CANN'S PRINCIPLES OF MOLECULAR VIROLOGY

CANN'S PRINCIPLES OF MOLECULAR VIROLOGY

EDWARD P. RYBICKI
Department of Molecular & Cell Biology
University of Cape Town
Cape Town, South Africa

Academic Press is an imprint of Elsevier
125 London Wall, London EC2Y 5AS, United Kingdom
525 B Street, Suite 1650, San Diego, CA 92101, United States
50 Hampshire Street, 5th Floor, Cambridge, MA 02139, United States
The Boulevard, Langford Lane, Kidlington, Oxford OX5 1GB, United Kingdom

ISBN 978-0-12-822784-8

For information on all Academic Press publications
visit our website at https://www.elsevier.com/books-and-journals

Cover illustration: SARS-CoV, with elements of its replication cycle. Copyright Russell Kightley Media, used with permission.

Publisher: Stacy Masucci
Acquisitions Editor: Kattie Washington
Editorial Project Manager: Sara Pianavilla
Production Project Manager: Sajana Devasi P K
Cover Designer: Mark Roger

Typeset by STRAIVE, India

Contents

8. Panics and pandemics

9. Subviral agents: Deltaviruses and prions

Preface

When I was seriously starting to structure my teaching efforts in second- and third-year virology at my university, I came across Alan J. Cann's first edition of *Principles of Molecular Virology* from 1993—and was immediately a fan because he organized things like I did.

That is, he described viruses and how they work in a comparative way, from first encountering a host cell, through replication and expression, to exiting the cell.

Moreover, his book was *affordable* in my developing country context, while many were not; accordingly, I recommended it to generations of virology students through the changing incarnations of our teaching, while I did not bother for other weightier offerings. We had a number of dealings over the years, formally with me writing a chapter on plant virus culture for a book he edited, but mainly via social media, where I adopted him as my trailblazing guru. Although I believe that I beat him on the World Wide Web (in 1994), it was he who subsequently introduced me to the concept of microbiology/virology blogging via his sadly defunct Microbiology Bytes (I do ViroBlogy on the WordPress platform), later to Twitter (@edrybicki), and also to news aggregators (Scoop.it Virology News). I owe him a lot, therefore, over quite a wide spectrum of social media reporting of academic and virological activities, over many years.

I was almost overcome, then, when Kattie Washington of Elsevier contacted me out of the blue to ask me to revise *Principles of Molecular Virology* for a seventh edition, as Alan was stepping away and had recommended that I take over. This was especially timeous, as I was considering launching my own textbook—and now I can instead set sail in an extremely well-crafted boat with a highly experienced crew! The unexpected advent of the SARS-CoV-2 pandemic in 2020 unavoidably delayed my finishing the book, but it also allowed me to considerably expand what had only been a notional chapter into a full-fledged (Chapter 8)!

I am also grateful to my long-time friend Russell Kightley, illustrator extraordinaire, for both making his pictures available, and enthusiastically altering them on demand. Thanks, Russell!

I am sincerely grateful to Alan for suggesting it, and the staff of Elsevier—in particular Kattie Washington, Sara Pianavilla, and Sajana Devasi—for their adoption of me as the revision author for the seventh edition of Cann's *Principles of Molecular Virology* and their patience with me as a new author.

Edward P. Rybicki
University of Cape Town, South Africa

Preface

CHAPTER

1

What are viruses, and how were they discovered?

INTENDED LEARNING OUTCOMES

- An understanding that viruses are organisms
- To be able to define how viruses are different from other biological organisms.
- To understand how the history of discovery of viruses led us to modern virology, and
- How the discovery and methods of investigation of viruses were coupled to technological advances in biology and molecular biology
- To understand how this has led to significant advances in molecular and cell biology and immunology

Introduction

This book is about "**molecular virology**," that is, the molecular basis of how viruses work. It looks at the astonishing **diversity** of viruses, the very different types of **genomes** viruses have, the **structure** of virus particles or **virions**, the ways viruses **infect cells**, how viruses **replicate themselves**, how they **spread**, and where they may come from, to infect us, and our crops and farmed animals—and literally everything else.

However, it is important to consider the basic nature of viruses first—and in order to understand how our present knowledge of viruses was achieved, it is useful to know a little about the history of virology, coupled to the technological developments that allowed their discovery and characterization. This helps both to understand the impact of viruses on our environment and to explain how we think about viruses, as well as what the current and future concerns of virologists are.

The principles behind some of the experimental techniques mentioned in this chapter may not be well known to all readers. That is why it may be helpful to you to use the Recommended Reading suggested at the end of this chapter to become more familiar with these methods, or you will not be able to understand the current research literature. In this and the subsequent

https://doi.org/10.1016/B978-0-12-822784-8.00001-5

chapters, terms in the text in ***bold italic print*** are defined in the glossary at the end of the book. Terms in **bold black print** are important terms to remember for understanding the literature on viruses.

Viruses as organisms

Viruses are **organisms** that are at the interface between molecules and cells; between what is usually termed "**living**" and "**dead.**" This creates problems for some traditional biologists; however, many of these can be cleared up with some concise definitions.

What are viruses?

Definition 1: Viruses

"Viruses are ***acellular*** organisms whose ***genomes*** consist of nucleic acid, which obligately ***replicate*** inside host cells using host metabolic machinery and ribosomes to form a pool of components which self-assemble into particles called ***virions***, which serve to protect the genome and to transfer it to other cells."

Definition 2: Organisms

"An organism is the unit element of a continuous **[genetic]** ***lineage*** with an individual evolutionary history."

SE Luria, JE Darnell, D Baltimore and A Campbell (1978). General Virology, 3rd Edn. John Wiley & Sons, New York, p4 of 578.

The concept of the virus as organism is contained within the concepts of individual viruses constituting continuous genetic **lineages**, and having independent **evolutionary histories**.

Thus, given this sort of lateral thinking, viruses become quite respectable as organisms:

- they most definitely **replicate**,
- their **evolution** can be traced quite effectively, and
- they are independent in terms of not being limited to a single organism as host, or even necessarily to a single species, genus or phylum of host.

Some virologists and microbiologists have recently suggested that the domain of "**life**" should be divided into two types of organisms: those that encode the full suite of protein-synthesizing machinery and make ribosomes (**cells**), and those that do not (**viruses**). The two fundamental types can thus be differentiated as "***ribocells***" and "***virocells,***" and the concept of drawing a line between them becomes rather arbitrary. These two terms will be used throughout this book.

While it was once true to say that known virus genomes were all smaller than those of bacteria, and that virions too were smaller, this is no longer the case. The largest viral genome is that of a **pandoravirus**, at 2.8 Mbp (million base pairs, or approximately 2500 genes). This is 20 times as big as smallest bacterial genome, from *Tremblaya princeps*: this is 139 kbp (thousand base pairs, with only 120 protein coding genes). The pandoravirus genome is in fact larger than the smallest ***eukaryotic*** genome known, which is currently the parasitic ***protozoan*** *Encephalitozoon* at just 2.3 Mbp. The largest virion known (currently that of Tupanvirus) is 1200 nm long, while the smallest bacteria (e.g., *Mycoplasma* spp.) are only 200–300 nm long.

Thus, the range of genetic complexity of viruses is now known to overlap with both ***prokaryotes*** and eukaryotes, though only at the highest bound for viruses.

In fact, while viruses mostly lack the genetic information that encodes the tools necessary for the generation of metabolic energy or for protein synthesis (i.e., ***ribosomes***), we are increasingly finding that the largest virus genomes may encode significant elements of their host's translational apparatus, such as tRNAs, rRNAs, and factors related to tRNA/mRNA maturation and ribosome protein modification. However, **no known virus has the biochemical or genetic means to generate the energy necessary to drive all biological processes,** and **all are absolutely dependent on their host cells for this function**. Lacking the ability to make ribosomes is therefore one factor which most clearly distinguishes viruses from all other organisms.

Viruses in the environment

In order to appreciate how important viruses are in the biosphere of this planet, one needs to understand that there is more genetic diversity among viruses than there is in all the rest of the bacterial, plant, fungal and animal kingdoms put together. For example, the species diversity of eukaryotes is estimated to be in the low tens of millions, and similar for prokaryotes—yet the diversity of viruses is only beginning to be determined, and cannot be accurately estimated.

In terms of number of organisms, given their much smaller size compared to most eukaryotes, there are probably at least **10^{30} individual prokaryotes on our planet**—everywhere from deep in the Earth's crust, to high in the atmosphere—making them the dominant cellular life form on the planet. More importantly, there are estimated to be at least **10 virus species infecting any single prokaryote**. As a result, **the greatest genetic diversity on planet Earth resides in virus genomes**. It is estimated that there are at least **10^{31} virus particles** on Earth—and that if one strung them end to end, the chain would stretch some **150 million light years**. This is the result of the success of viruses in parasitizing all known groups of living organisms, and particularly prokaryotes, and understanding this astonishing diversity is the key to comprehending the interactions of viruses with their hosts and how fundamental their importance is in certain ***biomes***. An extremely useful resource frequently used by the author for quick reference to just about any class of virus is **ViralZone**: this has encapsulated information on any given virus family and specific examples, with a structural description and brief account of how they replicate (see Recommended Reading below).

Virus genomes

Viruses have the most diverse types of genomes—the nucleic acids specifying their functions and structure—of any type of organism. While all cellular organisms—prokaryotes and eukaryotes, or **Bacteria, Archaea, and Eukarya** as they should be known—have **double-stranded DNA (dsDNA)** as their genetic material, viruses may have dsDNA or **single-stranded (ss) DNA**, and are unique in having **dsRNA** or **ssRNA** as well (Table 1.1).

Virus-like agents

Viruses are not alone in the acellular space. There are a number of other types of genomes—DNA and RNA—which have some sort of independence from cellular genomes. These

TABLE 1.1 Virus genome types.

DNA			
double-stranded		**single-stranded**	
Linear	Circular	Linear	Circular
Single component	Single or multiple	Single component	Single or multiple
RNA			
Double-stranded		**Single-stranded**	
Linear		Linear	
Single or multiple components		+ Sense[a]	− Sense
		Single or multiple	Single or multiple

a = mRNA-sense or translatable. (−)sense is complementary to (+)sense and must be transcribed to give mRNA.

include "**retrons**" or **retrotransposable elements**; bacterial and fungal (and eukaryotic ***organelle***) ***plasmids***; **satellite nucleic acids** and **satellite viruses** which depend on helper viruses for replication; and **viroids** and **virusoids (see below)**. A very different class of infectious agents—***prions***, associated with diseases such as **Creutzfeldt Jakob disease** in humans, **scrapie** in sheep, and **bovine spongiform encephalopathy** (**BSE**; "mad cow disease") in cattle—appear to be "**proteinaceous infectious agents,**" with no nucleic acid associated with them at all (see Chapter 9). Other virus-derived genomes and structural components (**polydnaviruses, tailocins**; see later) have been coopted by organisms as diverse as wasps and bacteria for their own survival.

Satellite viruses

There are viruses which depend for their replication on "helper" viruses: a good example is **tobacco necrosis satellite virus** (sTNV), which has a small piece of ssRNA which codes only for a **capsid protein (CP)**, and depends for its replication on the presence of TNV. Another example is **hepatitis delta virus (HDV):** this has a genome consisting of 1700 bases of negative sense **circular ssRNA**, which also codes for a structural protein (**delta antigen**). The 36nm diameter infectious particles consist of **hepatitis B virus** surface antigen (**HBsAg**) embedded in a cell-derived envelope, with **internal HDV nucleoprotein** (200 molecules of delta antigen complexed with the RNA) (see Chapter 9).

The **adeno-associated viruses (AAVs)** are also satellite viruses dependent on the much bigger linear dsDNA genome **adenoviruses** for replication, but which have **linear ssDNA genomes** and appear to be degenerate or defective **parvoviruses** (Fig. 1.1).

Much larger satellite viruses have been found parasitizing **giant dsDNA viruses**: the so-called "**Sputnik virophage**" was discovered associated with a giant **mimivirus** inside infected ***amoebae***, a class of unicellular eukaryotes. This has 74nm isometric particles with an internal lipid bilayer, and a 18kbp circular dsDNA genome capable of encoding up to 21 proteins—much bigger than many ***autonomous*** or **independently-replicating** viruses. Unlike other satellite viruses, the virophage uses the mimivirus's cytoplasmic virus factories for transcription, replication, and virion assembly, completely independent of the amoebal host's genes.

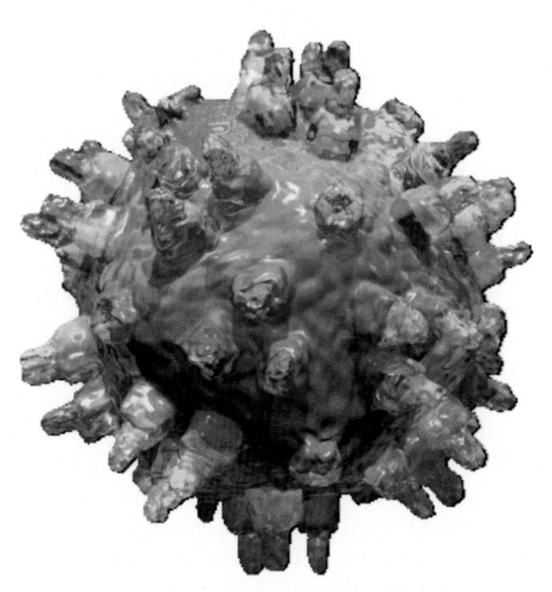

FIG. 1.1 Adeno-associated virus particle. This is a small particle (25 nm) with prominent surface spikes. *Courtesy of Russell Kightley Media.*

A unified system of **classification** for satellite viruses has recently been proposed, that regularizes the apparent discrepancy between regarding some (e.g., AAVs) as **defective viruses**, and others simply as **satellites**.

Satellite nucleic acids

Certain viruses have associated with them nucleic acids that are dispensable in that they are not part of the viral genome, and have no (or very little) sequence similarity with it, yet depend on the virus for replication, and are ***encapsidated*** by the virus. These are mainly associated with plant viruses and are generally **ssRNA**, both linear and circular: satellite tobacco necrosis virus (STNV) has a 1.2 kb ssRNA genome, which encodes only a coat protein which makes ***isometric*** 17 nm virions. However, several circular ssDNA satellites of plant-infecting **geminiviruses** have recently been found, in two separate classes—**alpha and beta satellites**.

Plasmids

Bacterial plasmids may share a number of properties with viral genomes—including modes of replication, as in single-strand circular DNA viruses and plasmids which replicate via **rolling circle mechanisms,** and circular dsDNA genomes replicating with a "**theta-like**" bidirectional replication forks—but they are not pathogenic to their host organisms. Some are transferred by ***conjugation*** between cells rather than by free extracellular particles, by means of *tra* genes encoding **pili**, or rod-like structures found on the surface of the cells. Plasmids generally encode some function that is of benefit to the host cell, to offset the metabolic load caused by their presence. There is even evidence that certain plasmids have evolved to encode **virus-like lipid membrane-derived particles** to transfer their genomes to other cells, and so resemble very primitive viruses.

Viroids

Viroids are **small naked circular ssRNA genomes** of 246–375 bp in length, which appear rodlike under the electron microscope due to their ***secondary structure*** and ***tertiary folding***, and are capable of causing diseases in plants. They code for nothing but their own structure, and are presumed to replicate by interacting with **host RNA polymerase II**, and to cause pathogenic effects by interfering with host DNA/RNA metabolism and/or transcription (Fig. 1.2). A structurally similar disease agent in humans is the **hepatitis D virus**, although this does encode a protein (Chapter 9; Fig. 9.1).

Retrons

Classic RNA-containing **retroviruses (HIV, HTLVs, bovine leukemia virus)**, and the DNA-containing **pararetroviruses** (**hepadnaviruses, caulimoviruses,** and **badnaviruses**) all share the unlikely attribute of the use of an enzyme complex consisting of a **RNA-dependent DNA polymerase/RNAse H/integrase (aka *reverse transcriptase*)** in order to replicate. They share this attribute with several **retrotransposons**, which are eukaryotic **transposable cellular elements** with striking similarities with retroviruses, that constitute more than 10% of the genome of eukaryotic cells. This includes well characterized entities such as the **yeast Ty element** and the ***Drosophila* copia element**, now classified as **pseudoviruses** and **metaviruses** respectively, in a taxon that includes retro- and pararetroviruses (see Chapter 2). **Retroposons** are similar in that they are eukaryotic elements which transpose via RNA intermediates, but they share no obvious genomic similarity with any viruses, other than the use of reverse transcriptase.

The human and other mammalian **LINE-1s** (**Long Interspersed Nuclear Elements**) are a group of **retrotransposable** elements which make up approximately **15% of the human genome**.

Bacteria such as *E. coli* also have reverse-transcribing transposons—known as **retrons**—but these are very different to any of the eukaryotic types, while preserving similarities in certain of the essential reverse transcriptase sequence motifs (see Chapter 3 for a detailed account of **mobile genetic elements, or MGEs**).

Polydnaviruses

Polydnaviruses are unusual in that they appear to be integral parts of their insect host genomes, yet have a virus-like stage and are used by their host to modify another insect's behavior and physiology. The two genera so far described—*Bracovirus* and *Ichnovirus*—of family *Polydnaviridae* (see Chapter 2 for taxonomy) contain viruses which have a **variable number of circular double-stranded DNA components**, with components ranging in size from **2 to > 31 kilobase pairs (kbp)**, for a total genome size of between **150 and 250 kbp**. Both

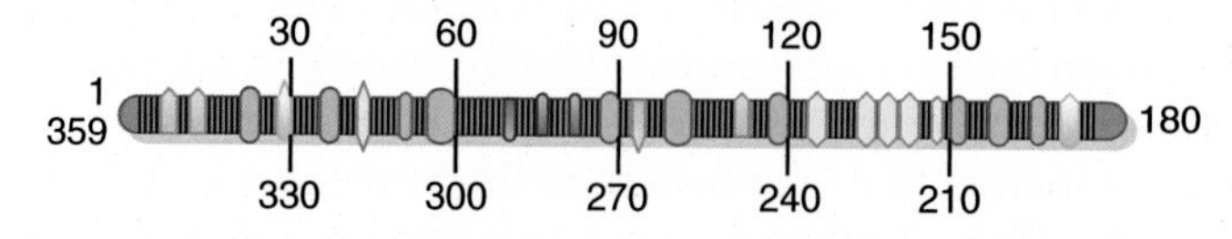

FIG. 1.2 Structure of a viroid. Viroids are naked, circular single-stranded RNA molecules (200–400 bases), with a high degree of secondary structure.

sets of viruses occur as **integrated *proviruses*** in the genomes of endoparasitic hymenopteran wasps. They replicate by amplification of the host DNA, followed by excision of ***episomal*** genomes by site-specific recombination, and only produce particles by budding from (ichnoviruses) or lysis of (bracoviruses) calyx cells in the oviducts of female wasps during the host pupal-adult transition. Moreover, the viruses in the two groups may well not be evolutionarily linked to one another, given that there is no antigenic or genome similarity, and the particles formed by the two groups are very different. Ichnoviruses make ellipsoidal particles with double membranes containing one nucleocapsid; bracoviruses make single-enveloped particles containing one or more cylindrical nucleocapsids. The latter may derive from **nudiviruses**, which appear to have contributed very substantially to wasp survival.

Particles are injected along with eggs into larvae of lepidopteran hosts; the DNA gets into secondary host cells and is expressed, but does not replicate—and this expression leads to some quite profound physiological changes, many of which are responsible for successful parasitism of the larva by the wasp. The association between wasp and virus has been termed an "**obligate mutualistic symbiosis,**" and appears to have evolved over more than 70 million years.

Viruses: Living or dead?

Viruses are simply **acellular organisms**, which find their full being inside host cells, where some measure of essential support services are offered in order to keep the virus life cycle turning. What most people consider as "viruses" are in fact **virus particles** or **virions**, the particles that viruses cause to be made in order to transport their genomes between cells, and to preserve them while doing so.

Thus, in a very real sense a virus IS the cell it infects—which is now increasingly becoming known as a **virocell**—because it effectively takes the latter over, and uses it to make **portable versions of the genome** (virions) that can infect other cells. Qualitatively, this is exactly what seeds and spores of plants and fungi do: they make specialized vehicles that preserve their genomes while not themselves replicating, and which can respond to changes in their environment to initiate a new organism.

While the debate on whether viruses are living or are indeed organisms gets almost theological in its intensity in certain biological circles, there is a very simple way around the problems—and that is to regard them as a **particle/organism duality**, much as physicists have learned to do with the dual **wave/particle nature of light**. Indeed, some virologists have gone as far as to suggest that the domain of "**life**" should be divided into two types of organisms: those that encode the full suite of protein-synthesizing machinery (**ribocells; ribosome-encoding**), and those that do not (**virocells; capsid/virion-encoding**) (see Recommended Reading). I agree with that sort of thinking—and recent evidence on the deep evolution of viruses appears to support it (see Chapter 3).

The history of the discovery of viruses

While people were aware of diseases of both humans and animals now known to be caused by viruses many hundreds of years ago, the concept of viruses as distinct entities dates back only to the very late 1800s. The word "**virus**" comes from a Latin word simply meaning "**slimy fluid,**" or poison: although the term had been used for many years previously to indiscriminately describe disease agents, the modern usage only started in the early 20th century.

While viruses were only discovered as a separate category of infectious organisms in the late 1800s, there are two notable cases of viral diseases for which vaccines were developed that predate their discovery: these were vaccines for **smallpox**, and for **rabies**. Both were dread diseases that were known to be transmissible—smallpox from human to human, and rabies by contact with infected animals—and which caused great fear in exposed human populations. The development of vaccines against them, therefore, constituted two very significant breakthroughs in medical science of the times, and moreover illustrated two very important principles that became fundamental to the later development of many more vaccines. These were the **use of a related agent** that did not cause severe disease in the case of smallpox, and the deliberate development of an **attenuated disease agent** in the case of rabies.

Smallpox vaccine

Smallpox was endemic in China by 1000 BCE. In response to observations that smallpox victims who recovered were subsequently immune, the practice of what became to be called "**variolation**" was developed, named in Europe after a Swiss Bishop in 570 CE who derived the term for smallpox infections from the Latin *varius* or *varus*, which respectively meant "**stained**" or "**mark on the skin.**" People inhaled the dried crusts from smallpox lesions like snuff or, in later modifications, inoculated the pus from a healing lesion into a scratch on the forearm. Variolation was practiced for centuries in China and India and later in Africa, before it was introduced to Europe in the 18th century CE, where it had been widely adopted by the 1750s. The practice of variolation was in fact introduced into North America prior to 1720 by a West African slave renamed Onesimus, in the household of the famous cleric Cotton Mather.

It was shown to be an effective method of disease prevention, although risky because the outcome of the inoculation was never certain: **2%–3% of variolated people died from it**, although this paled into insignificance compared to the **20%–60% who died of the natural infections**. Indeed, the man generally credited with developing the first smallpox vaccine, Edward Jenner, was nearly killed by variolation at the age of seven. However, it was an English farmer named Benjamin Jesty, in the village of Yetminster in North Dorset, UK, who in 1774 was the first to "***vaccinate***" anyone against smallpox: during a smallpox outbreak in the area, he ***inoculated*** his wife and two sons with **lymph** taken from lesions on the udder of an infected cow with a **cowpox** infection, on the strength of a folk tale about milkmaids who had previously been infected with the mild cowpox being immune to smallpox. It was only in May of 1796 that the country doctor Edward Jenner used cowpox-infected material obtained from the hand of Sarah Nemes, a milkmaid from his home village of Berkeley in Gloucestershire, England, to successfully vaccinate 8-year-old James Phipps. Jenner then inoculated young Phipps again, 2 months later, with material from a fresh smallpox lesion: the boy did not become infected, which allowed Jenner to surmise that the treatment resulted in **complete protection. (https://jennermuseum.com/.)**

While this would be regarded as completely unethical in today's world, Jenner went on to publish this and other cases, and eventually achieve quite considerable fame. He also coined the term ***vaccination***, derived from the Latin word *vacca* for cow, and his cowpox-derived inoculum became known as **vaccinia** after the name **Variolae Vaccinae,** which described the disease cowpox. Although initially controversial, vaccination against smallpox was almost universally adopted worldwide during the nineteenth century.

The nature of microorganisms

This early success, although a triumph of scientific observation and reasoning, was not based on any real understanding of the nature of infectious agents, which required a series of technological breakthroughs. The first of these was the invention of the **microscope** by Antoni van Leeuwenhoek (1632–1723), a Dutch merchant: he constructed the first simple microscopes, which Robert Hooke (1635–1703) in London copied and improved upon, and used to develop the first illustrations of **microfungi** in his book *Micrographia* in 1665. These were much later identified as probably being a **bread mold** or *Mucor* sp. By 1676 van Leeuwenhoek had observed and reported to the UK Royal Society on minute mobile "**animalcules**" in various specimens such as herb infusions and teeth scrapings: these were probably mainly bacteria. Both men became Fellows of the Royal Society, and communicated extensively. However, it was not until Robert Koch and Louis Pasteur in the 1880s jointly proposed the **"germ theory"** of disease, following the first **sterile culturing** of bacteria, that the significance of these organisms became apparent. Koch defined four famous criteria which are now known as **Koch's postulates**, which are still generally regarded as the best proof that an infectious agent—cellular or viral—is responsible for a specific disease:

1. The agent must be present in every case of the disease.
2. The agent must be isolated from the host and grown in vitro.
3. The disease must be reproduced when a pure culture of the agent is inoculated into a healthy susceptible host.
4. The same agent must be recovered once again from the experimentally infected host.

These principles are still central to the study of infectious diseases, but in the case of viruses that only infect humans, and which cannot be conveniently ***cultured***, it is regarded as being sufficient to simply show that **natural transmission of an identified agent causes the disease** (see Chapter 8).

Rabies vaccine

While he was involved in many discoveries of early microbiology, Pasteur also worked extensively on **rabies**, a lethal disease spread by bites from infected animals, which he identified as being caused by a "**virus**" (from the Latin for "**poison**") without knowing what the causative agent actually was. Working from 1881, he developed methods of producing ***attenuated*** virus preparations by progressively drying the spinal cords of rabbits experimentally infected with rabies which, when inoculated into other animals, would protect from disease caused by ***virulent*** rabies virus. He famously successfully used it in 1885 to treat the nine-year old Joseph Meister who had been bitten by a rabid dog. This was the **first artificially produced or attenuated virus vaccine**, as the ancient practice of variolation and the use of cowpox virus for vaccination had both relied on naturally-occurring viruses.

Porcelain filters and the discovery of viruses

Viruses were discovered as an excluded entity rather than by being seen or cultured, due to the invention of **efficient filters**: the fact that cell-free extracts from diseased plants and animals could still cause disease led people to theorize that an unknown infectious agent—a **"filterable virus"**—was responsible.

The invention that allowed viruses to be discovered at all was the **Chamberland-Pasteur filter.** This was developed in 1884 in Paris by Charles Chamberland, who worked with Louis Pasteur. It consisted of unglazed **porcelain "candles,"** with pore sizes of **0.1–1 μm (100–1000 nm)**, which could be used to completely remove all culturable bacteria or other cells known at the time from a liquid suspension (Fig. 1.3). Though this simple invention essentially enabled the establishment of a whole new science—**virology**—the continued development of the discipline required a string of technical developments, which I will highlight.

As the first in what was to be an interesting succession of events, Adolph Mayer from Germany, working in The Netherlands in 1886, showed that the "**mosaic disease" of tobacco** could be transmitted to other plants by rubbing a liquid extract, filtered through paper, from

FIG. 1.3 Chamberland filter for sterilizing liquids. This is a porcelain filter with very small pores that can completely remove culturable bacteria from liquids. *Image from https://tinyurl.com/33yx6scz: this file is made available under the Creative Commons CC0 1.0 Universal Public Domain Dedication and is in the public domain.*

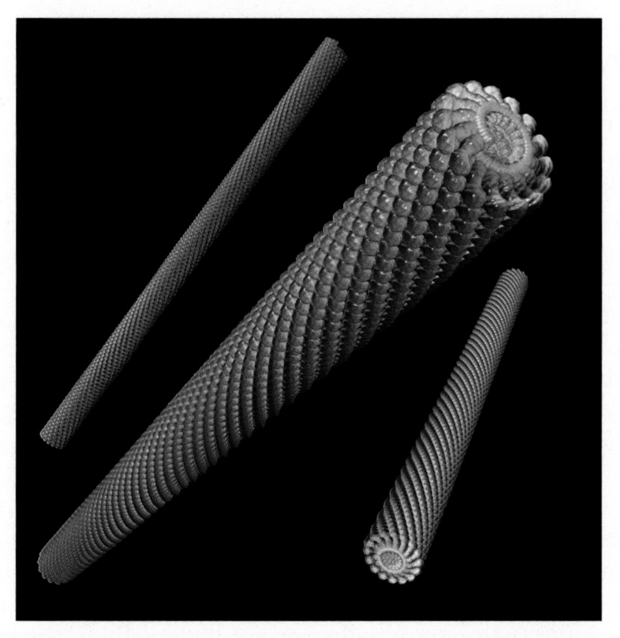

FIG. 1.4 Tobacco mosaic virus. Tobacco mosaic virus particles, showing RNA (*green*) embedded in a helix of protein (*purple*). *Courtesy of Russell Kightley Media.*

an infected plant onto the leaves of a healthy plant. However, he came to the conclusion it must be a bacterial disease.

The first use of porcelain filters to characterize what we now know to be a virus was reported by Dmitri Ivanowski in St Petersburg in Russia, in 1892. He had used a filter candle on an infectious extract of tobacco plants with mosaic disease, and shown that it remained infectious: however, he concluded the agent was either a ***toxin*** or bacterial in nature.

The Dutch scientist Martinus Beijerinck in 1898 described how he did similar experiments with bacteria-free extracts, but made the conceptual leap and described the agent of mosaic disease of tobacco as a "**contagium vivum fluidum,**" or **soluble living germ**. The extract was completely sterile, could be kept for years, but remained infectious. The term **virus** was later used to describe such fluids, also called **filterable viruses**, which were thought to contain no particles. The virus causing mosaic disease is now known as **tobacco mosaic virus (TMV)**: it makes rodlike (16×300nm) virions composed of a helix of coat protein and an ssRNA genome (Fig. 1.4).

Early filterable virus discovery timeline

The second virus discovered was what is now known as **foot and mouth disease virus (FMDV)** of farm and other animals, by the German scientists Friedrich Loeffler and Paul Frosch in 1899. This makes ssRNA-containing **isometric particles** around 30nm in diameter. Again, their sterile filtered liquid proved infectious in calves, providing the **first proof of viruses infecting animals**.

In 1898, G Sanarelli, working in Uruguay, described the smallpox virus relative and tumor-causing **myxoma virus** of rabbits as a virus—although he isolated it by ***centrifugation*** and not filtration.

The **first human virus** described was the agent that causes **yellow fever**: this was discovered and reported on in 1901 by the US Army physician Walter Reed, after pioneering work in Cuba by Carlos Finlay proving that **mosquitoes** transmitted the deadly disease.

A finding that was later to have great importance in **veterinary virology** was the discovery by Maurice Nicolle of France and Adil Mustafa in Turkey in 1902 that the legendary **rinderpest** or **cattle plague** was caused by a virus.

Sir Arnold Theiler, a Swiss-born veterinarian working in South Africa, had developed a crude vaccine against rinderpest by 1897, without knowledge of the nature of the agent: this consisted of blood from an infected animal, injected with serum from one that had recovered. This risky mixture worked well enough, however, to eradicate the disease in the region. He went on to do the same thing successfully for **African horse sickness virus** and other viruses.

The description in Annales de l'Institut Pasteur by Remlinger and Riffat-Bay from Constantinople in 1903 of the causative agent of **rabies** as a filterable virus was the culmination of many years of distinguished work in France on the virus, started by Louis Pasteur himself. Remlinger in fact credited Pasteur with having had the notion in 1881 that rabies virus was an ultramicroscopic particle. The same volume of the Annales which reported the rabies agent also has a discussion on whether or not the smallpox agent **variola virus** and the vaccine against it, **vaccinia virus**, were differently-adapted variants of the same thing, or were different viruses.

Adelchi Negri—who had previously discovered the **Negri bodies** in cells infected with rabies virus—showed in 1906 that the smallpox vaccine vaccinia virus was filterable. This was the final step in a long series of discoveries around smallpox, that started with Jesty and Jenner's use of what was supposedly **cowpox**, but may in fact have been **horsepox virus**, to protect people from the disease.

In 1904, Erwin Baur in Germany described an **infectious variegation of *Abutilon*** spp. that could only be transmitted by grafting, that was not associated with visible bacteria. This is now known to be due to **Abutilon mosaic virus**, a single-stranded DNA **geminivirus.** Interestingly, in 1906, Albrecht Zimmermann proposed that the agent of **mosaic disease of cassava** that had first been described from German East Africa (now Tanzania) in 1894, was a filterable virus—now known to be African cassava mosaic virus, and also a geminivirus (Fig. 1.5). Incidentally, the earliest recorded description of symptoms of a plant disease was probably in a poem in 752CE by the Japanese Empress Koken, describing **symptoms in eupatorium plants**. It was shown in 2003 that the striking yellow-vein symptoms were caused by a geminivirus infection, with **eupatorium yellow vein virus**.

Oluf Bang and Vilhelm Ellerman in Denmark were the first—in 1908—to associate a virus with an animal ***leukemia***: They successfully used a cell-free filtrate from chickens with **avian leukosis** to transmit the disease to healthy chickens.

In the same year, Karl Landsteiner and Erwin Popper in Germany showed that **poliomyelitis** or **infantile paralysis** in humans was caused by a virus. They proved this by injecting a cell-free extract of a suspension of spinal cord from a child who had died of the disease, into monkeys, and showed that they developed symptoms of the disease. The disease now known as **poliomyelitis** was first clinically described in England in 1789, as "a debility of the lower

FIG. 1.5 Mosaic disease in cassava and abutilon. Diseased cassava leaves (left) illustrating severe symptoms caused by African cassava mosaic virus infection. Ornamental abutilon leaf (right), showing symptoms of abutilon mosaic virus infection.

extremities." However, it had been known since ancient times, and had even been depicted clearly in an Egyptian painting from over 3000 years ago (Fig. 1.6).

The first **solid tumor-causing virus**, or virus associated with cancer, was described by Peyton Rous in the USA in **1911**. He showed that chicken **sarcomas, or solid connective tissue tumors,** could be transmitted by grafting, but also that a filterable or cell-free agent extracted from a sarcoma was infectious. The virus was named for him as **Rous sarcoma virus,** and is now known to be a **retrovirus**.

In 1915, Frederick Twort in the UK accidentally found a filterable agent that caused the bacteria he was growing to ***lyse***, or burst open. Although he showed that it could pass through porcelain filters, and could be transmitted to other colonies of the same bacteria, he was not sure whether or not it was a virus, and referred to it as "**the bacteriolytic agent**" (Fig. 1.7).

Subsequently, Félix d'Hérelle in Paris published in 1917 that he had discovered a virus that lysed a bacterial agent he was culturing that caused **dysentery**, or **diarrhea**. He named the virus "**bacteriophage,**" or **eater of bacteria**, derived from the Greek term "*phagein,*" meaning to eat. He showed a number of interesting properties of his shigella-specific bacteriophages, including that they could be adapted to other *Shigella* species or types by ***passaging*** them repeatedly, and that they protected rabbits against infection by lethal doses of bacteria. As a result of this, D'Hérelle's main interest in his new discovery was in using them as a **therapeutic agent for bacterial infections** in humans. In fact, d'Hérelle went to India in 1927, and put cholera phage preparations into wells in villages with cholera patients: apparently the death toll went down from 60% to 8%. Sadly, this idea did not take off in Europe or the Americas, largely due to the unreliability of the ill-understood phage preparations, although it was extensively exploited in the former USSR (see Recommended Reading). Indeed, he mentored George Eliava who went on to found the **Eliava Institute** in Tbilisi, Georgia, which became a major center for the use of bacteriophage cocktails against persistent bacterial infections in humans. A review on what has become known as **phage therapy** from the Institute was recently published to mark the centenary of Twort's discovery in 1915 (see Recommended Reading).

FIG. 1.6 Stele from an Egyptian tomb. Representation of a polio victim, Egypt 18th Dynasty (1403–1365 BC). https://commons.wikimedia.org/wiki/File:Polio_Egyptian_Stele.jpg

Possibly the worst human plague the world has ever seen swept across the planet in the period 1918–1922: this was known as the **Spanish Flu**, from where it was first properly reported, and it went on to **kill between 50 and 100 million people** all over the world. Most medical authorities at the time thought the disease was caused by bacteria, despite this contemporary comment in the British Medical Journal of 1918:

> *"Of the suggestion that the true virus is a filter-passer it has been said that it is only a cloak to our real ignorance as to its nature, but the results of some recent experiments by Nicolle and Lebailly...**afford good evidence that a filter-passing virus does exist.**"*

Further evidence that the 1918 influenza was filterable was provided in 1919 in The Lancet by T Yamanouchi and colleagues in Japan: they showed the agent passed a porcelain filter, and elicited immunity in volunteers. We now know it to have been **H1N1 influenza type A**: this is an ssRNA virus with a segmented genome.

The virus probably infected over one third of the humans alive at the time—1.5 billion—with a **case fatality rate of 10%–20%**. Some regions, like Alaska and parts of Oceania, had death rates of **over 25%** of the total population. By contrast, the normal mortality rate for seasonal flu is less than 0.1% of those infected.

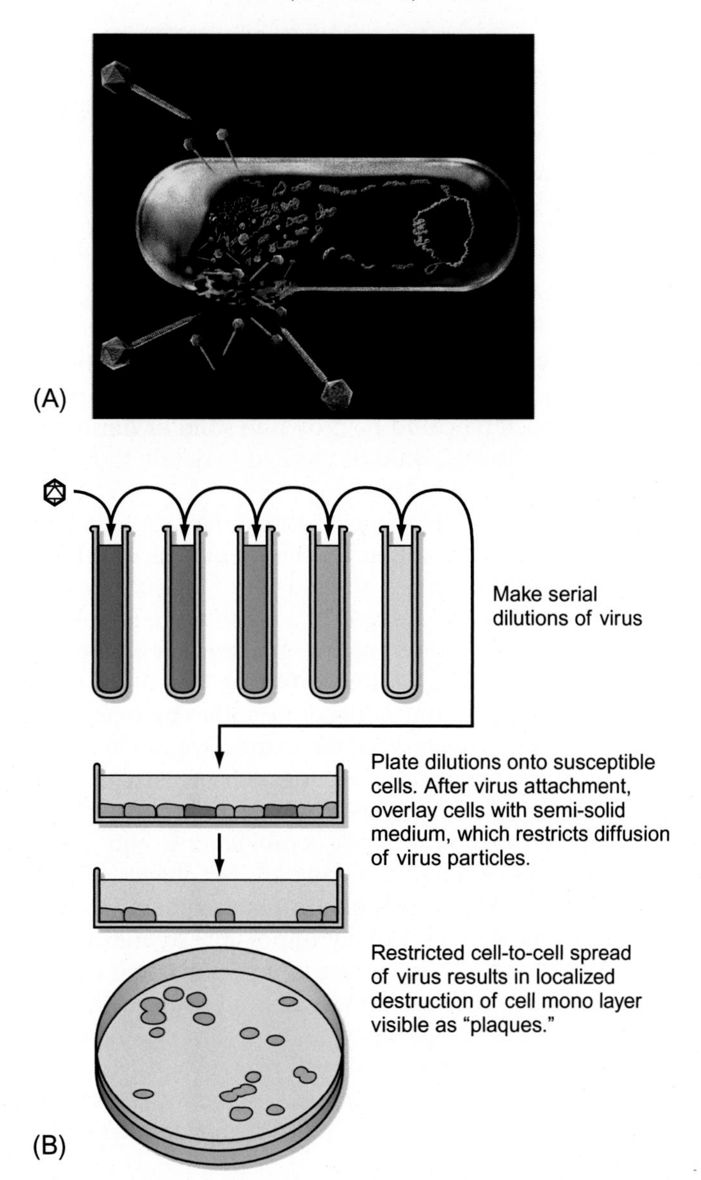

FIG. 1.7 **Bacterial lysis and plaque assays.** (A) *Escherichia coli* bacterium being burst by infecting bacteriophage λ. *Courtesy of Russell Kightley Media.* (B) Plaque assays are performed by adding a suitable dilution of a virus preparation to an adherent monolayer of susceptible cells. After allowing time for virus attachment to the cells, a semisolid culture medium containing a polymer such as agarose or carboxymethyl cellulose, which restricts diffusion of virus particles from infected cells, is added. Only direct cell-to-cell virus spread occurs, resulting in localized destruction of cells. After incubation, the medium is removed and the cells stained to make the holes in the monolayer (plaques) more visible. Each plaque results from infection by a single plaque-forming unit (p.f.u.) allowing the original number of virus particles to be estimated.

Modern reconstruction of the virus from archived tissue samples and frozen bodies found in permafrost has shown it probably jumped directly into humans from **birds**, as all influenza A viruses appear to originate in birds. The virus persisted as seasonal flu until 1958.

Agents of many other diseases were found to be "filterable viruses" in the 1920s, including **yellow fever virus** by Adrian Stokes in 1927, in Ghana. Indeed, the US bacteriologist and virologist Thomas Rivers in 1926 counted some **65 disease agents** that had been identified as viruses.

Culturing viruses

The discovery of bacterial viruses, or **phages** as they are colloquially known, was a landmark in the history of virology, as it meant that for the first time it was relatively easy to work with viruses: many kinds of bacteria could be grown in solid or liquid **culture media** quite easily, and the **life cycle** of the viruses could be studied in detail. In fact, this later led to the **birth of molecular biology**, as described later.

However, the beauty of working with phages was that they could be ***assayed***—or counted in terms of infectious units—so easily, either by the **plaque technique** or by infections of liquid cultures. This was not true of viruses of plants or of animals in the absence of similar culture techniques; these could only be assayed in a much more crude method using whole organisms. One such method was by determining infection endpoints by serial dilution of inoculum, such as the now-famous **ID_{50}**, or **dose infecting 50%** of the experimental subjects, which was done for FMDV by inoculating cattle, or for rabies by inoculating dogs.

This changed in 1929 for **plant viruses**, with the demonstration by the plant virus pioneer Francis O Holmes that **local lesions** caused by infection of particular types of tobacco by TMV could be used as a means of accurately assaying the infectivity of virus stocks (Fig. 1.8).

This was then extended to other virus/host combinations, and allowed the rapid and quantitative assay of virus stocks—which, as it had done for phages, allowed the study of the properties of plant viruses, and led to their **biological isolation** and then **purification**.

However, viruses of humans were effectively impossible to culture unless they infected animals, and whole animal culture of viruses was cumbersome and expensive. Accordingly, another important technological development was needed to allow the new science of **virology**, or **the study of viruses**, to develop: this was the use of **organs of small animals**, and most importantly, the **brains of live mice**, to culture viruses.

In 1929, Howard Andervost, working at Harvard University, showed that **human herpes simplex virus**—a large dsDNA virus—could be reliably cultured by injection into the brains of live mice.

By 1930 the South African-born Max Theiler—son of Sir Arnold, and also at Harvard—showed that **yellow fever virus** could also be similarly cultured. This discovery allowed the development of ***attenuated*** or **weakened strains** of virus by serial passage or repeated transmission of the virus between mice, and the successful animal testing of vaccine candidates and of protective **antisera.** Theiler was awarded the Nobel Prize in 1951 for this work, which **until 2008 was the first and only recognition of virus vaccine work by the Nobel Foundation.**

A landmark in medical virology was the development in **1931** of human vaccines against yellow fever virus, by Wilbur Sawyer in the USA: this followed on Theiler's mouse work in

FIG. 1.8 Local lesions in plant leaves. Local lesions caused by infection of *Nicotiana glutinosa* with tobacco mosaic virus.

using brain-cultured virus plus human immune serum from recovered patients to immunize humans—very similar to Theiler Senior's strategy with rinderpest.

Also in 1931, Richard E Shope in the USA managed to recreate **swine influenza** by intranasal administration of filtered secretions from infected pigs. Moreover, he showed that the classic severe disease required coinoculation with a bacterium—*Haemophilus influenza suis*—originally thought to be the only agent involved in the disease. He also found that people who had survived infection during the 1918 pandemic had antibodies protecting them against the swine flu virus, while people born after 1920 did not, which showed that the 1918 human and swine flu viruses were very similar if not identical. This was a very relevant discovery for what happened much later, in the **2009 influenza pandemic,** when the same virus apparently came back into the human population from pigs after circulating in them continuously since 1918.

Shope went on in 1932 to discover, with Peyton Rous, what was first called the **Shope papillomavirus** and later **cottontail rabbit papillomavirus**: this causes benign cancers in the form of long hornlike growths on the head and face of the animal. This was **the first proof that a virus could cause a cancer in mammals**—and may also explain the sightings in the US Southwest of the near-mythical "jackalope."

Patrick Laidlaw and William Dunkin, working in the UK at the National Institute for Medical Research (NIMR), had by **1929** successfully characterized the agent of **canine distemper**—a relative of **measles, mumps** and **distemper morbilliviruses**—as a virus, proved it infected dogs and ferrets, and in 1931 got a vaccine into production that protected dogs.

Christopher Andrewes, Patrick Laidlaw, and Wilson Smith reported in 1933 that they had isolated a virus from humans infected with influenza from an epidemic then raging. They had done this by infecting ferrets with filtered extracts from infected humans—after the fortuitous observation that infected ferrets could apparently transmit influenza to investigators, by sneezing on them! The "**ferret model**" was very valuable, as strains of influenza virus could now be clinically distinguished from one another without infecting people.

Chickens, eggs and viruses

Possibly the next most important methodological development in virology after the discovery of phages was the proof that **embryonated or fertilized hen's eggs** could be used to culture a variety of important animal and human viruses. Ernest Goodpasture, working at Vanderbilt University in the USA, showed in 1931 that it was possible to grow **fowlpox virus**—a relative of **smallpox**—by inoculating the **chorioallantoic membrane of eggs**, and incubating them further.

While tissue culture had in fact been practiced for some time—for example, as early as the 1900s, investigators had grown "**vaccine virus**" or the smallpox vaccine now called vaccinia virus in **minced up chicken embryos suspended in chicken serum**—this technique represented a far cheaper and much more scalable technique for growing pox- and other suitable viruses.

Frank Macfarlane Burnet in 1936 used **embryonated egg culture of viruses** to demonstrate that it was possible to do "pock assays" on chorioallantoic membranes that were very similar to the plaque assays done for bacteriophages, with which he was also very familiar (Fig. 1.9).

Also in 1936, Burnet started a series of experiments on culturing **human influenza virus** in eggs: he quickly showed that it was possible to do pock assays for influenza virus, and that:

"It can probably be claimed that, excluding the bacteriophages, egg passage influenza virus can be titrated with greater accuracy than any other virus."

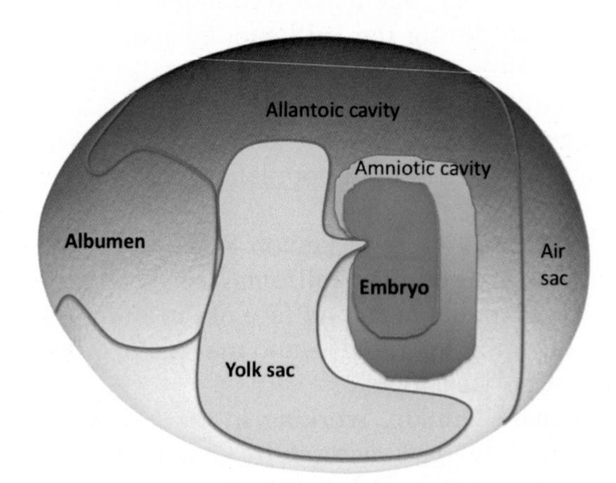

FIG. 1.9 Chorioallantoic culture. Embryonated hen's egg at 12 days, displaying anatomical features used for virus assay and multiplication. Chorioallantoic membrane lines the allantoic cavity.

By 1937 Max Theiler and colleagues in the USA adapted the French strain of **yellow fever virus (YFV)** he had previously grown in mouse brains to being grown in **chick embryos**, and showed that he could attenuate the already weakened strain even further—but it remained ***"neurovirulent,"*** as it caused **encephalitis** or brain inflammation in monkeys. He then adapted the first YFV characterized—the Asibi strain, from Ghana in 1927—to being grown in minced chicken embryos lacking a spinal cord and brain, and showed that after more than 89 passages, the virus was no longer ***"neurotrophic,"*** did not cause encephalitis, and could now safely be grown in eggs. The new **17D strain of YFV** was successfully tested in clinical trials in Brazil in 1938: The strain remains in use today, and is still grown in eggs.

Virus purification and the physicochemical era

Given that the nature of viruses had prompted people to think of them as "chemical matter," researchers had attempted from early days to **isolate, purify and characterize** the infectious agents. An early achievement was the **purification of a poxvirus** in 1922 by Frederick O MacCallum and Ella H Oppenheimer.

Much early work was done with phages and plant viruses, as these were far easier to purify or extract at the concentrations required for analysis, than animal or especially human viruses.

CG Vinson and AM Petre, working with TMV, showed in 1931 that they could **precipitate the virus from suspension as if it were an enzyme**, and that **infectivity** of the precipitated preparation was preserved. Indeed, in their words:

"...it is probable that the virus which we have investigated reacted as a chemical substance."

The fact that bacterial viruses or **phages** adsorbed irreversibly to their hosts as part of the infection process was shown by AP Krueger and Max Schlesinger in 1930–1931. Schlesinger later showed between 1934 and 1936 that the bacteriophage he worked with consisted of approximately equal amounts of protein and DNA, which was incidentally the first proof that viruses might be **nucleoprotein,** or **protein associated with nucleic acid,** in nature.

However, it was the development of two major physical techniques—**ultracentrifugation and electron microscopy**—that were largely responsible for the rapid advances in virology in the 1930s.

Biophysics of viruses

Biophysical studies can be considered under four areas: **physical methods, chemical methods, electron microscopy** and **crystallography**. Physical measurements of virions began in the 1930s with the earliest determinations of their proportions by filtration through membranes with various pore sizes. Experiments of this sort led to the first (rather inaccurate) estimates of the size of virions. The accuracy of these estimates was improved greatly by studies of the sedimentation properties of viruses in **ultracentrifuges**.

The ultracentrifuge

A technical development that was to greatly advance the study of viruses reached fruition by the 1930s: This was the **ultracentrifuge**, invented and developed first by Theodor ("The") Svedberg in Sweden as a purely analytical tool, and later by JW Beams and EG Pickels in the USA as an analytical and preparative tool. The ultracentrifuge revolutionized first, the physical analysis of proteins in solution, and second, **the purification of proteins, viruses and cell components**, by allowing centrifugation at speeds high enough to allow **pelleting** of subcellular fractions.

Analytical centrifugation and calculation of molecular weights of particles gave some of the first firm evidence that **virions were large, regular objects**. Indeed, it came to be taken as a given that one of the fundamental properties of a virus particle was its **sedimentation coefficient**, measured in **svedbergs** (a unit of 10^{-13} s, shown as $S_{20,W}$). This is also how ***ribosomes*** of pro- and eukaryotes came to be named: these are known as **70S (prokaryote)** and **80S ribosomes**, respectively, based on their **different sedimentation rates** (Fig. 1.10).

An important set of biophysical discoveries started in 1935, when Wendell Stanley in the USA published the first proof that TMV could be crystallized, at the time the most stringent way of purifying molecules. He also reported that the "protein crystals" were contaminated with small amounts of phosphorus. An important finding too, using ultracentrifugation and later, **electron microscopy**, was that the infectious TMV "protein" had a very high molecular weight, and was in fact composed of large, regular particles. This was a highly significant discovery, as it indicated that some viruses at least really were very simple infectious agents indeed.

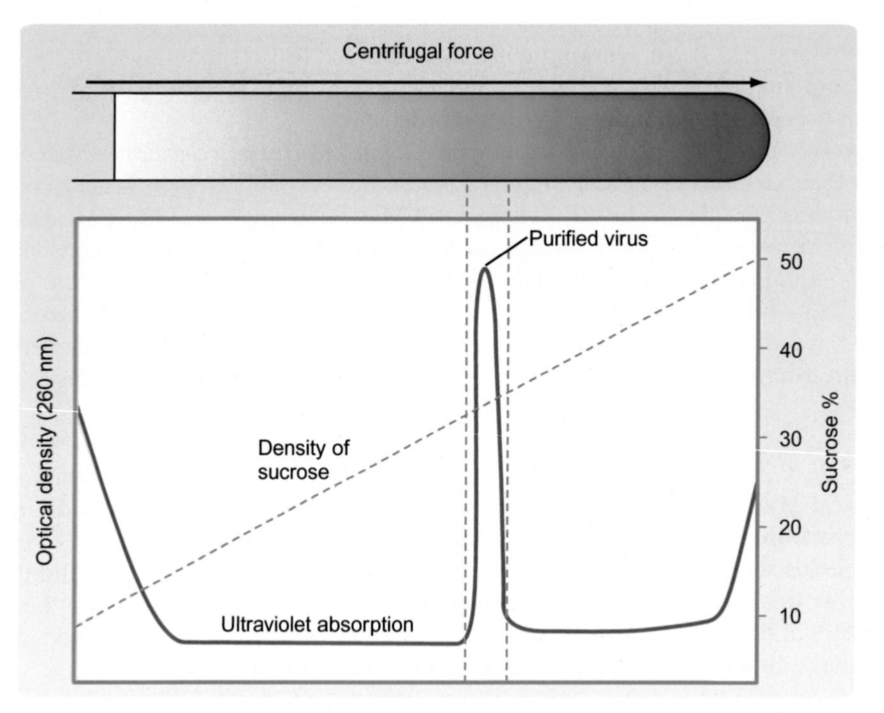

FIG. 1.10 Centrifugation of virus particles. A number of different sedimentation techniques can be used to study viruses. In rate-zonal centrifugation (shown here), virus particles are applied to the top of a preformed density gradient, that is, sucrose or a salt solution of increasing density from the top to the bottom of the tube. After a period of time in an ultracentrifuge, the gradient is separated into a number of fractions, which are analyzed for the presence of virus particles. The nucleic acid of the virus genome can be detected by its absorption of ultraviolet light. This method can be used both to purify virus particles or nucleic acids or to determine their sedimentation characteristics.

However, his first conclusion that TMV was composed only of protein was soon challenged, when Norman Pirie and Frederick Bawden working in the UK showed in 1937 that RNA—which consists of **ribose sugar molecules linked by phosphate groups**—could be isolated consistently from crystallized TMV as well as from a number of other plant viruses, which accounted for the phosphorus "contamination." This resulted in the realization that TMV and other plant virus particles—now known to be virions—were in fact "**nucleoproteins,**" or **complexes of protein and nucleic acid** (see Fig. 1.4).

Differential or preparative centrifugation, performed using parameters determined by analytical centrifugation, became of great value in obtaining purified and highly concentrated preparations of many different viruses, free of contamination from host cell components, which could be subjected to chemical analysis. The relative density of virions, measured in solutions of sucrose or CsCl or other solutes used to form density gradients, is also a characteristic feature, revealing information about the proportions of nucleic acid and protein in the particles.

Electron microscopy and virion structures

The development of the electron microscope, in Germany in the 1930s, represented a revolution in the investigation of virus structures: while virions of viruses like **variola** (smallpox) and vaccinia could just about be seen by light microscopy—and had been, as early as 1887 by John Buist and others—most viruses were far too small to be visualized in this way.

While Ernst Ruska received a Nobel Prize in 1986 for developing the electron microscope, it was his brother Helmut who first imaged virions—using beams of electrons deflected off virions coated in heavy metal atoms. From 1938 through the early 1940s, using his "supermicroscope," he imaged virions of poxviruses, TMV, **varicella-zoster herpesvirus**, and bacteriophages, and showed that they were all **regular and sometimes complex particles**, and were often very different from one another. Indeed, Wendell Stanley recounted in his 1946 Nobel Prize acceptance speech that TMV was known to be **rodlike** (15 × 280 nm); vaccinia virus particles were **bricklike** (210 × 260 nm); **poliovirus** particles were isometric or apparently spherical (diameter 25 nm); phages like **T2** were known to have a **head-and-tail** structure.

Electron microscopy became a routine tool in 1959, when Sydney Brenner and Robert Horne published *"A negative staining method for high resolution electron microscopy of viruses."* This method involved the staining of virions deposited on copper grids with **heavy-metal salts** such as **phosphotungstic acid (PTA)** or **uranyl acetate** (Fig. 1.11).

This simple technique revolutionized the field of electron microscopy, and within just a few years much information was acquired about the **architecture of virions**. Not only were the overall shapes of particles revealed, but also the details of the **symmetrical arrangement of their components**. Using such data, Francis Crick and James Watson (1956) were the first to suggest that virus ***capsids*** are composed of numerous identical protein subunits arranged either in **helical** or **cubic (icosahedral)** symmetry. In 1962, Donald Caspar and Aaron Klug extended these observations and elucidated the **fundamental principles of virion symmetry**, which allow repeated protomers to self-assemble to form capsids, based on the principle of **quasi-equivalence** (see Chapter 2). This combined theoretical and practical approach, with **crystallographic and electron microscopic image reconstruction** (see below), has resulted in our current understanding of the structure of virions (Fig. 1.12).

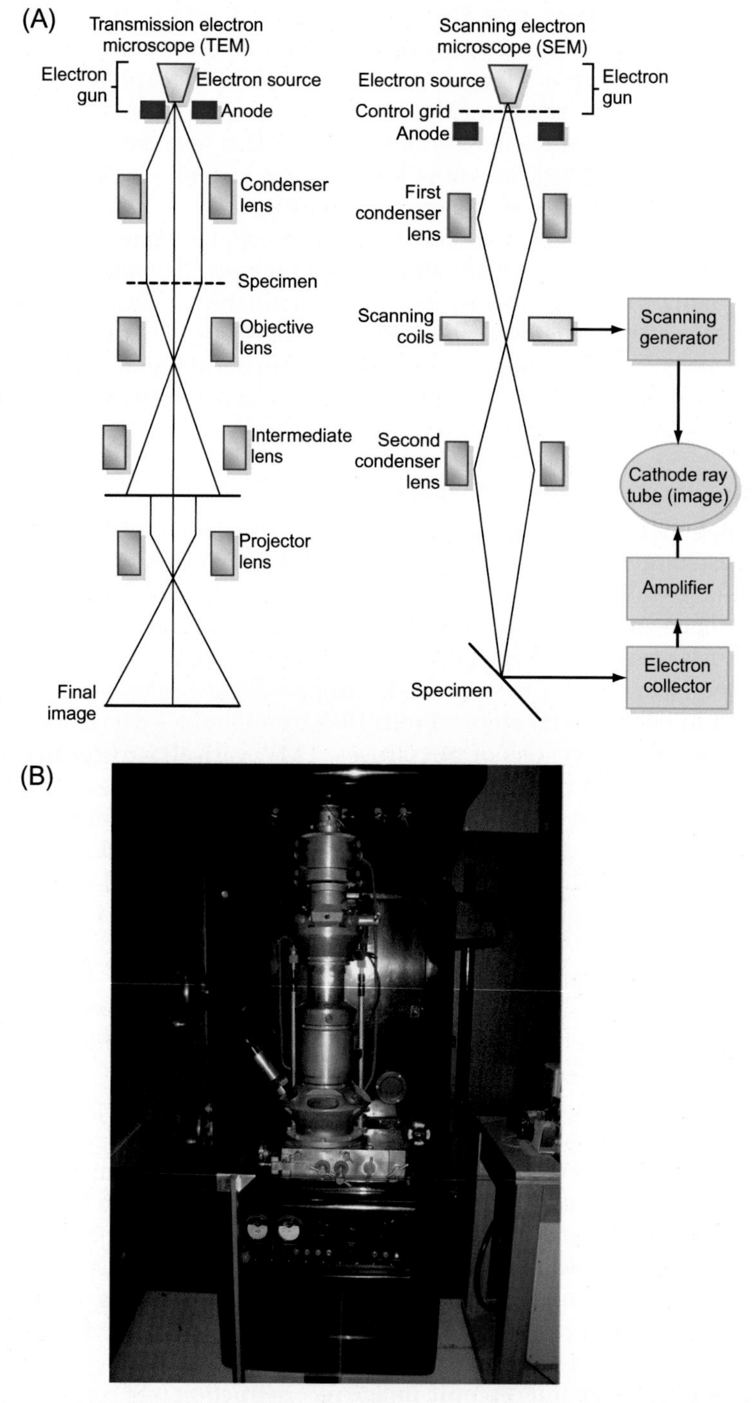

FIG. 1.11 Electron microscopy. (A) This figure shows the working principles of transmission and scanning microscopes. (B) Originally posted to Flickr as Electron Microscope Deutsches Museum by J. Brew. Creative Commons Attribution-Share Alike 2.0 Generic license.

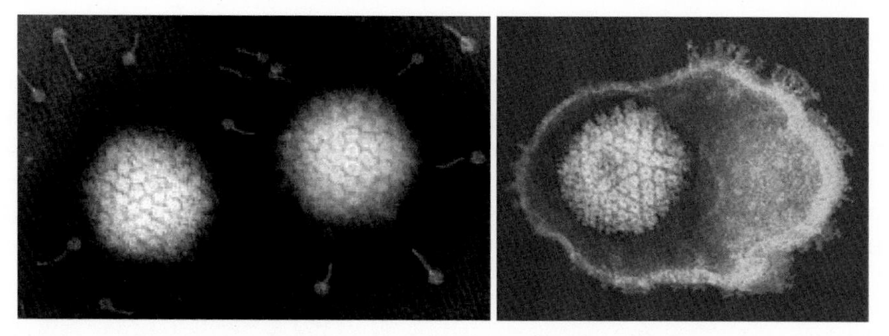

FIG. 1.12 Negatively-stained virus particles. Negatively-stained human adenovirus (left) and human herpesvirus type 1 (right). Images copyright LM Stannard.

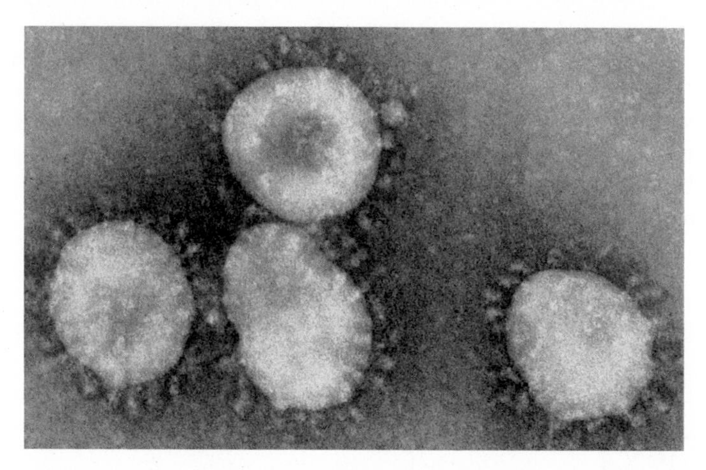

FIG. 1.13 Negatively-stained human coronavirus 229E particles (F Murphy, S Whitfield, CDC 1975).

An important discovery made possible by electron microscopy that has a particular resonance in these times of pandemic COVID-19, was the discovery by June Almeida in 1964 of the first **coronavirus**. This was a common cold-associated virus which had a similar appearance in electron micrographs—a halo or **corona** (crown-like) of envelope proteins surrounding essentially spherical enveloped particles—to **mouse hepatitis** and **infectious bronchitis virus** virions, also first seen by Dr. Almeida (Fig. 1.13).

Crystallography of virions

The most important early method for the detailed investigation of virion structures was the use of **X-ray diffraction** by crystals of purified virions, following the demonstration that it was possible to crystallize plant virus virions. This permits determination of their structure at an atomic level. TMV (first crystallized by Wendell Stanley in 1935) and **turnip yellow mosaic virus (TYMV)**, were among the first virus structures to be determined during the 1950s. It is significant that these two viruses represent the two fundamental types of virus nucleoprotein particle: **helical** in the case of TMV and **icosahedral** for TYMV (see Chapter 2). It is also often overlooked that Rosalind Franklin, of DNA structure fame, was the first to locate the RNA within TMV virions by X-ray diffraction.

The phage age

Since d'Herelle's work with bacterial viruses after 1918, phage biology leapt ahead of studies of other viruses, because of the ease, speed and convenience of the culture system. One of the more important aspects was the ease of **titrating** the phages: the "**plaque assay**" for counting infectious virions done on lawns of bacteria in a Petri dish was accurate, reproducible, and allowed everything from inactivation assays using radiation or chemicals, to growth cycle determinations, much more easily than could be done with animal viruses before the advent of egg-based culture.

In 1939, the former physicist Max Delbrück, working with the biologist Emory Ellis at Caltech, elucidated the growth cycle of a sewage-isolated Escherichia coli bacteriophage in a now-classic paper simply entitled *"The Growth of Bacteriophage"* (see Recommended Reading). This used the simple technique of counting plaques in a bacterial lawn following infection of a standard bacterial inoculum with a dilution series of a phage preparation. Their principal finding was that **viruses multiply inside cells in one step**, and not by **division and exponential growth** like cells. This was determined using the so-called "**one-step growth curve,**" which allowed the accurate determination of the titers of viruses released from bacteria that had been synchronously infected. This allowed calculation of not only the time of multiplication of the virus, but also the "**burst size**" from individual bacteria, or the number of viruses produced in one round of multiplication. This was a fundamental discovery, and allowed the rapid progression of the field of bacterial and phage genetics (Fig. 1.14).

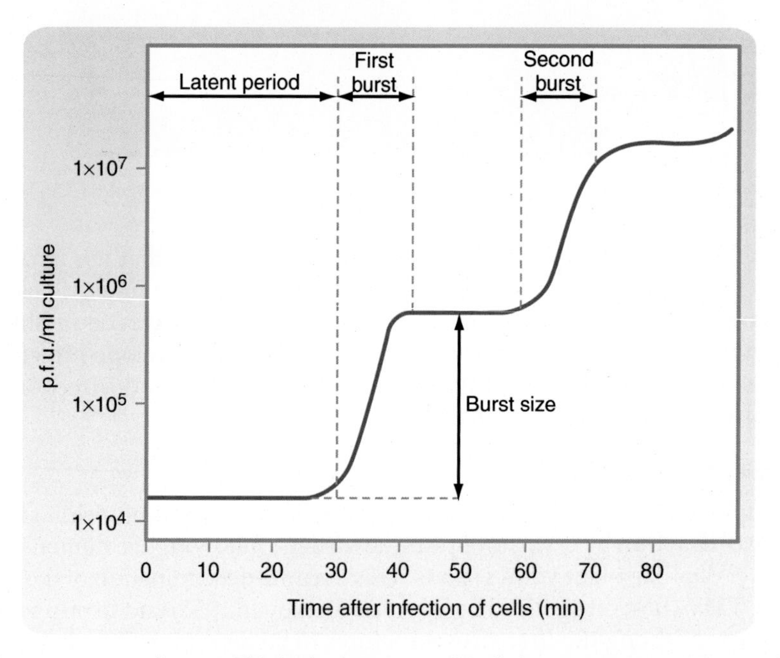

FIG. 1.14 The one-step growth curve. This is also known as a "single-burst experiment." First performed by Ellis and Delbruck in 1939, this classic experiment illustrates the true nature of virus replication. Details of the experiment are given in Chapter 4. Two bursts (crops of phage particles released from cells) are shown in this particular experiment.

One of the most important facets of this work was that it showed that **infection could be caused by single virions**: the power of the plaque assay meant that even dilutions of phage preparations that contained only a single particle could produce a detectable plaque.

Alfred Hershey and his coworker Martha Chase performed the legendary Hershey-Chase or "Waring blender" experiment, that they reported on in 1952, in order to prove whether or not DNA was the genetic material of a bacterial virus. They grew up preparations of the *E. coli* **bacteriophage T2** separately in the presence of the radioisotopes ^{35}S and ^{32}P, to label the **protein and nucleic acid components** of the phage respectively. They allowed adsorption of phages to bacteria in liquid suspension for different times, then sheared off adsorbed phage particles from the bacteria using the blender. Pelleting the bacteria by centrifugation and assaying radioactivity allowed them to determine that **over 75% of the ^{35}S**—incorporated into cysteine and methionine amino acids—remained in the liquid, or outside the bacteria, whereas **over 75% of the ^{32}P**—incorporated into the **phage DNA**—was found inside the bacteria. Subsequent production of phage from the bacteria showed that **DNA was probably the genetic material,** and that protein was not involved in phage heredity—a fundamental discovery at the time (Fig. 1.15).

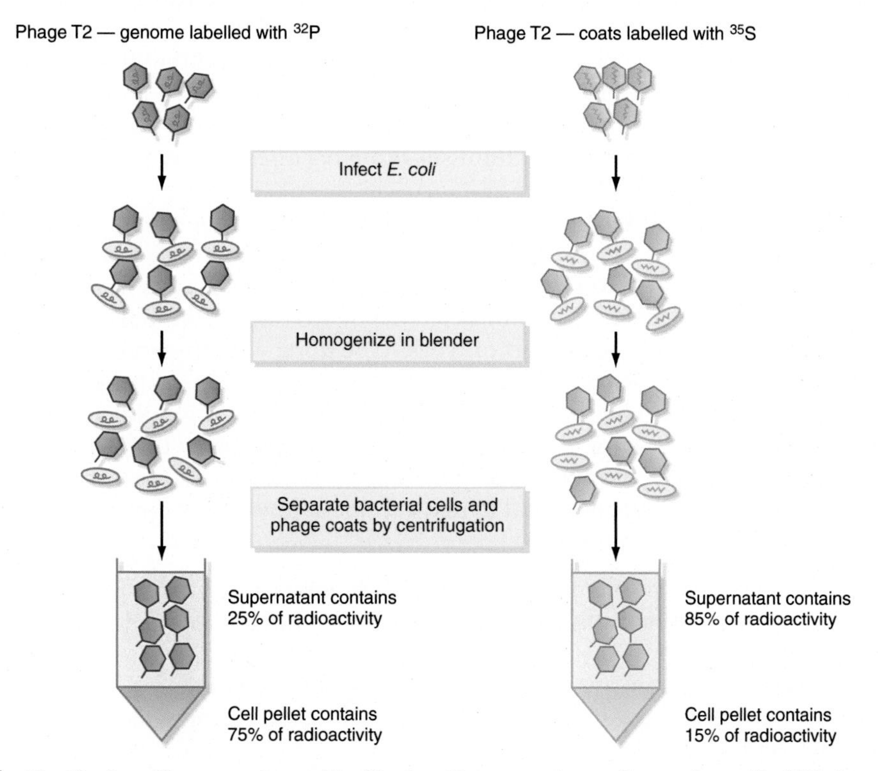

FIG. 1.15 The Hershey-Chase experiment. The Hershey Chase experiment, first performed in 1952, demonstrated that virus genetic information was encoded by nucleic acids and not proteins. Details of the experiment are described in the text.

Max Delbrück, Salvador Luria and Alfred Hershey received the 1969 Nobel Prize for Physiology or Medicine for their pioneering work on bacteriophages, which was fundamental in establishing the emerging field of **molecular biology**.

Animal cell culture

Cell culture began early in the twentieth century with **whole-organ cultures**, then progressed to methods involving individual cells, either **primary cell** cultures (cells from an experimental animal or taken from a human patient which can be maintained for a short period in culture) or **immortalized cell lines**, which, given appropriate conditions, continue to grow in culture indefinitely.

Possibly the most important development for the study of animal viruses since their discovery was the growing in 1949 of poliovirus in primary human cell cultures: this was reported by John Enders, Thomas Weller and Frederick Robbins from the USA, and was rewarded with a joint Nobel Prize to them in 1954. They did this around the same time as David Bodian and Isabel Morgan identified **three distinct types of poliovirus**. Previously, titration or assay of poliovirus, for example, required the injection of virus preparations into the brains of monkeys, or later, in the case of the **Lansing or Type II poliovirus strain**, into brains of mice.

Renato Dulbecco in 1952 adapted the technique to **primary cultures of chicken embryo fibroblasts** grown as monolayers in glass flasks. Using **Western equine encephalitis virus** and **Newcastle disease virus** of chickens, he showed for the first time that it was possible to produce **plaques** (see Fig. 1.8) due to an animal virus infection, and that these could be used to accurately assay **infectious virus titers**.

This figure shows a single-burst type of experiment for a picornavirus (e.g., poliovirus). This type of data can only be produced from ***synchronous infections*** where a high multiplicity of infection is used (Fig. 1.16).

He and Marguerite Vogt went on in 1953 to show the technique could be used to assay poliovirus—and went on to show that the principle of "**one virus, one plaque**" first established with phages, and later to plant viruses with local lesion assays in leaves, could be extended to animal viruses too. This heralded a major advance in animal and human virus studies, as it now allowed the same sort of work to be done with these viruses as had been done on phages for many years already. For example, it largely replaced earlier assays such as **endpoint dilution techniques**, which had been done in whole animals as infectious dose (ID50) assays, or later as **tissue culture infectious dose ($TCID_{50}$)** assays. While animal host systems are still used in modern virology—for example, to study pathogenesis of animal and some human viruses, and to test vaccine safety—they are increasingly being discarded for the following reasons:

- Breeding and maintenance of animals infected with pathogenic viruses is expensive.
- Animals are complex systems in which it is sometimes difficult to isolate the effects of virus infection.
- Results obtained are not always reproducible due to host variation.
- Unnecessary or wasteful use of experimental animals is increasingly viewed as being morally unacceptable.

The viral agent of **measles** was characterized by Thomas Peebles and Enders via tissue culture in 1954; **adenoviruses** were discovered in 1953 by Wallace Rowe and Robert Huebner and

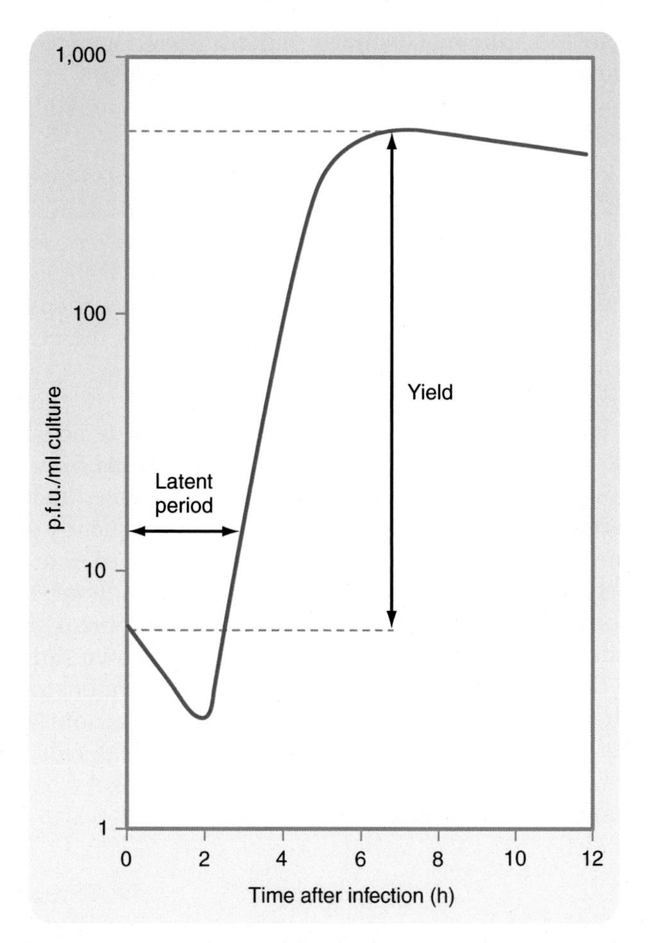

FIG. 1.16 Replication of lytic eukaryotic viruses. This can be assayed in a very similar way to that of bacterial viruses, except in cultured eukaryotic cells (see Fig. 1.13).

shown to be associated with **acute respiratory disease** soon afterward, by Maurice Hilleman and others. In the 1950s and 1960s, tissue culture led to the identification and isolation of many viruses and their association with human diseases—for example, common viruses such as the ssRNA **enteroviruses**, related to polioviruses. Widespread virus isolation led to the realization that ***subclinical*** virus infections—infections with no obvious symptoms—were very common; for example, even in ***epidemics*** of the most virulent strains of poliovirus there are approximately **100 subclinical infections for each paralytic case of poliomyelitis**. This is explored further in Chapter 8.

Virus infection has often been used since to probe the working of "normal" (i.e., uninfected) cells—for example, to look at macromolecular synthesis. This is true of the applications of phages in bacterial genetics, and in many instances where the study of eukaryotic viruses has revealed fundamental information about the cell biology and genomic organization of higher organisms. **Polyadenylation of messenger RNAs** (mRNAs) (1970), **chromatin**

structure (1973), and **mRNA splicing** (1977) were all discovered in viruses before it was realized that these phenomena could also be found in uninfected cells.

RNA as a genetic material

While it had been known since Bawden and Pirie's work in 1937 that TMV particles contained RNA, followed later by a number of other viruses, it must be remembered that DNA had only really been accepted as the genetic material of cells and viruses after the Hershey-Chase experiment in 1952 and the demonstration of the nature of DNA in 1953. Moreover, the way in which the information in DNA was used to make proteins was still very obscure in the 1950s, given that the proof that RNA was used as a template for the production of proteins was only provided in 1961 by Marshall Nirenberg.

It was regarded as a major breakthrough in molecular biology, therefore, when Heinz Fraenkel-Conrat, his wife Bea Singer, and Robley C Williams demonstrated in 1955–1957 that it was possible to **reconstitute fully infectious TMV virions** from separately-purified preparations of coat protein and RNA. At the time it was assumed that neither of the two components was infectious on its own; however, it was subsequently shown by Fraenkel-Conrat and Singer, and separately by Alfred Gierer and Gerhard Schramm, that **purified TMV RNA was in fact infectious**—albeit several hundred times more weakly per unit mass than the native or reconstituted particles. This was also a ground-breaking achievement, as it was the **first proof that RNA could be a genetic material in its own right** (Fig. 1.17).

Further important developments with TMV included the demonstration by Gierer and Karl-Wolfgang Mundry in 1958 that TMV mutants with altered genomes could be produced by treatment of virions with **nitrous acid**, which only alters nucleic acids, and the sequencing of the TMV coat protein in 1960 by two groups including Fraenkel-Conrat and Stanley and Knight in one, and Schramm in the other.

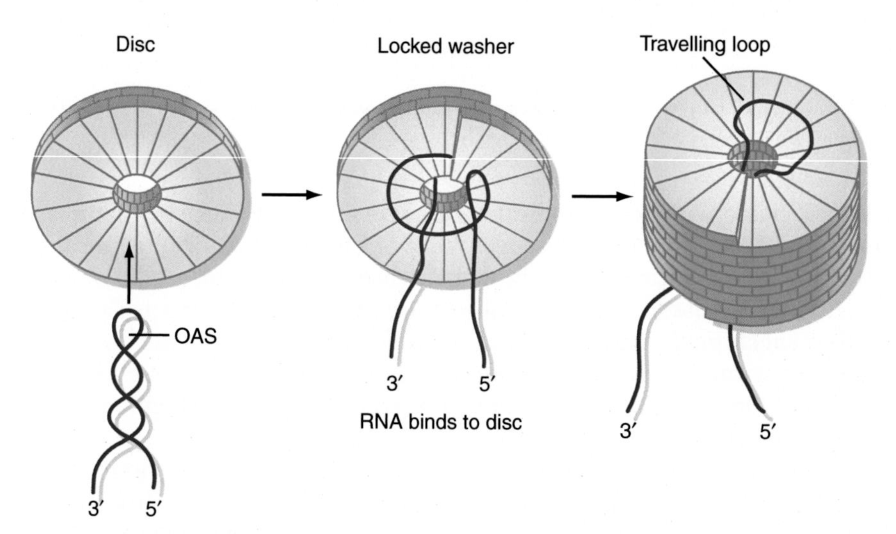

FIG. 1.17 TMV assembly. Purified coat protein and RNA can spontaneously self-assemble to form an infectious virion. The capsid proteins bind first to an "origin of assembly," then progressively trap and protect the fragile RNA genome of the virus as the helical particle forms.

Working from the example of Gierer and Scramm with TMV, Hattie Alexander and colleagues demonstrated in 1958 that RNA extracted from concentrated, partially purified preparations of polioviruses types I and II, that was free of protein and DNA, was infectious in cultured HeLa cells and human amnion cell monolayers. Moreover, the RNA produced progeny virus characteristic of that used to produce the RNA. The concept of RNA genomes was still new enough at the time to prompt their conclusion that the virus RNA was the essential infectious agent—**the first time this had been shown for an animal virus**. In the same year, the same was shown by Fred Brown for another distantly related picornavirus, **foot and mouth disease virus (FMDV)**. By 1968, it was known—thanks to Michael F Jacobson and David Baltimore and others—that several picornaviruses related to poliovirus appeared to have just one **open reading frame (ORF)** encoding a single polypeptide in their genomic RNA, and did not make a smaller mRNA (see Chapters 4 and 5).

An interesting class of new viruses was discovered in humans, birds, and later in other animals too, in the period 1953–1964. These were dubbed "**respiratory enteric orphans**" based on where they were found, and the fact they were not associated with any disease—which gave rise to the name **reovirus**, and their description as a distinct group of viruses by Albert Sabin in 1959. By 1962, the unique double-layered capsid morphology had been seen and the virions shown to contain RNA, and then in 1963 Peter J Gomatos and Igor Tamm showed using physical and chemical techniques that the viruses as well as the similar **wound tumor virus** isolated from plants had a genome consisting of **double-stranded (ds) RNA**—a finding unprecedented in biology at the time. Gomatos and Walther Stoeckenius went on to show by electron microscopy in 1964 that the reovirus genome was also segmented—another unprecedented finding for viruses.

A major highlight in molecular biology was Marshall Nirenberg and Heinrich Matthaei's 1961 demonstration of

"...an assay system in which RNA serves as an activator of protein synthesis in E. coli extracts,"

or the proof in an in vitro translation system that **RNA was the "messenger"**—now known as mRNA—that conveyed genetic information into proteins.

In 1962, A Tsugita, Fraenkel-Conrat, Nirenberg, and Matthaei used the still extremely novel in vitro translation system with purified TMV genomic RNA, and were able to show that:

"The addition of TMV-RNA to a cell-free amino acid incorporating system derived from E. coli caused up to 75-fold stimulation in protein synthesis (^{14}C-incorporation)."

Part of the protein synthesized formed a specific precipitate with anti-TMV serum, indicating that TMV coat protein had been made.

This was the first demonstration of in vitro translation from *any* specific mRNA, and incidentally also direct proof that the single-stranded TMV genome was **"messenger sense" RNA**. They concluded that their result showed that the newly-determined ***"genetic code"***—the nucleotide triplets that code for individual amino acids—was probably universal, given that it was a tobacco virus RNA being translated by a bacterial system (Fig. 1.18).

Later in 1962, D Nathans and colleagues used **coliphage f2 RNA** as template for translation in the same type of bacterial extract. They showed that polypeptides corresponding to the coat as well as other proteins were made, showing that it was the input virion RNA that was responsible.

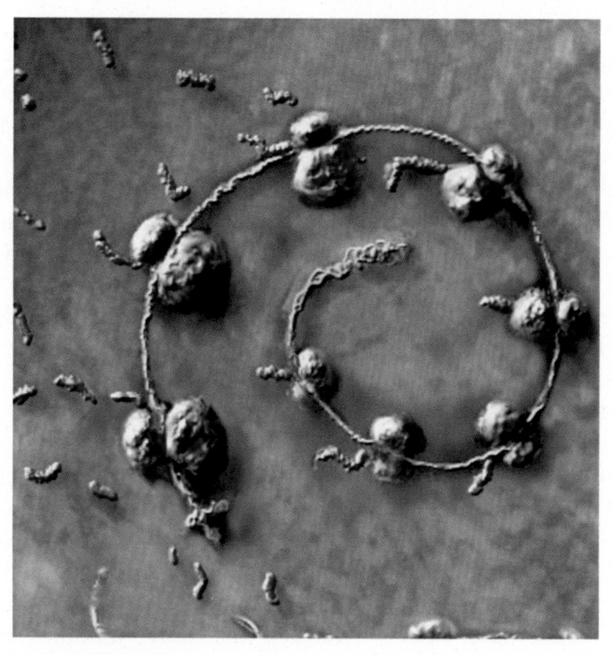

FIG. 1.18 Translation of viral RNA. Ribosomes translating protein from a messenger RNA molecule in the cell cytoplasm. *Courtesy of Russell Kightley Media.*

The dawn of molecular virology

The proof that RNA was both the messenger that conveyed information from DNA to be made into protein, and was in fact a genetic material in its own right, made possible a revolution in virology that transformed it into the science we know today. The new molecular biology together with well-established physical and biochemical techniques for molecular characterization, coupled with the ability to reliably culture bacterial, plant and now animal viruses as well, enabled an explosion of discovery that continues to this day.

A *tour de force* experiment in the modern molecular biological era was the 1965 in vitro synthesis of an infectious phage RNA genome by S Spiegelman and coworkers, using only purified Qß E. coli phage single-stranded virion RNA and the purified **viral replicase**. They remarked:

> *"For the first time, a system has been made available which permits the unambiguous analysis of the molecular basis underlying the replication of a self-propagating nucleic acid."*

There followed the demonstration in 1967 that the same could be done for a **ssDNA virus**: M Goulian and colleagues reported in that they had successfully made a completely synthetic and infectious **PhiX174 phage** genome, by means of a series of syntheses using purified virion ssDNA, *E. coli* DNA polymerase and a "**polynucleotide-joining enzyme,**" or **DNA ligase**.

Naked nucleic acids as infectious agents: Viroids

A potato disease that had been known in the New York and New Jersey state areas in the USA since the 1920s was the source of an exciting discovery by Theodor (Ted) Diener and WB Raymer, reported in 1967. The **potato spindle tuber** disease agent had proved recalcitrant over many years to being characterized or isolated; all that was known was that it could be transmitted mechanically using sap, or via grafting, and that no fungi, bacteria or viruses could be isolated from diseased material. Diener and Raymer speculated that the extractable infectious agent could be a **double-stranded RNA**.

By 1971 Diener had determined that the infectious RNA was several species with molecular weights ranging from **2.5×10^4 to 1.1×10^5 Da**—unprecedently small for a viral genome. No evidence for the presence in uninoculated plants of a latent helper virus was found. Thus, potato spindle tuber "virus" RNA was too small to contain the genetic information necessary for self-replication, and must rely for its replication mainly on plant mechanisms.

This was a revolutionary concept: **an infectious, pathogenic entity in the form of a naked RNA** that was too small to encode a replicase or any other protein. He proposed the term "**viroid**" to designate this and similar agents, a term that persists up to today. By 1979, they were known to be single-stranded circular RNA molecules with a high degree of sequence self-complementarity, which results in them appearing as "highly base-paired rods" (Fig. 1.19).

Reverse transcription and tumor viruses

In 1963, Howard Temin showed that the replication of **retroviruses**, whose particles contain RNA genomes, was inhibited by actinomycin D, an antibiotic that binds only to DNA. The replication of other RNA viruses is not inhibited by this drug. These facts were largely ignored until 1970, when two back-to-back papers in Nature magazine, by Howard Temin and S Mituzami, and David Baltimore respectively, revealed that "**RNA tumor viruses**" such as the agents found by Ellerman and Bang and Peyton Rous contained an enzyme activity named "**reverse transcriptase"** or more correctly **RNA-dependent DNA polymerase** in their virions, which converted the single-stranded RNA genomes into double-stranded DNA. Later this was shown to result in resulted in insertion of the DNA of the viral genome into the host cell genome, vindicating Howard Temin's 1960 proposal that.

> *"...a RNA tumor virus can give rise to a DNA copy which is incorporated into the genetic material of the cell."*

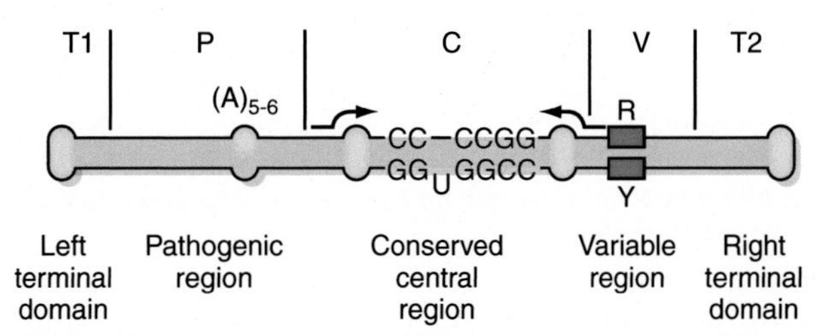

FIG. 1.19 Structure of a viroid. All viroids share common structural features despite having very different sequences. The central region is, however, highly conserved, and is involved in replication.

Baltimore and Temin both received a share of the Nobel Prize in Physiology or Medicine in 1975 for their discovery of reverse transcriptase—and shared it with Renato Dulbecco, who was credited with clarifying the process of infection and of cellular transformation by **DNA tumor viruses**.

Dulbecco used the double-stranded (ds) DNA **mouse polyomavirus PyV:** this was shown from 1959 onward to cause a variety of tumors in hamsters, hence the name "**poly-oma.**" The virus had been discovered in 1953 and tissue cultured by 1957; it was shown in 1959 that its purified DNA was infectious, and by 1963 by Dulbecco and others to have a **small, circular double-stranded DNA genome**.

He and colleagues showed that polyomavirus grew and could be assayed normally in certain cell cultures, but caused **tumor-like *transformation* of cells** in others in which it did not grow. They showed that transformed cell chromosomes contained covalently integrated viral DNA termed a **provirus**, which was active in producing mRNA which made virus-specific proteins. Thus, his work was the first to show how **DNA viruses might cause cancer**, and he and his colleagues deserved their award.

"...for their discoveries concerning the interaction between tumour viruses and the genetic material of the cell."

The related **simian vacuolating virus 40,** or **SV40**, was discovered in 1960 by Ben Sweet and Maurice Hilleman as a contaminant of live attenuated polio vaccines made between 1955 and 1961: this was as a result of use of vervet or "African green monkey" kidney cells that were inadvertently infected with SV40 to grow up the polioviruses. As a consequence, between 1955 and 1963, **up to 90% of children and 60% of adults—98 million people**—in the USA were inadvertently inoculated with live SV40. Given the demonstration by Bernice Eddy and others in 1962 that hamsters inoculated with simian cells infected with SV40 developed **sarcomas and ependymomas**, the class of viruses including SV40 and MPyV described earlier became known as "**polyomaviruses,**" and **DNA tumor viruses**. However, and despite considerable concern over many years, **SV40 has not been shown to cause or to definitively be associated with any human cancers**.

Viral genome cloning

The techniques of **recombinant DNA technology**—or the artificial introduction of genetic material from one organism into the genome of another—were pioneered between 1971 and 1973 by Paul Berg, Herbert Boyer and Stanley Cohen. In 1971, Berg performed an in vitro exercise in which a segment of the **lambda phage genome** was ligated into the purified DNA of SV40, which had been linearized using the then-new **restriction endonuclease, *Eco*RI**. Cohen, Annie Chang, Boyer and Robert Helling took the technology further in 1973 by showing that:

"The construction of new plasmid DNA species by in vitro *joining of restriction endonuclease-generated fragments of separate plasmids is described. Newly constructed plasmids that are inserted into Escherichia coli by transformation* ***are shown to be biologically functional replicons that possess genetic properties and nucleotide base sequences from both of the parent DNA molecules."***

Thus, **molecular cloning** had arrived—made possible in large part by use of viruses. The fundamental nature of this advance of molecular biology was rewarded by a half share of the 1980 Nobel Prize in Chemistry to Paul Berg.

Viral genome sequencing

Nucleotide sequencing, or the determination of the order of bases in nucleic acids, started with laborious, difficult techniques such as the two-dimensional fractionation of enzyme digests of **^{32}P-labeled RNA** described by Frederick Sanger and colleagues in 1965. DNA sequencing followed in 1970: Ray Wu described the use of *E. coli* **DNA polymerase** and radioisotope-labeled nucleotides to sequence the single-stranded ends of phage lambda DNA. He and colleagues followed this with a more general method in 1973, using extension of synthetic oligonucleotide ***"primers"*** annealed to target DNA—the so-called "**primer extension**" technique.

Walter Gilbert and Allan Maxam published in **1977** an immediately popular paper entitled *"A new method for sequencing DNA."* This became known as **Maxam-Gilbert sequencing**, or the chemical method, as it entailed sequencing by chemical degradation. Also in 1977, however, Frederick Sanger and colleagues adapted the Wu technique to come up with the so-called **Sanger method**, or "**DNA sequencing with chain-terminating inhibitors**": this soon became the industry standard for at least the next 20 years, because it was considerably easier and cheaper than the chemical method.

Gilbert and Sanger were awarded a share of the Nobel Prize in Chemistry in 1980, *"for their contributions concerning the determination of base sequences in nucleic acids."*

The first virus sequenced: MS2 coliphage

Walter Fiers and his coworkers completed the genome sequencing of the **ssRNA *E. coli* phage, MS2**, in 1976. They had previously also been responsible for **the first ever gene sequence, in** 1972: This was of the **coat protein gene** from the same virus. This was a highly significant publication, because it completed the work of years by their group by achieving the landmark that.

> ***"MS2 is the first living organism for which the entire primary chemical structure has been elucidated."***

PhiX174 phage sequencing

The next complete viral genome sequenced was that of the **circular single-stranded DNA coliphage PhiX174**, in 1977 by Sanger and his team in Cambridge, using the new sequencing technique invented by them. This was a DNA sequence of 5375 nucleotides.

This was the **first complete genome sequenced for any DNA-containing organism**, and a satisfying conclusion to many decades of work on the virus. One of the most interesting features of the sequence was the fact that several of the 11 genes are highly overlapping: that is, the **same DNA sequence is used to encode completely different genes in different open reading frames**. This represented an economy of use of genetic information that was hitherto unknown.

SV40 sequencing

SV40 had become an object of considerable interest as mentioned earlier in connection with Renato Dulbecco, and it was accordingly the next virus to be completely sequenced. This was by Walter Fier's group in 1978: They determined by Maxam-Gilbert sequencing that the circular dsDNA genome comprised 5224 base pairs, and had an interesting organization. In their words:

"...the T antigen is coded by two non-contiguous regions of the genome; the T antigen mRNA is spliced in the coding region. In the late region the gene for the major protein VP1 overlaps those for proteins VP2 and VP3 over 122 nucleotides but is read in a different frame."

This was the first time that **RNA splicing** had been demonstrated for an entire genome; indeed, it had only been discovered in 1977 when two separate groups of researchers showed that adenovirus-specific mRNAs made late in the replication cycle in cell cultures were **mosaics**, being comprised of **sequences from noncontiguous or separated sites in the viral genome**. This was subsequently found to be a common feature in eukaryotic but not prokaryotic mRNAs.

Sequencing of a viroid

In 1978, Heinz Sänger's group published the sequence and the predicted secondary structure of **potato spindle tuber viroid**. This was the first RNA genome to be sequenced using the still relatively new method of **generating complementary DNA (cDNA) from RNA** by use of reverse transcriptase. They stated in their Nature paper abstract that:

"PSTV is the first pathogen of a eukaryotic organism for which the complete molecular structure has been established."

This was true despite earlier whole genome sequence determinations, because in this case **the genome was the whole organism**, and they could predict a secondary structure for it (see Fig. 1.2) .

Sequencing of the first human polyomavirus

Two groups in 1979 published the complete nucleotide sequences of two strains of the **polyomavirus** known as **BKV**: this had been first isolated from a renal transplant patient with those initials in 1971, and **found to be present in about 80% of healthy blood donors**. It causes only mild infections—fever and respiratory symptoms—on first infection, and then subsequently infects cells in the kidneys and urinary tract, where it can remain causing no symptoms for the lifetime of infected individuals. In retrospect, this was one of the first demonstrations that viruses could be ***commensal*** **organisms**: that is, they could multiply in their host, but cause no harm.

The hepatitis B virus genome

Hepatitis B virus (HBV) was discovered more or less accidentally during serological studies in the 1960s by Baruch Blumberg and colleagues. However, a transmissible agent had been implicated in "**serum hepatitis**" as early as 1885, when A Lurman showed that "contaminated lymph" (serum) was to blame for an outbreak in a shipyard in Bremen, Germany, after a smallpox prevention exercise. Subsequently, reuse of hypodermic needles first introduced in 1909 was shown to be responsible for spreading the disease. By 1968 the "**Australia antigen**" was seen to consist of **22 nm empty particles**; by 1970 a **42 nm DNA-containing "Dane particle"** was found, which is now known to be the virion. By 1975 Blumberg and others had also implicated HBV in the causation of **primary hepatic carcinoma**, now known as **hepatocellular carcinoma (HCC),** and a serious complication of chronic infection with HBV, especially if acquired in early life.

The complete genome sequence of an *E. coli* genomic clone of the subtype ayw strain of HBV was reported by Francis Galibert and coworkers in 1979. This consisted of **3182** nucleotides, arranged as what had previously been shown by physicochemical techniques on DNA isolated from virions to be a circular structure with a "-" strand with a short gap, and an incomplete "+" strand of varying length (see Fig. 5.24). The reason for this would have to wait a few years, and would result in a new class of viruses being recognized that would reflect **very deep evolutionary links between viruses and cellular elements** (see Chapter 3).

Cauliflower mosaic virus

The first of what are now known to be **caulimoviruses**—family *Caulimoviridae*—was described in 1933 as dahlia mosaic virus. **Cauliflower mosaic virus (CaMV)** was described in 1937, and shown to have particles containing a DNA genome in 1968. This was visualized by electron microscopy by two groups in 1971 as relaxed open circles or also as linear forms—unlike the **supercoiled DNAs** of papilloma- or papovaviruses. By 1977 it was known that the "**nicked circular form**" was infectious, and fragments of the genome had been cloned in *E. coli*. Physical and biochemical characterization of the genome in 1978 showed that it consisted of three discrete lengths of single-stranded DNA—alpha, beta, and gamma, with alpha being the full genome-length—that annealed to one another to give a circular double-stranded form about 8000 nucleotides in length. By 1979 it was known that only the alpha strand was transcribed to give mRNA.

The complete sequence of the viral genome was published in 1980 (see Fig. 5.25), and was predicted to encode six proteins: this has since been upped to eight. The genome is transcribed into only two mRNAs—named **35S** and **19S**, on the basis of their sedimentation properties—and contains one discontinuity in the alpha or coding strand, and two in the noncoding sequence. Cloned DNA was also shown to be infectious in 1980, even if excised as a linear molecule.

Infectious, cloned poliovirus RNA

Naomi Kitamura and coworkers published the complete nucleotide sequence of **poliovirus type I** in 1981: this was a **ssRNA+ molecule 7433 nucleotides long, polyadenylated** at the 3′ terminus, and covalently linked to a small protein (VPg) at the 5′ terminus. An open reading frame of 2207 consecutive triplets spanned over 89% of the nucleotide sequence. Also in 1981, Vincent Racaniello—working in David Baltimore's lab—also sequenced the poliovirus I genome.

Later in **1981,** Racaniello used the three clones to construct one contiguous cDNA clone in pBR322, which he successfully used to transfect vervet monkey (=African green monkey) kidney cell cultures, producing infectious wild-type virus with which he produced characteristic plaques in HeLa cells. **This was the first proof that an infectious cDNA construct could be made for an RNA virus.**

Tobacco mosaic virus sequenced

P Goelet and five coworkers published the complete nucleotide sequence of the Vulgare (=common) strain of **tobacco mosaic virus** in 1982. They did this by using reverse transcriptase and synthetic oligonucleotide primers to generate a set of short, overlapping complementary DNA fragments covering the whole TMV genome. They reported a **6395 nucleotide sequence**, with just three major ORFs. They also found a unique **hairpin loop-encoding sequence region for assembly initiation**—nicely rounding out pioneering work by P Jonathan

Butler and colleagues and Genevieve Lebeurier and others—both published in January 1977, incidentally—on the physical mechanism of assembly, which had shown a **single site for assembly nucleation**. This has subsequently been used for making **virus nanoparticles (VNPs)** for use in various biotechnological fields.

The circle closes: TMV understood

The sequencing of the TMV genome almost, but not quite, brought to a complete circle the journey of discovery that started with its description as a "contagium vivum fluidum" in 1898. Closing the circle required a complete molecular understanding of the virus particle, which was achieved with the publication from Gerald Stubbs' group in 1986 of the **3.6 ångström (Å) resolution structure for TMV**, and then the refined **2.9 Å structure** in 1989.

This allowed the building of molecular models that helped explain the chemical and physical basis for virion self-assembly, as well as accounting for the positioning in particles of the **whole RNA and the whole of the more than 2000 individual protein subunit sequences** (Fig. 1.20).

The structure also brought to a conclusion a journey that had started as early as 1936, after Stanley's demonstration that TMV could be crystallized. Stubbs celebrated his group's subsequent refinements of the structure with this statement in the centenary book on TMV:

"...in 1989, Bernal and Fankuchen's remarkable patterns finally yielded the fulfillment of ***[Rosalind] Franklin's vision*** *with the publication of the 2.9 Å resolution structure...".*

A modest statement for a landmark achievement: the finalizing of the complete molecular understanding of the first virus discovered. **The circle was finally closed, 91 years after Beijerinck first showed how a filter could define a new class of organisms** (Box 1.1).

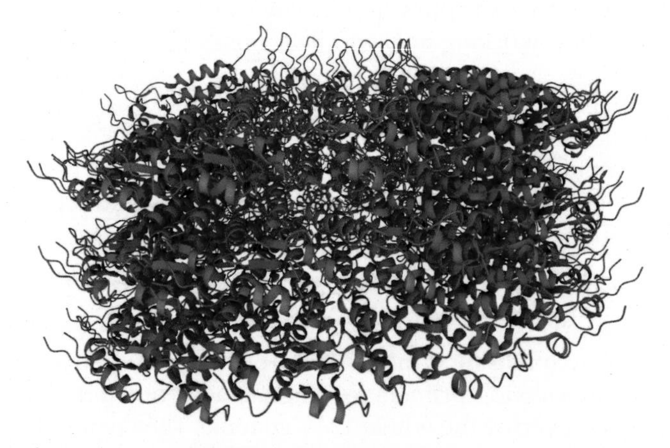

FIG. 1.20 Atomic-resolution structure of TMV. This figure depicts the 3-D reconstruction from X-ray crystallographic data of the 2.9 Å resolution structure of TMV virions. Protein is shown in *brown*, RNA in *green*. *From Protein Database (DB) accession 2TMV, https://doi.org/10.2210/pdb2TMV/pdb. Deposition Authors: Stubbs, G., Pattanayek, R., Namba, K., 1989.*

BOX 1.1

The timeline of TMV

The history of TMV has been well covered elsewhere—with an excellent free text written by Karen-Beth Scholtof, available online (see Recommended Reading)—but it is worth recapitulating how many times TMV appears in the history of virology, as new technological developments allowed new findings about its properties.

1898: Beijerinck shows that a bacteria-free filtered extract from mosaic diseased tobacco is infectious, and coins a new term—"**contagium vivum fluidum.**"

1929: FO Holmes shows that **local lesions** caused by infection of tobacco by TMV could be used as a means of assaying the infectivity of virus stocks.

1935: Stanley crystallizes TMV, and also shows by centrifugation and electron microscopy that it consists of **large, regular rodlike particles**.

1936: Bernal and Fankuchen publish X-ray diffraction pictures from gels of purified TMV.

1937: Bawden and Pirie show using chemical methods that TMV is a **ribonucleoprotein**.

1954: James Watson shows by X-ray studies that TMV is probably **helical**.

1956: Fraenkel-Conrat and Gierer and Schramm showed that **TMV RNA was infectious**, and provided all the information necessary to produce virions in plants.

1956: Rosalind Franklin locates the RNA within TMV virions by X-ray diffraction.

1958: Franklin and Holmes show that TMV virions are **helical**.

1958: Gierer and KW Mundry show that **TMV mutants with altered genomes** could be produced by treatment of virions with nitrous acid, which only alters nucleic acids.

1962: Tsugita et al. provided the **first demonstration of in vitro translation** from a specific mRNA, and also direct proof that the single-stranded TMV genome was "messenger sense."

1975: Holmes and Stubbs publish a 6.7Å map of the structure of TMV.

1977: Holmes, Stubbs & Warren publish 4 ÅA structure for TMV.

1977: Butler, Lebeurier and colleagues publish models for the **self-assembly of TMV.**

1982: Goelet and colleagues sequence the TMV genome.

1986: Sleat and colleagues showed that in vitro transcribed mRNAs containing a 5′ ORF and a 3′ TMV origin of assembly sequence could be efficiently encapsidated by purified TMV CP **to form pseudovirions.**

1989: Stubbs' group publishes a 2.9Å resolution map of TMV structure.

Transgenic hosts and viruses

The advent of molecular cloning unlocked the possibility of permanently inserting genes of our choosing—a technique known as ***"transgenesis,"*** or the making of a **transgenic organism**—into all cellular life forms. This obviously started with the invention of the technique using recombinant bacterial plasmids in the 1970s, and was then extended to plasmids of yeast, then to insertion of genes into the main genomes of bacteria and yeast, to making transgenic plants and plant cells in the 1980s, and then to animals (Box 1.2).

One unique advantage when using plants is that it is possible to insert foreign genes into cell genomes using an engineered form of a naturally-occurring soil bacterium: this was

BOX 1.2

What's wrong with transgenics?

For thousands of years farmers have transferred genes from one species of plant into another by crossing two or more species. This is the way that wheat was created over 10,000 years ago, as well as other food crops such as bananas. There was no control, other than trial and error, over which genes were transferred or the properties the resulting offspring possessed. In the 1980s it became possible to genetically modify plants and animals by transferring specific genes or groups of genes from another species. And so the controversy over genetically modified organisms (GMOs) arose: were they the saviors of humanity, feeding the starving and reducing pollution, or heralds of environmental doom? At about the same time, the first transgenic.

mice were made. Although there was an outcry at the time, this was dwarfed by the controversy over the first transgenic monkey in 2001. Genetically modified versions of our human relatives seemed too close to home for some people as this reminds them of eugenics, the selective breeding of humans with its very negative political and moral associations. In truth, science and technology are neutral, and it is societies who ultimately decide how they are used. Should we use these new technologies to feed the world and cure disease, or abandon them for fear of misuse? **It's not the technology, it's what we do with it that matters.**

formerly known as *Agrobacterium tumefaciens*, and is now named *Rhizobium radiobacter*. The organism naturally causes "**crown gall disease**" in plants that it colonizes: these are fleshy tumorous growths resulting from multiplication of cells into which the bacterium has inserted bacterial genes that produce metabolites beneficial to the bacteria. The mechanism of gene transfer was elucidated in 1974 in Rob Schilperoort and colleagues in Belgium—and was linked to a large plasmid, subsequently named the "**Ti plasmid,**" for "tumor-inducing." A specific segment of the Ti plasmid called the **"transferred DNA" (T-DNA)** was found by Mary-Dell Chilton's lab in the USA in 1977 to be mobilized into plant cells as a **protein-coated single-stranded linear DNA**, enabled by various factors encoded in the Ti plasmid. This DNA would be conveyed to into plant cells via bacterial-induced pores in plant cell walls and membranes, and then into the nuclei of these cells via **"nuclear localization sequences"** in the protein associated with it. In the nuclei that DNA is converted to linear double-stranded DNA by host repair polymerases. This can exist for some considerable time as an **episome** or independent subgenomic DNA—or more rarely, be incorporated covalently into the host genome (see Recommended Reading). By 1983 several groups of researchers including Chilton and the Marc van Montagu/Jeff Schell group in Belgium in had announced the use of *R. radiobacter* carrying engineered Ti plasmids to ***transform*** plant cells. The maturation of the technology came in **1984** from the Schell/van Montagu lab, with the regeneration of whole plants from transformed patches of tissue by means of selection for cells expressing antibiotic genes from the recombinant integrated T-DNA.

The generation of transgenic plants was very well established by the mid-1980s—and one of the first applications of the technology from 1986 onward was investigation of the potential for **engineering plants for resistance to plant viruses**, by Roger Beachy and coworkers. This started with the presumption that if transgenic plants expressed TMV coat protein (CP), virions entering their cells could not "**uncoat**" due to the abundant presence of TMV CP, and would therefore not be expressed and initiate an infection. This was in fact successful, and was copied for a number of other plant/virus combinations, including the transgenic papayas expressing **papaya ringspot virus CP**, which have played a major role in saving the Hawaiian papaya industry. However, later work showed that most resistance in transgenic plants was in fact due to "RNA interference" (RNAi) mechanisms (see Chapter 6).

In the 1980s, the first transgenic animals were also produced. Inserting all or part of a virus genome by means of **cell transfection** and ***illegitimate recombination*** into the nuclear DNA of an embryo (typically of a mouse) results in expression of virus mRNA and proteins in the resulting animal. This allows the pathogenic effects of virus proteins, individually and in various combinations, to be studied in living hosts. "**Humanized**" mice have been constructed from immunodeficient animals transplanted with human tissue. These mice form an intriguing model to study the pathogenesis of **human immunodeficiency virus (HIV)** in particular, as there is no real alternative to study the properties of HIV in vivo. Similarly, transgenic mice have proved to be vitally important in understanding the biology of prion genes. While these techniques raise the same moral objections as the more old-fashioned experimental infection of animals by viruses, they are immensely powerful new tools for the study of virus pathogenicity. A growing number of plant and animal viruses genes have been analyzed in this way, but the results have not always been as expected, and in some cases it has proved difficult to equate the observations obtained with those gathered from experimental infections. Nevertheless, this method has become quite widely used in the study of important diseases where few alternative models exist.

Viruses and serological/immunological methods

In 1941, George Hirst observed **hemagglutination** of **erythrocytes** or red blood cells by **influenza virus** (see Chapter 4): this is caused by virion surface proteins binding to particular chemical groups—such as **sialic acid**, in many **glycoproteins**—that are particularly common on erythrocytes, which are relatively simple to isolate from animal blood. This proved to be an important tool in the study of not only influenza but also several other groups of viruses—for example, **rubella virus**. In addition to measuring the **hemagglutination (HA) titer** (i.e., relative amount) of virus present in any preparation, this technique can also be used to determine the **antigenic type** of the virus. Hemagglutination will not occur in the presence of antibodies that bind to and block the virus hemagglutinin, or binding protein. If an antiserum is titrated against a given number of hemagglutinating units, the **hemagglutination inhibition titer** (HAI or HI) and specificity of the antiserum can be determined. Also, if antisera of known specificity are used to inhibit hemagglutination, the antigenic type of an unknown virus can be determined. This technique is still routinely used to investigate influenza viruses.

In the 1960s and subsequent years, many improved ***serological*** detection methods for viruses were developed, such as:

- Complement fixation tests
- Radioisotope immunoassays (**radioimmunoassays**)

- **Immunofluorescence** (direct detection of virus antigens in infected cells or tissue)
- **Immunoelectron microscopy** (use of antibodies in electron microscopy to identify virus-related structures)
- Enzyme-linked immunosorbent assays (**ELISAs**)
- **Western blot** assays or **immunoelectroblotting**

These techniques are sensitive, relatively simple and quick in the case of radio- and enzyme immunoassays, and are usually quantitative (Fig. 1.21). The **enzyme-linked immunoassays**—ELISA, western blots—were a very important development, as they offered sensitivity of detection of antigen or of antibodies that had only hitherto been possible with radioimmunoassays—first applied in 1960—without the expense and inconvenience in terms of use and disposal associated with radioisotopes. ELISA, for example, was developed in 1971, and offered quick and highly sensitive detection of viruses. Its first application was in detecting antibodies to **rubella virus** in 1975; plant virologists were very quick to pick up on the technique, with a first description of plant virus detection in 1976. Western blotting—using enzyme immunoassays to probe membranes with proteins "blotted" onto them after separation by SDS-polyacrylamide gel electrophoresis—was invented in 1979 as an ***autoradiographic*** and enzyme immunoassay method, but the use of enzymes rapidly became predominant. The first use of the technique with viruses was published in 1982, with two papers on its use with plant viruses (including one by Rybicki), and another on antigenic relationships among **coronavirus** structural proteins. It rapidly became the test of choice for demonstration of antibodies in humans to **HIV-1 and -2** proteins.

In 1975, George Kohler and Cesar Milstein isolated the first **monoclonal antibodies** from clones of cells selected in vitro to produce an antibody of a single specificity directed against a particular antigen. This enabled virologists to look not only at the whole virus, but at specific regions—**epitopes**, or antibody-binding structures or sequences—of individual virus antigens. This ability has greatly increased our understanding of the function of individual virus proteins. Monoclonal antibodies (**mAbs**) have found increasingly widespread application in other types of serological assays (e.g., ELISAs) to increase their reproducibility, sensitivity, and specificity. They are also increasingly being used as therapeutics for treatment of a number of diseases—such as rheumatoid arthritis and breast cancer—as well for viral diseases such as the pneumonia caused by **respiratory syncytial virus** in children (Fig. 1.22).

Modern molecular virology

The history of discovery of viruses most certainly does not end with the determination of the complete molecular structure of TMV virions in the 1980s: Indeed, there have been many exciting discoveries not covered, including of novel **hemorrhagic fever viruses** such as the various **Ebola** and **Marburg viruses, HIV-1 and -2** and their fiendishly complex interference in host immune responses, and **giant eukaryotic and prokaryotic viruses.** These and other discoveries will be dealt with in later Chapters. What is necessary to finish this Chapter, however, is a description of the modern and classical methods that evolved with the early virologists, that we presently use to characterize and to investigate viruses and their particles. While many methods have fallen into disuse with the rapid rise and application of methods

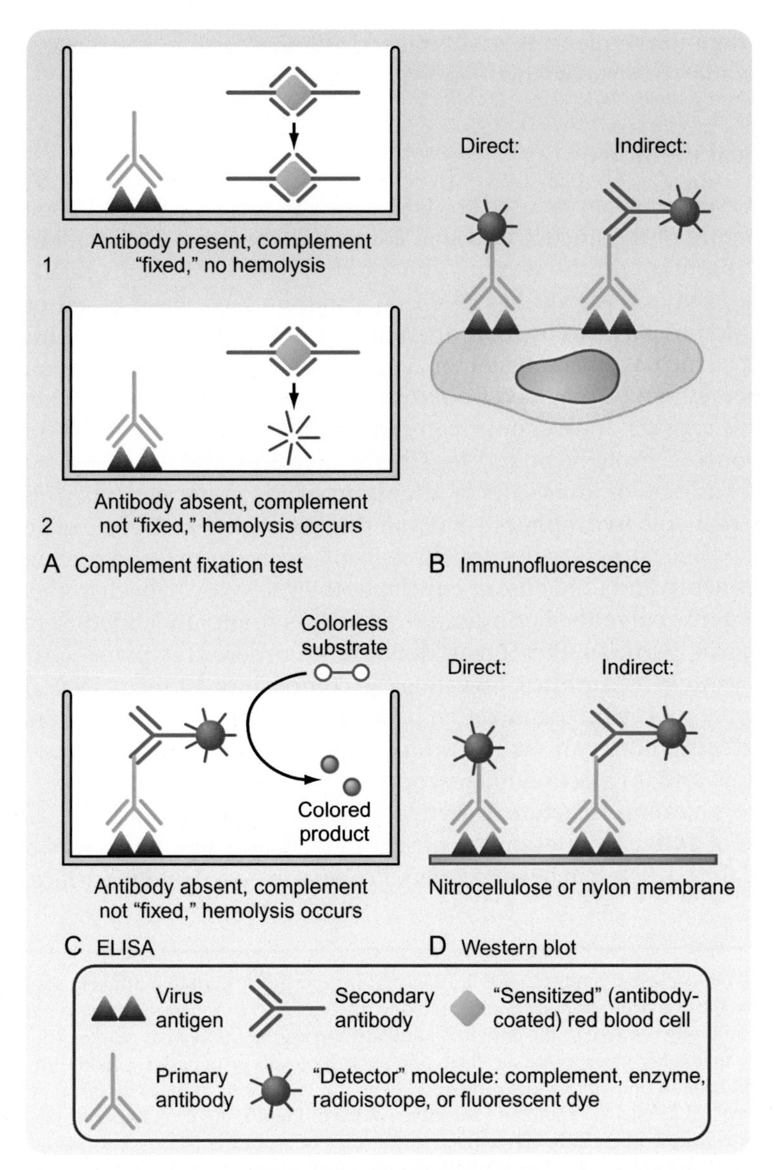

FIG. 1.21 Serological techniques in virology. The four assays illustrated in this figure have been used for many years and are still of widespread value in many circumstances. They are used to test both viruses and for immune responses against virus infection. (A) The complement fixation test works because complement is bound by antigen antibody complexes. "Sensitized" (antibody-coated) red blood cells, known amounts of complement, a virus antigen, and the serum to be tested are all added to the wells of a multiwell plate. In the absence of antibodies to the virus antigen, free complement is present which causes lysis of the sensitized red blood cells (hemolysis). If the test serum contains a sufficiently high titer of antivirus antibodies, then no free complement remains and hemolysis does not occur. Titrating the test serum through serial dilutions allows a quantitative measurement of the amount of antivirus antibody present to be made. (B) Immunofluorescence is performed using modified antibodies linked to a fluorescent

(Continued)

such as next-generation sequencing and "**transcriptomics**" and modern mass spectroscopy, basic characterization of viruses still often requires the classical methods.

Physicochemical methods

Chemical investigation can be used to determine not only the overall composition of virions and the nature of the nucleic acid that comprises the virus genome, but also the construction of the particle and the way in which individual components relate to each other in the capsid(s). Many classic studies of virion structure have been based on the gradual, stepwise disruption of particles by slow alteration of pH or the gradual addition of protein-denaturing agents such as urea, phenol, or detergents. Under these conditions, valuable information can sometimes be obtained from relatively simple experiments. The reagents used to denature virus capsids, for example, can indicate the basis of the stable interactions between its components. Proteins bound together by **electrostatic or charge interactions** can be disrupted by addition of **ionic salts** or **alteration of pH** under generally mild conditions; those bound by nonionic **hydrophobic** and **van der Waal's interactions** can be disrupted by "**chaotropic**" or protein structure-destroying reagents such as urea; and proteins that interact with lipid components, and lipid bilayer envelopes themselves, can be disrupted by **nonionic detergents** or **organic solvents**. For example, as urea is gradually added to preparations of purified **adenovirus** particles, they break down in an ordered, stepwise fashion which releases **subvirus protein assemblies**, revealing the composition of the particles. In the case of TMV, similar studies of capsid organization have been performed by **renaturation of the purified capsid protein** under various conditions (Fig. 1.23, and 1.16), and study by analytical ultracentrifugation and/or electron microscopy.

In addition to revealing structure, progressive denaturation can also be used to observe alteration or loss of **antigenic sites** on the surface of particles, and in this way a picture of the physical state of the particle can be developed. Proteins exposed on the surface of virions can

FIG. 1.21—Cont'd molecule that emits colored light when illuminated by light of a different wavelength. In direct immunofluorescence, the antivirus antibody is conjugated to the fluorescent marker, whereas in indirect immunofluorescence a second antibody reactive to the antivirus antibody carries the fluorescent marker. Immunofluorescence can be used not only to identify virus-infected cells in populations of cells or in tissue sections but also to determine the subcellular localization of particular virus proteins (e.g., in the nucleus or in the cytoplasm). (C) Enzyme-linked immunosorbent assays (ELISAs) are a rapid and sensitive means of identifying or quantifying small amounts of virus antigens or antivirus antibodies. Either an antigen (in the case of an ELISA to detect antibodies) or antibody (in the case of an antigen ELISA) is bound to the surface of a multiwell plate. An antibody specific for the test antigen, which has been conjugated with an enzyme molecule (such as alkaline phosphatase or horseradish peroxidase), is then added. As with immunofluorescence, ELISA assays may rely on direct or indirect detection of the test antigen. During a short incubation, a colorless substrate for the enzyme is converted to a colored product, amplifying the signal produced by a very small amount of antigen. The intensity of the colored product can be measured in a specialized spectrophotometer (plate reader). ELISA assays can be mechanized and are suitable for routine tests on large numbers of clinical samples. (D) Western blotting is used to analyze a specific virus protein from a complex mixture of antigens. Virus antigen-containing preparations (particles, infected cells, or clinical materials) are subjected to electrophoresis on a polyacrylamide gel. Proteins from the gel are then transferred to a nitrocellulose or nylon membrane and immobilized in their relative positions from the gel. Specific antigens are detected by allowing the membrane to react with antibodies directed against the antigen of interest. By using samples containing proteins of known sizes in known amounts, the apparent molecular weight and relative amounts of antigen in the test samples can be determined.

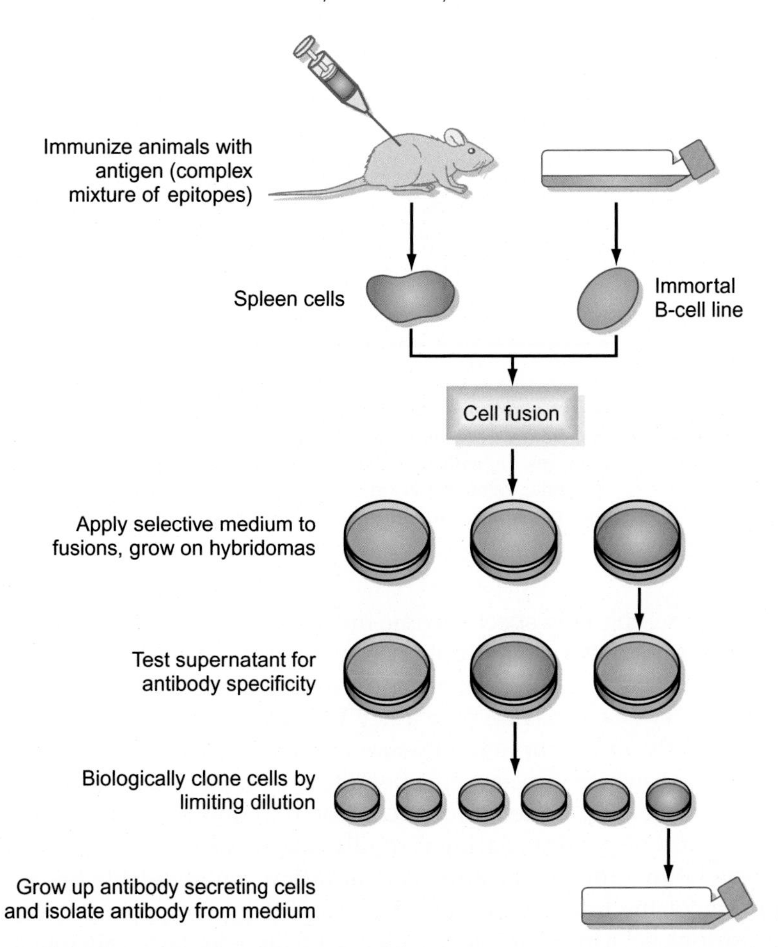

FIG. 1.22 Monoclonal antibodies. Monoclonal antibodies are produced by immunization of an animal with an antigen that usually contains a complex mixture of epitopes. Immature B-cells are prepared from the spleen of the animal, and these are fused with a myeloma cell line, resulting in the formation of transformed cells continuously secreting antibodies. A small proportion of these will make a single type of antibody (a monoclonal antibody) against the desired epitope. Recently, in vitro molecular techniques have been developed to speed up the selection of monoclonal antibodies.

be labeled with various compounds (e.g., **iodine, various organic chemicals, radioisotopes**) to indicate which parts of the protein are exposed and which are protected inside the particle or by lipid membranes. **Cross-linking reagents** are used to determine the **spatial relationship of proteins**, and of **nucleic acids and protein subunits**, in intact virions.

The physical properties of viruses can be determined by **spectroscopy** of various kinds: this includes using **ultraviolet light** to examine the **nucleic acid** and/or **protein content** of the particle or virion, **visible light** to determine **light-scattering properties** and particle sizes—now again a cutting-edge technique, given new technological developments—and various more modern techniques such as **rotational, vibrational, electronic, photoelectron** and **Auger spectroscopy** to study structural and other parameters.

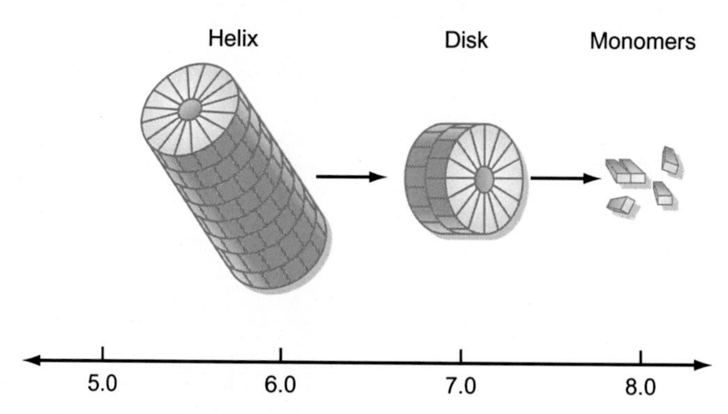

FIG. 1.23 Denaturation of TMV. The structure and stability of virus particles can be examined by progressive denaturation or renaturation studies. At any particular ionic strength, the purified capsid protein of TMV spontaneously assembles into different structures, dependent on the pH of the solution. At a pH of around 6.0, the particles formed have a helical structure very similar to infectious virus particles. As the pH is increased to about 7.0, disk-like structures are formed. At even higher pH values, individual capsid monomers fail to assemble into more complex structures.

One of the most exciting new spectroscopic methods for molecular virology, however, is **mass spectroscopy** (MS) in its various applications: this separates **particles of different mass to charge ratios**. The technique behind mass spectrometry dates back about a hundred years, but in the past few decades it has become a key technology in molecular virology. MS can now be used to directly and accurately measure the mass of viral proteins and particles, to identify and to even sequence virus-specific peptides in a total protein extract by comparing a peptide sequence to known protein sequences stored in online databases. It is also highly useful in investigation of **posttranslational modifications** of virus proteins such as **glycosylation, myristoylation, phosphorylation** and **sulfation,** and **disulfide bridging.**

Electrophoresis, or the fractionation of virions and their components in electrical fields in liquid or in various gel formulations, has long been an extremely valuable tool in routine molecular virology. Electrophoresis of intact virions has historically yielded and still yields information on **surface charge** and of **relative sizes** of **virions and virus-like particles** and whether or not they contain nucleic acids; however, it is electrophoretic analysis of individual virion proteins and of whole or partial nucleic acid genomes by gel electrophoresis (see Chapter 3), that has been and continues to be far more valuable. Most modern virology papers will still contain **sodium dodecyl sulfate-polyacrylamide gel electrophoresis (SDS-PAGE)** electropherograms describing viral proteins; many others will have nucleic acid fractionations to characterize virus genomes or messenger RNAs, in addition to sequence data.

It is not generally known that the most cited (35,000+ citations; Laemmli, 1970) methodological paper for SDS-PAGE was in fact about "*Cleavage of Structural Proteins during the Assembly of the Head of Bacteriophage T4,*" with the method described as an incidental to the main work.

Physical methods

X-ray crystallographers, using the ever more powerful radiation sources that have become available recently, can collect good data from even very small crystals, which is the normal result of attempts to crystallize virions or their proteins. Powerful **synchrotron** sources that generate intense beams of radiation have been built during the last few decades, and are now used extensively for this purpose. However, a number of important virions refuse to crystallize: this is a particularly common problem with irregularly-shaped viruses, and especially those with an outer lipid bilayer **envelope**. One further limitation is that some of the largest virions, such as poxviruses, contain hundreds of different proteins and are at present too complex to be analyzed using these techniques. Modifications of the basic **diffraction technique,** such as electron scattering by membrane-associated protein arrays, have helped to provide more information.

A technique still very much in use for the characterization of viruses and their virions is electron microscopy (EM)—this can be **transmission EM (TEM)**, which uses negative staining of suspensions of proteins or particles (see Figs. 1.11 and 1.12), or **EM sectioning**, where very thin sections of fixed cell culture material or even animal tissue is stained and observed, or **cryoelectron EM**, which makes use of extremely thin sections ("**cryosections**") of virions flash-frozen by exposure to liquid nitrogen. While **scanning EM** (SEM)—where samples are coated with metals by spray deposition—can be used, the resolution is generally much too low for ultrastructural investigations of viruses.

Two fundamental types of information can be obtained by TEM of virions: these are the absolute number of particles present in any preparation (total count), and the appearance and structure of the virions. TEM can provide a rapid method of virus detection and diagnosis, but may give misleading information—mainly because many things can resemble "**virus-like particles (VLPs)**," particularly in crude preparations, and for small virions. This difficulty can be overcome for crude preparations by using antisera specific for particular virus antigens, either alone, or conjugated to electron-dense markers such as the iron-containing protein ferritin or colloidal gold suspensions. This highly specific technique, known as **immunoelectron microscopy**, is useful in some cases as a rapid method for diagnosis, and for differentiating related virions and their proteins from one another.

Sectioning EM has been used for many years to study viruses in cells (virocells). While the resolution obtained from observation of thin tissue or cell sections is not as good as for TEM, the very clear association of virions with the outside of cell membranes during cell entry, or of newly-made or "**nascent**" particles with the inside of membranes during **virion budding** (see Fig. 2.12), was only made possible by this technique.

The relatively new technique of cryoelectron microscopy (cryoEM) is still developing in leaps and bounds, and can increasingly be used to supplant X-ray crystallography in virology. This uses extremely low temperatures, achieved by immersing a specimen suspended in dilute solution into liquid nitrogen or helium, to flash-freeze virions and even individual proteins in their native conformation in an amorphous ice matrix. This allows the operator to increase the amount of radiation falling on the specimen and hence the resolution, without disrupting the structure. Optical **3-D image reconstruction** by computer from thousands of individual particles or subunit proteins has allowed the determination of virion structures down to resolutions previously only attainable by X-ray crystallography. For example, in early 2020 the 2.56 Å resolution structure of an **adeno-associated satellite virus** was determined by cryoEM. A 3.5 Å structure was similarly obtained in 2020 for SARS-CoV-2 spike protein. For comparison, a typical atomic diameter is 0.25 nm, a typical protein alpha-helix is 1 nm in cross-section, and a DNA double helix 2 nm: this means that **atomic-level resolution** is possible for viruses using cryoEM. (see Recommended Reading).

Cryoelectron tomography allows a computerized three-dimensional reconstruction of structures ranging from virions down to single molecules, from a series of two-dimensional images taken at different tilt angles. Using these techniques it is possible to resolve structures in individual proteins down to 3Å (0.3nm) or less, as has been done for HIV-1 (Schur et al., 2016).

One of the most recent developments in physical methods to study viruses and their proteins is **atomic force microscopy** (AFM), only invented in 1986, which has been used to provide new insight into the structure of some viruses. The resolution ranges from subnanometer scales to 50nm or so, depending on sample type and instrument. While it remains a niche technology, recent successes have included determinations of structures of free virions adsorbed in liquid onto a mica support, of small isometric plant virus virions and even nucleoprotein cores of very large DNA viruses such as **irido- and herpes- and phycodnaviruses**. The latest results indicate that AFM may also be able to resolve molecular structures for viruses.

Nuclear magnetic resonance (**NMR**) has been widely used to determine the atomic structure of all kinds of molecules, including proteins and nucleic acids. The limitation of this method is that only relatively small molecules can be analyzed before the signals obtained become so confusing that they are impossible to decipher with current technology. At present, the upper size limit for this technique restricts its use to molecules with a molecular weight of less than about 50,000, considerably less than even the smallest virions. Nevertheless, this method may well prove to be of value in the future, certainly for examining isolated virus proteins if not for intact virions (Fig. 1.24).

Small-angle neutron scattering is another physical technique used to good effect to study virus structures: this depended on the invention of high-intensity, convenient neutron sources, and was first used in 1975 to study ribosome structure in liquid suspensions. By 1977 it had been applied to the study of small spherical RNA plant virus virions, but lack of

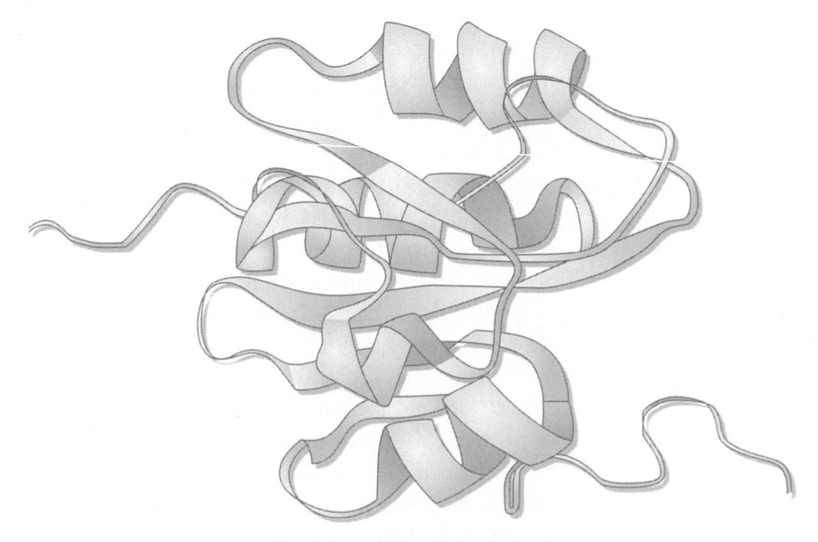

FIG. 1.24 Three-dimensional structure of a virus protein. This ribbon diagram of SV40 large T antigen was reconstructed from NMR data using a computer.

resolution limited its application to more detailed structures. Recently, however, technological advances in the technique have allowed much more detailed investigations. Recent applications include combining small-angle neutron scattering with atomic resolution data from X-ray crystallography and TEM to yield detailed quaternary structures for the nanoparticles (Krueger et al., 2021).

Molecular biology

Although protein chemistry has advanced greatly, the biggest impact from new technologies has come via nucleic acid–centered approaches: these have revolutionized virology and, to a large extent, shifted the focus of attention from particles and proteins toward the virus genome, the nucleic acid comprising the entire genetic information of the organism.

It is instructive that many fundamental discoveries in cell molecular biology were made by using viruses. Examples of note are that **vaccinia virus mRNAs** were found to be polyadenylated at their 3′ ends by Kates in 1970—**the first time this observation had been made in any organism**. Similarly, **"split genes,"** or sequences containing noncoding ***introns***, protein-coding ***exons***, and **spliced mRNAs**, were first discovered in adenoviruses by Roberts and Sharp in 1977, and later discovered to be the norm in eukaryotic cells. Introns in prokaryotes were first discovered in the genome of **bacteriophage T4** in 1984, a finding later extended to other prokaryotic viruses and also prokaryotes themselves.

Initially, molecular biology techniques focused on direct analysis of nucleic acids extracted directly from virions, using gel electrophoresis: this was a limiting procedure because of low sensitivity and the requirement for relatively large amounts of virus to be available. However, DNA virus genomes can be analyzed directly by restriction endonuclease digestion without resorting to molecular cloning: this approach was achieved for the first time with **SV40 DNA** in 1971.

The first pieces of DNA to be molecularly cloned were restriction fragments of **bacteriophage λ DNA** which were cloned into the DNA genome of SV40 by Berg and colleagues in 1972. This means that **virus genomes were both the first cloning vectors and the first nucleic acids to be analyzed** by these techniques. In 1977, the genome of bacteriophage ϕX174 was the first ***replicon*** to be completely sequenced (see earlier).

Molecular cloning of virus genomes, as DNA or cDNA, removed this limitation by enabling the production of essentially unlimited quantities of cloned virus nucleic acids. Increasing the sensitivity of nucleic acid detection using methods such as **Southern blotting** for DNA—**nucleic acid hybridization** done on immobilized target DNA on a membrane after separation by electrophoresis (Fig. 1.25)—**northern blotting** for RNA and western blotting for proteins (Fig. 1.20), allowed further advances to be made. Subsequently, phage genomes such as **M13** inovirus were highly modified for use as vectors in DNA sequencing. The enzymology of RNA-specific nucleases was comparatively advanced at this time, such that a spectrum of enzymes with specific cleavage sites could be used to analyze and even determine the sequence of RNA virus genomes (the first short nucleotide sequences of tRNAs having been determined in the mid-1960s). However, direct analysis of RNA by these methods was laborious and notoriously difficult. RNA sequence analysis did not begin to advance rapidly until the widespread use of reverse transcriptase (isolated from retroviruses) to convert RNA into cDNA in the 1970s.

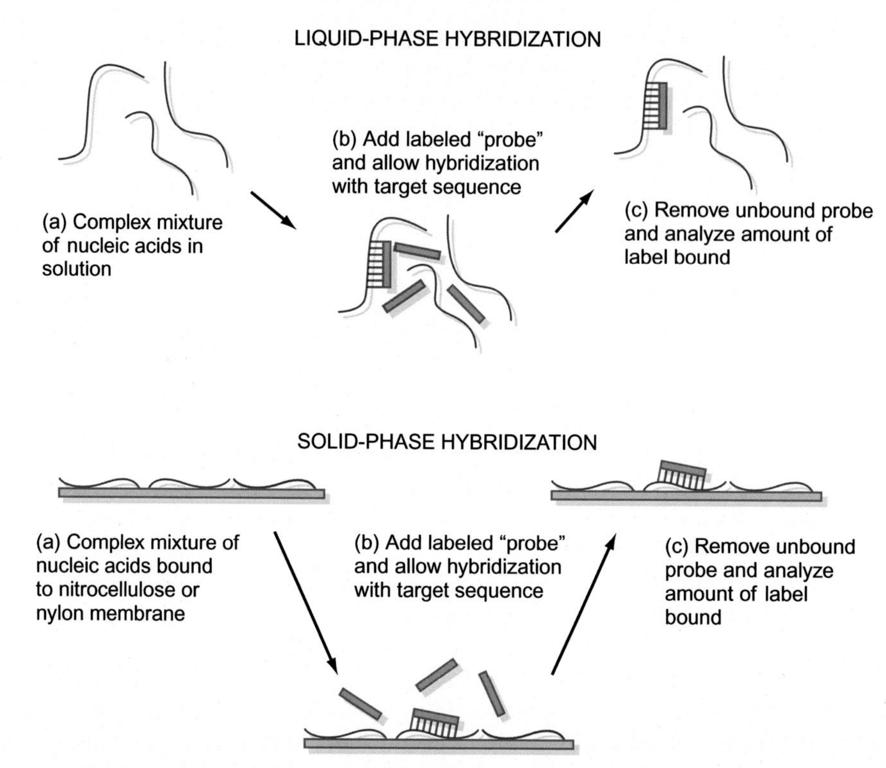

FIG. 1.25 Nucleic acid hybridization. Nucleic acid hybridization relies on the specificity of base-pairing which allows a labeled nucleic acid probe to pick out a complementary target sequence from a complex mixture of sequences in the test sample. The label used to identify the probe may be a radioisotope or a nonisotopic label such as an enzyme or photochemical. Hybridization may be performed with both the probe and test sequences in the liquid phase (top) or with the test sequences bound to a solid phase, usually a nitrocellulose or nylon membrane (below). Both methods may be used to quantify the amount of the test sequence present, but solid-phase hybridization is also used to locate the position of sequences immobilized on the membrane. Plaque and colony hybridization are used to locate recombinant molecules directly from a mixture of bacterial colonies or bacteriophage plaques on an agar plate. Northern and Southern blotting are used to detect RNA and DNA, respectively, after transfer of these molecules from gels following separation by electrophoresis (c.f. western blotting, Fig. 1.20).

Polymerase chain reaction amplification of virus genomes

Possibly the most important recent development in molecular biology was the development of the **polymerase chain reaction (PCR)** technique for amplification of defined sequences of DNA in the mid-1980s (Fig. 1.26). PCR allows **in vitro amplification** and detection of tiny quantities of nucleic acid from virions, or from complex mixtures found in virus-infected cells. One of the most important advances was the combination of PCR amplification with nucleotide sequencing methods, which revolutionized virus genome sequencing before the advent of **next-gen sequencing** (NGS; see below). PCR has continued to be developed: for example, **reverse-transcription (RT-PCR)** first copies RNA into DNA and then amplifies it; **quantitative PCR (qPCR)** allows spectrophotometric quantitation of DNA or cDNA amplification, either during or post-PCR—and may be done in real time, where concentrations of reaction products can be visualized as the reaction proceeds. More recently, **digital PCR**

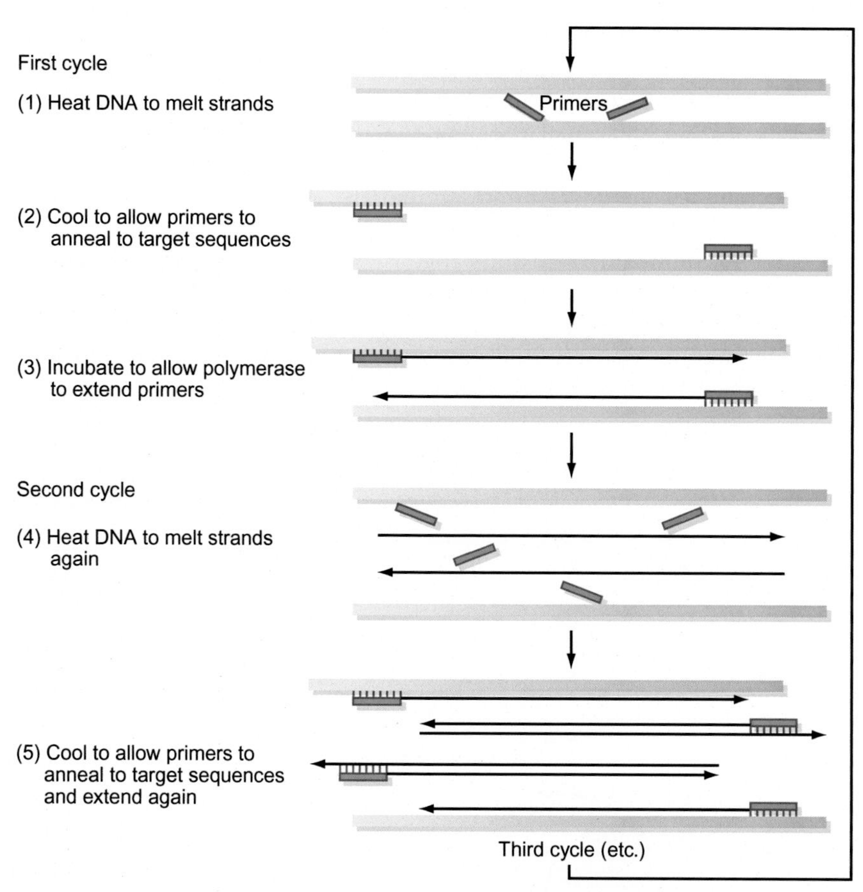

FIG. 1.26 DNA amplification by polymerase chain reaction (PCR). The PCR relies on the specificity of base-pairing between short synthetic oligonucleotide probes and complementary sequences in a complex mixture of nucleic acids to prime DNA synthesis using a thermostable DNA polymerase. Multiple cycles of primer annealing, extension, and thermal denaturation are carried out in an automated process, resulting in a massive amplification of the target sequence located between the two primers (2^n-fold increase after n cycles of amplification, i.e., a thousand copies after 10, and over a million copies after 20 cycles).

(dPCR), an amazingly sensitive technique, allows measurement of tiny quantities of viral nucleic acid in clinical samples. Routine diagnosis of SARS-CoV-2 infections, for example, are done by real-time RT-PCR, and virus loads in HIV-infected people are often assessed similarly.

Viromics

"**Genomics**" is the study of the composition and function of the genetic material of organisms. **Virus genomics** or "**viromics**" began with the first complete sequence of a virus genome in 1976, as described earlier. Initially, sequencing nucleic acids was a manual operation and extremely tedious; additionally, it used a radioactive isotope (^{32}P), which made it potentially hazardous. Gradually the technology improved, via PCR sequencing and the use of nonradioactive markers and capillary sequencing, and eventually this evolution led to what are sometimes called "**next generation sequencing**" (NGS) or **massively parallel** or **deep sequencing** methods. Essentially, most of the methods allow the sequencing of millions of small fragments of DNA in parallel after some form of amplification, or even of many single long molecules, to generate billions of base reads for thousands or even millions of DNA fragments. The techniques change fast enough that any detailed description would become redundant very quickly—however, a continuously updated description of current techniques can be found in the Recommended Reading under EMBL-EPI Training. **Sequence assembly** is done using a rapidly-evolving set of specialized software that allows assembly of many fragmented and partially overlapping sequences into **contigs**, or contiguous sequences, with a high depth of **coverage**, or repeats of the consensus sequence. This is in fact how the Wuhan isolate of SARS-CoV-2 was originally sequenced (Fig. 1.27).

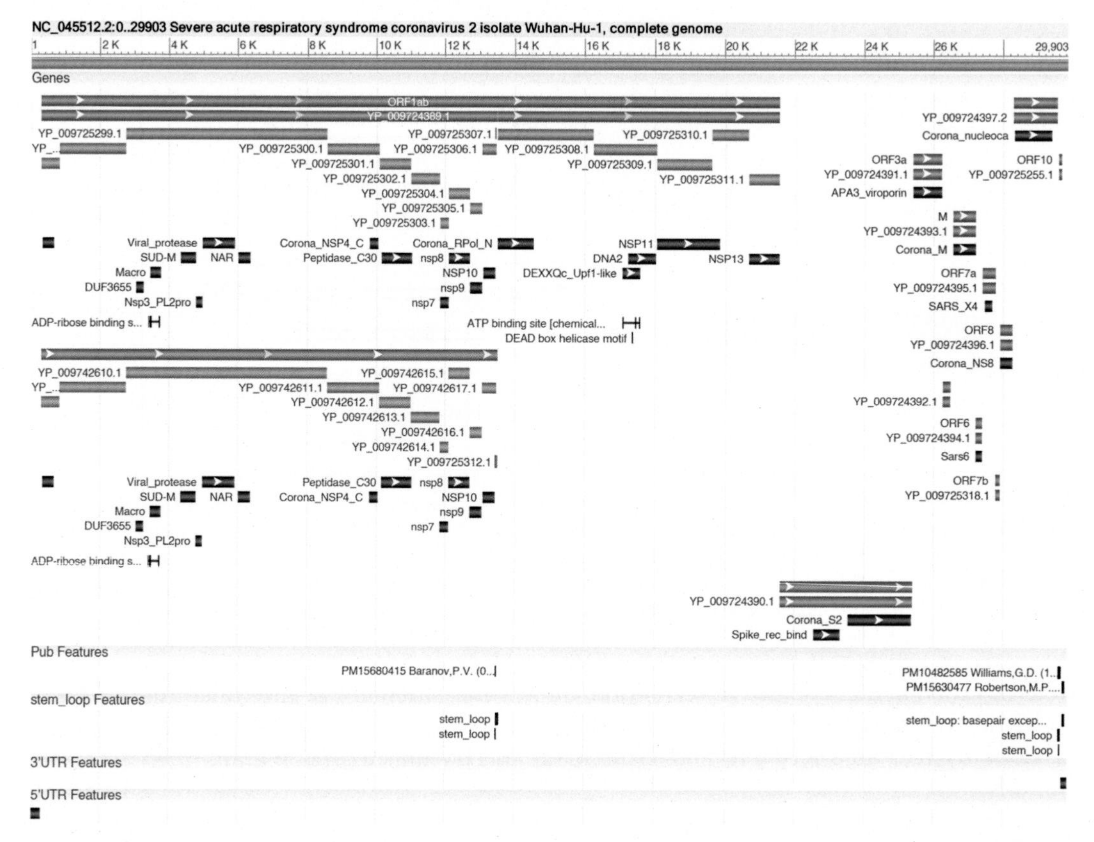

FIG. 1.27 Bioinformatics of viruses. This is a graphic depiction of the information contained in the whole genome (29,303nt of ssRNA) of the **Wuhan-Hu-1** isolate—the first, or reference isolate—of **severe acute respiratory syndrome coronavirus type 2** (SARS-CoV-2). *Colored boxes* represent open reading frames (ORFs), with accession numbers for particular sequences shown alongside. This data was made available very shortly after the description of the virus, and annotated soon thereafter. It and other viral sequence information is freely available via the reference NC_045512.2 from the US National Institutes of Health (NIH) National Center for Biotechnology Information (NCBI) server (https://go.usa.gov/xsTCf).

There is a distinction between sequencing and analyzing all the DNA or cDNA in a sample—commonly termed **shotgun metagenomics**—or sequencing the product of amplification of DNA from a sample using a set of PCR primers directed to a highly conserved DNA sequence in a given set of organisms. The former potentially allows sequencing of genes from Bacteria, Archaea, viruses, fungi and other eukaryote communities all at the same time. The latter would be used to target bacteria and Archaea, for example, by use of **16S ribosomal RNA** gene-specific primers, eukaryotes by **18S ribosomal RNA** gene–specific primers, and viruses by primers targeting specific genes shared across related groups, such as **DNA polymerases**. This is known as **amplicon metagenomics**, as it focuses in on **PCR amplicons only**.

Virus bioinformatics

Vast international databases of nucleotide and protein sequence information exist, and these can be rapidly accessed by computers to compare newly determined sequences with the complete genome sequences of thousands of putative virus contigs that appear literally daily (Table 1.2).

However, the enormous quantities of data being generated—many times larger than any biologist has ever had to deal with before—has required the development of new parallel fields such as **bioinformatics**—a broad term first used in 1979 to describe the **application of statistics and computer science to molecular biology**. As well as simply storing and displaying information, by spotting patterns, bioinformatics can infer function from simple digital information and so is central to all areas of modern biology. For example, sequences of viruses must be annotated—that is, all of the coding regions and control sequences (e.g., promoters, terminators) must be located and marked, and their functions surmised—which allows comparisons of whole genomes in an informed way.

TABLE 1.2 Genomic comparison of different organisms.

Organism	Genome size	Number of genes
Porcine circovirus	1759 nt	2
Hepatitis B virus	3200 bp	4
SV40	5200 bp	6
Herpes simplex virus	152 kbp (152,000 bp)	77
Pandoravirus	2.4 Mbp (2,473,870 bp)	2500
Escherichia coli	4.6 Mbp (4,600,000 bp)	3200
Yeast	12.1 Mbp (12,100,000 bp)	6000
Caenorhabditis elegans	97 Mbp (97,000,000 bp)	19,000
Arabidopsis	100 Mbp (100,000,000 bp)	25,000
Drosophila	137 Mbp (137,000,000 bp)	13,000
Mouse	2600 Mbp	25,000
Human	3200 Mbp	25,000

The current trend is to lump related information together and to analyze it in a holistic way rather than one molecule at a time. This is often referred to as "-**omics**"—approaches such as **gen**omics or **prote**omics or **metabol**omics combine many of the above methods to examine all of the genes (genome) or proteins (**proteome**) or specific genes related to metabolic processes (**metabolome**) in a given organism or environment. For example, the vaccinia virus genome (the vaccinia virus nucleotide sequence and all its genes) and a mimivirus proteome (all the possible proteins produced by a mimivirus) can be studied in this way. Metagenomics enables the direct study of **all the uncultured genomes from complex environments** (e.g., the ocean or the human body), without requiring each one to be identified and cloned for fully sequenced. As well as looking at the complex mixture of viruses found in natural environments such as water or soil, work is currently underway to study the human virome—all of the genes from all of the viruses which infect humans. The new field of **virus ecology** has also emerged in recent years, and aims to analyze whole viral communities even though most of the viral sequences detected—which is the majority of sequences detected in any given metagenome—show no significant homology to previously known sequences. In fact, there is a 2020 description of an amoeba-infecting virus with a 44.9 kbp dsDNA genome containing 74 predicted genes, of which >90% had no homology to any known genes in any database at the time: this was the so-called Yaravirus, found in a Brazilian lake (see Boratto et al., 2020, below).

Functional viromics

The information stored in nucleic acids in digital archives is not of itself meaningful, unless the functions of the molecules and their interactions can be understood. **Functional viromics** attempts to go beyond the description or cataloging of viral genes and proteins to understanding the interactions between all of the components of a biological system—a virocell. This includes many investigational methods, such as **colocalization studies**; yeast and other **one- and two-hybrid systems** to study interactions between proteins; **microarrays**—two-dimensional grids of probes on a solid substrate such as a glass slide to allow high throughput screening—and **gene knockout experiments** through mutagenesis or **RNA interference (RNAi)** (see Chapter 6).

Overview

We have emerged from this history of the discovery of viruses as enabled by technological developments with a profound understanding of **viruses and virocells**, and how intimately related these are to ourselves and our environment. However, the current pace of research in virology tells us that there is still far more that we need to know. The rest of this book will explain in detail what we already understand about viruses, how we study them now, and how we may study them in future.

References

Boratto, P.V.M., Oliveira, G.P., Machado, T.B., Andrade, A.C.S.P., Baudoin, J.P., Klose, T., Schulz, F., Azza, S., Decloquement, P., Chabrière, E., Colson, P., Levasseur, A., La Scola, B., Abrahão, J.S., 2020. Yaravirus: a novel 80-nm virus infecting *Acanthamoeba castellanii*. Proc. Natl. Acad. Sci. U. S. A. 117 (28), 16579–16586. https://doi.org/10.1073/pnas.2001637117.

Krueger, S., Curtis, J.E., Scott, D.R., Grishaev, A., Glenn, G., Smith, G., Ellingsworth, L., Borisov, O., Maynard, E.L., 2021. Structural characterization and modeling of a respiratory syncytial virus fusion glycoprotein nanoparticle vaccine in solution. Mol. Pharm. 18 (1), 359–376. https://doi.org/10.1021/acs.molpharmaceut.0c00986.

Schur, F.K., Obr, M., Hagen, W.J., Wan, W., Jakobi, A.J., Kirkpatrick, J.M., Sachse, C., Krausslich, H.G., Briggs, J.A., 2016. An atomic model of HIV-1 capsid-SP1 reveals structures regulating assembly and maturation. Science 353 (6238), 506–508. https://doi.org/10.1126/science.aaf9620.

Recommended Reading

American Academy of Microbiology, 2013. Viruses Throughout Life & Time: Friends, Foes, Change Agents. http://academy.asm.org/index.php/browse-all-reports/5180-viruses-throughout-life-time-friends-foes-change-agents.

Brown, N.N., Bhella, D., 2016. Are Viruses Alive? Microbiology Society, 2016. https://microbiologysociety.org/publication/past-issues/what-is-life/article/are-viruses-alive-what-is-life.html.

Ellis, E.L., Delbrück, M., 1938. The growth of bacteriophage. J. Gen. Physiol. 22 (3), 365–384. https://www.ncbi.nlm.nih.gov/pmc/articles/PMC2141994/pdf/365.pdf.

Forterre, P., 2010. Defining life: the virus viewpoint. Orig. Life Evol. Biosph. 40 (2), 151–160. https://www.ncbi.nlm.nih.gov/labs/pmc/articles/PMC2837877/pdf/11084_2010_Article_9194.pdf.

Fraenkel-Conrat, H., Singer, B., 1999. Virus reconstitution and the proof of the existence of genomic RNA. Philos. Trans. R Soc. Lond. B Biol. Sci. 354 (1383), 583–586. https://doi.org/10.1098/rstb.1999.0409.

Gelvin, S.B., 2012. Traversing the cell: *Agrobacterium* T-DNA's journey to the host genome. Front. Plant Sci. https://doi.org/10.3389/fpls.2012.00052.

Hershey, D., Chase, M., 1952. Independent Functions of Viral Protein and Nucleic Acid In Growth of Bacteriophage. https://www.ncbi.nlm.nih.gov/pmc/articles/PMC2147348/pdf/39.pdf.

Hulo, C., de Castro, E., Masson, P., Bougueleret, L., Bairoch, A., Xenarios, I., Le Mercier, P., 2011. 2011. ViralZone: a knowledge resource to understand virus diversity. Nucleic Acids Res. 39 (Database issue), D576–D582. https://viralzone.expasy.org/.

Kutatelaze, M., 2015. Experience of the Eliava Institute in bacteriophage therapy. Virol. Sin. 30 (1), 80–81. https://www.virosin.org/fileup/PDF/20150115.pdf.

Murphy, F.A., 2020. A highly recommended, very detailed history of the early days of virology by a highly distinguished scientist. In: The Foundations of Virology, second ed. https://tinyurl.com/t2xtp4zf.

Press Release, Karolinska Institutet, The Nobel Prize in Physiology or Medicine for 1969. https://www.nobelprize.org/prizes/medicine/1969/press-release/.

Reardon, S., 2014. Phage Therapy Gets Revitalized. https://www.nature.com/news/phage-therapy-gets-revitalized-1.15348.

Saunders, K., Bedford, I.D., Yahara, T., Stanley, J., 2003. The earliest recorded plant viral disease. Nature 422, 831 (2003) https://www.nature.com/articles/422831a.

Scholthof, K.-B.G., 2008. Tobacco Mosaic Virus: The Beginning of Plant Virology. https://www.apsnet.org/edcenter/apsnetfeatures/Pages/TMV.aspx.

Temin, H., 1975. The DNA provirus hypothesis. In: Nobel Lecture. https://www.nobelprize.org/uploads/2018/06/temin-lecture.pdf.

CHAPTER

2

Viruses and their particles

INTENDED LEARNING OUTCOMES

On completing this chapter you should be able to:

- Understand the basis for the classification of viruses and classification criteria
- Explain the need for viruses to make virions.
- Understand the principles of virion structure and assembly.
- Describe examples of different types of virus particles, from simple to more complex forms.

Viruses and their particles

Viruses are both molecules and organisms; they are inert particles and actively replicating and expressing organisms. Virus particles, or **virions**, are "**nanoscale spacecraft**"—navigating the spaces between cells to keep their genomes safe and to deliver them to cells in order to multiply. This chapter describes how we classify viruses, and how viruses build their virions.

The classification of viruses

Summary

1. Viruses are classified just as other organisms are, but not using the same type of nomenclature
2. Their taxonomy attempts to reflect presumed evolutionary relationships, including the fact that viruses as a class of organisms have more than one origin
3. The most important criteria in their classification are host type, particle morphology, and genome type, sequence, and evolutionary relationships

Introduction

Given that viruses may be regarded as organisms, it is reasonable to expect that they be classified accordingly, although not necessarily as cellular organisms have been. A universal system for **classifying viruses**, and a unified **taxonomy**, or system of **classification** and **nomenclature**, has been established by the International Committee on Taxonomy of Viruses

Cann's Principles of Molecular Virology
https://doi.org/10.1016/B978-0-12-822784-8.00002-7

(ICTV) since 1966. The system makes use of a series of ranked **taxons**, or taxonomic ranks. While at first the taxonomy was largely based on the criteria of **host type, particle morphology,** and **genome type**—because complete genome sequencing was not feasible until recently—it has recently been brought into line with relationships that have been revealed by deep phylogenetic analysis of nucleic acid and predicted protein sequences (see Koonin et al., 2020). Analysis took account of the fact that three simple protein domains are present in the proteomes of most viruses, constituting ***"super virus hallmark genes"*** or **VHGs**. These are namely,

(i) the RNA recognition motif (RRM) forming the core of **all virus polymerases**;
(ii) the eight stranded antiparallel β-barrel "single jelly roll" (SJR) protein fold, that is the core domain of SJR- and "double jelly roll" (DJR) major capsid proteins;
(iii) the **"superfamily 3 helicase" (S3H)** that is the replicative **helicase** of diverse DNA viruses.

Additional genes linking virus lineages are the **HK97 major capsid protein (MCP)** module and **"HUH" endonucleases.** These motifs and domains link the vast majority of viruses: their involvement in the deep evolution of viruses will be covered in **Chapter 3**.

A major recent shake-up of virus taxonomy as a result of deep phylogenetic analysis of VHGs (see Koonin et al. 2020) reflects the deep evolutionary relationships revealed by that analysis, and is shown in Table 2.1 below. Taxon names are italicized, with the appropriate suffixes as shown. Taxons in bold are the ones most used by virologists. See this Chapter 3 for a more in-depth discussion of hallmark genes.

TABLE 2.1 Summary of current virus taxons.

Rank	Total, MSL-35[a]	Taxon suffix	Example	Includes
Realm	4	*-viria*	*Riboviria*	All viruses with a RNA stage in the replication cycle
Kingdom	9	*-virae*	*Orthornavirae*	RNA viruses with no DNA stage in the life cycle
Phylum	16	*-viricota*	*Pisuviricota*	"Picornavirus supergroup"
Subphylum	2	*-viricotina*		
Class	36	*-viricetes*	*Pisoniviricetes*	
Order	55	*-virales*	*Picornavirales*	Viruses with picornavirus-like replication strategy
Suborder	8	*-virineae*		
Family	168	*-viridae*	*Picornaviridae*	Viruses more closely related to generic picornaviruses
Subfamily	103	*-virinae*		
Genus	1421	*-virus*	*Aphthovirus*	Viruses related to foot-and-mouth disease virus
Subgenus	68	*-virus*		
Species	6590	*-virus*	*Foot-and-mouth disease virus*	Second virus discovered, causes disease in cattle and swine

a = Master Species List 35, ICTV 2020.

Virus realms

Riboviria: viruses replicating via **RNA-dependent RNA polymerases (RdRps)** or **DNA-dependent RNA polymerases (RT)**—all RNA viruses (ss+, ss-, ds) and RT-using retro- and pararetroviruses and reverse transcribing transposon-related viruses.

Duplodnaviria: viruses of prokaryotes and eukaryotes with **dsDNA genomes** and **HK97-fold capsid proteins.**

Monodnaviria: viruses of prokaryotes and eukaryotes with **ssDNA genomes** or **small circular dsDNA genomes** that utilize rolling-circle replication and encode replication initiation endonucleases of the **HUH superfamily.**

Varidnaviria: viruses of prokaryotes and eukaryotes with dsDNA genomes and single vertical jelly-roll or double jelly-roll capsid proteins.

Adnaviria: viruses with filamentous virions that infect Archaea with linear A-form dsDNA genomes, and characteristic major capsid proteins unrelated to those encoded by other known viruses.

Ribozyviria: this is a realm of with one family (*Kolmioviridae)* and eight genera, based on similarity of sequences found mainly by metaviromic screens to hepatitis D virus (HDV; genus *Deltavirus*). These are essentially satellite viruses, all with circular ssRNA genomes encoding a single protein.

Taxonomic criteria

The most important criteria to establish taxonomic placement are:

- **Host organism(s)**: eukaryote; prokaryote; vertebrate, fungi, plants
- **Particle morphology**: filamentous; isometric; tailed; naked; enveloped; complex
- **Genome type**: RNA; DNA; ss- or ds-; circular; linear; segmented
- **Genome sequence** and evolutionary relationships of hallmark genes

While a number of classical criteria such as disease symptoms, antigenicity, protein profile and host range are also important in precise identification, consideration of the above four criteria—and in many cases, just morphology—are sufficient in most cases to allow identification of a virus down to familial if not generic level. In this age of **next-generation sequencing** (NGS; Chapter 1), of course, it is often sufficient to simply isolate or even just concentrate virions from environmental or diseased host samples and then extract and sequence all or part of their genomes, in order to accurately identify viruses down to the **species level**—often without culturing them. One important caveat is that **a sequence is not a virus**: without some indication that a purportedly complete viral sequence is in fact complete and its virions infectious, it may be premature to call it a virus.

Virus orthography

Orthography in the sense of virus taxonomy is a **set of conventions for writing names**. The names of all **taxons** or taxonomic groups are written in *italics* with the first letter *Capitalized* (see examples above). Proper nouns in species names should be capitalized as well: e.g., *Semliki Forest virus, West Nile virus.*

This format refers only to **official taxonomic entities**: these are **concepts**, while viruses—the things that infect cells, and cause diseases—are real. **Vernacular** or common-use forms of names are **neither capitalized nor written in italics**. Thus, while *Tobacco mosaic virus* refers to the **species** of the first virus discovered, and *Tobamovirus* to the **genus, tobacco mosaic viruses** or **tobamoviruses** are the entities that you work with.

It is important to distinguish **common or colloquial names** of viruses, and the formal taxonomic designations given to **species** of viruses. This serves to highlight the difference between a virus as a thing—that is, something that causes infections, and which has properties that can be measured, such as genome size and type—and the **taxonomic concept** of a virus species. The former gets a common name, like **poliovirus type 1** or **tobacco mosaic virus**; the latter gets a formal name that should only be used when specifically describing the taxonomic entity, which exists only as a concept.

The ICTV is involved only with the naming of virus taxa, rather than giving viruses common names, or naming diseases.

For an outbreak of a new viral disease there are three names to be decided: these are the **name of the disease**, a **name for the virus** itself, and a **formal name for the species**. The World Health Organization (WHO) is responsible for the first; expert virologists for the second, and the ICTV for the third. In the case of the pandemic caused by what was at first called **nCoV-19**—for "novel coronavirus from 2019"—the WHO named the disease **coronavirus disease 19**, or **COVID-19**, and the ICTV ended up naming it **severe acute respiratory syndrome coronavirus type 2**, or **SARS-CoV-2**.

There are often discrepancies between the taxonomic descriptions and the ordinary names of viruses: for example, while **Ebola virus disease (EVD)** may be caused by a spectrum of related viruses including **Ebola, Sudan, Taï Forest,** and **Bundibugyo viruses**, the agent causing the **2014–2016** West Africa outbreak was **Ebola virus**, which is classified as Species *Zaire ebolavirus* in Genus *Filovirus*, Family *Filoviridae*, Order *Mononegavirales*. This gets worse with the **papillomaviruses**, where human papillomaviruses type 16 and 18, which together cause over 70% of cervical cancers in women, are designated as Species *Alphapapillomavirus* 9 and 7 respectively, in Genus *Alphapapillomavirus*, Subfamily *Firstpapillomavirinae*, Family *Papillomaviridae*.

Virions

Summary

1. Virions are virus-specified particles that contain **all or part of the genome** of the virus.
2. Virions may be simple nucleocapsids—**protein associated with nucleic acid only**—or more complex.
3. Basic elements of simple virions include **nucleic acid**—various forms of RNA or DNA—and associated **capsid protein**. This may assemble as helical or isometric nucleoproteins.
4. More complex forms may have **more than one shell of protein**, or **one or more membranes** surrounding the particle or internal components, or a **head-and-tail** arrangement like many prokaryotic viruses.

What are virions?

Virions are virus particles: they are the inert carriers of the viral genome, and are assembled inside cells, from **virus-specified components**: they do not grow, and do not form by division.

Fig. 2.1 shows an illustration of the approximate shapes and sizes of different families of viruses. Virus particles may range in size by nearly 100-fold, from around 17 nm in diameter (**porcine circovirus**) to 1200nm in length ("**Pithovirus sibericum**"). The protein subunits in a virus **capsid**—the protein shell that nearly all viruses make—are **redundant**, that is, there are many copies in each particle. Damage to one or more capsid subunits may make that particular subunit nonfunctional, but rarely does such limited damage destroy the infectivity of the entire virion: this helps make the capsid an effective barrier. Capsids are very tough, about as strong as a hard plastic such as Perspex or Plexiglas, although they are so small. They are also elastic, and are able to deform by up to a third of their diameter without breaking.

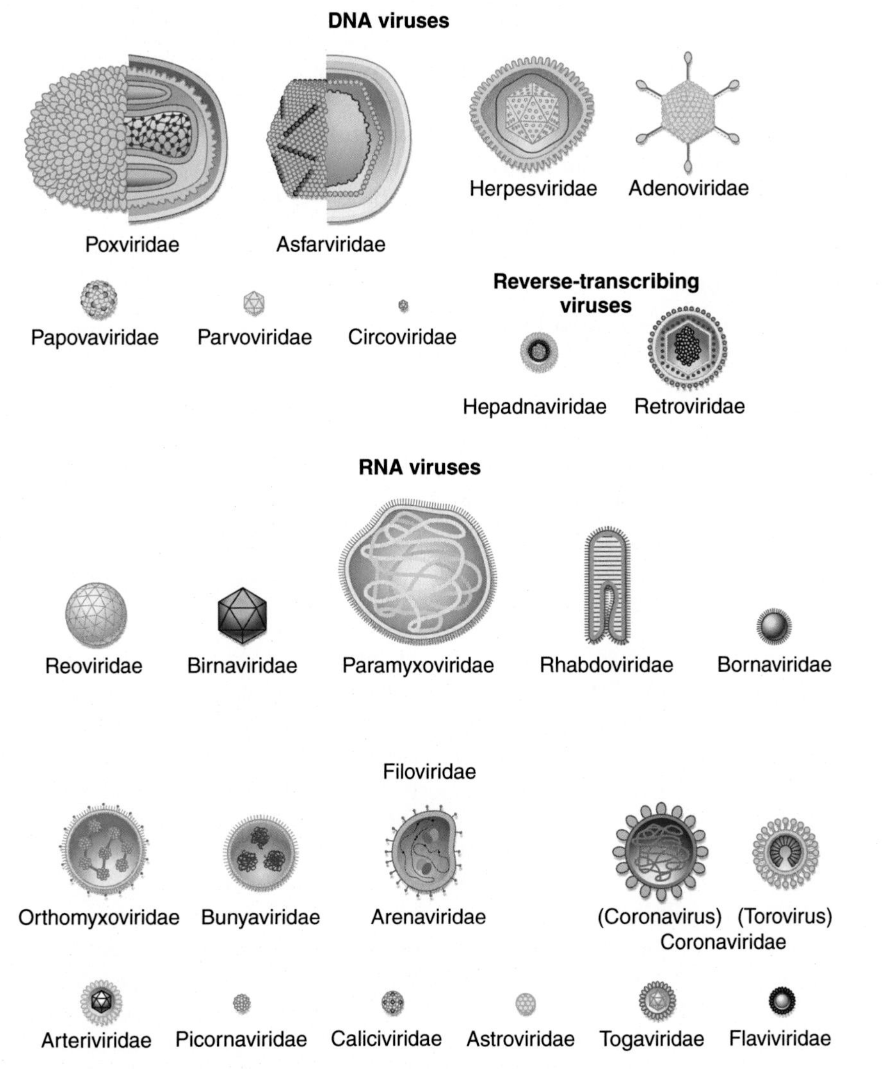

FIG. 2.1 Shapes and relative sizes of virions. A diagram illustrating the shapes and sizes of viruses of families. The virions are drawn approximately to scale, but artistic license has been used in representing their structure. In some, the cross-sectional structures of capsid and envelope are shown, with a representation of the genome. For the very small virions, only their size and symmetry are depicted. *From F.A. Murphy.*

This combination of strength, flexibility, and small size means that it is physically difficult (although not impossible) to break open virus particles by physical pressure—and indeed, capsids may be able to resist considerable internal pressure due to dense genome packing (Box 2.1).

A major function of most virions is **specific recognition of suitable host cells**, and binding to these and delivering the genome into them. The outer surface of the virion is thus responsible for recognition of and interaction with the host cell. Initially, this takes the form of specific binding of a **virus attachment protein** to a **cellular receptor molecule,** followed by virus- and cell-specific processes that deliver at least the virus genome, and often all or some of the rest of the virion, inside the host cell. These processes will be covered in more detail later, in Chapter 4.

Note the strong resemblance in Fig. 2.2 between a bacterial virus—T4 phage or *Enterobacteriophage* T4 of *Escherichia coli*, which evolved possibly billions of years ago—and the type of only human-crewed spacecraft to have landed on another planet: this is a case of **macroscale design mimicking the nanoscale**, where genetic material is protected from the environment until it can be delivered to a new habitat—with specific signals from "landing gear" responsible for allowing this delivery.

Composition of virions

Basically, all virions have a **capsid structure**, which encloses a **genome.** Starting from the inside going out, the virion consists of:

1. **Genomic nucleic acid**: this may be RNA or DNA, single- or double-stranded, and may be all or part of the genome
2. **Protein associated with the genome**: in the simplest cases, this may be in the form of a protein or proteins helically-assembling or isometrically-assembling to form

BOX 2.1

Why do viruses make virions?

Why do viruses bother to form a particle to contain their genome? Some unusual and infectious agents such as viroids (see Chapter 8) don't. The fact that viruses pay the genetic and biochemical cost of encoding and assembling the components of a particle shows there must be some benefits. Virions may be regarded as the **extracellular phase** of the virus: they are exactly analogous to **spacecraft** in that they take viral genomes from a cell to another cell, and they protect the genome in inhospitable environments in which the virus cannot replicate. After leaving the host cell, the virion enters environments that could quickly inactivate an unprotected genome, by inflicting physical, chemical or enzymatic damage. Nucleic acids are susceptible to shearing by mechanical forces, and to chemical modification by radiation in the ultraviolet range in sunlight. Mild acid environments such as encountered in animal stomachs may also inactivate nucleic acids, whereas virions are often highly resistant. The natural environment is also full of **nucleases**, either from dead cells, or which have been deliberately secreted as defense against infection—and single-stranded RNA is especially labile. As an extreme example from the author's lab of the value of a capsid, purified TMV virions have a half-life of 30 min at 80°C and are still infectious after 40 years at 4°C.

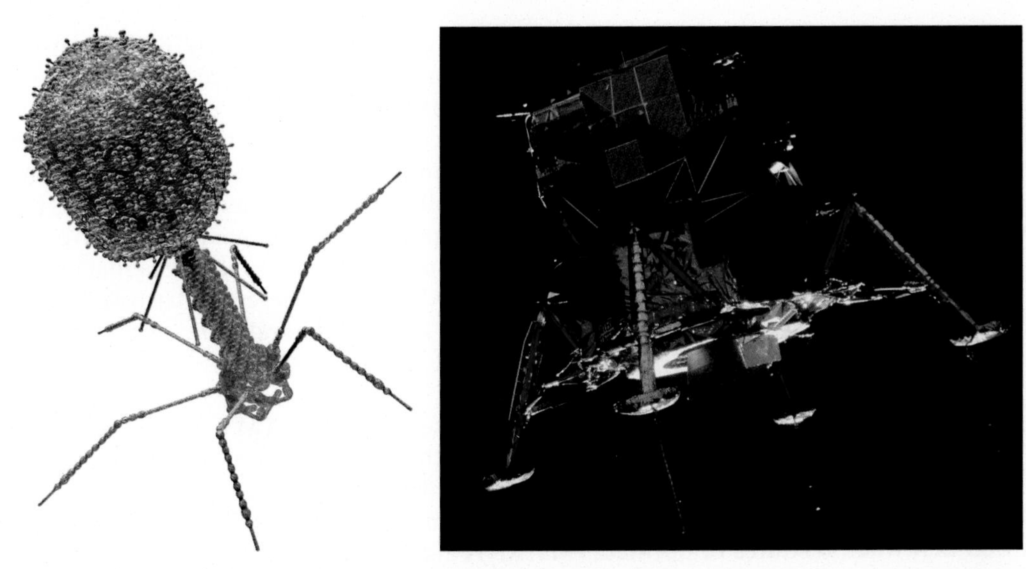

FIG. 2.2 Comparison of a tailed phage and a human spacecraft. (A) T4 bacteriophage. *Courtesy of Russell Kightley Media.* (B) the Apollo 11 Lunar Excursion Module (LEM), photographed in lunar orbit (NASA).

nucleocapsids. In more complex particles, enzymes such as **replicases** may also be associated with the genome.

3. Other virions may have more **complex** structures with two or more layers of capsid proteins, or possibly **tail structures**, with nucleocapsid innermost, possibly also with one or more **lipid bilayer membranes** with inserted viral proteins.

If the virions are simple **nucleoproteins**—that is, contain only nucleic acid and protein—then they are usually composed only of virus-specified components. However, certain host components may be incorporated within isometric virions, such as **polyamines**: these are **polycationic** (many + charges) compounds which serve to neutralize charge on the viral nucleic acid as it is packed into isometric capsids. Some small DNA viruses (e.g., **papilloma-** or **polyomaviruses**) may in addition encapsidate host **histones** associated in **nucleosome** complexes with virus genomic DNA. The ssRNA– **arenaviruses** have virions that appear granular when viewed by EM, as they contain **host ribosomes**. Larger enveloped virions, such as those of **pox- or herpes- or baculoviruses**, may incorporate many host proteins inside virions, and also have many host proteins embedded in their outer **envelopes**, or membranes.

In order to form virions, viruses must overcome two fundamental problems. First, they must assemble the particle using only the information available from the components that make up the particle itself. This is the same kind of **molecular self-assembly** that is used to make a large number of complicated intra- and extracellular structures, such as the helical **flagella** and **pili** in bacteria, and the various **cytoskeletal filaments—microfilaments, intermediate filaments and microtubules**—found in eukaryote cells. Second, virion capsids generally have regular geometric shapes, even though the proteins from which they are made are irregular. How do these simple organisms solve these difficulties? The solutions to both problems lie in the rules of **symmetry**, elucidated very elegantly from the 1950s onward as a result of X-ray crystallography and then 3-D image reconstruction from electron micrographs.

For excellent representations of virus structure, go to the University of Wisconsin's **VirusWorld** (http://www.virology.wisc.edu/virusworld/tri_number.php), where there is a gallery of 3-D image reconstructions of different viruses, together with explanations of structure. The forces that drive the assembly of virus particles include **hydrophobic** and **electrostatic** interactions, as well as **van der Waal's interactions** to a lesser extent: only rarely are covalent bonds involved in holding together the subunits. In biological terms, this means that **protein–protein, protein–nucleic acid,** and **protein–lipid interactions** are involved. We now have a good understanding of general principles and repeated structural motifs that appear to govern the construction of many diverse, unrelated viruses. These are discussed below under the two main classes of virus structures: **helical** and **icosahedral** symmetry.

Helical nucleocapsids

This is one of the simplest forms of viral capsid, exemplified by **TMV** and other **rodlike** or **filamentous** virions of plant viruses (see Fig. 1.4): the protein is "wound on" to the viral nucleic acid (generally ssRNA, though M13 and other filamentous bacteriophage virions contain circular ssDNA) in a simple **helix,** like a screw. In 1957, Fraenkel-Conrat and Williams showed that when mixtures of purified TMV RNA and coat protein (CP) are incubated together, rodlike virions formed (Chapter 1, see Fig. 1.17). The discovery that these could form spontaneously from purified subunits without any extra information indicates that the particle is in the **free energy minimum state**, and is the most energetically favored structure of the components. This inherent stability is an important feature of virions: although some are very fragile and are unable to survive outside the protected host cell environment, many are able to persist for long periods, for many years as in the case of TMV. While purified TMV CP can also assemble into rodlike structures, this occurs at a higher, nonphysiological pH than for virion assembly—because not only protein–protein, but also protein–nucleic acid interaction is required for authentic virion formation.

In the TMV structure protein subunits are all in **equivalent crystallographic positions**, related by a **right-hand "screw translation."** TMV virions are 300 nm long and 18 nm across (see Fig. 2.3). They are built from a **17.5 kDa protein** (blue) that is arranged in a right-handed helix that forms a long hollow rigid tube. The **6.4 kb ssRNA+** sense viral genome is wound within this helix (see also Fig. 1.16).

In the case of TMV, this is the entire virion: this is also the case for all rodlike and **filamentous** virions where no membranes are involved, and assembly of nucleoproteins follows the same principles of protein–protein and protein–nucleic acid interactions. Such viruses include ssRNA+ plant viruses like the rodlike hordeiviruses (e.g., **barley stripe mosaic virus**), and the filamentous potyviruses (e.g., **potato virus Y**), potexviruses (**potato virus X**), and closteroviruses (**grapevine leafroll viruses**), and the ssRNA− **tenuiviruses,** that resemble inner nucleocapsids of enveloped **bunyaviruses**. The ssDNA bacteria-infecting **inoviruses** (M13) also have long filamentous particles, but the CP is initially associated with the cytoplasmic membrane of gram-negative bacteria and assembly requires interaction of the circular ssDNA with this complex, and concomitant loss of a cytoplasmic DNA-binding protein (Fig. 2.4).

One important principle for helical capsids for virions is that there **is theoretically no limit on the size of genome** that can be encapsidated, because the helix can carry on growing as long as there is nucleic acid to interact with. However, **flexibility** of helical nucleocapsids

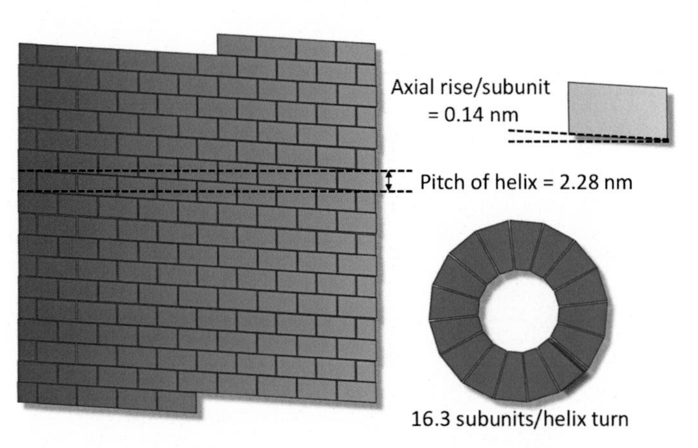

FIG. 2.3 Tobacco mosaic virion symmetry.

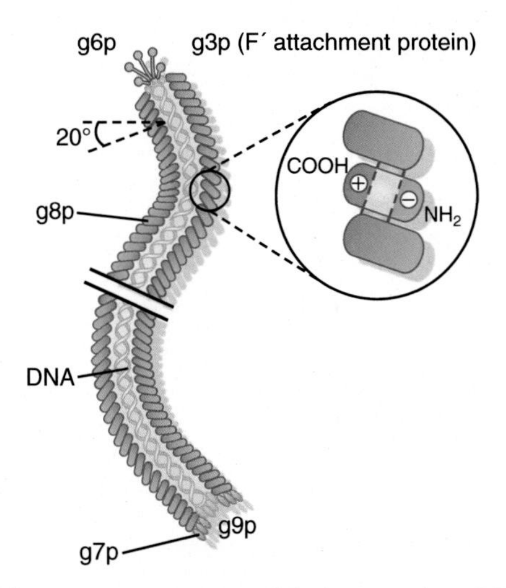

FIG. 2.4 Bacteriophage M13. Schematic representation of the bacteriophage M13 particle (*Inoviridae*). Major coat protein g8p is arranged helically, the subunits overlapping like the scales of a fish. Other capsid proteins required for the biological activity of the virion are located at either end of the particle. Inset shows the hydrophobic interactions between the g8p monomers (*shaded region*).

is an important property: long helical particles are increasingly likely to be subject to shear forces in the environment, and the ability to bend reduces the likelihood of breakage or damage. This is probably the reason that all ssRNA+ viruses with larger genomes than TMV have **flexuous filamentous** virions rather than rodlike ones.

Essentially all of the rodlike- and filamentous-capsid naked ssRNA+ viruses—all plant-infecting viruses—have CPs that are based on a "**four-helix bundle**" capsid protein (see Koonin et al., 2015; References/Recommended Reading), that probably have a common deep evolutionary origin. Most of the other filamentous nucleocapsid (NC) proteins, however, are not as similar to one another, and almost certainly have diverse origins. Interestingly, several of these appear to have been recruited from among **cellular endonucleases**: a

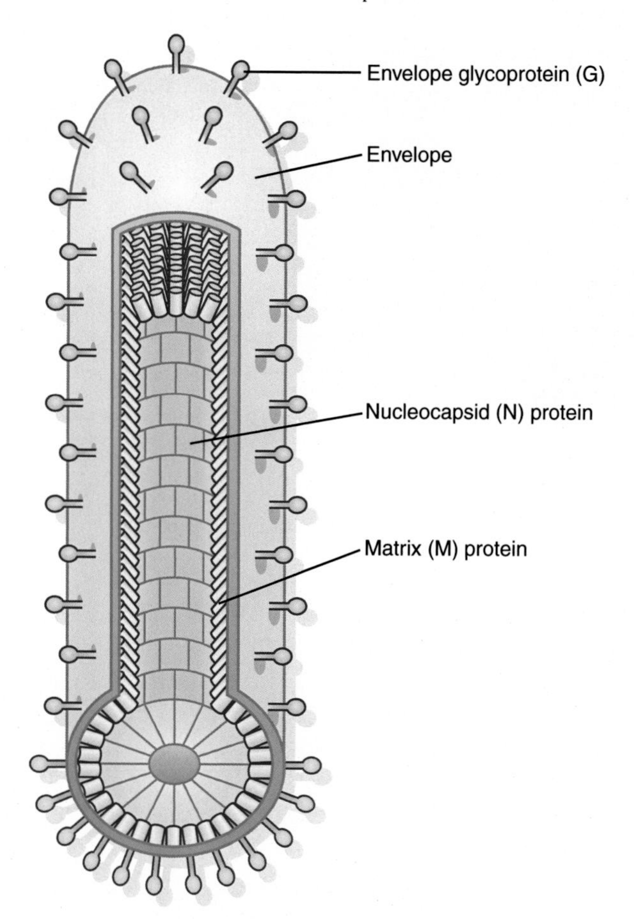

FIG. 2.5 Rhabdovirus particle. Rhabdovirus particles, such as those of VSV, have an inner helical nucleocapsid surrounded by an outer lipid envelope and its associated glycoproteins.

comprehensive analysis of viral proteins (Krupovic and Koonin, 2017) reports that the enveloped ss RNA− **Crimean-Congo hemorrhagic fever bunyavirus** (CCHFV) NC has a metal-dependent DNA-specific endonuclease activity, whereas the structurally-related NC of the ssRNA− Lassa fever arenavirus—a virus only very distantly related in some respects to CCHFV—has a similar head domain, but completely different tail, which has dsRNA-specific 3′-5′ exonuclease activity (see Chapter 3).

In the case of all animal viruses with helical nucleocapsids, the helical structures are enclosed within **enveloped** particles, and are formed by what is normally termed the **nucleocapsid** or **N protein**. Most of these also have a layer of **matrix protein** associated with the inner surface of the lipid bilayer envelope, and the envelope outer surface is studded with virus-encoded proteins, that are usually glycoproteins. Excellent examples of such virions are those of the ss-RNA− **paramyxoviruses** such as **measles virus** and **mumps virus,** the **parainfluenza viruses** which are human respiratory pathogens, and the related **rabies virus** and **vesicular stomatitis virus,** both **rhabdoviruses** (see Fig. 2.5). The helical nucleoproteins are also **flexuous,** because of genome size and packing constraints. The matrix (M) protein is usually the most abundant protein in the virion: for example, in VSV particles there are approximately 1800 copies of the M protein, 1250 copies of the N protein, and 400 G glycoprotein trimers.

Isometric nucleocapsids

Simple capsids

These are built up according to simple structural principles, as outlined at the VirusWorld site (n.d.). Put simply, nearly all **isometric virions** are constructed around **icosahedral symmetry**: **icosahedra** are regular solids with 20 equilateral triangles for faces. Early in the 1960s, direct examination of a number of small "spherical" viruses by electron microscopy, followed later the by 3-D image reconstruction from electron micrographs that earned Aaron Klug a Nobel Prize, revealed that they appeared to have icosahedral symmetry. At first sight, it is not obvious why this pattern should have been chosen by diverse virus groups; however, although in theory it is possible to construct virus capsids based on simpler symmetrical arrangements such as **tetrahedra** (4 faces), **cubes** (6 faces) or **dodecahedra** (12 faces), there are practical reasons why this does not occur. It is more economical in terms of genetics to design a capsid based on a large number of identical, repeated protein subunits rather than fewer, larger subunits. In order to construct a capsid from repeated subunits, a virus must "know" the rules that dictate how these are arranged. For an **icosahedron**, the rules are based on the rotational symmetry of the solid, known as 2–3–5 symmetry, which has the following features (Fig. 2.6):

- An axis of twofold rotational symmetry through the center of each edge.
- An axis of threefold rotational symmetry through the center of each face.
- An axis of fivefold rotational symmetry through the center of each corner (vertex).

Because protein molecules are irregularly shaped and are not regular equilateral triangles, the simplest icosahedral capsids are built up by using **three identical subunits to form each triangular face**. This means that **60 identical subunits** are required to form a complete capsid (3 subunits per face, 20 faces). In most cases, analysis reveals that icosahedral virus capsids contain more than 60 subunits, which causes a theoretical problem. A regular icosahedron composed of 60 identical subunits is a very stable structure because all the subunits are equivalently bonded (i.e., they show the same spacing relative to one another and each occupies the minimum free energy state). With more than 60 subunits it is impossible for them all to be arranged completely symmetrically with exactly equivalent bonds to all their neighbors, as a true regular icosahedron consists of only 20 identical subunits. To solve this problem, in 1962, Caspar and Klug proposed the idea of **quasi-equivalence**. Their idea was that subunits in nearly the same local environment form nearly equivalent bonds with their neighbors, permitting self-assembly of **"quasi-icosahedral"** capsids from multiples of 60 subunits. In these higher order icosahedra, the symmetry of the particle is defined by the **triangulation number** of the icosahedron (Fig. 2.6). An excellent account of the basics of triangulation numbers and quasi-equivalence is given at the VirusWorld site; accordingly, it will not be covered in detail here.

The quasi-icosahedral capsid is possibly nature's most popular means of enclosing viral nucleic acids. They come in many sizes, from tiny ***T*=1 structures** with only 60 identical subunits (nanoviruses, e.g., **banana bunchy top virus**; circoviruses, **porcine circovirus**-1 and -2; ~18 nm diameter; microviruses, **φX174**), ***T*=3 structures** with 180 structural units (many plant

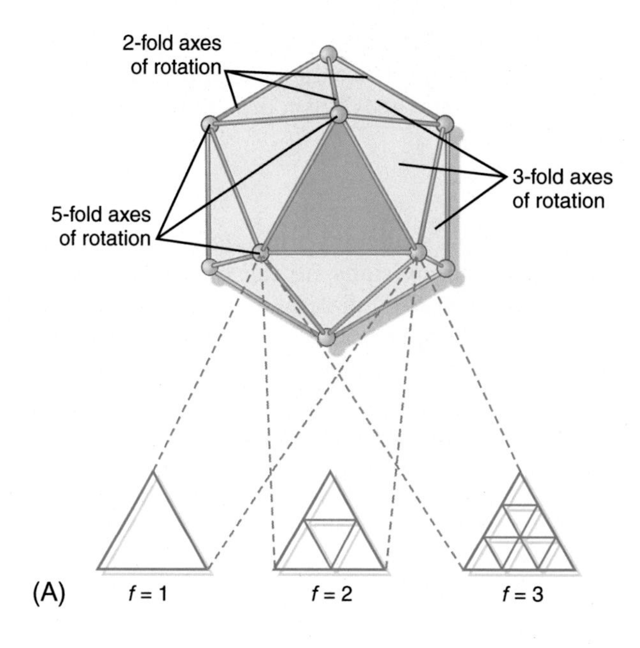

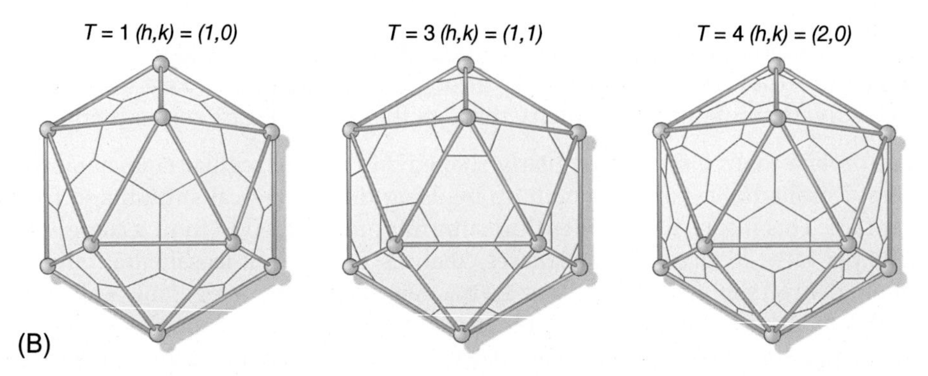

FIG. 2.6 Icosahedral symmetry and triangulation numbers. (A) Illustration of the 2 3 5 symmetry of an icosahedron. More complex (higher order) icosahedra can be defined by the triangulation number of the structure, $T=f^2.P$. Regular icosahedra have faces consisting of equilateral triangles and are formed when the value of P is 1 or 3. All other values of P give rise to more complex structures with either a left-hand or right-hand skew (see VirusWorld, http://www.virology.wisc.edu/virusworld/tri_number.php). (B) Icosahedra with triangulation numbers of 1, 3, and 4.

viruses; **polioviruses** and **foot-and-mouth disease virus**), ***T*****=4**240 subunit structures (animal-infecting enveloped ssRNA+ **togaviruses**), ***T*****=7**420 subunit structures such as many **tailed phage head structures**, through to huge structures such as those of **iridoviruses** or **mimiviruses** (up to 400nm diameter; ***T*****=972–1200**), with an internal and an external lipid membrane.

The simplest isometric virions are those of the viruses with the smallest genomes: these are virions such as those of the ssRNA+ **satellite tobacco necrosis virus** (1239nt), the linear ssDNA **canine parvovirus** (CPV, 5323nt; *Protoparvovirus, Parvoviridae*) and the defective **adeno-associated viruses** (AAVs, ~4700 nt; *Dependoparvovirus, Parvoviridae*); circular ssDNA-containing **porcine circovirus** (PCV, 1768 nt; *Circovirus, Circoviridae*), and **microviruses** infecting *E. coli* and other bacteria (e.g., φX174, 5386 nt; *Microvirus, Microviridae*). These all have a **simple icosahedral $T=1$ surface lattice structure**. All structural subunits of these capsids are in the same positional state; that is, they all have the same interactions with their neighbors. The packing capacity of the virions has an upper bound of just greater than 5 kb. All of these small virions have **single-stranded genomes** in their virions: this is a consequence of the limited space available for the genome, and the fact that ssRNA and DNA can fold more compactly than dsRNA or DNA, which are more rigid molecules.

A unique derivative structure is that of the ssDNA-containing **geminivirus** virions—denoting twins, after Castor and Pollux from Greek mythology—which have **two incomplete $T=1$ icosahedra,** with 55 subunits each, joined at the missing vertex with a small rotational twist. The virions are termed **geminate**, or twinned, and each double particle contains one molecule of circular single-stranded DNA (Fig. 2.7).

Examples of bigger (e.g., $T=3$, 4, 7) and slightly more complex nonenveloped isometric structures are common, as many phages, plant viruses and a number of insect and animal viruses have quasi-icosahedral capsids of 30–50 nm in size. It is interesting to see how viruses have evolved to encapsidate genomes larger than can be accommodated in a $T=1$ shell with 60 subunits: a simple rule appears to be that most **particles have 12 vertices at 5-fold rotational axes of symmetry**, with **pentameric** (=5 subunit) groupings of subunits in these locations, and then **hexameric** (=6 subunit) structures in the faces of the particles.

It is also interesting that viruses with similar-sized particles, with the same basic structure, may use one or two or several types of proteins to make their virions. The ssRNA+ brome mosaic virus (BMV; *Bromovirus, Bromoviridae*) has ~30 nm particles composed of a single 20 kDa **coat protein (CP)**. These have a very similar $T=3$ structure to the **picornaviruses** (family *Picornaviridae*), which have three structural proteins—resulting from gene duplication of a single precursor—filling the roles of the three quasi-equivalent positions required by the $T=3$ structure. All of these are **eight-stranded antiparallel β-barrel**, or **"single jellyroll fold proteins"** (Fig. 2.8; see also Chapter 3). Cowpea mosaic comovirus (CPMV; *Comovirus, Secoviridae*), which is evolutionarily related to picornaviruses—they are all in the Order *Picornavirales*—has two capsid proteins: one occupies the fivefold rotational axes of symmetry, and the other—which looks like the product of a **fusion of a gene duplication** of the first—occupies two of the positions the picornavirus proteins do (i.e., a **DJR** or **double jelly roll protein**) (Fig. 2.9).

SJR-fold CPs are the most prevalent among isometric virions, representing ~28% of all CPs analyzed. These are structurally similar to four groups of host cell-derived SJRs, spread across all three cellular Kingdoms (Bacteria, Archaea and Eukarya, referred to as "**ribocells**") (see Chapters 1 and 3). The next most common classes of CPs are the **DJR fold** proteins of isometric viruses, found in ~10% of taxa. These are almost exclusively dsDNA viruses, infecting all ribocells, and include mid-sized viruses (e.g., adenoviruses, PRD1-related phages) as well as giant-sized viruses (e.g., mimiviruses). The DJR fold is not found in cellular proteins. The next most common structural fold found in **major CPs** (MCPs) of dsDNA viruses is the **HK97-like-fold**, named for the bacteriophage HK97 gp5 protein. This is found in bacteria- and archaea-infecting tailed viruses of the taxonomic Order *Caudovirales*, one of the most abundant and diverse group of viruses on planet Earth, as well as in the animal-infecting herpesviruses (Order *Herpesvirales*, Family *Herpesviridae*).

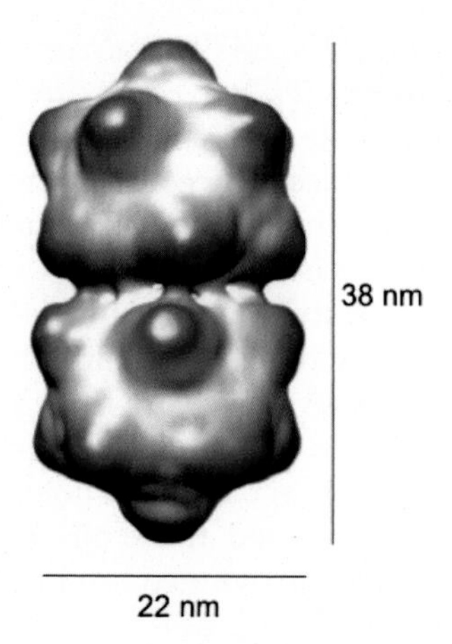

FIG. 2.7 Geminivirus structure. Three-dimensional reconstruction of a geminate maize streak virus particle from cryoelectron microscopy data. *Image copyright Kyle Dent and University of Cape Town.*

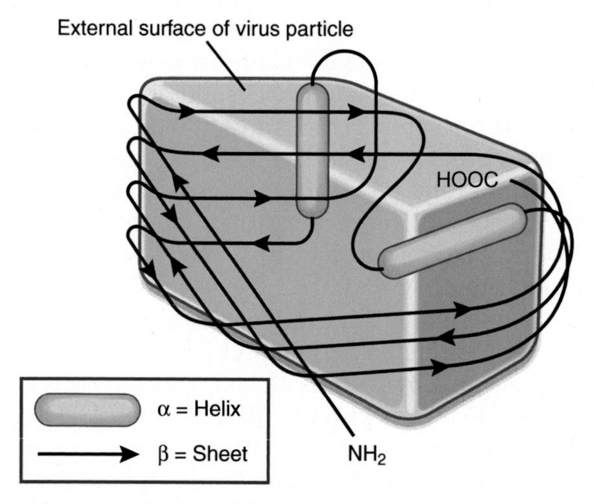

FIG. 2.8 The eight-stranded antiparallel β-barrel or SJR-fold. The subunit structure found in most isometric capsids.

More complex isometric capsids

More complex capsids are generally found for viruses with larger genomes, whether composed of RNA or DNA. These include virions such as those of the small dsDNA papillomaviruses and polyomaviruses (Families *Papillomaviridae* and *Polyomaviridae*), dsRNA rotaviruses (Family *Reoviridae*), dsDNA adenoviruses (Family *Adenoviridae*) and various similar-sized bacterial viruses (e.g., phage PRD1; *Tectivirus, Tectiviridae*).

While many—but not all—simpler isometric particles can self-assemble from subunits, even in vitro, a number of viruses with isometric particles or isometric heads for many phages, cannot. For example, assembly of the $T=1$ particle of φX174 starts with self-assembly of an empty precursor particle called the procapsid, with the aid of two **scaffolding proteins**: these

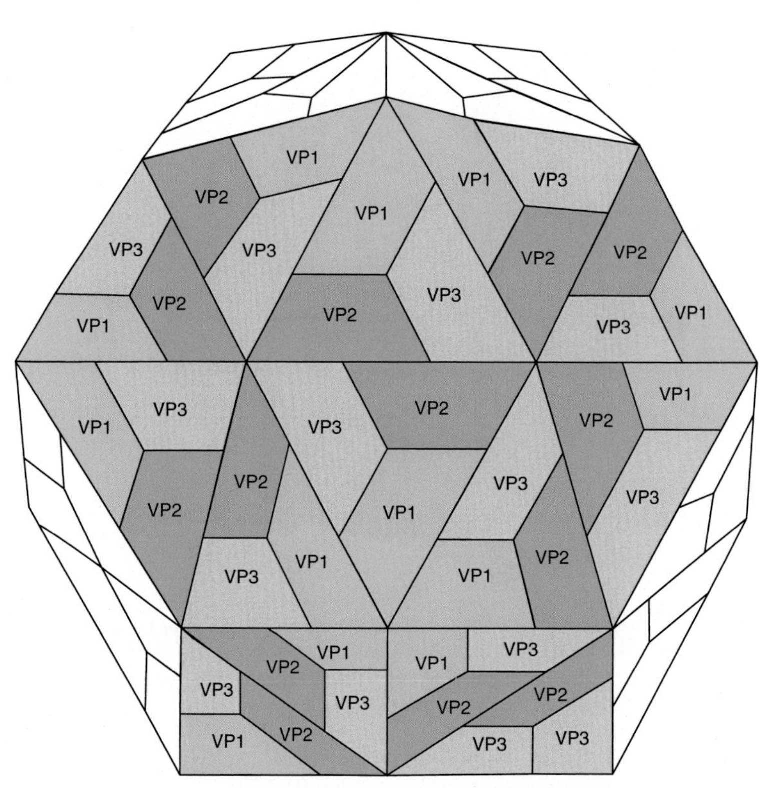

FIG. 2.9 Quasi-icosahedral $T=3$ particles. Bromovirus, comovirus and picornavirus particles are icosahedral structures with triangulation number $T=3$. In bromovirions one capsid protein is found in all 3 positions shown, with subunit 1 at fivefold rotational axes of symmetry (VP1), and subunits 2 and 3 in the equilateral triangle faces (VPs 2 and 3). In picornavirions, three different virus proteins (VP1, 2, and 3) comprise the surface of the particle, with VP1 at position 1 and VPs 2 and 3 at the positions shown. A fourth protein, VP4, is not exposed on the surface of the virion but is present in each of the 60 repeated units that make up the capsid. In comovirions, particles have a pseudo $T=3$ symmetry as there are only 120 subunits: however, the SJR-fold CPS (=small) protein occupies position 1, and the DJR CPL (=large) protein occupies both positions 2 and 3.

exit the particle upon entry of the genome. Virus particles with large triangulation numbers use different kinds of subunit assemblies for the faces and vertices of the icosahedron, and also use internal scaffolding proteins which act as a framework. These direct the assembly of the capsid, typically by bringing together preformed **subassemblies** of proteins. Some of these types of assembly will be discussed in Chapter 4.

Adenoviruses are possibly the poster children for "**nanoscale spacecraft**": their resemblance to early orbiting satellites is almost uncanny, with the "aerials" or **surface spikes** serving the purpose of communicating with cells by recognition of surface receptor proteins to enable delivery of their genomes. It took more than a decade to obtain a X-ray crystallographic structure to 3.5Å resolution, one of the most complex structures ever solved by this technique (Fig. 2.10; also see References/Recommended Reading). Particle assembly also employs scaffolding proteins which are displaced by import of the linear dsDNA genome into the procapsid via a **molecular motor**, similarly to the structurally and evolutionarily related PRD1 phage (see also Chapter 3).

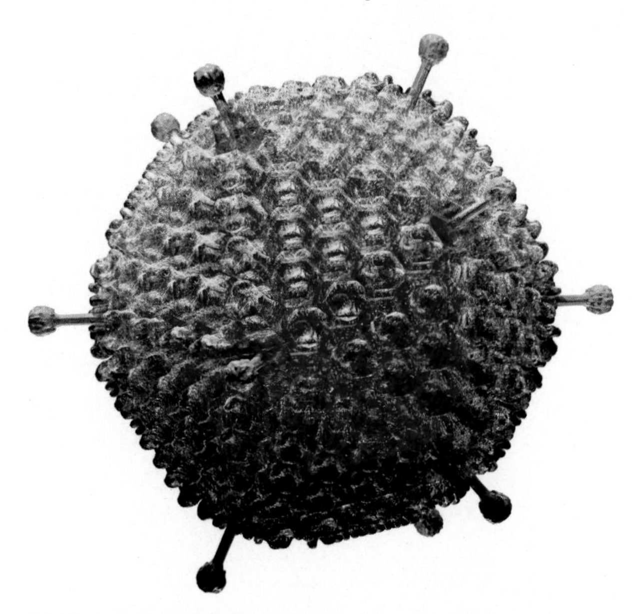

FIG. 2.10 Adenovirus 36. The translucent *yellow* outer layer represents the viral hexon proteins (arranged as an icosahedral capsid). At the 12 5′-fold rotational axes (vertices) lie the pentons (*purple*) from which radiate the fibers (*blue*). The fiber proteins are responsible for mediating entry into host cells. The *red* inner ball represents the linear double-stranded DNA genome. *Courtesy of Russell Kightley Media.*

Enveloped virions

Some virions additionally have **envelopes**, or **lipid bilayer membranes**: these are derived from host cells, and always have viral proteins inserted in them (Fig. 2.11). These are almost always **glycoproteins** in eukaryote viruses and may be heavily **glycosylated**, or covered in sugars that are added in the **endoplasmic reticulum**. These are often referred to as **spikes**, and are anchored into the envelope by **hydrophobic transmembrane anchor domains** that traverse the membrane, with a terminal **cytoplasmic domain**. Their main function is to recognize receptor molecules on host cells so as to allow entry. There is usually an inner layer of protein (generally called the **matrix**) inside the envelope; this is also in contact with the inner nucleocapsid protein (Fig. 2.12A). Matrix proteins are not usually glycosylated and are often very abundant: for example, in retroviruses they comprise approximately 30% of the total weight of the virion. Some matrix proteins contain transmembrane anchor domains. Others are associated with the membrane by hydrophobic patches on their surface or by protein–protein interactions with the cytoplasmic domain of envelope glycoproteins. While most enveloped virions have a matrix, ssRNA+ **coronaviruses** such as SARS-CoV and SARS-CoV2 and **flaviviruses** such as West Nile and yellow fever viruses, as well as ssRNA− **bunyaviruses**, do not. This is probably because their envelopes with closely-packed viral membrane proteins are sufficiently structured as to provide an effective capsid.

The **human immunodeficiency virus type 1 (HIV-1)** virion is a good example of an enveloped virion. Trimeric assemblies of **Env**, which is also called **gp160** and which is processed to form the **gp120** head and the stalk protein **gp41**, are embedded in a membrane which surrounds a regular isometric structure composed of the **matrix protein** (**MA, p17**), within which is the **capsid** (**CA, p24**) which contains the two-component ssRNA+ genome associated with

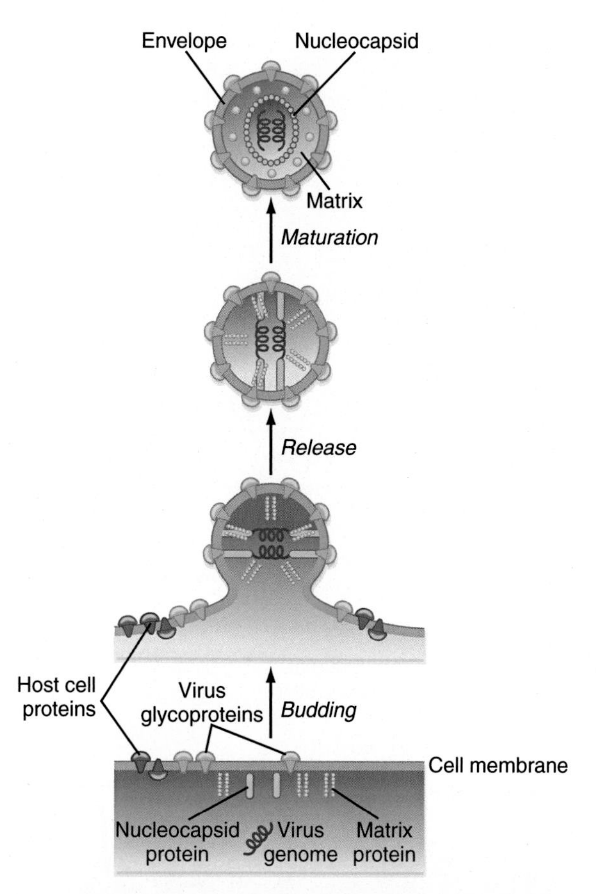

FIG. 2.11 Budding of enveloped particles and envelope proteins. Enveloped virus particles are formed by budding through a host cell membrane, during which the particle becomes coated with a lipid bilayer derived from the cell membrane. For some viruses, assembly of the structure of the particle and budding occur simultaneously, whereas in others a preformed core pushes out through the membrane. Several classes of proteins are associated with virus envelopes. Matrix proteins link the envelope to the core of the particle. Virus-encoded glycoproteins inserted into the envelope serve several functions. External glycoproteins are responsible for receptor recognition and binding. Host-cell-derived proteins are also sometimes found to be associated with the envelope, usually in small amounts.

a **nucleocapsid protein** (**NC**), along with associated **reverse transcriptase** and other minor nonstructural proteins. Another good example is the ssRNA− **influenza virus type A** (family *Orthomyxoviridae*) (Fig. 2.12B): the often elongated particles of around 100 nm in diameter have a tightly-packed array of two envelope glycoproteins—trimeric **hemagglutinin (HA)** and tetrameric **neuraminidase (NA)**—embedded in a membrane on their surfaces, a layer of **matrix protein (M1)**, and eight helical nucleoprotein complexes composed of the genomic ssRNA− molecules and the **nucleoprotein (N)**, complexed with the three components of the **viral replicase**. A minor protein called M2 is a **transmembrane protein** and acts as a proton (H^+ ion) channel to allow acidification of the interior of the virion during disassembly and during maturation in the *trans*-Golgi apparatus (see Chapter 4).

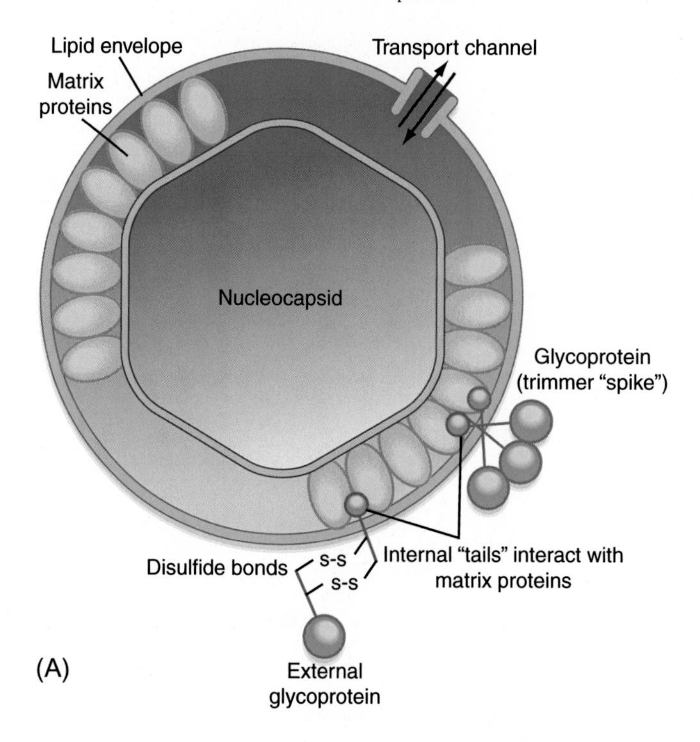

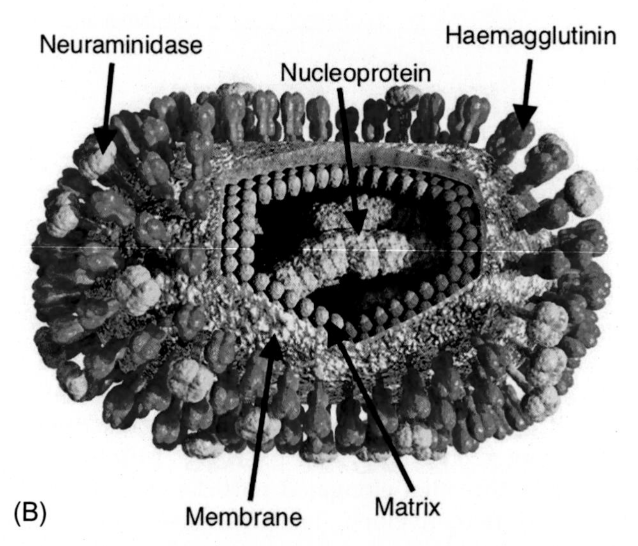

FIG. 2.12 Enveloped virions. (A) Several classes of proteins are associated with virion envelopes: matrix proteins beneath the membrane link it to the core or nucleocapsid; viral glycoproteins on the membrane outer surface are usually virion attachment proteins (VAPs); transmembrane proteins can be transport channels. (B) Influenza A virus: hemagglutinin trimers (*orange*) are densely packed in the envelope (*green-blue*), with lower concentrations of tetrameric neuraminidase (*yellow*). The matrix shell (*light purple*) is in contact with the underside of the membrane, and contains the helical nucleocapsids (*yellow*, 8 per virion) that cover the ssRNA− genome. *Courtesy of Russell Kightley Media.*

FIG. 2.13 Ebola virus particle. The cutaway illustration shows trimeric glycoprotein (*pink*, GP) spikes, embedded in the envelop (*greenish*), with the adherent matrix (VP40) overlaying the helical nucleocapsid (N protein) containing the single-component ssRNA− genome (*yellow*). Virions also contain virus-encoded polymerase (L), polymerase cofactor VP35 and transcription factor VP30. *Courtesy of Russell Kightley Media.*

The nucleoprotein inside the envelope and matrix can be **helical** (paramyxoviruses, rhabdoviruses, bunyaviruses) or **isometric** (flaviviruses, hepadnaviruses, retroviruses). An unusual virion morphology is that of **filoviruses** (family *Filoviridae*; e.g., Ebola, Marburg viruses), which have long flexuous enveloped particles, with a helical nucleoprotein (Fig. 2.13).

The prokaryote-infecting **cystoviruses** (e.g., **Pseudomonas virus φ6**, genus *Cystovirus*, family *Cystoviridae*) are highly unusual in that they are the only dsRNA viruses to have enveloped virions. These are probably ancestrally related to animal **reoviruses** (see Chapter 3), and like them, have a multilayer isometric capsid with incorporated RdRp (see below). Unlike reoviruses, however, they have a lipid membrane obtained within their bacterial hosts that has embedded in it the virus attachment protein. There are however a number of enveloped virus families of prokaryotes that have dsDNA genomes, such as the *Fuselloviridae*, *Lipothrixviridae*, and *Plasmaviridae*.

The only enveloped plant virus virions are those of viruses that also either infect insects themselves, or are closely related to viruses that do. These include **rhabdo- and bunyaviruses** (e.g., lettuce necrotic yellow rhabdovirus; tomato spotted wilt tospovirus). This is almost certainly related to host biology, as infecting animal cells requires receptor-binding proteins whereas infecting plant cells does not (see Chapter 4). Interestingly, the highly filamentous and nonenveloped ssRNA− **tenuiviruses** (genus *Tenuivirus*, family *Phenuiviridae*, and order *Bunyavirales*) look like bunyaviruses that have lost their envelope and associated proteins, and have become completely adapted to being plant viruses.

Complex or compound virions

While the majority of virions can be fitted into one of the three structural groups outlined above (i.e., helical symmetry, icosahedral symmetry, or enveloped viruses), there are many whose structure is more complex. In these cases, although the general principles of symmetry already described are often used to build parts of the virion shell (this term being appropriate here because such viruses often consist of several layers of protein and lipid), the larger and more complex virions cannot be simply defined. Because of the complexity of some of these particles, until very recently they largely defied attempts to determine detailed atomic structures using the techniques described in Chapter 1, such as X-ray crystallography or EM 3-D image reconstruction or cryoEM.

Most prokaryote viruses have a more complex structure than the simple isometric and helical nucleocapsids typical of many plant and animal viruses. Even viruses with relatively small genomes may have more than one type of architecture associated with their virions, normally in the shape of some kind of **tail structure** attached to an **isometric head** at one vertex. All of the larger viruses have tail assemblies that include a **molecular motor** that both packs DNA into preformed capsids, and can propel the viral genome into cells.

Bacterial viruses with tails

The **T-even viruses**—part of the ***"T4-like virus"*** genus, family *Myoviridae*, Order *Caudovirales* (tailed phages)—have the general structure of an **elongated isometric head**, containing a tightly-packed linear ~170 kbp dsDNA genome. The head is attached via a **connector assembly** or **collar** to a **contractile helically-constructed tail** with a rigid helical **core**, that terminates in a **baseplate** with **tail fibers** that bind specifically to the **bacterial host cell surface** to start the infection process (see Chapter 4). The baseplate has **lysozyme-like enzymes** associated with it whose function is to dissolve the bacterial cell wall to allow tail core penetration during infection (Fig. 2.14).

λ-like viruses in family *Siphoviridae* (also *Caudovirales*) have thinner **noncontractile tails** with tail fibers and isometric heads, with circularly permuted linear dsDNA genomes of ~48 kbp (phage λ; see Chapters 3 and 4) (Fig. 2.15).

The **P22-like** and **Φ29-like** viruses in family *Podoviridae* (*Caudovirales*) have isometric heads and short, thick tails (Fig. 2.16). P22 phages have ~44 kbp linear dsDNA genomes: the tail bases contain both the attachment proteins and the molecular motor to drive DNA into virions during assembly, and into cells during infection. The viruses attack bacteria in genus *Salmonella*.

Large dsDNA-containing virions

A negatively-stained electron micrograph showing an enveloped herpesvirus virion was shown in Fig. 1.12B: this shows clusters of glycoprotein B spikes in the envelope, a granular tegument composed of virus-coded proteins within it, surrounding a characteristic icosahedral mega-capsid. Herpes virions are similar to short-tailed phages: there is evidence from **cryoelectron microscopy** that virions may have **a very short tail structure with a DNA motor** that is probably involved with DNA entry into the nucleus of infected cells (Fig. 2.17). The entry of DNA is a highly regulated process, and the DNA inside capsids is very tightly packed—which state has been described as packaging a micrometer-long DNA into a nanometer-scale capsid. This results in neighboring regions of DNA electrostatically repelling each other due to their negatively-charged backbones, resulting in an internal pressure of 20 atm inside capsids. A similar situation exists for all tailed bacteriophages.

Very large enveloped virions may have a highly complex structure: for example the dsDNA-containing **orthopoxvirus** virions—like **variola virus**, the agent of smallpox in humans, and cow-, camel- and monkeypox viruses—may have **more than one membrane**, with both an internal lipid layer as well as an outer envelope that is not necessarily present, depending on whether they are produced by budding (with outer membrane), or by cell lysis (without). Poxviruses and other viruses with complex structures such as **African swine fever virus** (*Asfarviridae*) and **iridoviruses** (*Iridoviridae*) obtain their membranes in a different way from "simple" enveloped viruses such as retroviruses or influenza virus. Rather than budding at the

FIG. 2.14 Phage T4. Virions of a typical myovirus, with elongated isometric head, thick contractile tail and prominent tail fibers. The one on the left has just bound the bacterial surface by means of the tips of tail fibers; the one on the right has injected DNA into the cell after baseplate binding and conformational change leading to tail contraction. *Courtesy of Russell Kightley Media.*

cell surface or into an intracellular compartment, thus acquiring a single membrane, these complex viruses are wrapped by the endoplasmic reticulum, thus acquiring two layers of membrane (Fig. 2.18, see also Chapter 4).

The outer surface of the virion is composed of lipid and protein, if it buds. This layer surrounds the **core**, which is biconcave (dumbbell-shaped), and two "**lateral bodies**" whose function is unknown. The core is composed of a tightly compressed **nucleoprotein**, and the double-stranded DNA genome is wound around it.

Arthropod-infecting **baculoviruses** have been investigated for their potential investigation as biological control agents for insect pests. In addition, **recombinant** (=genetically manipulated) baculoviruses are very commonly used as expression vectors in cultured insect cells to produce large amounts of recombinant proteins, with their best-known application being in the manufacture of the **blockbuster** human papillomavirus vaccine Cervarix by

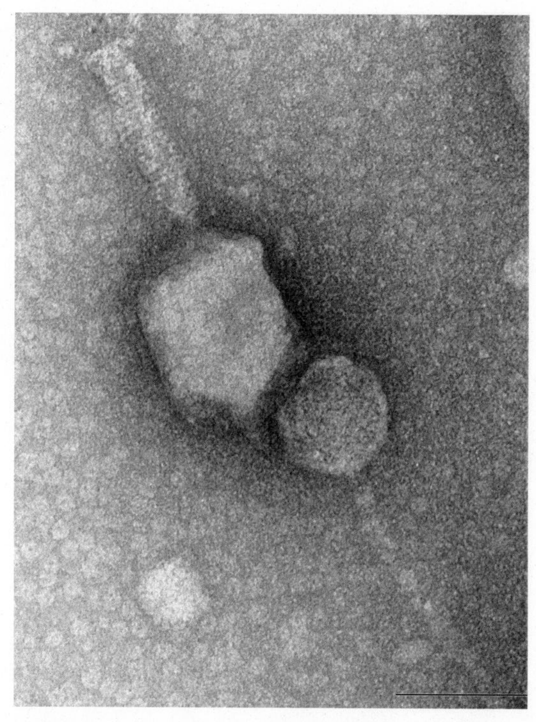

FIG. 2.15 Myovirus and siphovirus. Negatively stained electron micrograph from a mixed phage infection of *E. coli* (own collection). T4-like myovirus (top), λ-like siphovirus (bottom).

FIG. 2.16 P22 podovirus. Large, regular isometric head with stud-like protrusions, and short, thick tail structure with attachment spikes. *Courtesy of Russell Kightley Media.*

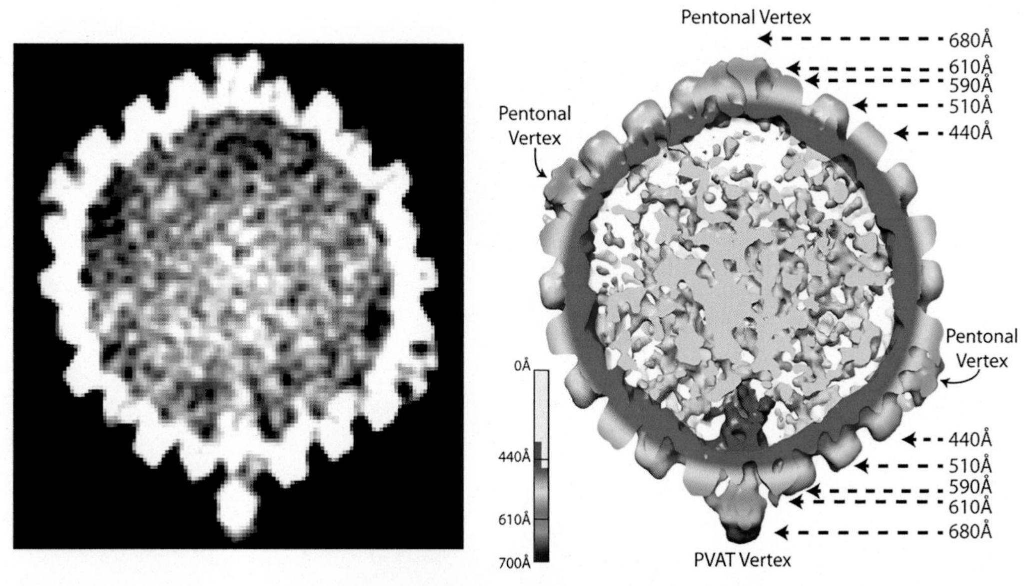

FIG. 2.17 Herpesviruses. CryoEM reconstruction of a herpesvirus virion. The left-hand panel shows a gray-scale density map of herpesvirus type 1. In the right-hand panel, the capsid shell is radially colored, with internal DNA density shown in cream, and portal density in purple (see also Fig. 1.12B). *Schmid, M.F., Hecksel, C.W., Rochat, R.H., Bhella, D., Chiu ,W., et al., 2012. A Tail-like Assembly at the Portal Vertex in Intact Herpes Simplex Type-1 Virions. PLoS Pathog. 8(10), e1002961. https://doi.org/10.1371/journal.ppat.1002961.*

GlaxoSmithKline (see Chapter XX). These complex viruses contain 12–30 structural proteins and make rodlike (hence "**baculo**") nucleocapsids that are 30–90nm in diameter and 200–450nm long, and contain the 90- to 230-kbp dsDNA genome (Fig. 2.19). The nucleocapsid is surrounded by an envelope, which may in some genera of *Baculoviridae* be contained in a **crystalline protein matrix** composed of the proteins composed of proteins **polyhedrin, polyhedron envelope protein** (PEP) and **P10.** This matrix and virion(s) within it is referred to as an **occlusion body**, and virions within such bodies are **occluded** (see Chapter 4). Nonoccluded baculovirus virions (BV) bud from cell surfaces; occluded forms (ODV) acquire their envelopes from the nuclear membrane. ODVs are responsible for environmental infection of arthropods, while BV spread infection once inside the arthropod (Chapter 4).

Mimiviruses (family *Mimiviridae*), which were originally presumed to be Gram-positive bacteria infecting an amoebal species because of their size—have a very large (~500nm diameter) isometric capsid, with internal membranes and other structures. The most striking feature of mimivirus virions is the dense coating of fibrils, ~50nm long, on their surfaces. The structure of the mimivirus capsid has recently been determined by a combination of electron tomography and **cryo-scanning electron microscopy**—sophisticated image reconstruction techniques based on cryoEM—and atomic force microscopy, which has a resolution of 1nm (see Chapter 1). They apparently have **two specialized portals** for both insertion of genome DNA into their huge particles, and eventual delivery of the genome into cells. The latter has been termed the "**stargate,**" and includes the whole of a **fivefold vertex** that opens like a flower to deliver DNA, and is opposite the smaller **DNA insertion portal** in the center of a triangular face (Fig. 2.20).

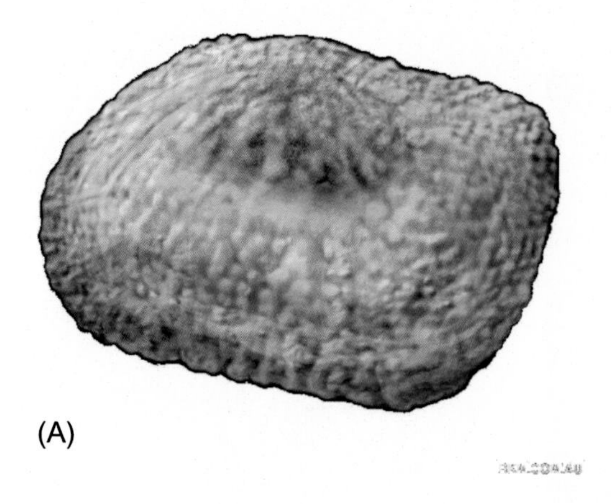

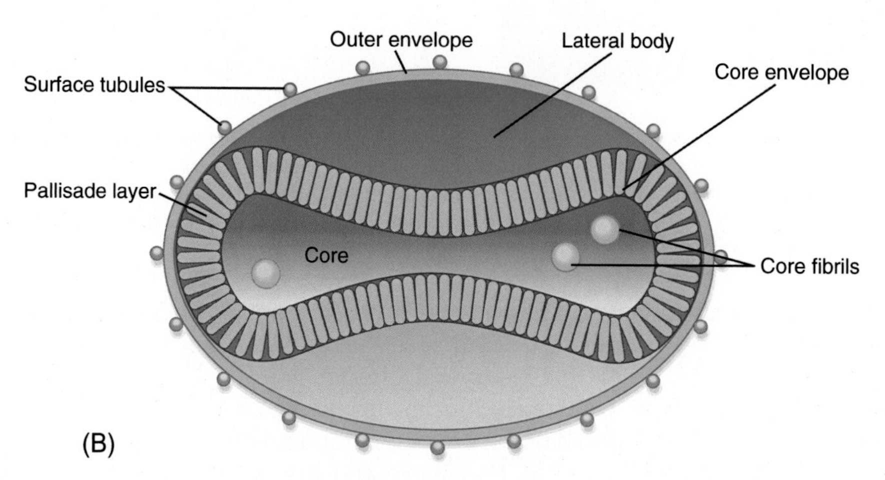

FIG. 2.18 Poxvirus particle. (A) Depiction of an enveloped smallpox virus particle. *Courtesy of Russell Kightley Media.* (B) Poxvirus particles are the most complex virions known and contain more than 100 virus-encoded proteins, arranged in a variety of internal and external structures.

Nonstructural and host proteins in virions

All dsRNA and ssRNA− viruses and retroviruses contain **polymerases** within their virions: this is **RNA-dependent RNA polymerase (RdRp)** for ds- and ssRNA− viruses, and **RNA-dependent DNA polymerase/DNA polymerase** (reverse transcriptase, RT) for retroviruses. ssRNA+ virus genomes are translated to give RdRp upon cell entry. The large dsDNA poxviruses and mimiviruses do not use the nucleus as a replication site (see Chapter 4): these take their own DNA-dependent DNA polymerases into cells in the virion, and establish "**viral factories**" in the cytoplasm, where their genomes are replicated.

Enveloped virions regularly incorporate **host membrane proteins** in their envelopes: while this may be a side effect of the "budding" process by which viral nucleoprotein complexes acquire envelopes, it may also in some cases be a mechanism for avoidance of host immune systems (e.g., hepatitis B virus incorporates **human serum albumin** into its envelope; HIV virions may incorporate host **major histocompatibility complex or MHC proteins**).

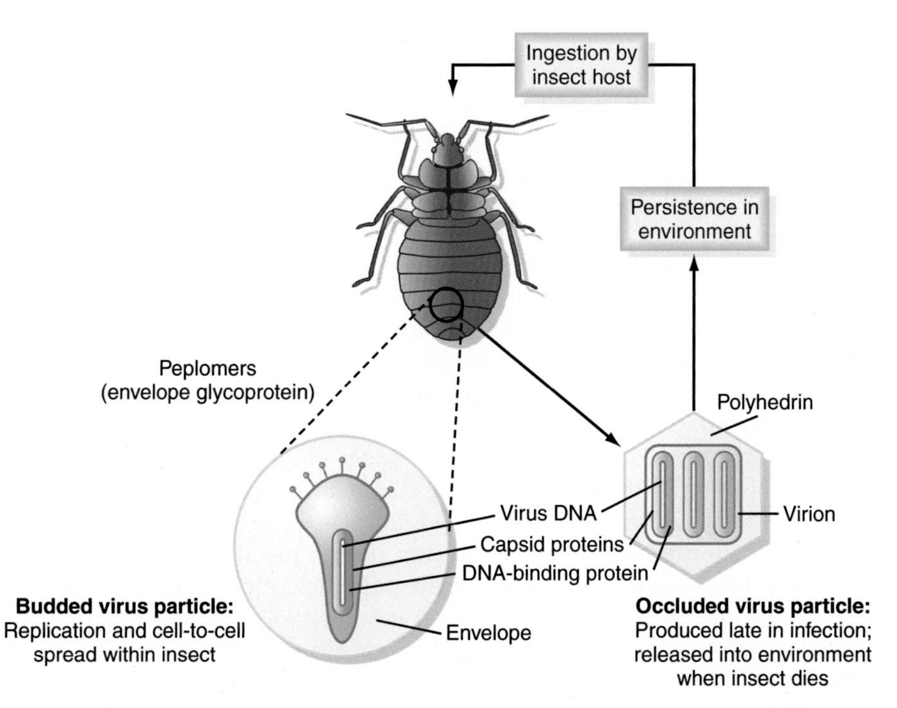

FIG. 2.19 Baculovirus particles. Some baculovirus particles exist in two forms: a relatively simple "budded" form found within the host insect, and a crystalline, protein-occluded form (occlusion body) responsible for environmental persistence.

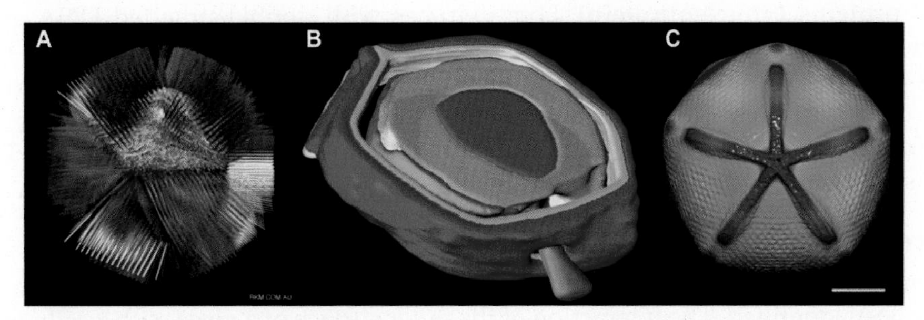

FIG. 2.20 Mimivirus structure. (A) Particle with surface spikes cleared away, showing outer and inner capsids and membranes. (B) Tomographic slice through a mimivirus particle undergoing DNA packaging, showing the orifice in an icosahedral face through which DNA (*green*) is packaged, which spans the outer and inner capsid shells (*red, orange*), and the internal membrane (*blue*). The protein core underlying the membrane is *gray*. (C) Surface-shaded rendering of cryoEM reconstruction of untreated Mimivirus. Looking down the *starfish-shaped* feature associated vertex. *Panel (B): From Fig. 7B, Zauberman, N., Mutsafi, Y., Halevy, D.B., Shimoni, E., Klein, E., Xiao, C., et al., 2008. PLoS Biol. 6(5), e114. https://doi.org/10.1371/journal.pbio.0060114.; Panel (C) Fig. 5 A. Xiao, C., Kuznetsov, Y.G., Sun, S., Hafenstein, S.L., Kostyuchenko, V.A., Chipman, P.R., et al., 2009. Structural studies of the giant mimivirus. PLoS Biol 7(4), e1000092. https://doi.org/10.1371/journal.pbio.1000092.*

If the virions are simple **nucleoproteins**—that is, contain only nucleic acid and protein—then they are usually composed only of virus-specified components. However, certain host macromolecules may be incorporated within virions, such as **polyamines**: these are **polycationic** (many + charges) compounds which serve to neutralize charge on the viral nucleic acid as it is packed into the capsid. Some small DNA viruses may in addition encapsidate **host histones** associated in **nucleosome** complexes with virus genomic DNA.

Protein–nucleic acid interactions and genome packaging

The primary function of the virus particle is to contain and protect the **genome** before delivering it to the appropriate host cell. To do this, the proteins of the **capsid** must interact with the nucleic acid genome. While this is not a problem for helical nucleocapsids, whose length is proportional to genome of genome component size, there are significant physical constraints to **encapsidating** (packaging) relatively large nucleic acid molecules into relatively small isometric capsids. In most cases, the linear virus genome when stretched out is at least an order of magnitude longer than the diameter of the capsid. Simply folding the genome in order to stuff it into such a confined space is quite a feat in itself, but the problem is made worse by the electrostatic repulsion by the negative electrostatic charges on every phosphate group of every nucleotide in the polynucleotide backbone. This means that the **polyanionic** genome resists being crammed into a small space. Viruses with isometric nucleocapsids overcome this difficulty by packaging, along with the genome, various positively charged molecules to counteract this negative charge repulsion. These include small **cations** or positively charged ions (Na^{+}, Mg^{++}, K^{+}, etc.) or polyamines, and various nucleic acid-binding proteins. Some of these proteins are virus-encoded and are enriched in amino acids with **basic side-chains,** such as **arginine and lysine**, which interact with the genome. This category includes all virion **nucleocapsid** proteins. There are many examples of such proteins—for example, **retrovirus nucleocapsid (NC)** and **rhabdovirus N** and **influenza virus NP proteins (=nucleoprotein)**. Some viruses with double-stranded DNA genomes have basic **histone-like** molecules closely associated with the DNA. Again, some of these are virus-encoded (e.g., **adenovirus** polypeptide VII). In other cases, however, the virus may use cellular proteins: for example, the polyomavirus genome assumes a condensed **chromatin**-like structure in association with four **cellular histone proteins** (H2A, H2B, H3, and H4), similar to that of the host cell genome.

Virus receptors: Recognition and binding

Cellular **receptor** molecules used by a number of different viruses from diverse groups have now been identified. The interaction of the outer surface of a virus with a cellular receptor is a major event in determining the subsequent events in replication and the outcome of infections. It is this binding event that activates inert extracellular virus particles and initiates the replication cycle. Receptor binding is considered in detail in Chapter 4.

Other interactions of the virus capsid with the host cell

As described earlier, the function of the virus **capsid** is not only to protect the **genome** but also to deliver it to a suitable host cell, and more specifically, to the appropriate compartment of the host cell (in the case of **eukaryote** hosts) in order to allow replication to proceed. One example is the **nucleocapsid** protein of viruses which replicate in the nucleus of the host cell.

These molecules contain within their primary amino acid sequences *"**nuclear localization signals**"* that are responsible for the migration of the virus genome plus its associated proteins into the nucleus where replication can occur. Again, these events are discussed in Chapter 4.

Virions are not inert structures. Many virions contain several enzymatic activities: in most cases these are not active outside the biochemical environment of the host cell, although an exception may have to be made for certain viruses of Archaea, which undergo significant morphological changes outside their host cells. I said at the beginning of this Chapter that virions are **nanoscale spacecraft**; that their function is to deliver their precious genomic cargo to an environment—the cell—in which it can be replicated. Virions are often not inert in this respect, as will become clear in Chapter 4: even very simple virions, with one or only a couple of capsid proteins, are programmed by their structures to safeguard the genome in hostile environments, and release it in congenial sites. Virions that enter cells via interaction with cellular receptors undergo morphological changes that are quite profound, and may even involve use of "**nanomachines**" such as molecular motors, or active injection mechanisms. We now understand many of these mechanisms in great detail—and this fundamental structural understanding of the structure and function of virus particles is beginning to have benefits in a number of ways. Some of these have been long anticipated, such as rational vaccine design based on structural data rather than trial and error; however, others are new, such as the role this knowledge is playing in **nanotechnology** and the design of molecular machines (see Rybicki, 2019; Recommended Reading).

References

Koonin, E.V., Dolja, V.V., Krupovic, M., 2015. Origins and evolution of viruses of eukaryotes: the ultimate modularity. Virology 479–480, 2–25. https://doi.org/10.1016/j.virol.2015.02.039.

Koonin, E.V., Dolja, V.V., Krupovic, M., Varsani, A., Wolf, Y.I., Natalya Yutin, F., Zerbini, M., Kuhn, J.H., 2020. Global organization and proposed megataxonomy of the virus world. Microbiol. Mol. Biol. Rev. 84, e00061-19. https://doi.org/10.1128/MMBR.00061-19.

Krupovic, M., Koonin, E.V., 2017. Multiple origins of viral capsid proteins from cellular ancestors. Proc. Natl. Acad. Sci. U. S. A. 114 (12), E2401–E2410. https://doi.org/10.1073/pnas.1621061114.

Rybicki, E.P., 2019. Plant molecular farming of virus-like nanoparticles as vaccines and reagents. Wires Nanomed. Nanobiotechnol. 12 (2), e1587. https://doi.org/10.1002/wnan.1587.

VirusWorld site: [http://www.virology.wisc.edu/virusworld/tri_number.php/].

Recommended reading

Ahi, Y.S., Mittal, S.K., 2016. Components of adenovirus genome packaging. Front. Microbiol. 7, 1503. Available from https://www.frontiersin.org/articles/10.3389/fmicb.2016.01503/full.

Andrade, A.C., Rodrigues, R.A.L., Oliveira, G.P., Andrade, K.R., Bonjardim, C.A., La Scola, B., Kroon, E.G., Abrahão, J.S., 2017. Filling knowledge gaps for mimivirus entry, uncoating, and morphogenesis. J. Virol. 91, e01335-17. Available from: https://doi.org/10.1128/JVI.01335-17.

Häring, M., Vestergaard, G., Rachel, R., Chen, L., Garrett, R.A., Prangishvili, D. (Eds.), 2005. Independent virus development outside a host. Nature 436, 1101–1102.

Klug, A., 1999. The tobacco mosaic virus particle: structure and assembly. Philos. Trans. R. Soc. Lond. B Biol. Sci. 354 (1383), 531–535.

Luque, D., Caston, J.R., 2020. Cryo-electron microscopy for the study of virus assembly. Nat. Chem. Biol. 16, 231–239.

McDonald, S.M., Patton, J.T., 2011. Assortment and packaging of the segmented rotavirus genome. Trends Microbiol. 19 (3), 136–144.

Anon., Phage T4 structure. Available from https://talk.ictvonline.org/ictv-reports/ictv_9th_report/dsdna-viruses-2011/w/dsdna_viruses/70/myoviridae-figures.

Reddy, V.S., Kundhavai Natchiar, K., Stewart, P.L., Nemerow, G.R., 2010. Crystal structure of human adenovirus at 3.5 Å resolution. Science 329, 1071–1075.

Rohrmann, G.F., 2008. Baculovirus Molecular Biology. NCBI Bookshelf. http://www.ncbi.nlm.nih.gov/bookshelf/br.fcgi?book=bacvir.

Sgro, J.-Y., 2011. Triangulation Numbers. Virus world Server, Institute for Molecular Virology, University of Wisconsin-Madison. Available from http://www.virology.wisc.edu/virusworld/tri_number.php.

CHAPTER

3

Virus origins and genetics

INTENDED LEARNING OUTCOMES

On completing this chapter, you should be able to:

- Describe the range of structures and compositions seen in virus genomes.
- Explain how the composition and structure of a genome affects the genetic mechanisms which operate on it.
- Discuss representative examples of virus genomes to illustrate the range of genetic diversity seen in viruses.

The structure and complexity of virus genomes

As **obligate intracellular parasites**, the forms and compositions of virus genomes are necessarily affected by the genetic mechanisms operating in their host cells—although unlike the genomes of all cells, which are composed of DNA, virus genomes may have their genetic information encoded in **either DNA or RNA**. The forms and composition of viral genomes are also more varied: While prokaryotes have mainly single-component circular and occasionally multiple-component or sometimes linear dsDNA genomes (e.g., *Streptomyces*, *Helicobacter*), and all eukaryotes have multicomponent linear dsDNA, virus genomes may be as varied as shown in Table 3.1.

Thus, the chemistry and structures of virus genomes are more varied than any of those seen in ribocells from the entire bacterial, archaeal, or eukaryote realms. In particular, viruses are the only organisms on this planet to still have **RNA as their sole genetic material**. They are also the only autonomously replicating organisms to have **single-stranded DNA genomes**.

Single-stranded RNA virus genomes may be **positive-sense** (i.e., the same polarity or nucleotide sequence as the **mRNA;** designated as **ssRNA+**), **negative-sense** (opposite polarity; **ssRNA−**, needs to be transcribed for mRNA), or **ambisense** (a mixture of the two, with an ssRNA− replication strategy).

Virus genomes range in size from approximately 1800 nucleotides (nt) (e.g., certain ssDNA **circoviruses**) to 2.8 million base pairs (bp) of dsDNA in the case of **Pandoravirus**, which is nearly five times as large as the smallest bacterial genome (*Mycoplasma genitalium*

Cann's Principles of Molecular Virology
https://doi.org/10.1016/B978-0-12-822784-8.00003-9

TABLE 3.1 Types of virus genomes.

DNA			
Double-stranded		**Single-stranded**	
Linear	Circular	Linear	Circular
Single component	Single[a]	One or two components	Single or multiple
RNA			
Double-stranded	**Single-stranded**		
Linear	Linear		Circular (*Deltaviridae* only)
Single or multiple components	+sense or −sense[b]		−sense
	Single or multiple		Single

a While polydnaviruses may have multiple circular dsDNA components, they are now classed as "viriforms," and not viruses.
b +sense RNA can be translated; genomic −sense has to be transcribed for mRNA and +sense.

at 580,000 bp). The smallest and least complex viral genomes may have as few as only two genes (e.g., circoviruses), while the genomes of the largest double-stranded DNA viruses (e.g., mimi-, irido-, herpes-, pox- and mimi- and pandoraviruses) are sufficiently complex to have mostly escaped complete functional analysis yet (annotation; see Chapter 1), even though the complete nucleotide sequences of a large number of examples have been known for several years. Satellite viruses or genomes—which require coinfection with an autonomously-replicating virus in order to multiply—may have just a single gene (e.g., **satellite tobacco necrosis virus**, ssRNA+; **alphasatellites** of nano- and geminiviruses; ss circular DNA).

The evolutionary origins of viral genomes define their **replication strategies** to quite a fundamental degree: this will be discussed below. Chapter 4 explains how these genomes **replicate**, and Chapter 5 deals in more detail with the mechanisms that regulate the **expression** of virus genetic information.

Whatever the size or composition of a virus genome, because they are intimately intertwined with the host cell for their very existence, the genome must contain information encoded in a way that can be recognized and decoded by the particular type of host cell. Similarly, the control signals that direct the expression of virus genes must be appropriate to the host. Accordingly, many DNA viruses of both prokaryotes and eukaryotes closely resemble their host cells in terms of the mechanisms used to replicate, and to express and translate their genomes. The purpose of this chapter is to describe the diversity of virus genomes, and to consider how and why this variation may have arisen.

Origins of viruses

Summary

1. Viruses have more than one origin, in capture of structural genes by "**mobile genetic elements,**" and some lineages are truly ancient.

2. Many modern virus genomes result from the exchange of "**cassettes**" of essential genes or groups of genes.
3. Many ssRNA− viruses infecting plants and vertebrates probably originated in insects.
4. The "**retroid cycle,**" or use of reverse transcriptase in replication, is a characteristic of cellular mobile genetic elements as well as of four major groupings of viruses.
5. **Rolling circle replication** is a feature of bacterial plasmids and diverse families of small ssDNA viruses and some small dsDNA viruses.

"The ubiquity of viruses in the extant biosphere and the results of theoretical modeling indicating that emergence of selfish genetic elements is intrinsic to any evolving system of replicators together imply that virus-host coevolution had been the mode of the evolution of life ever since its origin". **Koonin et al. (2015)**

From what did viruses evolve?

The answer to this question is not simple, because while viruses all share the characteristics of being obligate intracellular parasites that use host cell machinery to make their components, some of which then self-assemble to make particles which contain their genomes, they most definitely **do not have a single origin**. Indeed their origins may be spread out over a considerable period of geological and evolutionary time.

The graphic (see Fig. 3.1) depicts a possible simple scenario for the evolution of viruses: "wild" genetic elements could have escaped, or even been the agents for transfer of genetic information between, both RNA-containing and DNA-containing "protocells," to provide the precursors of retroelements and of RNA and DNA viruses. Later additions from Bacteria, Archaea, and their progeny Eukarya would complete the virus zoo.

Viruses infect all types of ribocells, from Bacteria through Archaea to Eukarya; from *E. coli* to mushrooms; from amoebae to human beings—and **virus particles may even be the**

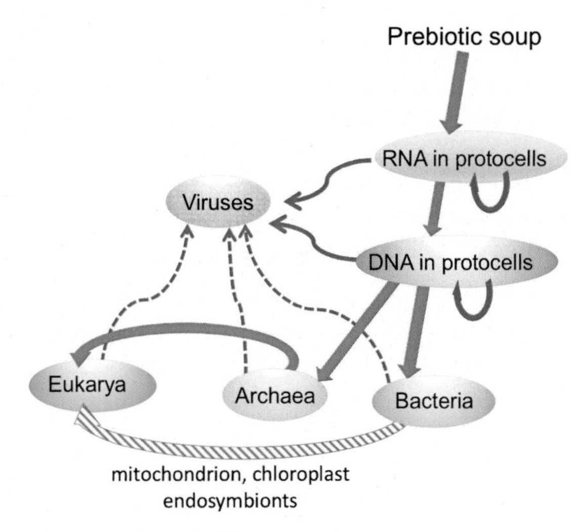

FIG. 3.1 A simple scheme depicting the probable origins of viruses.

single most abundant and varied organisms on the planet, given their abundance in all the waters of all the seas of planet Earth (see Chapter 1).

Given this diversity and abundance, and the propensity of viruses to **swap and share successful modules** between very different lineages and to pick up bits of genome from their hosts or to donate to their hosts (**horizontal genetic transfer; HGT**), it was previously very difficult to speculate sensibly on their deep origins—however, massive recent metagenomic studies and phylogenomic comparisons of viral and cellular sequences have allowed significant breakthroughs in this regard, and I shall outline some of the probable deep evolutionary scenarios.

Historically there have been three scenarios proposed to explain the origins of viruses: these are

- descent from primordial and precellular genetic elements;
- reductive evolution from cellular precursors; and
- escape of genes or genomes from cells.

These three scenarios could all be true when considering different collections of viruses; however, groups of viruses that are related phylogenetically must have arisen largely via only one of these scenarios. However, recent deep analysis of **core viral genes**—those involved in genome replication, and in building virions, respectively—has led to the conclusion that **virus replication modules** arose from a pool of genetic elements—probably "**mobile genetic elements,**" many of which are still associated with all cellular lineages—that arose very early in cell evolution, while **structural protein modules** were acquired from cellular hosts at various times thereafter. This is a "**chimeric scenario,**" where already-existing genetic parasites consisting of a core collection of genes specifying a replication mechanism became viruses by recruiting cellular proteins as structure-forming units, which made them much more easily transmissible between cells (see Krupovic et al., 2019).

Virus hallmark genes

As introduced in Chapter 2, virus hallmark genes can be used to deduce deep phylogenetic relationships between evolutionary lineages of viruses. For example, the **RNA recognition motif** (RRM) is a very common component of **RNA-binding domains** in all ribocells. Structurally related domains are also widespread in many viruses and mobile genetic elements, as seen in Table 3.2, and these can be grouped according to which particular derivative of the RRM motif they employ for replication.

Thus, **six different families** of replication modules are found in virus genomes:

- RdRp is in all RNA viruses replicating via RNA intermediates;
- RT is in all viruses that have both DNA and RNA phases in their replication;
- RCRE is in most ssDNA viruses and some small dsDNA viruses;
- larger dsDNA viruses may have
  - PolA,
  - PolB/protein-primed PolB, or
  - AEP replication machinery.

TABLE 3.2 Viral and cellular replication modules deriving from an ancestral RRM motif.

Replication enzyme	Present in:
Reverse transcriptase (**RT**)	**Cells**: eukaryote telomerase and RT-related enzymes **Mobile elements**: group II introns, various retrons, retroposons, retroplasmids, retrotransposons (pro- and eukaryotes) **Viruses**: reverse-transcribing viruses (retro-, pararetroviruses)
DNA-dependent DNA polymerase family A (**PolA**)	**Cells**: bacteria and eukaryotic mitochondria **Mobile elements**: none **Viruses**: some dsDNA viruses
DNA-dependent DNA polymerase family B (**PolB**)—RNA primed[a]	**Cells**: bacteria, archaeab and eukaryotes **Mobile elements**: prokaryote plasmids and **casposons** (see Box 3.1) **Viruses**: all large and many medium-sized dsDNA viruses
pPolB—protein-primed[a]	**Mobile elements**: linear cytoplasmic and mitochondrial plasmids of fungi and plants **Viruses**: DNA viruses with small-to-medium-sized genomes (e.g., adenoviruses, PRD1 phage)
DNA-dependent DNA polymerase archaeo-eukaryotic primase (**AEP**)	**Cells**: bacteria, archaea, and eukaryotes **Mobile elements**: many prokaryotic plasmids **Viruses**: many dsDNA viruses
RNA-dependent RNA polymerase (**RdRp**): template-dependent	**Cells**: none **Mobile elements**: none **Viruses**: all RNA viruses replicating via RNA intermediates
Rolling circle replication endonuclease (**RCRE**)	**Cells**: none **Mobile elements**: numerous plasmids, some transposons **Viruses**: most ssDNA and some small dsDNA viruses

a Homologous polymerases, but PolB replication is primed by RNA fragments, and protein-primed PolB works by having a protein covalently attached to the termini of linear dsDNA genomes as a primer for opposite strand synthesis.
Adapted from Krupovic, M., Dolja, V.V., Koonin, E., 2019. Origin of viruses: primordial replicators recruiting capsids from hosts. Nat. Rev. Microbiol. 17, 450–458.

By contrast, in ribocells no **mobile genetic element** (MGE) machineries use RdRp; RCRE is only used by prokaryotic **plasmids** and pro- and eukaryotic **transposons** (**helitrons**, see below); all ribocells and various plasmids use PolB and AEP, but only bacteria and the bacterial-descended **mitochondria** use PolA (Box 3.1).

These replication modules have picked up a wide variety of cellular proteins that can form structures over the course of coevolution with ribocells, to create viruses. A good example of this are the "**single jelly-roll fold**" (SJR) proteins (see Fig. 2.8, Chapter 2): these are also called **jelly-roll capsid** (**JRC**) or Swiss roll fold proteins, and are more formally known as eight stranded antiparallel β-barrel proteins. SJR fold proteins are the most common virus capsid proteins, found in around one-third of virus families: these are mainly eukaryotic ssRNA and ssDNA viruses. These proteins have structural homologues in all ribocells, especially in **carbohydrate-binding proteins**: in fact, clustering of viral and cellular SJR fold proteins by structural similarity indicates that there may be **two or three separate origins** for

BOX 3.1

Mobile genetic elements

There are a wide variety of mobile genetic elements (MGEs) found in all types of cells. These include viruses of all kinds, virus-like agents, plasmids, DNA transposons, Polintons (named after DNA POLymerase and INTegrase), group II introns, and reverse-transcribing elements such as **retrons, retroposons, and retrotransposons**. There is considerable overlap between some of these agents, and all share the property of being "**selfish replicators,**" or agents whose molecular imperative is first and foremost to make more copies of themselves. Most are integrated into the genome of the cell harboring them, but retain mobility by encoding machinery that allows them either to replicate mobile versions of themselves that subsequently reintegrate, or can excise themselves and reintegrate elsewhere. Viruses and plasmids are special cases in that they are extrachromosomal. However, certain plasmids can integrate into prokaryote cell genomes and then excise again; retroviruses have incorporated chromosomal integration into their life cycle, and retrotransposons (reverse-transcribing; **class I transposons**) can make particles and occasionally be transmitted horizontally. Given their ubiquity, it is probable that most of these elements emerged early on in cellular evolution—which makes them good candidates for being "parents" of many lineages of viruses.

MGEs and their more numerous nonfunctional remnant sequences may constitute a large proportion of many organisms' genomes: this is up to two-thirds in humans, and nearly 90% in maize. The **DNA transposons (class II)** can move via a "cut and paste" mechanism, for transposons with **inverted terminal repeats** (whole transposon is excised by **transposase** enzyme, migrates and reinserts, creating short target site duplications); via a "**rolling circle**" mechanism very similar to ssDNA viruses (**Helitrons**); and as **Polintons**, or **self-synthesizing transposons**.

Transposases, classically considered as selfish genes, have been found to outnumber cellular or "housekeeping" in sequence database analyses. This suggests that they must offer selective advantages to the organisms in which they are found. These include movement of transposons within and even between cellular genomes, which promote the mutations and larger genome rearrangements that can accelerate biological diversification and evolution.

the viral proteins, possibly among carbohydrate-binding cellular proteins. These comparisons also indicate that while eukaryotic ssRNA and ssDNA viruses acquired their SJR fold protein through one event (=**monophyletic origin**), bacterial ssDNA micro- and the much larger dsDNA sphaerolipoviruses acquired their SJR fold capsid proteins independently of one another (=**polyphyletic origins**), and of eukaryotic viruses. It is suggested that the SJR capsid protein was ancestral to all eukaryotic RNA viruses, but that in many cases the original protein was replaced by other unrelated capsid and nucleocapsid (=binds to the RNA) proteins.

The second most common isometric capsid protein type are the **double jelly-roll fold** (DJR) proteins (see Chapter 2): these are found in about 10% of known viruses with isometric

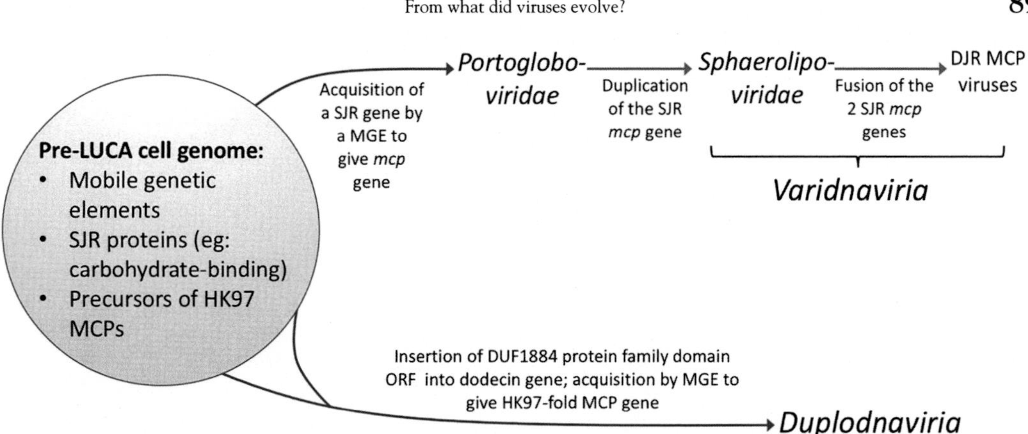

FIG. 3.2 Early evolution of dsDNA viruses. Acquisition of precursor major capsid protein (MCP) genes by mobile genetic elements (MGEs) is postulated to have given rise to the two major realms of prokaryote DNA viruses. Dodecin: a protein assembling as a **dodecamer**, or group of 12 subunits. *Modified from Krupovic, M., Dolja, V., Koonin, E., 2020. The LUCA and its complex virome. Nature Reviews Microbiology 18(11), 661–670, doi: 10.1038/s41579-020-0408-x.*

particles, particularly dsDNA viruses infecting all three main lineages of ribocells. Additionally, metagenomic trawls through marine environments have unearthed a huge diversity of such protein genes linked to assembled putatively viral genomes, but not any in cellular genomes. This suggests an ancestral origin by **gene duplication and fusion** in a viral genome. Given the presence of SJR and DJR fold capsid proteins in viruses infecting bacteria, archaea, and eukarya, it is suggested that these emerged very close to the time of the **last universal cellular ancestor** of all ribocells—**LUCA** (Fig. 3.2).

While just a few years ago there was only a handful of families officially recognized of viruses with HK97-like MCPs, literally dozens are now being created each year just in the realm of tailed phages—which are probably the most widespread and diverse group of viruses on this planet.

LUCA and its viruses

The concept of a LUCA comes from study of the three primary domains of cellular life—bacteria, archaea, and the latter's effective subdomain, Eukarya—given that these had to come from somewhere. Given that the primary evolutionary division among ribocells was first between bacteria and archaea (see Fig. 3.1), looking at the present distribution of viruses across these two domains should allow some inferences as to what the virome of LUCA looked like (Fig. 3.3).

The hypothetical virome of LUCA is highly diverse and complex, and dominated by dsDNA viruses. This indicates that LUCA was a community rather than being as related as a species or even genus of organisms; what "LUCA" had that is important from what descended from it, was a shared **core set of genes** inherited by all ribocells, and a "**pangenome**" that provided genes and even defense systems against viruses, to various lineages of descendant organisms.

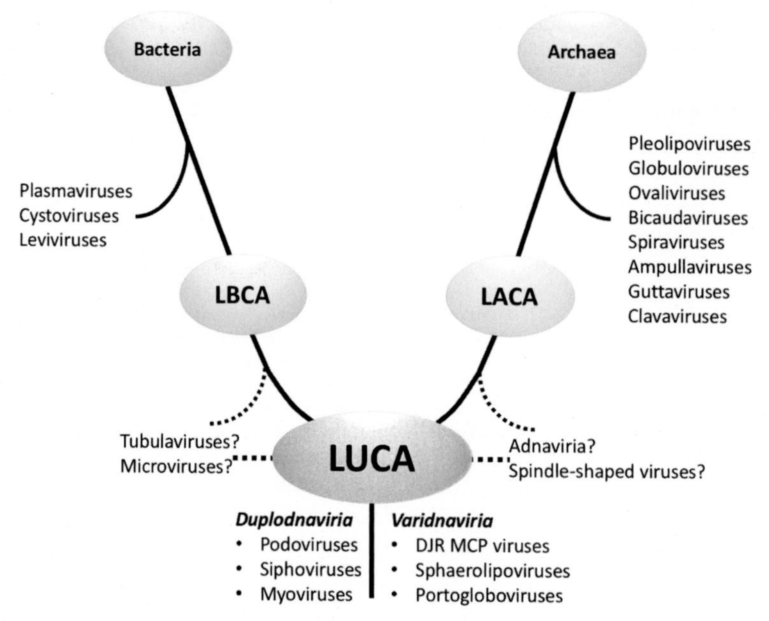

FIG. 3.3 The hypothetical LUCA virome. This shows the possible virome of the last universal cellular ancestor (LUCA), as well as the distribution of viral taxa in extant bacteria and archaea. The figure shows the possible split of the LUCA virome into the viromes of the last bacterial common ancestor (LBCA) and the last archaeal common ancestor (LACA). Dotted lines indicate uncertainty about whether the virus groups shown might have been part of the LUCA virome, or were divided into viruses of the LBCA and LACA, respectively. *DJR*, double jelly-roll fold; *MCP*, major capsid proteins. *Modified from Krupovic et al. (2020).*

One important feature of the postulated LUCA virome is that there are only two lineages of RNA viruses: these are ssRNA+ viruses in family *Leviviridae*, and dsRNA viruses in family *Cystoviridae*, both presently exclusive to bacteria. Thus, although the original protocells (see Fig. 3.1) may well have had RNA genomes, and would be expected to have had an RNA virome, by the time of LUCA DNA viruses had already evolved and diversified to the point that they had almost completely supplanted the possibly more primitive and less efficient RNA viruses. This of course begs the question, where did all the RNA viruses of eukaryotes come from? This is explained below, but it is necessary first to introduce the context that allowed their development.

The origins of modern eukaryote viruses

Eukaryote DNA viruses

I will first discuss the origins of eukaryote DNA viruses as this is much simpler to explain than for RNA viruses, given that major lineages of especially dsDNA viruses have clear similarities to certain families of prokaryotic viruses. It is postulated that eukaryote ssDNA viruses have a **polyphyletic** or diverse origin in several distinct lineages of prokaryote plasmids that replicated via rolling circle mechanisms, and therefore had RCRE modules (see Table 3.2). By means of multiple recombination events involving ssRNA viruses, these rolling circle Reps (RC-Reps) picked up SJR-fold capsid proteins (CPs; see Fig. 3.4), over a long period of geological time. Subsequent evolutionary development include swapping out the CP and/or RC-Rep, including picking up a pPolB module; switching to having dsDNA genomes by Rep modification; integrating into arthropod genomes, and even becoming CP-less RC transposons (see Koonin et al., 2015).

Most double-stranded DNA genomes with intermediate to large genomes have two distinct origins within the bacterial virome that probably coexisted with LECA, the last eukaryote common ancestor. One group—viruses in the Order *Herpesvirales*—probably evolved from an ancestral caudovirus, or a tailed phage of the *Caudovirales*, Realm *Duplodnaviria* (see Fig. 3.3). These viruses share major capsid proteins with the HK97 fold that are structurally homologous with MCPs of the *Caudovirales* phages (Fig. 3.4).

The other is a much more varied group with a more complex origin, that probably originated very early in eukaryogenesis from ancestral tectiviruses (family *Tectiviridae*) infecting the α-proteobacterial endosymbiont that was the ancestor of mitochondria (see Fig. 3.5). Indeed, it is striking that larger DNA viruses appear to derive only from ancestors infecting Bacteria, whereas Eukarya arose from a **mesophilic** clade of Archaea. However, given that both mitochondria (descended from gram-negative **Proteobacteria**) and chloroplasts (**Cyanobacteria** or blue-green algae) derive from within Bacteria, it is not surprising that bacterial viruses or transposons are a likely source for larger eukaryotic dsDNA viruses.

It has been proposed that the giant viruses among what are known unofficially as the "**Megavirales**"—and in particular mimi- and pandoraviruses—constitute a new domain of life, that derived by "**reductive evolution**" from a ribocell. However, several independent studies showed that the viruses probably evolved from viruses with much smaller genomes by acquisition of genes from both their host and from other viruses, given that the closest relatives of many genes in particular viruses were within the repertoire of their particular ribocell hosts (Box 3.2).

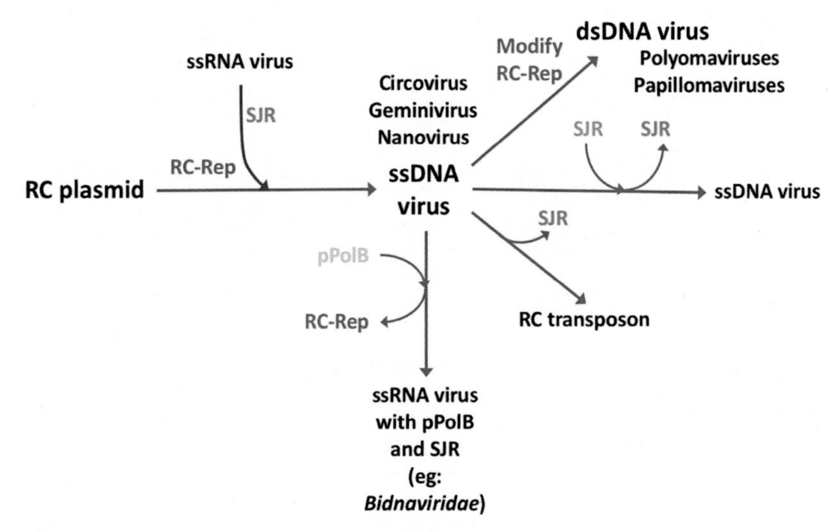

FIG. 3.4 Evolution of eukaryote ssDNA viruses. Evolution of ssDNA viruses of eukaryotes: polyphyletic origin from different plasmids and multiple cases of recombination with ssRNA viruses. Abbreviations: *JRC*, jelly roll capsid protein; *pPolB*, protein-primed DNA Pol B; *RC plasmid*, replicates via rolling circle; *RC-Rep*, rolling circle replication protein. *Different colors of SJR* denote distinct variants of the gene. *Modified from Koonin, E.V., Dolja, V.V., Krupovic, M., 2015. Origins and evolution of viruses of eukaryotes: the ultimate modularity. Virology 479–480:2–25. https://doi.org/10.1016/j.virol.2015.02.039.*

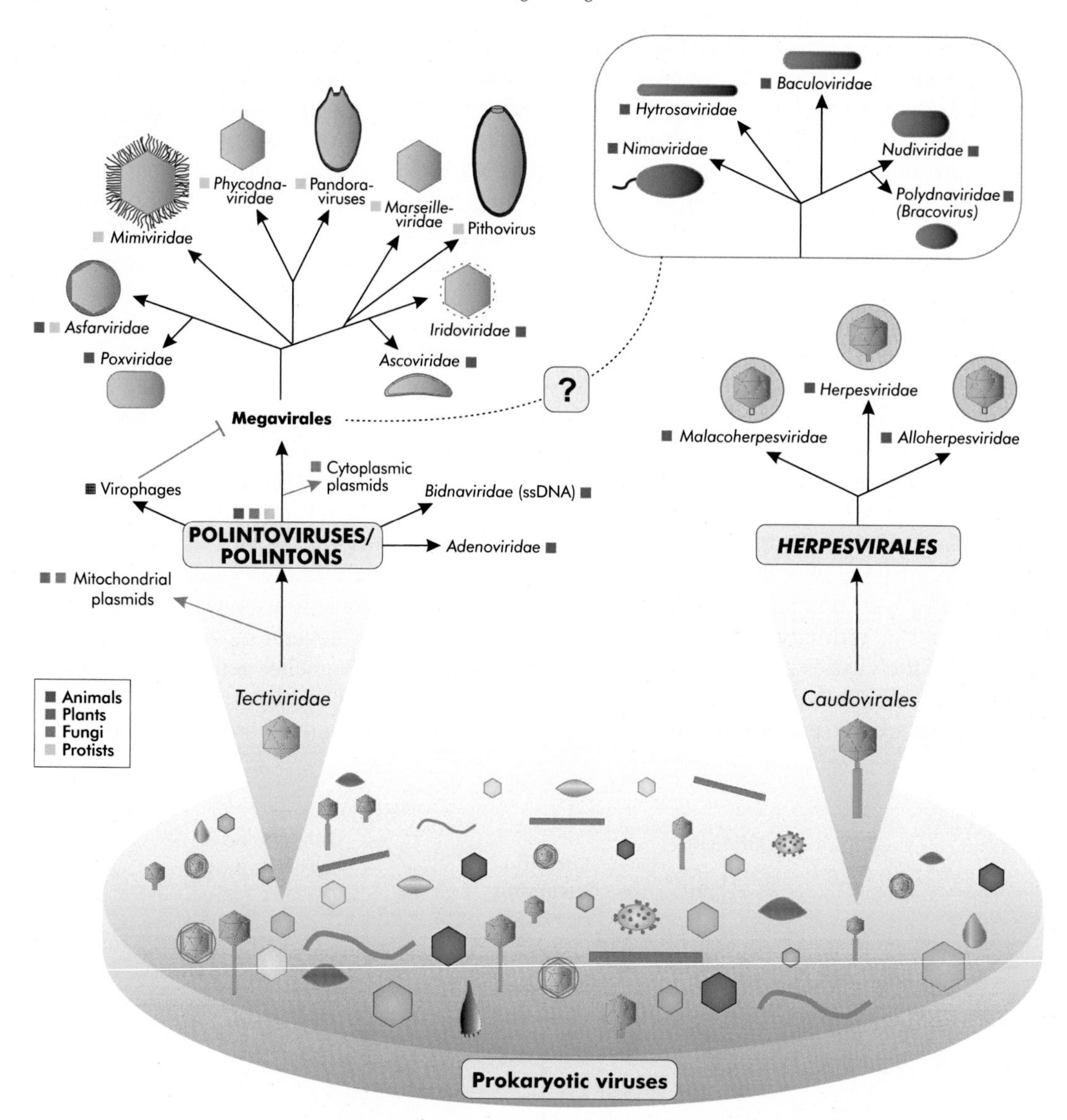

FIG. 3.5 Evolution of large dsDNA viruses of eukaryotes. These are shown as originating from two distinct groups of bacteriophages. The *dotted line with a question mark* shows a tenuous evolutionary relationship. The host ranges of the eukaryotic virus groups are *color-coded as shown in the inset*. The *hatched yellow square for the virophages* indicates that these viruses parasitize on the giant viruses of the family *Mimiviridae* which themselves infect amoeba and other protists. For each family of large eukaryotic viruses, a simplified schematic depiction of the virion structure is included. *From Koonin, E.V., Dolja, V.V., Krupovic, M., 2015. Origins and evolution of viruses of eukaryotes: the ultimate modularity. Virology 479–480, 2–25. https://doi.org/10.1016/j.virol.2015.02.039.*

BOX 3.2

Polintons: Transposons or viruses?

The large complex satellite viruses popularly known as "**virophages**" that are associated with mimi- and other giant viruses are interesting, in that they more closely resemble the tectiviruses they probably descend from than any of the other viruses in that lineage, and appear very similar to the self-synthesizing DNA transposons known as Polintons. Moreover, virophages can in fact integrate like transposons into the genomes of both their "parent" giant viruses, and their **protist** hosts. There is also evidence that integration into the host genome can confer adaptive immunity in the host to the giant virus. Metagenome and comparative genome analysis has revealed an extensive family of Polinton-like viruses (PLVs), and has shown that PLVs and related smaller transposons called transpovirons derive from Polintons. These elements have the largest genomes of known nonrecombinant* transposons, of 15–20 kb, with flanking long terminal inverted repeats (TIRs). All Polinton genomes (and those of PLVs) encode a pPolB and a retrovirus-like (RVE) integrase. Most also encode proteins homologous to major and minor capsid proteins of the tectivirus lineage. Additionally, many encode a DNA-packaging ATPase and a maturation protease: these are evolutionarily related to the enzymes responsible for virion assembly in the giant viruses in the same lineage. Accordingly, it is speculated that Polintons are descended from viruses, but have now adapted wholly (no capsid protein genes) or partially (could still make capsids) to being transposons.

*A 180 kb transposon named *Teratorn* found in teleost fish genomes appears to have originated by fusion of a piggyBac transposon and a herpesvirus (Inoue et al., 2017).

Retroelements and retroviruses

One of the most common types of selfish replicators or MGEs in eukaryotes are reverse-transcribing or **retroelements (REs)**, described as class I transposons (see Box 3.1). These have a hallmark gene, as do all nonreverse transcribing viruses: This is RNA-dependent DNA polymerase, or reverse transcriptase (RT). Aside from this single hallmark component, the REs can behave as RNA or DNA plasmids, as MGEs that move around in host genomes via reverse transcription and insertion (retroposons, retrotransposons), and as viruses that may pack either DNA or RNA into isometric capsids. While they occur less frequently in prokaryotes, there are similar elements in Archaea and Bacteria: these include Group II introns, retrons, diversity-generating retroelements and retro-plasmids (see Table 3.3 and Fig. 3.6), but no viruses. An important feature of all the nonviral elements is that they are **vertically transmitted**, or inherited along with the cell genome by descendants.

Construction of phylogenetic trees from RT sequences found in all cellular organisms has allowed a hypothetical reconstruction of the evolution of eukaryotic REs: this is shown in Fig. 3.6. Prokaryote RTs are much more diverse in sequence than those found in eukaryotes: thus, it is speculated that all eukaryotic REs are derived from one group of prokaryotic REs, and possibly from **Group II introns**. Given that eukaryotic RNA splicing mechanisms also appear to have originated with Group II introns, it is possible that these functions entered

TABLE 3.3 Prokaryote retroelements.

Retroelement	Components	Description
Retrons/ retron-like	Reverse transcriptase (RT) gene and multicopy single-stranded DNA (msDNA)	Vertically inherited in bacteria, suggestive of some "normal" function(s). Makes multiple copies of a branched RNA-DNA hybrid. Function as antiphage defense systems, causing abortive infections
Retro-plasmids	Reverse transcriptase gene, circular or linear plasmids	Replicate in fungal mitochondria, most likely to be of bacterial origin, derive from retrons
Diversity-generating REs (DGRs)	Bacteriophage-derived RT gene, major tropism determinant (*mtd*); accessory tropism determinant (*atd*). MTD is a receptor-binding protein	Unusual retroelements in some phage and bacterial genomes; use RT to modify specific target genes to change phage receptor specificity to help phage evade bacterial resistance
Group II introns	RT gene, X/D/E maturase gene (function in the splicing of introns)	A large class of self-catalytic ribozymes and REs found in all three domains, which site-specifically mobilize to new DNA sites. Most common REs in archaea and bacteria (>70% of RTs), and only group that can spread horizontally. Commonly present in mitochondrial genomes. **Probable ancestors of eukaryotic REs/RT**

Adapted from Koonin, E.V., Dolja, V.V., Krupovic, M., 2015. Origins and evolution of viruses of eukaryotes: the ultimate modularity. Virology 479–480:2–25. https://doi.org/10.1016/j.virol.2015.02.039.

eukaryotes via endosymbiotic bacteria, and possibly those that became mitochondria. This is a "**symbiogenetic**" scenario.

Simple eukaryote REs—also known as **retroposons**, or "**nonlong terminal repeat** (**non-LTR**) REs"—are widely distributed, and some have important functions—particularly with relevance to "**endogenous viral elements**" or EVEs (see below). They may be classed as the following (see Fig. 3.6 for structure):

- **Long interspersed nuclear elements (LINEs)**: these are the most common REs in eukaryotes. They actively transpose due to an endonuclease gene (END), and constitute up to ~17% of human genome.
- **RVT**: possibly LINE-related elements; only identifiable gene is RT; present as single copies and not known to be mobile; some members also found in bacterial genomes.
- **"Penelope-like" REs (PLEs)**: these are simple REs encoding a fusion of RT with a GIY-YIG endonuclease (allows cleavage/movement); found in invertebrates and certain vertebrates.
- **Telomerase RT (TERT)**: these are related to PLEs. They constitute a pan-eukaryotic enzyme **essential for the replication of the ends of linear chromosomes**.

PLE-like REs antedate the LECA, although the fusion with endonuclease happened later. Recruitment of a PLE-related RT for the **telomerase** function was an important early event during evolution of the eukaryotic cell (it allowed big linear genomes). Several groups of eukaryotes, in particular insects, have lost the original TERT gene, and instead use a variety of non-LTR retrotransposons as telomeric repeats.

Dictyostelium intermediate repeat sequence **(DIRS)** elements are more complex REs which have nonstandard LTRs, and are found in slime molds and other simple eukaryotes. They are also known as tyrosine recombinase retroelements, for the enzyme they have instead of an integrase (INT) or endonuclease (END).

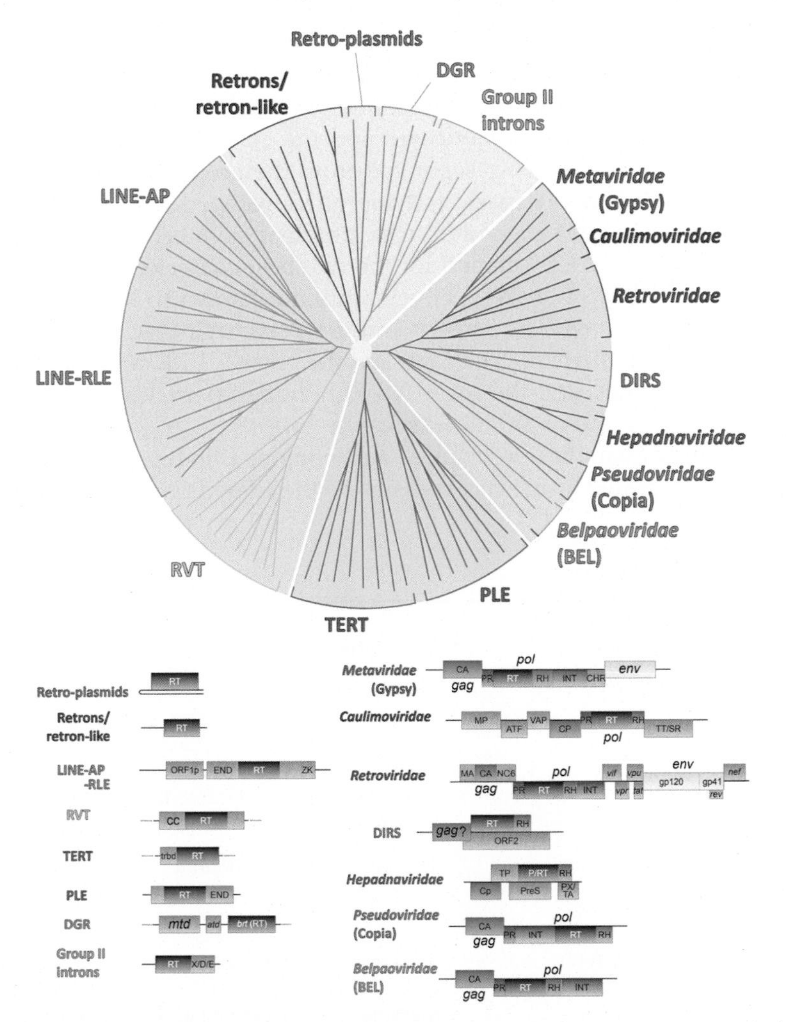

FIG. 3.6 Suggested evolution of REs and reverse-transcribing viruses. Genomic organizations of selected representatives of the major groups of retroelements overlay the phylogenetic tree of the reverse transcriptases. Abbreviations: *DGR*, diversity-generating retroelements; *X/D/E*, maturase, DNA binding, and endonuclease domains, respectively, of the intron-encoded protein; *mtd*, major tropism determinant; *atd*, accessory tropism determinant; *brt*, bacteriophage reverse transcriptase; *LINE*, long interspersed nucleotide elements; *END*, endonuclease; *ZK*, zinc knuckle; *gag*, group-specific antigen; *env*, envelope; *pol*, polymerase; *PR*, aspartate protease; *RT*, reverse transcriptase; *RH*, RNase H; *INT*, integrase; *CHR*, chromodomain; *MA*, matrix protein; *CA/Cp*, capsid protein; *NC*, nucleocapsid; *6*, 6-kDa protein; *vif, vpr, vpu, tat, rev, and nef*, regulatory proteins encoded by spliced mRNAs; *gp120 and gp41*, the 120- (surface) and 41-kDa (transmembrane) glycoproteins; *ATF*, aphid transmission factor; *VAP*, virion-associated protein; *TT/SR*, translation *trans*-activator/suppressor of RNA interference; *TP*, terminal protein; *P*, polymerase; *PreS*, presurface protein (envelope); *PX/TA*, protein X/transcription activator; *trbd*, telomerase RNA-binding domain; *cc*, coiled-coil. *From Koonin, E.V., Dolja, V.V., Krupovic, M., 2015. Origins and evolution of viruses of eukaryotes: the ultimate modularity. Virology 479–480, 2–25. https://doi.org/10.1016/j.virol.2015.02.039.*

It is very striking that the groups of "real" viruses—those in families *Retroviridae, Caulimoviridae, Hepadnaviridae*—are evolutionarily interspersed among **retrotransposons**, or REs with LTR sequences. Moreover, the two families with dsDNA-containing virions are not each other's closest relatives: **hepadnaviruses** and the envelope-gene-free **pseudoviruses** and **belpaoviruses** are very distinctly separated from the group containing the Env-gene-encoding retroviruses and **metaviruses** and the nonenveloped **caulimoviruses**. The REs classed as viruses in fact exhibit an evolutionary gradient from cellular REs through to autonomous viruses. Belpaoviruses—named for the two best characterized REs in the family, found in fruit flies and silkworms, respectively—are not known to be infectious, although they make intracellular capsids. The same is true for pseudoviruses, whose type members (see Table 3.4) are the Copia RE from fruit flies, and the yeast Ty1 RE. Metaviruses include the Gypsy RE found in fruit flies, and the yeast Ty3 RE. Budded particles with membranes and Env proteins are infectious, although this is unusual.

Retroviruses sit midway in the spectrum: as **autonomous** (=independent) viruses they are infectious, but their integrated form acts exactly as do the related Gypsy REs in terms of being

TABLE 3.4 Reverse-transcribing viruses.

Virus family; Examples of genera	Nucleic acid in capsid	Description	Type viruses
Caulimoviridae; Caulimovirus, Badnavirus	Open circular "gapped" dsDNA	These are all plant viruses which package the DNA form of the genome into the virions as circular but not covalently-closed molecules, and transcribe LTR-RNA genomes	Cauliflower mosaic caulimovirus; banana streak badnavirus
Hepadnaviridae[a]; *Orthohepadnavirus, Avihepadnavirus*	Open circular partially dsDNA	These are found in vertebrates	Hepatitis B virus; duck hepatitis B virus
Belpaoviridae; Semotivirus	ssRNA+	The second-most abundant class of LTR REs, occur in multiple phyla including basal metazoans but not mammals, suggesting they arose early in animal evolution. Not infectious	*Drosophila melanogaster* Bel virus; *Bombyx mori* Pao virus
Metaviridae[a]; *Errantivirus, Metavirus*	ssRNA+, 2 molecules?	The most abundant class of LTR REs, first found in arthropods but present in all eukaryotes. Can be infectious, due to presence of Env	*Drosophila melanogaster* Gypsy virus; *Saccharomyces cerevisiae* Ty3 virus
Pseudoviridae; Hemivirus, Pseudovirus	ssRNA+, 2 molecules	First described in fungi; now known from many eukaryotes, especially plants. Not infectious	*Drosophila melanogaster* Copia virus; *Saccharomyces cerevisiae* Ty1 virus
Retroviridae[a]; *Lentivirus; Simiispumavirus*	ssRNA+, 2 molecules		HIV-1 and -2; simian foamy virus

[a] *MGE virions have envelopes and envelope proteins and are transmissible.*
From ViralZone Reverse transcribing viruses (https://viralzone.expasy.org/295); ICTV Taxonomy (https://talk.ictvonline.org/taxonomy/).

transposable, and if they integrate into germ line cells, they can be vertically transmitted and become **endogenous retroviruses** (see below and Chapter 4).

Caulimoviruses come next on the spectrum: while these viruses replicate via reverse transcription in the cytoplasm, then **transcription in the nucleus** of infected plant cells to give LTR-containing ssRNA forms of the genome that complete the cycle to dsDNA in the cytoplasm (see Chapter 4), they do not use integration into the genome as part of the replication cycle. However, integrated forms of the genome can be found, mainly for the bacilliform **badnaviruses** and for **two other genera**, and integrated tandem copies of banana streak virus and other badnaviruses can create infectious RNA forms of the genome.

Hepadnaviruses follow the same replication mode as caulimoviruses, and integrate quite often into the host cell genome—often engaging REs such as LINEs to do so—but integrated copies do not appear to **produce infectious virus**. They do, however, cause disruptions of genome and cell cycle regulation that can lead to cell transformation and possibly to cancers (see Chapter 6). Their evolution is interesting, as they are related in terms of core replication machinery to the nonenveloped **nackednaviruses** found in fish—albeit across a 400 million year divide. Both employ protein-primed reverse transcription that utilizes a characteristic RNA "epsilon element," and share a structurally homologous capsid protein—although hepadnaviruses have acquired an envelope protein in addition (see Beck et al., 2021).

Caulimoviruses and hepadnaviruses are obviously highly derived forms from different RE lineages that have lost and/or displaced several genes of the ancestral RT virus, with the exception of RT and RH, and also a protease module (PR) for caulimoviruses. Other key features of their evolution include:

- **Capsid protein (CAP)** of caulimoviruses shares a C2HC Zn-knuckle motif with the **NCs** of **retroviruses, pseudoviruses** and **metaviruses**.
- The **movement protein (MP)** and **aphid transmission factor** (ATF) of **caulimoviruses** must have been acquired from **plants and aphids**, respectively, to allow cell-to-cell movement in plants, and enhanced transmissibility by aphids.

At first sight, it does not appear likely that two very different groups of viruses—those that have **ssRNA+ genomes in particles but replicate via dsDNA**, and those that have **dsDNA in particles but replicate via ssRNA+**—could possibly be related in any way. However, once one understands that these two genome types are simply on **opposite sides of a cycle**—which I have termed the "**retroid cycle**"—it becomes much easier to understand. This is shown in Fig. 3.7. Further details on replication—for example, how the respective types of viruses derive **long terminal repeat sequences** in **DNA** forms of the genome (**retro-** and related viruses) or RNA forms (**caulimo-** and **hepadnaviruses**)—are discussed below, and in Chapter 5.

Retroviruses are worth discussing more, given their importance in human and animal diseases—and the fact that **endogenous retrovirus** sequences (**ERVs**; see below)—retrovirus genomes that have integrated into germline cells, and been **vertically inherited**—are very useful in determining breakpoints in host evolution by molecular archaeology, and often play important roles in their hosts after being coopted during cell development. As much as 10% of the human and other vertebrate genomes are composed or ERVs: Many of these sequences are degraded or fragmented, however, with this tendency increasing with the age of probable insertion. All nonendogenous retroviruses have three gene functions in common, although some—e.g., HIV-1 and -2, human T-cell leukemia viruses (HTLV)—may have several more (see Fig. 3.6). The core genes encode polyproteins that are processed into several other entities:

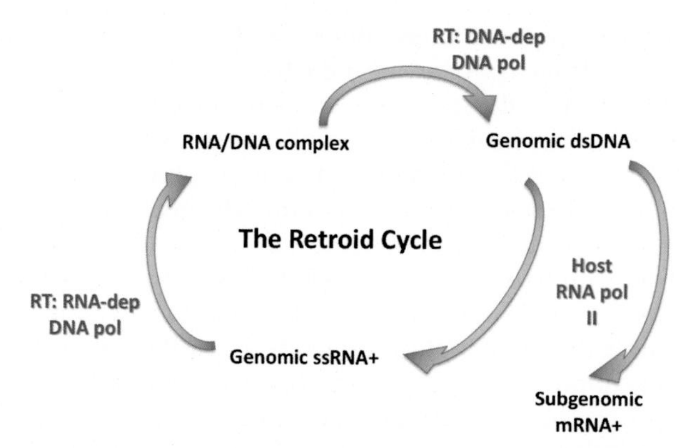

FIG. 3.7 The retroid cycle of replication via reverse transcription. The cycle depicts the common features of going from RNA to DNA and back again, and vice-versa.

these are ***pol***, ***gag***, and ***env***. *Pol* encodes a protease, Pro; reverse transcriptase, RT; RNase H, RH, and an integrase, INT. *Gag* (for group-specific antigen) encodes the structural polyprotein, processed into matrix (MA), capsid (CA), and nucleocapsid (NC) proteins. Env (for envelope protein) contains information to produce a polyprotein that is processed into a two-component Class I fusion glycoprotein: in HIV-1 this is Env, or gp160, that is cleaved by a protease to give a gp41 stalk with a transmembrane domain, and a gp120 head. While Gag and Pol proteins are homologous among retroviruses—they each descend from a common ancestor—Env proteins of different retroviruses appear to have been obtained from a number of other RNA and DNA viruses during their evolution, just as for retrotransposons.

During the process of reverse transcription (Fig. 3.8), the two ssRNA+ molecules that comprise the retrovirus genome are converted into a dsDNA molecule that is longer than the RNA templates due to the **duplication of direct repeat sequences** at each end—the long terminal repeats (LTRs) (Fig. 3.8). Complete conversion of retrovirus RNA into dsDNA only occurs in a **partially uncoated core particle** and cannot be duplicated accurately in vitro with the same reagents in vitro. This indicates that the conformation of the two RNAs inside the nucleocapsid dictates the course of reverse transcription—the 3′ ends of the two ssRNA+ strands are probably held adjacent to one another inside the core.

Reverse transcription has important consequences for retrovirus genetics. First, it is a **highly error-prone process**, because RT does not carry out the **proofreading** functions performed by **cellular DNA-dependent DNA polymerases**. This results in the introduction of many mutations into retrovirus genomes and, consequently, **rapid genetic variation**. In addition, the process of reverse transcription promotes **genetic *recombination***. Because two RNAs are packaged into each virion and used as the template for reverse transcription, recombination can and does occur between the two strands, and especially if superinfection occurs, with **more than one** retrovirus infecting the same cell. Although the mechanism responsible for this is not clear, if one of the RNA strands differs from the other (e.g., by the presence of a mutation) and recombination occurs, then the resulting virus will be **genetically distinct** from either of the parental viruses. This happens frequently with **HIV-1** genomes, for example, where **circulating recombinant forms** are common.

After reverse transcription is complete, the double-stranded DNA migrates into the nucleus, still in a core particle (see Chapter 4). This contains the RT complex: this is the product

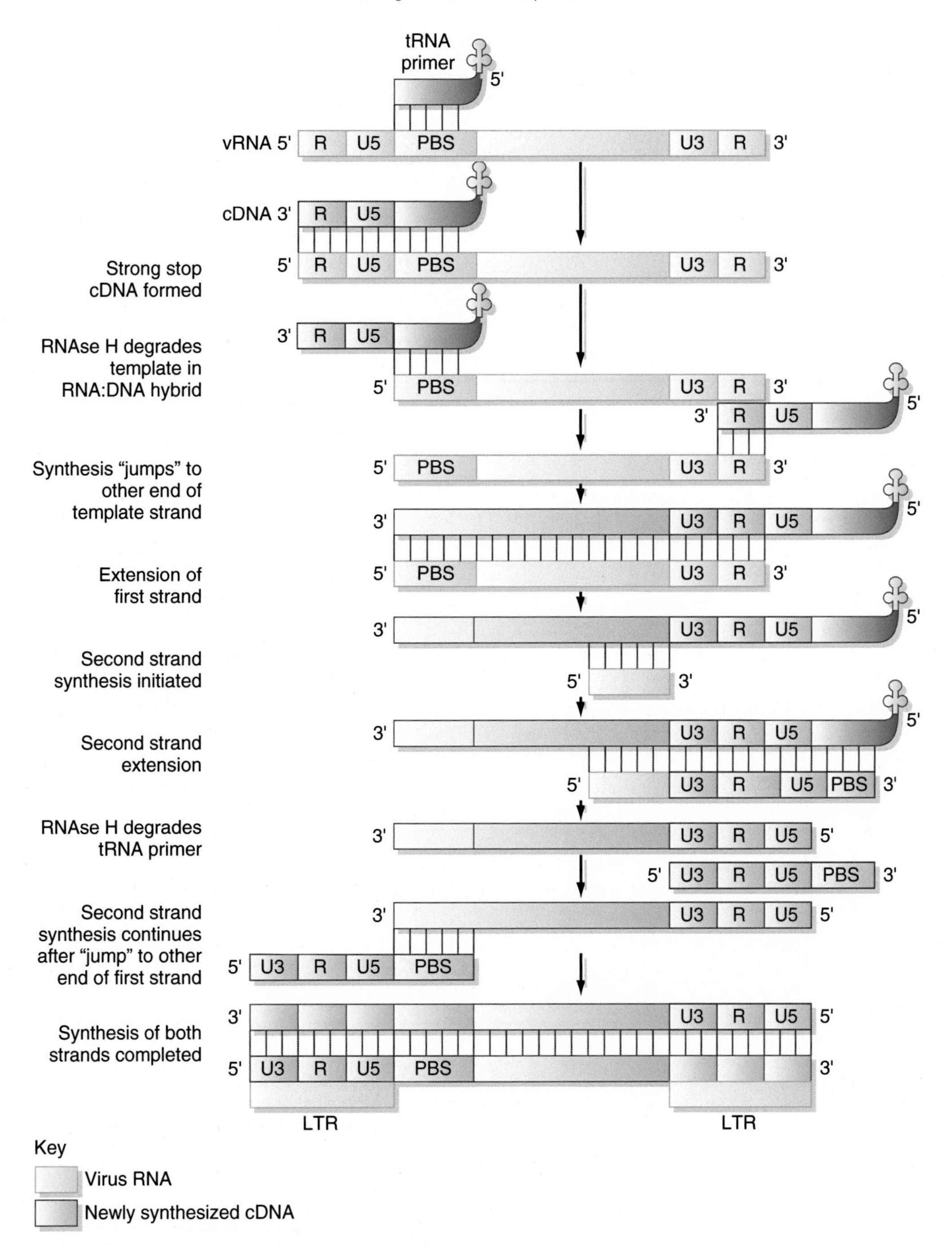

FIG. 3.8 Reverse transcription and generation of DNA forms of retroviruses. The highly complex mechanism of reverse transcription of retrovirus RNA genomes, in which two molecules of RNA are converted into a single (terminally redundant) double-stranded DNA provirus, is depicted.

of the ***pol*** gene, and has three distinct enzymatic activities: these are **RNA-dependent DNA polymerase (RT)**, **RNase H**, which degrades **RNA templates base-paired to DNA**, and **integrase**, which catalyzes **integration of virus DNA** into the host cell genome, after which it is known as the **provirus** (Fig. 3.9). Integration of retrovirus DNA involves a linear form, although circular forms can be found in infected cells. The net result of integration is that 1–2 bp are lost from the end of each LTR and 4–6 bp of cellular DNA are duplicated on either

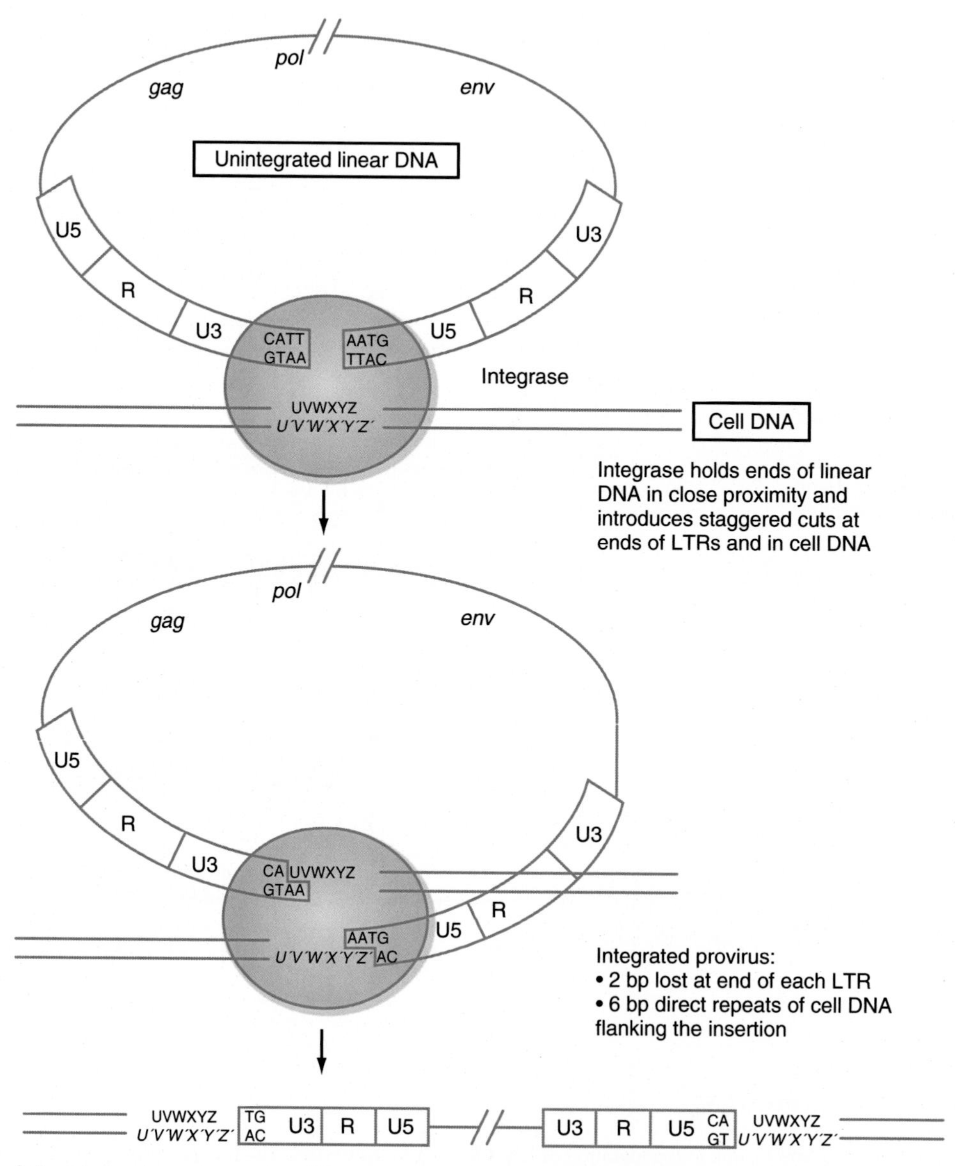

FIG. 3.9 Mechanism of retrovirus-like genome integration. This depicts the mechanism of integration of retrovirus-like genomes into the host cell chromatin: this will be true for retro-, meta-, pseudo-, and belpaoviruses.

side of the provirus. There may be numerous regions or sites in the eukaryotic genome that are more likely to be integration sites than others, although there is no clear specificity for any given virus. Following integration, the DNA provirus genome for most retroviruses becomes essentially a collection of cellular genes and is at the mercy of the cell for expression (see Chapter 5). Viruses with complex genomes such as **HIV-1 and -2**, and the **human T-cell leukemia viruses (HTLVs)**, however, do encode proteins with *trans*-acting capability that can specifically **repress or activate** provirus expression.

Eukaryote RNA viruses

Eukaryotes are unique among cellular organisms in having a rich and diverse RNA virome, or viruses with RNA in virions that replicate via RNA intermediates, nearly all of which have no relatives in any prokaryote, and the most diverse of which have ssRNA+ genomes. The only hallmark gene they all share is **RNA-dependent RNA polymerase, or RdRp**, which is absolutely required for their replication. An enzyme or complex with this specificity—**RNA template-dependent RNA synthesis**—is not present in any ribocell. In fact, while eukaryotes do possess a polymerase involved in making small interfering RNAs (siRNAs), this is distantly related to DNA-dependent RNA polymerases, while viral RdRps are distantly related to RTs. Thus, RNA viruses all encode RdRp, and viruses with nontranslatable genomes include the enzyme in their virions. Moreover, the enzymes in all three classes of RNA viruses—ssRNA+, ssRNA−, and dsRNA—are obviously **homologous** (deriving from the same source) at the structural level. The RNA viruses epitomize the concept of "modular evolution," where distinct replication modules may pair up with many varieties of structural modules, to give a patchwork type of evolution.

The only contributions of the prokaryote RNA virome to eukaryotic RNA viruses are thought to be from the ssRNA **leviviruses** (family *Leviviridae*), and the dsRNA-containing enveloped **cystoviruses** (family *Cystoviridae*). The RdRps of leviviruses are most closely related to those in the fungus-infecting "**naked RNA virus**" *Narnaviridae*, and the isometric ssRNA+ plant-infecting **ourmiaviruses** (genus *Ourmiavirus*; no familial relatives). Narnaviruses are capsidless and not infectious; they are vertically transmitted as RNA plasmids of fungal mitochondria—which indicates an endosymbiogenetic origin from phages of the mitochondrial ancestor. Ourmiaviruses have movement and capsid proteins encoded in their genome that are related to plant-infecting **tombusviruses**, from a completely different lineage (see Fig. 3.10), presumably picked up by recombination in a plant infected by a fungus and a tombusvirus. The only viruses with **RdRps** directly derived from cystoviruses are the **reoviruses** (family *Reoviridae*; many genera). These are not enveloped, and infect many vertebrates, arthropods, molluscs, fungi, plants, and a green microalga. While reoviruses and cystoviruses share a capsid architecture with a variety of other eukaryotic dsRNA viruses—$T=1$ or "pseudo T = 2" capsids (see Chapter 2) composed of 60 dimers of 2 identical proteins—the other viruses do not have a related RdRp. Thus, they probably also evolved by recombination, to give the same structural but different replication modules.

All other RNA viruses in Eukarya—ssRNA+, ssRNA−, and other dsRNA viruses—appear to derive from a **common ssRNA+ ancestor**. This was probably assembled at the dawn of eukaryogenesis from components derived from prokaryotes: these included a RdRp derived from retron RT, a Superfamily 3 helicase from a rolling-circle (RC) plasmid, a "high temperature requirement A" (HtrA) family serine protease (found in bacteria and mitochondria), and a SJR-fold structural protein. Given the mitochondrial affinity of the protease module, this

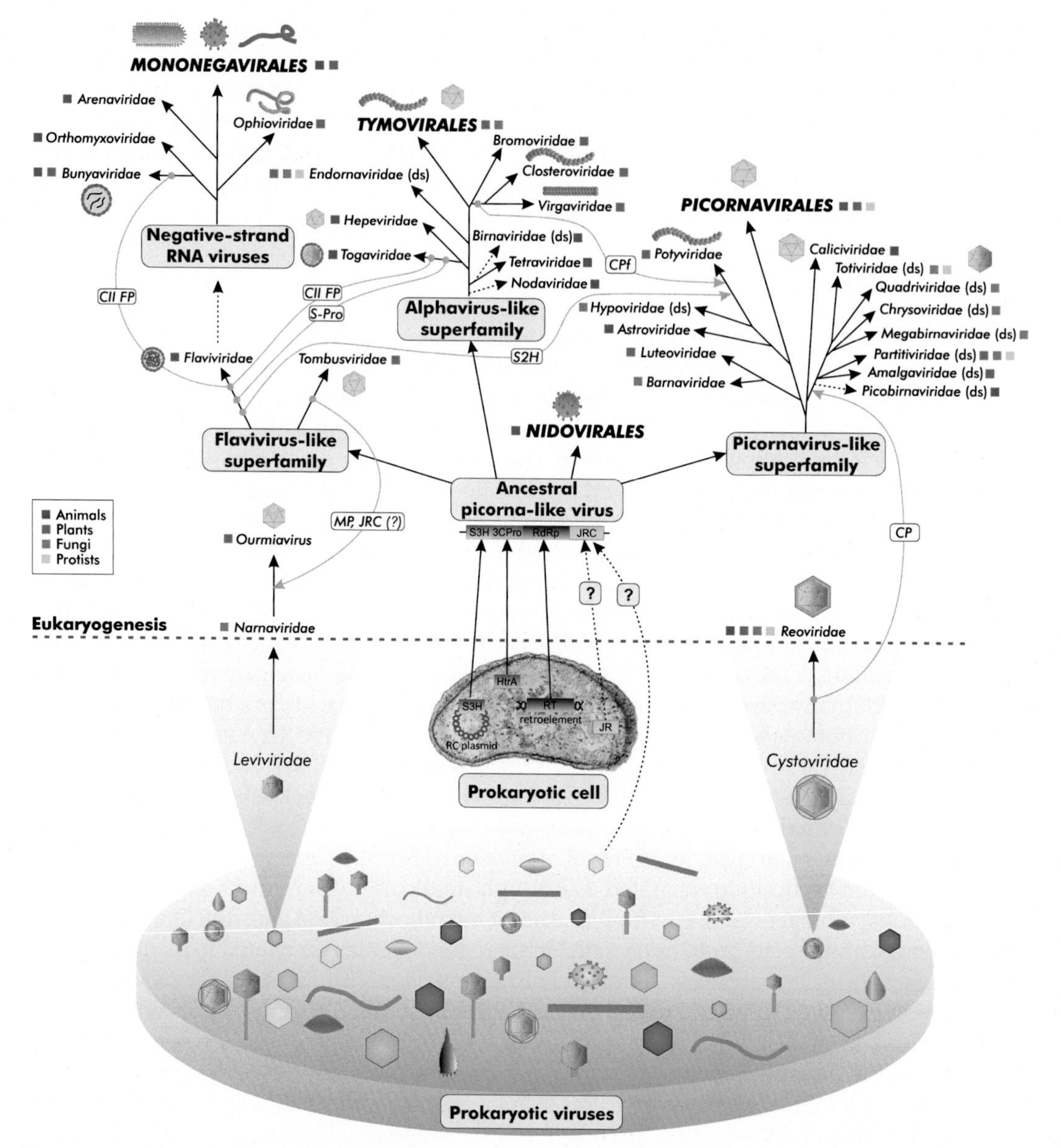

FIG. 3.10 Origin of major eukaryotic RNA virus groups. Origin of the major groups of RNA viruses of eukaryotes, based on the symbiogenetic scenario of eukaryogenesis. The host ranges of viral groups are *color-coded as shown in the inset.* Icons of virion structures are shown for selected groups. Ancestor-descendant relationships that are considered tentative are shown with *dotted lines,* and particularly weak links are additionally indicated by *question marks* (see text for details). Key horizontal gene transfer events are shown by *grey, curved arrows.* Abbreviations: *CII FP,* class II fusion protein; *CP,* capsid protein; *CPf,* capsid protein of filamentous viruses; *JRC,* jelly roll capsid (protein); *MP,* movement protein; *RT,* reverse transcriptase; *S2H,* superfamily 2 helicase; *S3H,* superfamily 3 helicase. *From Koonin, E.V., Dolja, V.V., Krupovic, M., 2015. Origins and evolution of viruses of eukaryotes: the ultimate modularity. Virology 479–480:2–25. https://doi.org/10.1016/j.virol.2015.02.039*

may well all have happened in the mitochondrial ancestor. As can be seen in Fig. 3.7, this postulated ancestral picorna-like virus had a genome that encoded a helicase (S3H), an **endoproteinase** (3CPro), RdRp and a SJR structural protein (labeled SJR, for single jelly-roll fold). This then diverged to give rise to **picornavirus-like**, **flavivirus-like**, and **alphavirus-like** superfamilies of viruses, grouped mainly based on their replication module similarities.

Viruses not fitting into these superfamilies are members of the Order *Nidovirales*: nidoviruses are characterized by having the largest known RNA genomes, not encoding SJR capsid proteins, and all having enveloped virions. It is supposed that the large genomes are enabled by their large and complicated replication apparatus including a **proofreading 3′-to-5′ exoribonuclease** (ExoN). This is related to **DEDDh** exonucleases, part of the DnaQ-like (DEDD) exonuclease superfamily found in all ribocell types, and probably acquired from a host cell by the ancestral nidovirus. Family *Coronaviridae* are nidoviruses, and their most famous member is SARS-CoV-2, in genus *Betacoronavirus*. Referring to coronaviruses in fish, Eddie Holmes of the University of New South Wales recently wrote:

> *"There are also clear phylogenetic links to invertebrate viruses, so I suspect that some of these families (and where we designate these on the tree is largely arbitrary) may be as old as the* ***Metazoa*** *[multicellular animals]"* https://twitter.com/edwardcholmes/status/1365413669547380737?s=20

The presently-existing picornaviruses, alphaviruses, and flaviviruses are all **ssRNA+** viruses, as their postulated predecessor must have been. However, the formerly-described "picorna-like superfamily" now known as phylum *Pisuviricota* includes five families of dsRNA viruses; order *Tymovirales* also has two dsRNA families: all of these are distinct from the prokaryote cystovirus-descended *Reoviridae*, although those families in the *Pisuviricota* appear to have acquired their structural protein module from reoviruses. It is suggested that these diverse lineages of dsRNA viruses independently converged on a "**dsRNA lifestyle,**" which is just an adaptation of the ssRNA+ replication cycle (see Chapter 4), given that replication complexes of these viruses contain **dsRNA replication intermediates**.

Detailed investigations RdRp relatedness among RNA viruses showed that dsRNA viruses apparently evolved from ssRNA+ ancestors as at least two independent lineages, whereas ssRNA− viruses apparently derive from dsRNA viruses, and not from ssRNA+ viruses as had previously been assumed (see Wolf et al., 2018). This was also obviously a more recent event than the separation of the major ssRNA+ and dsRNA virus lineages, which is borne out by proteins and even genetic organization being more similar between many of the ssRNA− viruses, than is the case among the ssRNA+ lineages. The ssRNA− viruses are also limited to infecting animals and green plants—with evidence pointing to the terrestrial plant viruses being derived from animal-infecting viruses, probably by long exposure of plants to colonizing arthropods or nematodes in which the viruses replicated, as diversity of ssRNA− viruses is far greater in invertebrates and vertebrates, than in plants, and more diverse in marine than in terrestrial environments. It is entirely likely, therefore, that the viruses originated in arthropods in marine environments, and were transmitted to vertebrates there, before first arthropods and then vertebrates—and their viruses—emerged onto dry land to follow bacteria, fungi, and plants.

Terrestrial viruses

One may contend that it was the coevolution of insects and plants—because what else were insects going to eat?—that has driven much of terrestrial viral evolution. Basically, the only terrestrial organisms around some 450 million years ago were primitive green plants, fungi, and bacteria. Bacteria and fungi would both have parasitized plants as the only organic food source available apart from each other, which would have led to fungal viruses also infecting plants, and vice-versa. As an example, **partitiviruses** (phylum *Pisuviricota*, family *Partitiviridae*, segmented dsRNA genome) are found in fungi and **protists** and plants, with a wider spectrum of similar viruses found only in fungi. After the emergence of insects, these ate plants (and each other), and viruses in insects entered plants, and vice-versa, as a result of close and prolonged association. Here viruses in the **Alphavirus-like superfamily** are instructive, as viruses infecting only insects or only plants, and some infecting both, have very similar replication machinery, if different structural modules. The emergence of amphibians—vertebrates derived from fish—over 250 million years ago brought a whole new virome onto dry land—and a new niche for insects to colonise and swap viruses with. An interesting example here would be herpesviruses: these are known to infect a wide variety of marine animals, including the invertebrate coral polyps—yet they are only found in vertebrates on land, because that was the only host that migrated there. Relevant to this is the speculation that tobamoviruses related to tobacco mosaic virus have been cospeciating with their angiosperm host plant lineages for at least 100 million years—and may have originated in algae, or even in association between fungi and algae, before the emergence of land plants (Fig. 3.11).

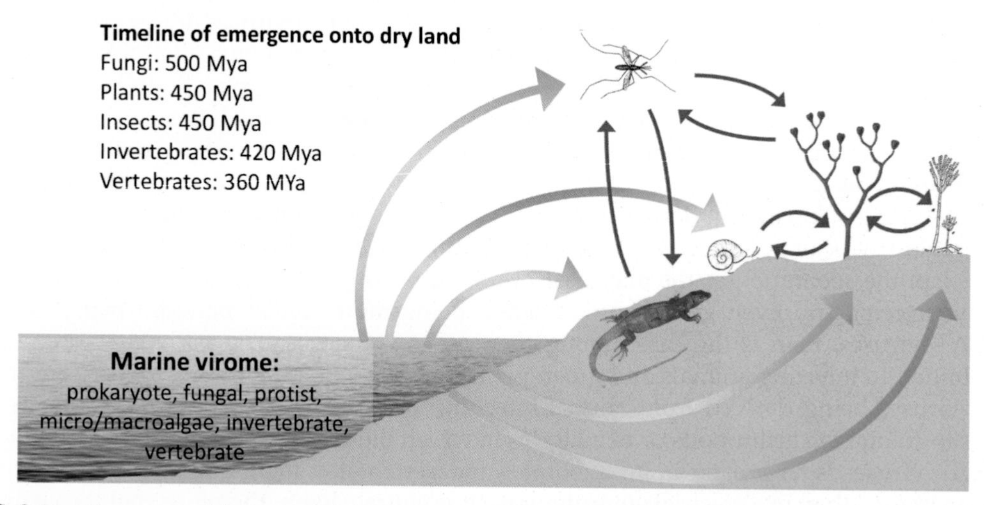

FIG. 3.11 Emergence of the terrestrial virome. In this depiction, the terrestrial virome emerged onto land with the organisms that harbored them: this is shown by the blue-brown arrows for each major group. This would have been a subset of prokaryotes (emerged at least 1.8 billion years ago), followed by representatives of fungi, green plants (all land plants derive from a subset of marine algae), insects and other invertebrates, and finally, vertebrates. Red arrows depict the interchange of viruses between terrestrial organisms, with plants being central to many of these. Insects were also an important bridge between viruses of invertebrates, plants, and vertebrates.

Of ERVs and EVEs

Endogenous retroviruses (ERVs)

Endogenous retrovirus sequences (ERV) are numerous in vertebrates, and constitute around 8% of the human genome (HERVs). These are all the remnants of retroviruses that have inserted into the germline of humans and their ancestors, as none of them are still viable, and the oldest sequences may be very fragmented, with no intact ORFs. Once ERVs have been inserted, they are replicated along with the host genome—and are subject to the same accuracy (or fidelity) of copying as other host DNAs, which is far more accurate than their own RT. This allows them to be used for "**molecular paleontology,**" as comparison of ERVs between progressively less related species can give an excellent idea of which of them entered ancestral host species, and which are new entrants, in only one or more lineages.

While some human ERVs (HERVs) are predicted to be truly ancient—up to 100 million years old, predating the divergence of all primates—it has been difficult to assess the relatedness of these sequences past ~100 million years. Use of foamy virus sequences, however—genus *Spumaretrovirus*—has enabled far deeper dives into evolutionary antiquity, given that these viruses have a very stable history of **cospeciation** with their hosts. Recent evidence from detection of foamy virus ERV (FLERV) sequences in fish, amphibians, and shark genomes—all species that diverged >100 million years ago—suggest that foamy viruses originated at least as long ago as 450 million years, in jawed fish, and that retroviruses as a class are even older. The antiquity of retrotransposon-like viruses (metaviruses, pseudoviruses) is obvious if you note their range of hosts: these include all eukaryotes surveyed (metaviruses), and eukaryotes including yeasts and fungi, arthropods, and plants (pseudoviruses). It is therefore likely that these were the origin of retroviruses as we know them now.

While this is fascinating to some, more fascinating has been the revelation that certain endogenous retrovirus sequences—including the LTR-containing retrotransposons that are pseudoviruses and metaviruses—have been "**exapted**" by the host. This term denotes recruitment of a biological entity for a new function unrelated to the original one. In this case, an exogenous gene or part of it, or even a promoter sequence, ends up being used by the host to provide a function that may become essential.

One of the most important of these events must be the repeated exaptation throughout the evolution of mammals, of genes encoding "**syncytins,**" or membrane proteins derived from retrovirus Envs. These play a central role in the development of the **placenta** that is central to mammals bearing live young after internal incubation of fetuses. While the original event that helped give rise to the mammalian ancestor involved one Env, at least 10 mammalian lineages have subsequently domesticated different *env* genes to replace the original—and several different versions playing related roles in certain of them—to provide the function of forming **syncytia**, or multinucleate giant cells. These arise as a result of neighboring **trophoblast** cells expressing the virus-derived fusion protein that is expressed at a particular stage of their development, and fusing together to form the **syncytiotrophoblast**, which acts as a shield between maternal tissues and the developing embryo: this limits exchange of migratory immune cells between embryo and mother, and prevents immune rejection of the fetus. Moreover, syncytiotrophoblasts do not express ***Major Histocompatibility Complex*** **(MHC)** molecules, so stay hidden to T cells (see Chapters 4 and 6).

Another set of exaptations that appears central to the development of higher nervous systems in insects and mammals is the parallel recruitment of **Gag-related proteins** from **Ty3/Gypsy-like metaviruses** (mice) and **Copia-like pseudoviruses** (flies), respectively. A gene called *arc* in both organisms helps neurons build connections with each other to enable long-term memory formation and other neurological functions. **Extracellular vesicles** (EVs) released by neurons in mice and flies contain a capsid made from Arc, which carry mRNA for Arc: these bud through the cell membrane to produce EVs which fuse with neighboring neuronal cells to release mRNA, which is translated. Flies lacking *arc* form fewer neuron-muscle connections; mice have problems forming long-term memories. The human genome contains at least 85 similar genes, many present also in other mammals, which have been grouped into the Mart, PNMA and SCAN families. Proteins from all three are involved in various types of tissue differentiation, but it is not known if this is via formation of **virus-like particles**, as for Arc. This is a very fast-developing field of research, however.

The cooption of *gag* genes in mammals also includes use of an ERV *gag* dating from an insertion event 45–50 Myr ago as the "rodent restriction factor" Fv1, which is a postentry inhibitor of **murine leukemia virus** infections. Similarly, a coopted *gag* produces Gag as a trans-dominant release blocker for **jaagsiekte sheep retrovirus**, which cause an infectious lung cancer in sheep and relatives in the genus Ovis. An exciting finding in humans is that a HERV *gag* dating back at least 43 Myr—and so present in all simian primates, including Old World and New World monkeys, apes, and humans—is active in the placenta, as are associated *env* genes. Thus, both *gag* and *env* from at least one HERV play some as yet undefined role in placental development. However, despite their expression being tightly regulated and limited to the placenta in normal animals, transcripts are also upregulated in diffuse large B-cell lymphomas, the most common of the **non-Hodgkin's lymphomas**—meaning they could be a useful **biomarker** for the disease (see Boso et al., 2021).

While the two *arc* exaptations are ancient—dating to before the development of mammals for syncytins, and in the early development of insects and mammals for Arc—a more recent recruitment from retroviruses involves the expression in the bovine embryo of genes of the two bovine endogenous retroviruses **BERV-K1** and **BERV-K2**. These are both expressed throughout different stages of bovine embryo preimplantation development. It is suggested that ERVs may be involved in the establishment of **totipotent/pluripotent states** (capability to differentiate into any of the cells found in the organism) in cells of the early-stage bovine embryo, which is similar to recent discoveries for human and mouse embryos as well.

It is known that HERV LTRs are involved in the transcriptional regulation of various cellular genes, and that HERVs and their products (RNA, cytosolic DNA proteins) modulate the host immune system. HERV sequences have helped shape and expand the network of **interferon** responsive genes (see Chapter 6), and modulate **innate immunity** effectors. This may result in inflammatory and autoimmune disorders on the one hand, while on the other, can help control excessive immune activation through immunosuppressive effects. HERVs have also been proposed more generally to protect against exogenous retroviral infections, as has been shown for mouse and sheep exaptations (see above).

Most ERVs that have been well characterized are at least several million years old, which means that their hosts have productively survived the event. One negative manifestation of recent retroviral integration into germline cells, however, was very recently found in koalas. Here, researchers were able to study the ongoing endogenization—a process that probably started around 50,000 years ago—of **koala retrovirus (KoRV)** in its early stages in live animals. Sampling tumors and normal tissue from 10 koalas allowed identification of 1002 unique integration sites (ISs), with "hotspots" of integration in the vicinity of known cancer genes. The endogenized elements are transcriptionally active, meaning koalas have high levels of circulating and infectious virions. Their cancers arise from reinfection and reinsertion into new sites in the genomes of nongermline cells, upregulation by promoters in KoRV LTRs of "**proto-oncogenes**" (genes whose unregulated expression could cause cell transformation), and then clonal expansion of transformed cells to form tumors. The study indicated that the host suffers a "tremendous mutational load" during KoRV invasion of germline cells—which suggests that this is a process that has been repeatedly experienced, and repeatedly overcome, during the evolution of vertebrate lineages.

Endogenous viral elements (EVEs)

It has recently been recognized that sequences deriving from **certain nonreverse transcribing RNA and DNA viruses** can actually be inserted into host cell genomes—and can potentially afford them protection against future infection by the same or similar viruses, among other functions. This phenomenon had been discovered previously, in the shape of "**pseudogenes**": these are gene-like sequences in animal genomes that have no introns, may have no promoter associated with them, may even contain poly-A sequences characteristic of mRNAs, and are obviously derived by reverse transcription from cellular mRNAs. For viruses, this may have been happening over aeons of evolutionary time, and has involved hosts as diverse as plants (integrated poty- and geminivirus sequences), honeybees (integrated Israeli bee paralysis virus), and integrated filovirus-like elements in the **genomes of bats, rodents, other mammals, and marsupials**. In the case of mammals, transcribed fragments have been found that are homologous to a fragment of the filovirus genome, which when expressed, can interfere with the assembly of Ebola virus, indicating an advantage to the organisms in having the sequence preserved in their genomes. The case of the bee EVE is interesting, as researchers showed that 30% of bee populations in Israel had EVEs derived from the ssRNA+ **dicistrovirus** Israeli acute paralysis virus (IAPV) coat protein genes—and were resistant to infection by the virus, which is implicated in mass mortality of bees.

A fascinating recent example of molecular virus "**paleovirology**" was the demonstration that endogenous viral elements (EVEs) derived from a **wide variety of RNA viruses (ssRNA+, ssRNA−, dsRNA)** could be found in the genomes of mammals and insects, as well as **ssDNA circo- and parvoviruses** in mammal and bird genomes, and a **family of hepatitis B-like pararetrovirus** sequences in birds and a tick (see Katzourakis and Gifford, 2010). In mammals, but not in insects, complete genes for RNA virus proteins were typically flanked by the "**target site duplications**" (TSDs) and 3′ poly-A tails that are characteristic of LINE-mediated retrotransposition of viral mRNAs. Thus, endogenous RT activity provided by LINEs is inadvertently providing cDNAs derived from genomic and mRNAs for integration into cell genomes—which will be inherited if those cells are in the germline (see Fig. 3.8).

The finding of related families of virus-derived insertions in widely-diverged animal species allows the deduction that **the viruses have been diverging for at least as long as their hosts**—pushing back the possible origin of the EVEs to more than 30 million years in the case of filoviruses, by the same kind of molecular paleontology that was used to look at

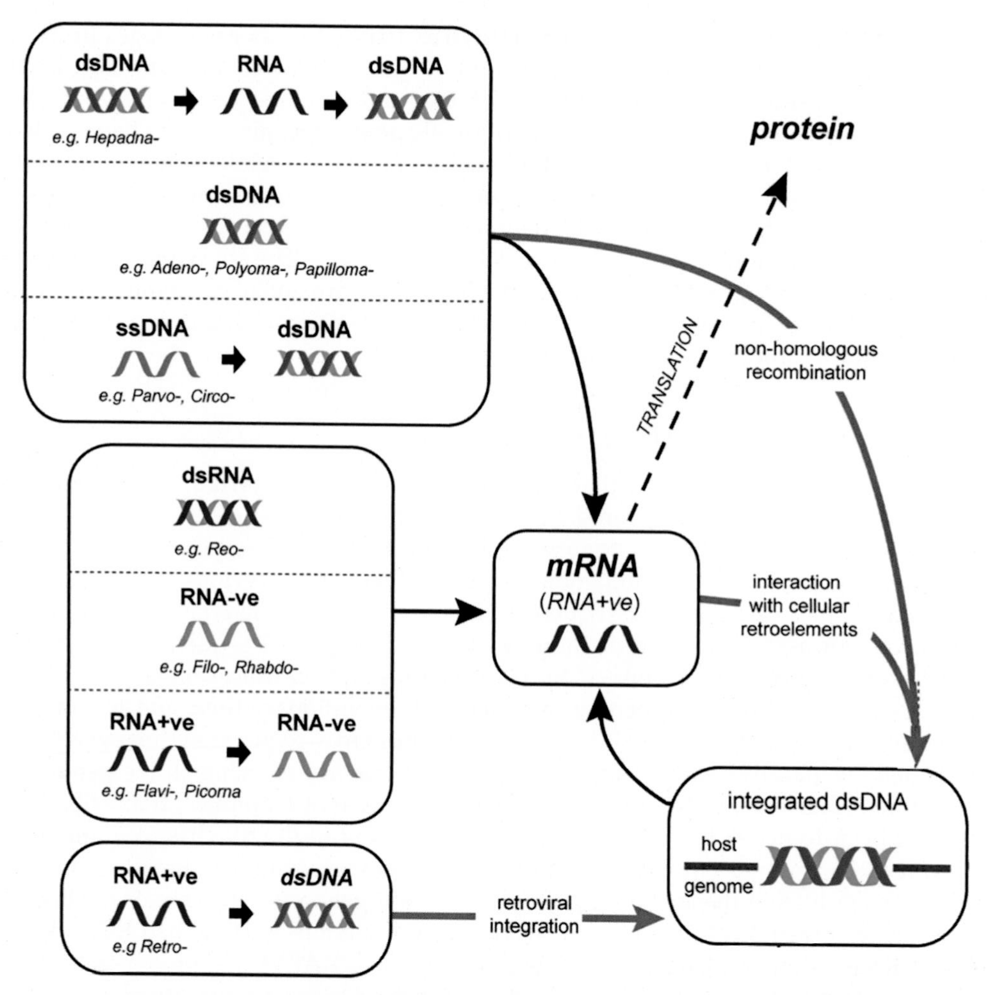

FIG. 3.12 Origin of viral sequences inserted into eukaryotic cell genomes. Examples of the known animal virus replication strategies are shown to the left of the figure, with the representative families listed for each case. *Arrows* indicate steps in replication. *Red lines* indicate pathways that lead to viral genetic material becoming integrated into the nuclear genome of the host cell. For retroviruses, integration occurs as an obligate step in replication. For all other animal viruses integration occurs anomalously, through interaction with cellular retroelements such as LINEs, or via nonhomologous recombination with genomic DNA. If integration occurs in a germ line cell that goes on to develop into a viable host organism, an EVE is formed. Abbreviations: *dsDNA*, double-stranded DNA; *ssDNA*, single-stranded DNA; *dsRNA*, double-stranded RNA; *RNA−ve*, negative sense, single-stranded RNA; *RNA−ve*, negative sense, single-stranded RNA; *RNA+ve*, positive sense, single-stranded RNA.

retrovirus origins. More impressive, however, was a recent study that determined that true **endogenous hepatitis B viruses (eHBVs)** are only present in **saurians** (all reptiles, birds), and that metahepadnaviruses (which infect fish) must have originated over 300 million years ago (Figs. 3.12 and 3.13).

Interesting insights into the evolution of life on land, and its virome, was provided by evidence that giant virus (NCLDV) genes are present in both fungi and in early-diverging land

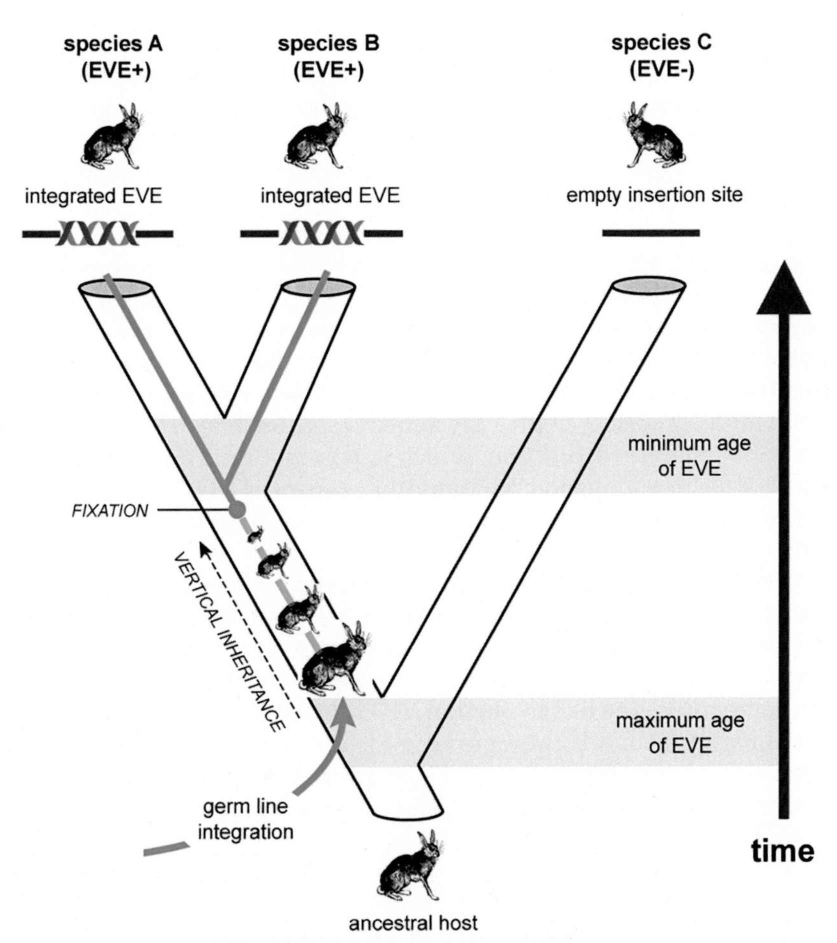

FIG. 3.13 Evolution of EVEs in a host lineage. If integration occurs in a germ line cell that goes on to develop into a viable host organism, an EVE is formed. Green lines show the evolution of an EVE in its host lineage: the EVE reaches genetic fixation at the point indicated, and is inherited by all descendant hosts thereafter. Assuming that insertion occurs randomly, the presence of related EVEs at the same locus in both descendant species A and B indicates that insertion occurred prior to their divergence, allowing a minimum age for the insertion to be inferred from the estimated timescale of their evolution. Conversely, the presence of an empty insertion site in species C provides a maximum age for the insertion. *Both Figs. 3.12 and 3.13 adapted from Fig. 1 of Katzourakis, A., Gifford, R.J., 2010. Endogenous viral elements in animal genomes. PLoS Genet. 6 (11), e1001191. https://doi.org/10.1371/journal.pgen.1001191.*

plants—despite there being no known NCLDVs infecting either terrestrial fungi or plants. In fungi, two distinct groups of NCLDV genes were found in seven fungal phyla, constituting around one-third of the known phyla. While infections of terrestrial fungi by NCLDVs is not ruled out, it is possible that the acquisition occurred when fungi were still limited to aquatic environments. A scan of land plant genomes for NCLDV sequences turned up candidates in only the genomes of the **bryophyte** *Physcomitrella patens* (a moss) and the **lycophyte** *Selaginella moellendorffii*. These two lineages diverged before 400 million years ago; mosses do not have true vascular tissue, while lycophytes do. This indicates a possibly very ancient origin of these sequences, possibly before true emergence onto dry land between 510 and 630 million years ago. Evidence that NCLDVs are still being inserted into marine plant genomes bolsters this contention: for instance, genome segments from various members of the *Phycodnaviridae* have been found as presumed recent inserts in the nuclear genomes of algal groups as diverse as the brown algae *Ectocarpus siliculosus* and some *Feldmannia* spp., the haptophyte *Emiliania huxleyi*, and the green microalga *Chlorella variabilis*.

EVEs that do not affect or only have slightly negative effects on their hosts may be "fixed" in the genome (i.e., not under negative selection), and will accumulate mutations at the same rate as host genes. However, if EVEs confer an advantageous phenotype on the host, they may spread through the population by positive selection. These are then **exapted** sequences, and will be selected so as to maintain their functionality. One such is the ssRNA-**bornavirus**-derived EBLN-1 element (N for nucleoprotein gene-derived), which is intact in several monkey species, with a presumed insertion date of >54 million years. This strongly indicates a conserved function, which is possibly resistance to bornavirus infection. The same is true for endogenous geminivirus sequences (eGVs) found in *Camellia* species: the rep genes are transcribed, and may well act as a defense mechanism (via siRNA, see Chapter 6).

One of the most interesting exapted sequences, in light of the involvement of **bats** in so many **zoonotic human diseases**, including **SARS-CoV-2** (see Chapter 8), is the presence in bats of the genus *Myotis* (microbats; mouse-eared bats) of a gene apparently derived from a filovirus. This is a VP35 homologue, which in bats weakly inhibits interferon responses (see Chapter 6). In comparison, the native filoviral VP35 strongly inhibits interferon production, which promotes its survival. It is presumed the *Myotis* VP35 "damps down" interferon responses lessen the kind of inflammation that is often caused by virus infections, and so may lessen immune response-associated damage. This could be an important factor in their apparent tolerance to many virus infections.

The complexity of virus genomes

As discussed in Chapter 1, virus genome structures and nucleotide sequences have been intensively studied in recent decades because the power of recombinant DNA technology has focused much attention in this area, largely because viruses are such good model organisms for understanding cellular regulatory processes. This has resulted, particularly in recent years, in an explosion of raw data—which has had to be matched by an explosion in bioinformatics techniques to process and make sense of all this data, and create new knowledge.

A fundamental divide among ribocells as hosts to viruses is illustrated by the fact that prokaryote genomes are smaller and replicate faster than those of nearly all eukaryotes. Hence, they can be regarded as "streamlined" compared to the latter, which have introns in most of their genes, and generally also have large amounts of what has been termed "junk" DNA of unknown function. Viruses with prokaryotic hosts in particular must be able to replicate sufficiently quickly to keep up with their host cells: this is reflected in the compact nature of especially smaller bacteriophages. **Overlapping genes** are common, and the maximum genetic capacity is compressed into the minimum genome size. The genome of the ssDNA phage ΦX174, for example, is a circle 5386 bases in size thought to encode 11 genes on one strand of

DNA, 4 of which are completely nested within other coding sequences. All control sequences such as promoters and terminators are contained within open reading frames—and it is noteworthy that despite this being the first DNA virus ever sequenced (in 1977), only in 2020 was a high resolution map of transcription from this organism generated (see Chapter 5).

The genome of the *E. coli*-infecting phage T4 is an interesting special case: while the genome consists of 160 kbp of dsDNA and is highly compressed, with **promoters** and translational control sequences nested within the coding regions of overlapping upstream genes, it was also the **first prokaryote genome found to contain introns** (see Chapter 1). The presence of introns in bacterial virus genomes, which are under constant ruthless pressure to exclude "junk sequences," suggests that these genetic elements must have evolved mechanisms to escape or neutralize this pressure and to persist as parasites within parasites.

In viruses with eukaryotic hosts there is also often pressure on genome size for smaller viruses, despite the lack of any such constraint on the host genome. Here the pressure is mainly from the packaging size of the virus particle (i.e., the amount of nucleic acid that can be incorporated into the virion). Therefore, these viruses commonly show highly compressed genetic information when compared with the low density of information in eukaryotic cellular genomes. This is exemplified by the dsDNA **polyomavirus SV40**, only the second DNA genome ever sequenced (Chapter 1). Here, a 5243 bp circular genome encodes up to nine proteins in bidirectional frames using two opposed promoters, with the use of several introns, one gene nested within another, and three proteins produced via differentially spliced mRNAs from the same ORF (see Chapter 5).

There are some quite important exceptions, however, to the rule of streamlining and reduction of virus genome sizes. Some well-known bacteriophages (e.g., the family *Myoviridae*, such as T4) were shown many years ago to have relatively large genomes, up to 170 kbp. However, quite recently there has been the discovery of "megaphages," with particles 200–300 nm in diameter and dsDNA genomes around 550 kbp in size, that infect gut bacteria in the genus *Prevotella* in primates, humans, and pigs. The viruses encode a suite of their own tRNAs, make use of an alternative genetic code to the normal cellular version, and some even encode a **gene editing enzyme called CasΦ**, a novel component of a CRISPR-Cas antiviral defense system used by the virus for superinfection exclusion or superinfection immunity at the cellular level, and incidentally to defend the host against other viruses (see Chapter 6).

Mimiviruses and the distantly related **megaviruses** (order *Imitervirales*) are amoeba-infecting discoveries of the early 2000s. These dsDNA-containing isometric virions are around 400 nm in diameter and are covered with protein filaments up to 100 nm long. The genome of mimivirus, at approximately 1.2 Mbp, contains around 1200 open reading frames (ORFs), only 10% of which show any similarity to proteins of known function. Pandoraviruses, also isolated from amoebae, have oval virions around 1 μm in length, and dsDNA genomes of around 2.5 Mbp in size. These encode up to 2500 proteins, with around 93% of these not shared with any other organisms. It is notable that all these large eukaryote virus genomes contain many genes involved in their own replication, particularly enzymes concerned with nucleic acid metabolism. These viruses partially escape the restrictions imposed by the biochemistry of the host cell by encoding additional biochemical equipment. Viromics scans of large viruses of eukaryotes found in aquatic environments—all belonging to the **nucleocytoplasmic large DNA viruses (NCLDV)** supergroup, as do all the giant viruses—turned up thousands of virus genomes encoding genes involved in the cellular uptake and transport of a wide variety of substrates, in the translation machinery, and genes involved in photosynthesis. Many of these functions could serve to give a competitive advantage to the host cells, meaning that the viruses are more **commensals** than predators. This could be a major factor in their survival in ecosystems that are dominated by protists and resource-poor, such as seawater. It is thought that currently-known giant viruses evolved from smaller ancestral viruses with far fewer genes, and expanded their genomes via gene duplications and horizontal gene transfer from host cells and from coinfecting viruses.

Molecular genetics

As already described, the techniques of molecular biology have been a major influence on concentrating much attention on the virus genome—and it is worth taking some time here to illustrate how some of these techniques have been applied to virology, remembering that these newer techniques are complementary to and do not replace the classical techniques of virology. Initially, any investigation of a virus genome will usually include questions about the following:

- Composition—DNA or RNA, single-stranded or double-stranded, linear, or circular
- Size and number of segments
- Nucleotide sequence
- Genome terminal structures if linear
- Genome coding capacity—open reading frames, ORFs
- Regulatory signals—transcription **enhancers**, **promoters**, and terminators

It is possible to separate the molecular analysis of virus genomes into two types of approach: these are the physical analysis of structure and nucleotide sequence, essentially performed in vitro, and a more biological approach to examine the structure-function relationships of intact virus genomes and individual genetic elements, usually involving analysis of the virus phenotype in vivo.

The conventional starting point for the physical analysis of virus genomes was generally the isolation of nucleic acids from virus preparations of varying degrees of purity (see Chapter 1). More recently, the emphasis on extensive purification has declined as techniques of molecular cloning and amplification and direct sequencing have become much more advanced, allowing direct examination of viral genomes in complex mixtures of sequences without the need for prior purification. The use of **viral transcriptomics**—sequencing all transcripts from virus-infected cells (see Chapter 1)—is another rapidly-expanding field of investigation, as the power of next-gen sequencing allows very detailed analysis of how viruses and cells interact at the molecular level.

Phenotypic analysis of virus populations has long been a standard technique of virology. In modern terms, this might be considered as functional genomics. Examination of variant viruses and naturally occurring spontaneous mutants is an old method for determining the function of virus genes. Molecular biology has added to this the ability to design and create specific mutations, deletions, and recombinants in vitro. This **site-directed mutagenesis** is a very powerful tool. At one time genetic coding capacity of a viral genome could be examined in vitro by the use of cell-free extracts to translate mRNAs. Later, it was easier to express cloned fragments of genomes in appropriate cells, and analyses products by a variety of techniques, the simplest among these being **SDS-PAGE fractionation** and **western blotting** (see Chapter 1). Complete functional analysis of virus genomes can, however, only be performed on intact viruses, infecting appropriate host cells. Even here, though, the techniques of modern molecular biology can assist: the relative simplicity of most virus genomes compared with even the simplest cell allows us to "rescue" infectious virus from purified or cloned nucleic acids. Indeed, many virus genomes—including the whole **horsepox virus** genome of 212 kb (see Noyce et al., 2018)—have now been de novo synthesized in vitro and shown to infect cells. In some cases for simpler viruses, it has been shown that even cell extracts can support a complete virus replication cycle: David Baltimore and colleagues did this for poliovirus RNA as early as 1964; in 1991 Eckard Wimmer and his group both synthesized infectious poliovirus RNA in vitro from archived sequence data, and showed that it could be replicated and infectious particles made in HeLa cell lysates.

Virus genetics

Although nucleotide sequencing now dominates the analysis of virus genomes, functional genetic analysis of viruses was based largely on the isolation and analysis of mutants, naturally occurring or made chemically or by use of radiation, usually achieved using plaque purification (**biological cloning**) for viruses that could be cultured in this way. This included bacterial, yeast, insect, and other animal viruses. Some plant viruses could be similarly analyzed using local lesion assays. An impressive amount of genetic analysis was done on viruses from the earliest days of cell and other tissue and whole plant culture, starting with bacterial viruses after 1918, continuing through plant viruses from the 1920s, the use of chicken eggs for animal virus assays from the 1930s, and culminating in mammalian and insect cell cultures from the 1950s (see Chapter 1).

In the case of viruses for which no such systems exist—because they are not cytopathic, or do not replicate in culture, or did not cause local lesions—little genetic analysis was possible before the development of molecular genetics, whereafter progress was extremely rapid in many cases. This was particularly the case for **nonculturable viruses** such as **hepatitis B and C viruses** (HBV and HCV), and human and other **papillomaviruses** (HPVs) (see Chapter 1). In the case of HCV, the disease was originally termed "non-A, non-B hepatitis (NANBH)" from epidemiological and transmission studies, and the ssRNA+ virus was discovered as a result of screening a **phage lambda gt11 cDNA expression library** in *E. coli* with sera from an infected patient. This picked up a fragment of what was later found to be nonstructural protein 4: use of this cDNA segment for **molecular hybridization studies** on NANBH-infected samples picked up a ~10 kb ssRNA+, the sequence of which was distantly related to known **flaviriruses**. This and the epidemiological work that led up to it, and the proof that cloned virus genomes were infectious, was rewarded by a joint Nobel Prize in Physiology or Medicine in 2020.

However the genetics of viruses was studied, certain common concepts have been derived. An important one, which in the case of **HPVs** was initially arrived at by means of DNA:DNA hybridization studies (see Fig. 1.24), was that virus genomes derived from patient samples could be divided into distinct "**types,**" based on their degree of sequence homology as assessed by annealing efficiency. This is true of many groups of viruses, and was the basis of a **homology-based taxonomy** that was considerably in advance of similar efforts with plants and animals. In most cases these relationships were verified by partial sequencing of cloned and then later in vitro amplified DNA and cDNA fragments, and more recently by routine whole genome sequencing, including of whole viromes obtained from the vaginal tract in humans.

Another concept that was in part elicited by ingenious use of coinfection in cell cultures and recombination studies—or **reassortant** studies in the case of viruses with divided genomes—was **genetic mapping by biological cloning** (e.g., by plaque selection) of genetic variants: these could be naturally occurring, or resulting from repeated passage of viruses, or generated by use of chemicals and radiation. In any case, the use of the biology of virus isolates to characterize them—observing how the **genome determines the phenotype**—has become fundamental to our understanding of virus-host relationships, and of the **determinants of virulence, host range,** and **transmission**. To this end, it is useful to clear up some terminology used to describe viruses so as to avoid confusion—given that the terms do have specific meanings, and their misuse can lead to serious misunderstandings.

Virus variants

"Mutant," "strain," "type," "variant," and even "isolate" are all terms used rather loosely by virologists to differentiate particular viruses from each other and from the original "parental," "wild-type," or "street" isolates of that virus, the latter term frequently used in the case of rabies virus. More accurately, these terms are generally applied as follows:

- **Isolate**: a biological isolation of a virus from a disease case.
- **Mutant**: a virus whose genome is known to contain one or a very few nucleotide changes compared to a "type" isolate.
- **Variant**: a virus whose genotype differs in multiple locations from the original wild-type strain, for which there may be a phenotypic difference established—for example, in a new clinical isolate from a patient.
- **Strain**: different lines of the same virus (e.g., from different geographic locations or patients), usually with significant sequence and biological differences.
- **Type**: Different serotypes of the same virus (e.g. various antibody neutralization phenotypes that do not generally elicit cross-neutralizing antibodies).

Mutant viruses can arise spontaneously, generally as a result of copying errors during replication, or they can be induced by use of radiation or chemicals, or they can be artificially constructed. In some viruses, natural mutation rates may be as high as **10^{-3} to 10^{-4} per incorporated nucleotide** (e.g., in retroviruses such as HIV; ssDNA geminiviruses such as MSV), whereas in others they may be as low as **10^{-8} to 10^{-11}** (e.g., in herpesviruses), which is similar to the upper end of mutation rates seen in cellular DNA for simple eukaryotes. These differences are due to the mechanisms of genome replication, with error rates in RNA-dependent RNA polymerases (RdRp) generally being significantly higher than DNA-dependent DNA polymerases, because they mostly do not have **proofreading functions** like DNA polymerases do, which correct copying errors as they happen. RdRps of coronaviruses such as **SARS-CoV-2** do have proofreading functions—meaning coronaviruses mutate about **4× less than influenza viruses**—but in general mutation rates are higher in most RNA viruses than in most DNA, and especially dsDNA viruses.

For a virus, mutations are a mixed blessing. The ability to generate antigenic variants that can escape immune responses is a clear advantage, but mutation also results in many defective genomes, as most mutations are deleterious. In the most extreme cases (e.g., HIV), the error rate is 10^{-3} to 10^{-4} per nucleotide incorporated per replication cycle. This means the overall rate of nucleotide substitution in HIV is approximately **1 million times higher** than that of the human genome. The HIV genome is approximately 10kb long; therefore, there will be **1–10 mutations in every genome copied**. This means that the wild-type virus actually consists of a fleeting majority type that dominates the dynamic equilibrium (i.e., the population of genomes) present in all cultures of the virus, and that the rest of the sequences are a mixture of mutants/variants. These mixtures of molecular variants are known as a ***quasispecies***, and also occur in other RNA viruses (e.g., picornaviruses) as well as with ssDNA viruses (e.g., geminiviruses), because of the lack of proofreading by the host repair DNA polymerases that are involved in rolling circle replication of ssDNA viruses (see Chapter 4). However, the **majority of these variants will be noninfectious** or seriously disadvantaged, and are therefore rapidly weeded out of a replicating population. This mechanism is an important force in virus evolution. It is noteworthy that there were only 4–10 mutations accumulated for all the sequenced SARS-CoV-2 genomes that infected people in the United States in mid-2020, compared to the original virus found in Wuhan months earlier: thus, only a small proportion of the 24 mutations that could have been predicted to occur from the mutation rate, were in fact viable.

One penalty of having a quasispecies population of virus genomes with many of them being defective is that many virions may be needed to initiate an infection. Zika virus—an ssRNA+ mosquito-transmitted flavivirus—seems to have evolved to infect cells as aggregates of virions: a recent study found that individual plaques in cultured cells contained **10 infecting viral genomes** on average. Few plaques contained more than two dominant genomes, indicating most genomes in an aggregate were defective. However, these may still have a function in terms of assisting replication of other defective mutants, and having their replication facilitated by wild-type or replication-competent virus genomes (see Fig. 3.14). The same has been shown for plant viruses via local lesion assays (Chapter 1).

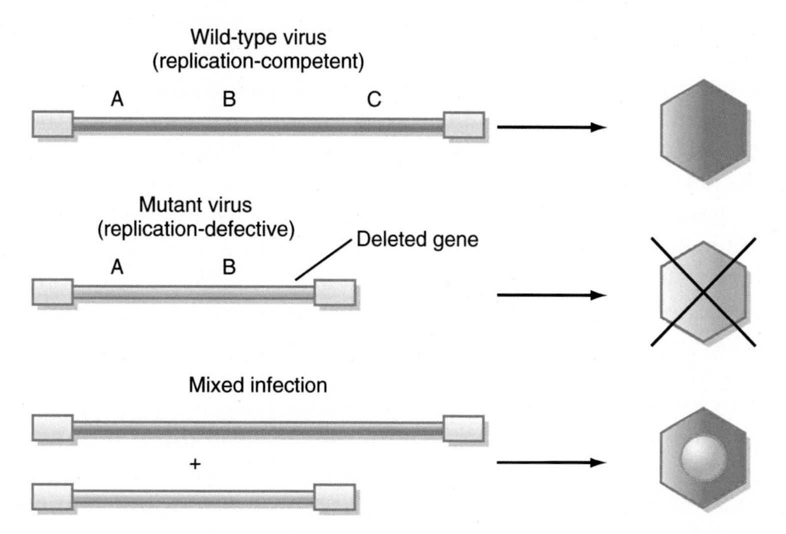

FIG. 3.14 Helper viruses. Helper viruses are replication-competent viruses that are capable of rescuing replication-defective genomes in a mixed infection, permitting their multiplication and spread.

A mistake that is sometimes made about especially RNA viruses involves conflating mutation with evolution: while most RNA and ssDNA genome viruses are subject to high mutation rates, this does not necessarily mean they evolve or change their general properties faster—it simply means that particular properties, such as antigenicity, can change fast. In fact, it has been pointed out that the percent genome sequence difference between HIV-1 and HIV-2 (60%–70%), which possibly diverged some **30,000 years ago**, is similar to that between all animal lineages since the **Cambrian explosion** period (~**500 million years ago**). However, this has led to no changes in morphology or gene function—unlike the phenotypic differences between jellyfish and humans, for example.

Historically, before the advent of modern molecular biological techniques, most genetic analysis of viruses was performed on virus mutants isolated from either from mutagen-treated populations, or which had occurred spontaneously. Currently, defined and highly specific mutations in particular genes or control sequences (e.g., single-nucleotide changes, specific insertions or deletions) are usually introduced using molecular methods such as PCR or one of the cutting-edge gene engineering methods such as **CRISPR/Cas**. This is known as **site-directed or site-specific mutagenesis**, and is far more precise and accurate than the older methods of mutant generation.

The disadvantage of simply determining nucleotide sequences as a primary tool for studying viruses is that the sequence alone frequently says little about the function of gene, apart from what can be guessed from comparisons with similar genes in other organisms. To understand how the virus actually works, **functional analysis** is required—and this is where mutations, whether naturally occurring or artificially introduced, are invaluable. The phenotype of a mutant virus—which in most cases means its biological properties—depends on the type of mutation(s) it has, and also upon the location of the mutation(s) within the genome. Each of the classes of mutations can occur naturally in viruses or may be artificially induced for experimental purposes:

- **Biochemical markers**: These include drug resistance mutations, mutations that result in altered virulence, polymorphisms resulting in altered electrophoretic mobility of proteins or nucleic acids, and altered sensitivity to inactivating agents.
- **Deletions**: Similar in some ways to nonsense mutants (below) but may include one or more virus genes and involve noncoding control regions of the genome (**promoters**, etc.). Spontaneous deletion mutants often accumulate in virus populations as defective-interfering (DI) particles, with significantly reduced genomes. These noninfectious but not necessarily genetically inert genomes are thought to be important in establishing the course and pathogenesis of certain virus infections (see Chapter 6). Genetic deletions can only revert to wild-type by **recombination**, which usually occurs at comparatively low frequencies, or **reassortment**, for viruses with multicomponent genomes such as reoviruses. Deletion mutants are very useful for assigning structure-function relationships to virus genomes, as they are easily mapped by physical analysis.
- **Host range**: This term can refer either to whole animal hosts or to permissive cell types in vitro. **Conditional mutants** of this class have been isolated using **amber-suppressor cells**—mostly for phages but also for animal viruses using in vitro systems—where "amber" stop codons in viral genomes are suppressed, so as to allow conditional expression (see below).

- **Nonsense**: These result from alteration of coding sequence of a gene to one of three translation stop codons (UAG, amber; UAA, ocher; UGA, opal). Translation is prematurely terminated in comparison to the wild-type gene, resulting in the production of an amino-terminal fragment of the protein. The phenotype of these mutations can be suppressed by propagation of virus in a cell (bacterial or, more recently, animal) with altered suppressor tRNAs. Nonsense mutations are rarely "leaky" (i.e., the normal function of the protein is completely obliterated) and can only revert to wild-type at the original site, so they usually have a low reversion frequency.
- **Plaque morphology**: This is a property that can be exploited for both prokaryote and eukaryote viruses, where these can be cultured in vitro. Mutants may be either large-plaque mutants which replicate more rapidly than the wild-type, or small-plaque mutants, which are the opposite. Plaque size is often related to a temperature-sensitive (t.s.) phenotype. These mutants are often useful as unselected markers in multifactorial crosses.
- **Temperature-sensitive (t.s.)**: This type of mutation is very useful as it allows the isolation of conditional-lethal mutations, a powerful means of examining virus genes that are essential for replication and whose function cannot otherwise be interrupted. Temperature-sensitive mutations usually result from mis-sense mutations in proteins (i.e., amino acid substitutions), resulting in proteins of full size with subtly altered conformation that can function at (permissive) low temperatures but not at (nonpermissive) higher ones. Generally, the mutant proteins are immunologically unaltered, which is frequently useful. These mutations are usually "leaky"—that is, some of the normal activity is retained even at nonpermissive temperatures. Protein function is often impaired, even at permissive temperatures, therefore a high frequency of reversion is often a problem with this type of mutation because the wild-type virus replicates faster than the mutant. This type of mutation can be useful in making vaccines: one live influenza A vaccine that cannot replicate outside of the upper respiratory tract was made this way, for example.
- **Cold-sensitive (c.s.)**: These mutants are the opposite of temperature-sensitive mutants and are very useful in bacteriophages and plant viruses whose host cells can be propagated at low temperatures, but are less useful for animal viruses because their host cells generally will not grow at significantly lower temperatures than normal.
- **Revertants**: Reverse mutation is a valid type of mutation in its own right. Most of the above classes can undergo reverse mutation, which may be either simple "back mutations" (i.e., correction of the original mutation) or second-site "compensatory mutations," which may be physically distant from the original mutation and not even necessarily in the same gene as the original mutation.
- **Suppression**: **Suppression** is the inhibition of a mutant phenotype by a second suppressor mutation, which may be either in the virus genome or in that of the host cell. This mechanism of suppression is not the same as the suppression of chain-terminating amber mutations by host-encoded suppressor tRNAs, which could be called "informational suppression." Genetic suppression results in an apparently wild-type phenotype from a virus which is still genetically mutant—a **pseudorevertant**. This phenomenon has been best studied in **prokaryotic** systems, but examples have been discovered in animal viruses—for example, reoviruses, vaccinia, and influenza—where

suppression has been observed in an attenuated vaccine, leading to an apparently virulent virus. Suppression may also be important biologically in that it allows viruses to overcome the deleterious effects of mutations and therefore be positively selected.

Mutant viruses can appear to revert to their original phenotype by three pathways:

1. **Back mutation** of the original mutation to give a wild-type genotype/phenotype (true reversion).
2. A second, **compensatory mutation** that may occur in the same gene as the original mutation, thus correcting it—for example, a second frameshift mutation restoring the original reading frame (intragenic suppression).
3. A **suppressor mutation** in a different virus gene or a host gene (extragenic suppression).

Genetic interactions between viruses

Genetic interactions between viruses often occur naturally, as host organisms are frequently infected with more than one virus. These situations were classically too complicated to be analyzed successfully; however, new nucleic acid sequencing and proteomics techniques have allowed significant advances in this area of research. Experimentally, genetic interactions can be analyzed by mixed infection (**superinfection**) of cells in culture. Two types of information can be obtained from such experiments:

- The assignment of mutants to functional groups known as **complementation** groups.
- The ordering of mutants into a linear genetic map by analysis of **recombination** frequencies.

Complementation results from the interaction of virus gene products during **superinfection**—or simultaneous infection with a high multiplicity of viruses, as in the case of Zika virus discussed above. This results in production of one or both of the parental viruses being increased, while both viruses remain unchanged genetically. In this situation, one of the viruses in a mixed infection provides a functional gene product for another virus which is defective for that function (Fig. 3.14). This kind of test allows functional analysis of unknown mutations if the biochemical basis of any one of the mutations is known. Complementation can be asymmetric—that is, only one of the mutant viruses is able to replicate. This can be an absolute or a partial restriction. When complementation occurs naturally, it is usually the case that a replication-competent wild-type virus rescues a replication-defective mutant. In these cases, the wild-type is referred to as a "**helper virus**," such as in the case of **defective transforming retroviruses** containing oncogenes (see Fig. 3.10 and Chapter 7).

Recombination is the generation of **hybrid genomes** with sequences derived from both of the two parents, resulting from mixing of virus genomes during superinfection. There are three mechanisms by which this can occur, depending on the nature and organization of the virus genome:

- **Intramolecular recombination via strand breakage and religation**: This process occurs in all DNA viruses and in RNA viruses that replicate via a DNA intermediate (retroviruses, pararetroviruses). It is believed to be caused by cellular enzymes, as no virus mutants with specific recombination defects have been isolated.

- **Intramolecular recombination by "copy-choice"**: This process occurs in both DNA and RNA viruses, probably by a mechanism in which the virus polymerase switches template strands during genome synthesis. There are cellular enzymes that could be involved (e.g., **splicing** enzymes), but this is unlikely and the process is thought to occur essentially as a random event. **Defective-interfering** (**DI**) particles in RNA virus infections are frequently generated in this way (see Chapter 6).
- **Reassortment**: In viruses with **segmented genomes**—such as **influenza** and other **myxoviruses**—the genome segments can be randomly shuffled during simultaneous or superinfection. Progeny viruses receive (at least) one of each of the genome segments, but probably not from a single parent. For example, influenza virus has eight genome segments; therefore, in a mixed infection, there could be a theoretical maximum of $2^8 = 256$ possible progeny virus genomes. Packaging mechanisms may be involved in generating reassortants.

Classically, intramolecular recombination was assessed by looking at the probability that breakage-reunion or strand-switching will occur between two specific markers in different genomes. This is proportional to the physical distance between them, and pairs of genetic markers were arranged on a linear map with distances measured in **"map units"** (i.e., percentage recombination frequency). This type of analysis has been very largely superseded in recent years by use of mutants generated by the new editing techniques, and next-gen sequencing.

Recombination occurs frequently in nature: for example, influenza virus reassortment has resulted in worldwide epidemics (pandemics) that have killed many millions of people (Chapter 6). Recombinations between **bat and other coronaviruses** have also probably contributed significantly to emergence of new coronaviruses capable of causing human disease (see Chapter 8). This makes these genetic interactions of considerable practical interest, and not merely an academic concern.

Reactivation is the generation of infectious (recombinant) progeny from noninfectious parental virus genomes. This process has been demonstrated in vitro and may be important in vivo. For example, it has been suggested that the rescue of defective, long-dormant HIV proviruses during the long clinical course of **acquired immune deficiency syndrome** (AIDS) may result in increased antigenic diversity and contribute to the pathogenesis of the disease. It is also a factor in **feline leukemia**, where exogenous infectious virus may reactivate latent endogenous viruses.

Nongenetic interactions between viruses

A number of nongenetic interactions can occur between viruses that can materially affect disease manifestation or even host range of viruses. **Eukaryotic** cells nearly all have a **diploid** genome with two copies of each chromosome, each bearing its own **allele** of the same gene. The two alleles on sister chromosomes may differ in many chromosomal locations, if the genome results from sexual reproduction, which involves the mixing of **haploid** genomes from two parent cell lines. This is known as **heterozygosity**. Among viruses only retroviruses are truly diploid, with two complete copies of the entire genome in every virion. Some DNA viruses, however, have long stretches of repeated sequences and are therefore partially heterozygous: good examples here are herpesviruses and poxviruses. In a few (mostly enveloped) viruses, aberrant packaging of multiple genomes may occasionally result in **multiploid** particles that are heterozygous (e.g., up to 10% of the ssRNA− Newcastle disease virus virions). This process is known as **heterozygosis,** and can contribute to the genetic complexity of virus populations.

Another nongenetic interaction between viruses that is commonly seen is **interference**. This process results from the resistance to superinfection in cells already infected by another virus. Many cases of interference in natural situations are mediated by molecules known as **interferons**, produced by cells in response to virus infection (see Chapter 6). **Homologous interference** (i.e., against the same virus) in eukaryotic cells often results from **defective interfering (DI) particles,** which compete for essential cell components and block replication of the wild-type genome. Plant ssRNA+ viruses often block reinfection of the same cells or even infection of neighboring cells via ***siRNA*-mediated genome degradation**, or by means of virus-specific proteins interfering with various stages of replication or spread.

Interference can also result from other types of mutation (e.g., dominant temperature-sensitive mutations), or by mechanisms such as **sequestration of virus receptors** due to the production of virus-attachment proteins by viruses already present within the cell (e.g., in the case of avian retroviruses).

Phenotypic mixing in virions (see Fig. 3.15) can vary from extreme cases, where the genome of one virus is completely enclosed within the capsid or envelope of another (called "**pseudotyping**"), to more subtle cases where the capsid/envelope of the progeny contains a **mixture of proteins** from both viruses. This mixing gives the progeny virus the phenotypic properties (e.g., types of cell infected or cell tropism, or even transmission to new hosts) dependent on the proteins incorporated into the particle, without any genetic change. Subsequent generations of viruses inherit and display the original parental phenotypes, so infections could be single-cycle. This process can occur easily in viruses with non-enveloped or naked capsids that are closely related (e.g., different strains of mammalian enteroviruses, or plan bromo- or geminiviruses), or in enveloped viruses, which need not be related to one another (Fig. 3.11). In this latter case, the phenomenon is due to the non-specific incorporation of different virus glycoproteins into the virion envelope, resulting in a mixed phenotype. Rescue of replication-defective ***transforming retroviruses*** by a helper virus is a form of pseudotyping.

Phenotypic mixing or pseudotyping has proved to be a very useful tool to examine biological or immunological properties of viruses, without having to work with what could be very dangerous pathogens. **Vesicular stomatitis virus (VSV)** readily forms pseudotypes containing retrovirus envelope glycoproteins, giving a plaque-forming virus with the properties of VSV but with the cell tropism of the retrovirus. This trick has been used to study the cell tropism of HIV and other retroviruses. It is also possible to construct **pseudoviruses** where a tissue culture-adapted harmless virus such as VSV or even engineered retroviruses such as HIV are used to display envelope proteins of nonculturable or highly pathogenic viruses: this is useful in studying interaction of antibodies and viruses by **pseudovirion neutralization assays**, for example, which is extensively used in the study of vaccine candidates for HIV, and for studying **SARS-CoV-2**.

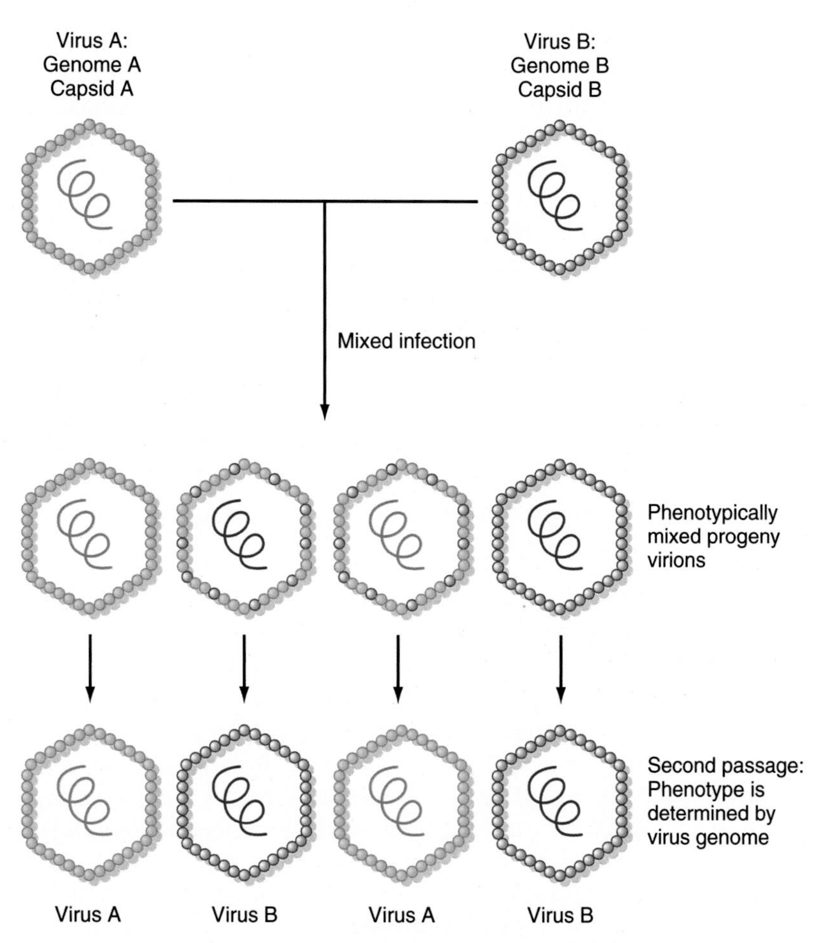

FIG. 3.15 Phenotypic mixing of viruses by capsid protein exchange. Phenotypic mixing occurs in mixed infections, resulting in genetically unaltered virus particles that have some of the properties of the other parental type due to sharing a capsid.

The global virome

Vast number of virus particles populate this planet. Our best estimate is that there are at least **10^{31} virus particles on Earth**, and that may have to be revised upward as viruses that often escape metagenomic scans (e.g., ssRNA and ssDNA viruses) become included. Viruses outnumber bacteria 10 to 1 in most ecosystems, including in our own bodies. Years ago we abandoned the idea that viruses are only associated with diseases: we have known that virus infection is much more prevalent than that—and indeed, is probably universal—for decades. But as the powerful techniques of molecular biology described in Chapter 1 have matured, we have come to realize how complex and abundant the virus world—the **global virome**—really is. This knowledge has resulted in the development of a new concept in recent years: this is the concept of the virome, or all of the viruses in a given environment. For example, the environment could be that of an entire ocean, of an isolated lake, or within particular locations of the human body. Our knowledge of viromes is possibly the fastest expanding area of virology, with the inexorable logarithmic annual increase of candidate genome sequences that become available as a result of the ever-evolving next-generation sequencing tools (see Chapter 1). Recent work has for instance established that 2 major groups of cultured microbial algae have 18 novel RNA viruses. These included members of the ssRNA+ *Tombusviridae* as well as dsRNA members of the family *Amalgaviridae*, previously known only from land plants. Viruses similar to the dsRNA fungus-infecting mitoviruses were also found, suggesting a horizontal transfer between algae and fungi similar to what happened with cryptovirus transfer between fungi and land plants. A highly novel finding was of a ssRNA– bunya-like virus in microalgae: these have hitherto been supposed to only occur in land plants as a result of transfer from insects. The authors made the comment that

> *"…these data suggest that the scarcity of RNA viruses in algae results from limited investigation rather than their absence".*

This is also undoubtedly true for most environments. Indeed, a landmark 2022 preprint (see He et al., 2022) that looked at 1941 game animals in China—a meta-transcriptomic survey from live animal samples—found 102 mammal-infecting viruses, 65 of them novel, and 21 considered to be of high risk to humans and domestic animals. Another example is the paper on diversity and evolution of the terrestrial animal virome (Harvey and Holmes, 2022). Given that the volume of information on global viromes is expanding so fast, however, it is perhaps better to give examples of further reading and links to useful sites, than to attempt to capture a quickly obsolete snapshot of a fast-moving field (see Recommended Reading).

Epidemiology

Epidemiology is concerned with the distribution of disease and the developing strategies to reduce or prevent it. Virus infections present considerable difficulties for this process. Except for **epidemics** where acute symptoms are obvious, the major evidence of virus infection available to the epidemiologist is the **presence of antivirus antibodies** in patients. This information frequently provides an incomplete picture, and it is often difficult to assess whether a virus infection occurred recently or at some time in the past. Techniques such as the isolation of viruses in experimental plants or animals are laborious and impossible to apply to large populations. Molecular biology provides sensitive, rapid, and sophisticated techniques to detect and analyze the genetic information stored in virus genomes and has resulted in a new area of investigation: **molecular epidemiology**, informed by our growing knowledge of the human and other viromes.

Knowledge drawn from taxonomic relationships allows us to predict the properties and behavior of new viruses or to develop drugs based on what is already known about existing viruses. It is believed these shared patterns suggest the **descent of present-day viruses from a limited number of ancestors**, with significant mixing of replication modules and multiple independent acquisitions of structural proteins. Although it is tempting to speculate on events that may have occurred before the origins of life as it is presently recognized, it would be unwise to discount the pressures that might result in viruses with diverse origins assuming **common genetic solutions** to common problems of storing, replicating, and expressing genetic information. This is particularly true now that we appreciate the plasticity of virus and cellular genomes and the mobility of genetic information from virus to virus, cell to virus, and virus to cell. There is no reason to believe that virus evolution has stopped, and it is dangerous to do so. The practical consequences of ongoing evolution and the concept of **emergent viruses** are described further in Chapter 8.

Summary

Molecular biology has put much emphasis on the structure and function of the virus genome. At first sight, this tends to emphasize the tremendous diversity of virus genomes. On closer examination, similarities and unifying themes become more apparent. Sequences and structures at the ends of virus genomes are in some ways functionally more significant than the unique coding regions within them. Common patterns of genetic organization seen in virus superfamilies and orders suggest either that many viruses have evolved from common ancestors, or that exchange of genetic information between viruses has resulted in common solutions to common problems. Increasingly, it is clear that there is much to learn about virus-cell interactions, and the profound effects of coevolution of viruses and their hosts—including exaptation of genes in both directions.

References

Beck, J., Seitz, S., Lauber, C., Nassal, M., 2021. Conservation of the HBV RNA element epsilon in nackednaviruses reveals ancient origin of protein-primed reverse transcription. Proc. Natl. Acad. Sci. U. S. A. 118 (13), e2022373118. https://doi.org/10.1073/pnas.2022373118. 33753499. PMCID: PMC8020639.

Boso, G., Fleck, K., Carley, S., Liu, Q., Buckler-White, A., Kozak, C.A., 2021. The oldest co-opted gag gene of a human endogenous retrovirus shows placenta-specific expression and is upregulated in diffuse large B-cell lymphomas. Mol. Biol. Evol. 38 (12), 5453–5471. https://doi.org/10.1093/molbev/msab245. 34410386. PMCID: PMC8662612.

Harvey, E., Holmes, E.C., 2022. Diversity and evolution of the animal virome. Nat. Rev. Microbiol. https://doi.org/10.1038/s41579-021-00665-x.

He, W.-T., Hou, X., Zhao, J., Sun, J., He, H., Si, W., Wang, J., Jiang, Z., Yan, Z., Xing, G., Lu, M., Suchard, M.A., Ji, X., Gong, W., He, B., Li, J., Lemey, P., Guo, D., Changchun, T., Holmes, E.C., Shi, M., Su, S., 2022. Virome characterization of game animals in China reveals a spectrum of emerging pathogens. Cell. https://doi.org/10.1016/j.cell.2022.02.014. ISSN 0092-8674.

Inoue, Y., Saga, T., Aikawa, T., et al., 2017. Complete fusion of a transposon and herpesvirus created the *Teratorn* mobile element in medaka fish. Nat. Commun. 8, 551. https://doi.org/10.1038/s41467-017-00527-2.

Katzourakis, A., Gifford, R.J., 2010. Endogenous viral elements in animal genomes. PLoS Genet. 6 (11), e1001191. https://doi.org/10.1371/journal.pgen.1001191.

Koonin, E.V., Dolja, V.V., Krupovic, M., 2015. Origins and evolution of viruses of eukaryotes: the ultimate modularity. Virology 479–480, 2–25. https://doi.org/10.1016/j.virol.2015.02.039.

Krupovic, M., Dolja, V.V., Koonin, E., 2019. Origin of viruses: primordial replicators recruiting capsids from hosts. Nat. Rev. Microbiol. 17, 450–458.

Noyce, R.S., Lederman, S., Evans, D.H., 2018. Construction of an infectious horsepox virus vaccine from chemically synthesized DNA fragments. PLoS One 13 (1), e0188453. https://doi.org/10.1371/journal.pone.0188453.

Wolf, Y.I., Kazlauskas, D., Iranzo, J., Lucía-Sanz, A., Kuhn, J.H., Krupovic, M., Dolja, V.V., Koonin, E.V., 2018. Origins and evolution of the global RNA virome. MBio 9 (6), e02329-18. https://doi.org/10.1128/mBio.02329-18. 30482837. PMCID: PMC6282212.

Recommended reading

Aziz, R.K., Breitbart, M., Edwards, R.A., 2010. Transposases are the most abundant, most ubiquitous genes in nature. Nucleic Acids Res. 38 (13), 4207–4217. https://doi.org/10.1093/nar/gkq140.

Campbell, S., Aswad, A., Katzourakis, A., 2017. Disentangling the origins of virophages and polintons. Curr. Opin. Virol. 25, 59–65. https://doi.org/10.1016/j.coviro.2017.07.011. Epub 2017 Aug 9 28802203.

Delwart, E., 2013. A roadmap to the human virome. PLoS Pathog. 9 (2), e1003146.

Domingo, E., Holland, J.J., 2010. The origin and evolution of viruses. In: Topley & Wilson's Microbiology and Microbial Infections. John Wiley & Sons.

Forterre, P., Prangishvili, D., 2009. The great billion-year war between ribosome- and capsid-encoding organisms (cells and viruses) as the major source of evolutionary novelties. Ann. N. Y. Acad. Sci. 1178 (2009), 65–77.

Krupovic, M., Koonin, E.V., 2017. Multiple origins of viral capsid proteins from cellular ancestors. Proc. Natl. Acad. Sci. U. S. A. 114 (12), E2401–E2410. https://doi.org/10.1073/pnas.1621061114. Epub 2017 Mar 6 28265094. PMCID: PMC5373398.

Legendre, M., Bartoli, J., 2014. Thirty-thousand-year-old distant relative of giant icosahedral DNA viruses with a pandoravirus morphology. Proc. Natl. Acad. Sci. U. S. A. 111 (11), 4274–4279.

Nguyen, M., Haenni, A.L., 2003. Expression strategies of ambisense viruses. Virus Res. 93, 141–150.

CHAPTER

4

Virus genomes and their replication

INTENDED LEARNING OUTCOMES

On completing this chapter, you should be able to:

- Discuss the different virus genome types and organizations, and how these may be grouped in seven replication strategies.
- Explain the phases of virus replication.
- Discuss how these processes may be blocked in order to combat virus infections.

Overview of virus replication

Understanding the details of virus replication is very important. This is not just for academic reasons, but also because this knowledge often provides the key to fighting virus infections. Once virus genomes enter cells, they need to be **expressed**, and then **replicated**. In prokaryotes this commences as soon as the genome—and accompanying proteins, if any—enters the cell. In eukaryotes this may also happen, if the virus replicates in the cytoplasm—however, in the case of most DNA viruses and even some RNA viruses, this involves **transport to the cell nucleus**. In all cases, soon after infection **the intact virion ceases to exist**, and the **virus is no longer infectious**—unless the genome alone is infectious, which is only true for a few types of virus (see Chapter 2). The old terms "**eclipse phase**" or "**latent period**" describe that part of a virus life cycle when **no infectious virus can be extracted** from cells which had just been exposed to infectious virions. An excellent illustration of the concept in terms of a virus assay experiment is shown by the Ellis and Delbrück single-burst experiment first published in 1939 (Fig. 1.14).

Virus replication can be divided into a number of stages, as shown in Fig. 4.1. These are purely arbitrary steps, used here for convenience in explaining the replication cycle of a non-existent "typical" virus. Regardless of their hosts, **all viruses must undergo each of these stages in some form to successfully complete their replication cycles**. Not all the steps described here are detectable as distinct stages for all viruses—often they blur together and appear to occur almost simultaneously. Some of the individual stages for some model viruses have been studied in great detail, and a tremendous amount of information is known about them. Other stages, and in particular, those of viruses that cannot be easily cultured in vitro,

Cann's Principles of Molecular Virology
https://doi.org/10.1016/B978-0-12-822784-8.00004-0

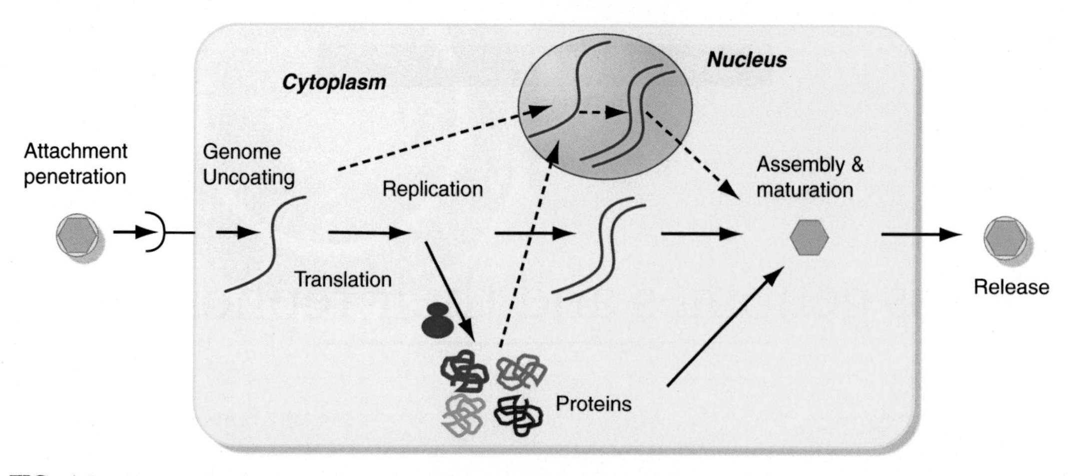

FIG. 4.1 A generalized scheme for virus replication. This diagram shows an outline of the steps which occur during replication of a "typical" virus. For prokaryotic viruses, all steps take place in the cytoplasm. For eukaryotic viruses, replication usually occurs in the nucleus for viruses with DNA genomes and in the cytoplasm for viruses with RNA genomes, although there are exceptions to both rules.

have been much more difficult to study, and considerably less information is available. Most of this chapter will examine the process of virus replication, and will look at some of the variations on the basic theme with examples of the virus genomes involved. Essentially, though, the process can be broken down into **entrance, expression,** and **replication** of the viral genome, and **exit of assembled virions** at the end of the process.

Entrance into cells and viral genome replication

The essential stages of all virus replication cycles can be reduced to the following (see Fig. 4.1, Box 4.1):

- Attachment to host cells and penetration into cells
- Partial or complete uncoating of the genome inside cells
- Expression of the genome
- Genome replication
- Assembly and maturation of virions
- Release of virions

Attachment of virions to cells

Viruses require to enter cells in order for their genomes to be expressed and replicated. The way this happens depends largely upon the **type of the cell, and of the virus**: the **cell type** has a great deal of influence on the strategy the virus uses to gain access; in turn, **specific virus types** may employ **different strategies to gain access to the same cell type**. However, the greatest

BOX 4.1

Entrance, entertainment, and exit

I have used the meme "**Entrance, Entertainment, and Exit**"—which happens to be the title of three Acts of a work entitled "The Grand Vizier's Garden Party" on the 1969 album "Ummagumma" by Pink Floyd—for the following reasons. First, students still know who Pink Floyd is/are, so they remember it better. Second, because it is a very simple encapsulation of the process. Third, because it neatly separates three crucial aspects of the virus life cycle. Fourth, it gives you the opportunity to describe three very different kinds of strategy for interfering with said life cycle.

Thinking of the fourth point, and just for HIV for example, those would be:
- entry inhibitors, like antibodies or fusion inhibitors,
- nucleoside analogue or nonnucleoside reverse transcriptase inhibitors, and
- protease inhibitors to prevent the polyprotein processing necessary for regulatory protein cleavage and particle assembly (see Chapter 5).

Just to say that it helps make virology fun. At least for me.

commonality in strategy is probably observed between **viruses infecting a single broad type of host**, defined by the nature of their cell walls. These may be defined as (see also Fig. 4.5):

- **bacterial** (e.g., bacteria, archaea; **rigid cell walls**),
- **animal** (e.g., all animal cells have only membranes; **no cell walls**), and
- **plant-like** (e.g., green multicellular plants, algae, microalgae, fungi; **thick rigid cell walls**).

Arguably, it is most logical to consider the first interaction of a virus with a new host cell as the starting point of the cycle. Technically, virus **attachment** consists of specific binding of a **virus-attachment protein** to a cellular **receptor** molecule. Many examples of virus receptors are now known (see Fig. 4.2 and Recommended Reading). However, the one factor that unifies all virus receptors is that **they did not evolve and are not manufactured by cells to allow viruses to enter cells**—rather, viruses have subverted molecules required for normal cellular functions.

Virus receptors fall into many different classes. For eukaryote cells, these include immunoglobulin-like superfamily molecules, membrane-associated receptors, and transmembrane (TM) transporters and channels. For prokaryotes, these receptor sites may be lipopolysaccharides, cell wall proteins, teichoic acid, or flagellar or pilus proteins. The target receptor molecules on eukaryote cell surfaces may be **proteins** (usually glycoproteins) or the **carbohydrate structures** (glycans) present on **glycoproteins** or **glycolipids**. The former are usually specific receptors in that a virus may use a particular protein as a receptor. Carbohydrate groups are usually less specific, because the same configuration of sugar side-chains may occur on many different **glycosylated** membrane-bound molecules: a good example here is the **sialic acid**—more properly, ***N-acetylneuraminic acid*** (Neu5Ac or NANA)—found in animals and some prokaryotes. Some complex viruses (e.g., **poxviruses, herpesviruses**) use more than one receptor, and therefore have alternative routes of uptake into cells.

Plant viruses face special problems initiating infection. The outer surfaces of plants are composed of protective layers of waxes and pectin, but more significantly, each cell is surrounded by a thick wall of cellulose overlying the cytoplasmic membrane, that is considerably thicker (200–300nm) than the dimensions of most virions. To date, no plant virus is known to use a specific cellular receptor of the type that animal and bacterial viruses use to attach to cells. Instead, plant viruses rely on a **breach of the integrity of a cell wall to introduce a virion directly into a host cell**. This is achieved either by the **piercing mouthparts** of the vector associated with transmission of the virus (e.g., aphids, whiteflies, nematodes, leafhoppers), or simply by **mechanical damage to cells** (e.g., beetles' jaws; pruning instruments) that transiently exposes the cell contents to virions. After replication in an initial cell, the lack of receptors poses further problems for plant viruses in recruiting new cells to the infection: These issues are discussed in Chapter 6.

Some of the best understood examples of virus-receptor interactions are from the non-enveloped ssRNA+ **picornaviruses** (family *Picornaviridae*). The virus-receptor interaction in picornaviruses has been studied intensively from the viewpoint of both the structural features of the virus responsible for receptor binding and those of the receptor molecule. The major **human rhinovirus** (HRV; causes the common cold) receptor molecule, **ICAM-1** (intercellular adhesion molecule 1 or CD54), is an **adhesion molecule** whose normal function is to **bind cells to adjacent structures**. Structurally, ICAM-1 is similar to an immunoglobulin molecule, with constant (C) and variable (V) domains homologous to those of antibodies and is regarded as a member of the **immunoglobulin superfamily** of proteins. Similarly, the poliovirus receptor, PVR, or CD155, is an integral membrane protein that is also a member of this family, with one variable and two constant domains which is involved in establishment of intercellular junctions between epithelial cells (Fig. 4.2).

Since the structure of a number of picornavirus capsids is known at a resolution of a few angstroms (Chapter 2), it has been possible to determine the features of the virus responsible for receptor binding. In HRVs, there is a deep cleft known as the "**canyon**" in the surface of each triangular face of the icosahedral capsid, which is formed by the flanking monomers, VP1, VP2, and VP3 (Fig. 4.3 and Fig. 2.9). Biochemical evidence from a class of inhibitory drugs that block attachment of HRV particles to cells indicates that the interaction between ICAM-1 and the virion occurs on the floor of this canyon. Unlike other areas of the virus

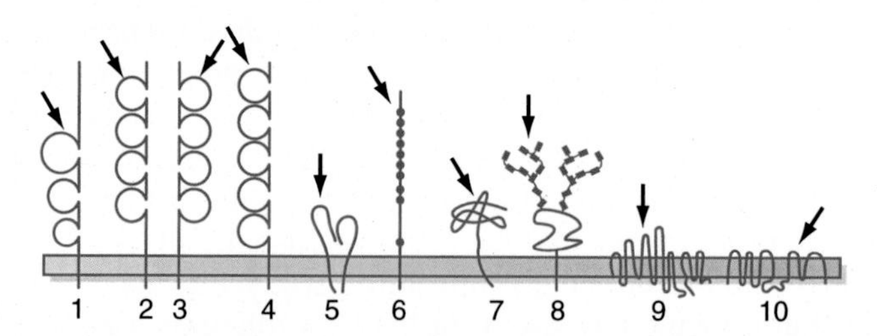

FIG. 4.2 Eukaryotic virus receptors. The arrows in this figure indicate approximate virus-attachment sites. (1) Poliovirus receptor PVR. (2) CD4: HIV receptor. (3) Carcinoembryonic antigen(s): MHV (coronavirus). (4) ICAM-1: most rhinoviruses. (Note that 1–4 are all immunoglobulin superfamily molecules.) (5) VLA-2 integrin: ECHO viruses. (6) LDL receptor: some rhinoviruses. (7) Aminopeptidase N: coronaviruses. (8) Sialic acid (on glycoprotein): influenza, reoviruses, rotaviruses. (9) Cationic amino acid transporter: murine leukemia virus. (10) Sodium-dependent phosphate transporter: Gibbon ape leukemia virus.

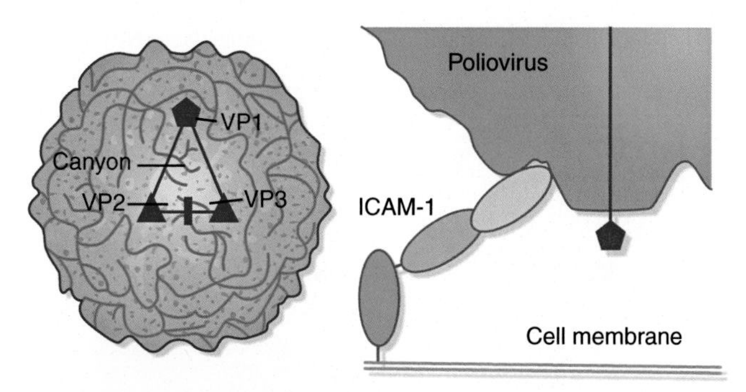

FIG. 4.3 Rhinovirus receptor binding. Rhinovirions have a deep surface cleft, known as the "canyon," between the three monomers (VP1, VP2, and VP3) making up each face of the particle.

surface, the amino acid residues forming the internal surfaces of the canyon are relatively **invariant**. It was suggested that these regions are protected from antigenic pressure because the antibody molecules are too large to fit into the cleft. This is important because radical changes here, although allowing the virus to escape an immune response, would **disrupt receptor binding**. Subsequently, it has been found that the binding site of the receptor extends well over the edges of the canyon, and the binding sites of neutralizing antibodies extend over the rims of the canyon. Nevertheless, the residues most significant for the binding site of the receptor and for neutralizing antibodies are separated from each other. In polioviruses, there is a similar canyon that runs around each fivefold vertex of the capsid. The highly variant regions of the capsid to which antibodies bind are located on the "peaks" on either side of this trough, which is again too narrow to allow antibody binding to the residues at its base. The invariant residues at the sides of the trough interact with the receptor (Fig. 4.3).

Even among picornaviruses, which are a structurally closely related family of viruses, there is considerable variation in receptor usage. Although 90 serotypes of HRV use ICAM-1 as their receptor, some 10 serotypes use proteins related to the low-density lipoprotein (LDL) receptor. **Encephalomyocarditis virus** has been reported to use the immunoglobulin molecule vascular cell adhesion factor or glycophorin A. Several picornaviruses use other ***integrins*** as receptors: Some **enteric cytopathic human orphan (ECHO) viruses** use VLA-2 or fibronectin, and **foot-and-mouth disease viruses** have been reported to use an unidentified integrin-like molecule. Other ECHO viruses use complement decay-accelerating factor (DAF, CD55), a molecule involved in complement regulation.

Another well-studied example of a virus-receptor interaction is that of enveloped ssRNA− **influenza viruses** (family *Orthomyxoviridae*). The influenza virus **hemagglutinin protein** (HA) forms one of the two types of glycoprotein spikes on the surface of the virions (see Chapter 2), the other type being formed by the **neuraminidase (NA)** protein. Each HA spike is composed of a trimer of three molecules, while the NA spike is a tetramer (Fig. 4.4). The HA spikes are responsible for binding the influenza virus receptor, which is **sialic acid** (*N*-acetyl neuraminic acid), a sugar group commonly found on a variety of glycosylated molecules in humans. As a result, **little cell-type specificity** is imposed by this receptor interaction, so influenza virions bind to a wide variety of different cell types (e.g., causing **hemagglutination** of red blood cells) in addition to the cells in which **productive infection** occurs.

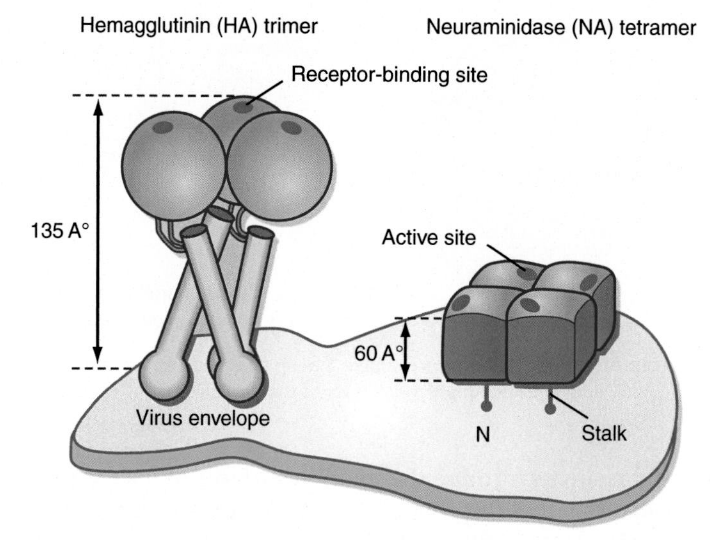

FIG. 4.4 Influenza virus glycoprotein spikes. The glycoproteins spikes on the surface of influenza virus (and many other enveloped viruses) are multimers consisting of three copies of the HA protein (trimer) and four copies of the NA protein (tetramer).

The NA molecule of influenza virus and HN protein of the distantly related **paramyxoviruses** (family *Paramyxoviridae*; e.g., measles and mumps viruses) illustrates another feature of this stage of virus replication. Attachment to cellular receptors is in most cases a reversible process—if **penetration** of the cells does not ensue, the virus can **elute or detach** from the cell surface. Some viruses have specific mechanisms for detachment, and the NA protein is one of these. NA is an **esterase** that **cleaves sialic acid from sugar side-chains** on glycoproteins. This is particularly important for influenza: because the receptor molecule is so widely distributed, the virus tends to bind inappropriately to a variety of cells and even cell debris. However, elution from the cell surface after receptor binding has occurred often leads to changes in the virus (e.g., loss or structural alteration of **virus-attachment protein**) which decrease or eliminate the possibility of subsequent attachment to other cells. Thus, in the case of influenza, cleavage of sialic acid residues on unoccupied receptors by NA leaves these groups bound to the active site of the HA, preventing that particular molecule from binding to another receptor (Box 4.2).

In most cases, the expression (or absence) of receptors on the surface of cells largely determines the **tropism** of a virus (i.e., the type of host cell in which it is able to replicate). In some cases, intracellular blocks at later stages of replication are responsible for determining the range of cell types in which a virus can carry out a productive infection, but this is not common. This is what makes it possible to **infect a wider range of cell types with naked nucleic acid** (ssRNA+, ss−, or dsDNA) than with virions, for example. Therefore, this initial stage of replication and the very first interaction between the virus and the host cell has a major influence on virus pathogenesis, and in determining the course of a virus infection.

In some cases, interactions with more than one protein are required for virus entry. These are not examples of **alternative receptor use**, as neither protein alone is a functional receptor—both are required to act together. An example is the process by which **adenoviruses**

BOX 4.2

Why do these details matter?

This chapter describes in quite a lot of detail the interactions of certain viruses with their receptors—and if you look at the research literature about virus receptors, you'll find it's huge. Why all the fuss? It's because this first interaction of a virion with a host cell is in some ways the most important step in replication—it goes a long way to determining what happens in the rest of the process. For one thing, if a cell has no receptors for a virus, it does not get infected. So **tropism**, the ability of a virus to infect a particular cell type, is largely controlled by **receptor interactions**, and less by events inside the cell. Going on from there, small changes can have big effects, so this process is important to understand in detail. At present, for example, the H5N1 type of influenza A virus can infect humans and when it does, it's likely to kill them—but it really struggles to do this because at the moment, it's really a bird (avian) virus. With a very small change in the receptor usage, H5N1 could become a far more transmissible and deadly human virus. In addition, when you understand these processes, you can use them against the virus. We've had antiinfluenza drugs for decades, but they were not very good. More modern influenza drugs such as Tamiflu and Relenza inhibit the NA protein involved in receptor interactions (although in release from the cell rather than uptake). If H5N1 ever does make the jump to being a human virus, we're going to need these and newer drugs to stay alive.

enter cells. This requires a two-stage process involving an initial interaction of the **virion fiber protein** with a range of cellular receptors, which include the major histocompatibility complex class I (MHC-I) molecule and the coxsackievirus-adenovirus receptor. Another virion protein, the **penton base** (see fig. 2.11), then binds to the ***integrin family*** of cell-surface proteins, allowing internalization of the particle by a process called ***receptor-mediated endocytosis***. Most cells express primary receptors for the adenovirus fiber protein; however, the internalization step is more selective, giving rise to a degree of cell selection.

A similar observation has been made with human immunodeficiency virus (HIV). The primary receptor for HIV is the **helper T-cell differentiation antigen, CD4**: while this is present on many cell types in humans, it is especially densely expressed on the surface of helper T-cells. Transfection of human cells that do not normally express or only sparsely express CD4 (such as **epithelial cells**) with **recombinant CD4-expression constructs** makes them **permissive for HIV infection**; however, transfection of rodent cells with human CD4-expression vectors does not permit productive HIV infection—something else is missing from the mouse cells. If HIV **provirus DNA** is inserted into rodent cells by transfection, virus is produced, showing that there is no intracellular block to infection. Thus, there must be one or more accessory factors in addition to CD4 that are required to form a functional HIV receptor. These are a family of proteins known as **β-chemokine receptors**. Several members of this family have been shown to play a role in the entry of HIV into cells, and their distribution may be the primary control for the tropism of HIV for different cell types (**lymphocytes, macrophages**, etc.). Furthermore, there is evidence, in at least some cell types, that HIV infection is not blocked by competing soluble CD4, indicating that in these cells a completely different receptor strategy may be being used. Several candidate molecules have been put forward to fill this role (e.g., **galactosylceramide** and various other candidate proteins). However, if any or all of these do allow HIV to infect a range of CD4-negative cells, this process is much less efficient than the interaction of the virus with its major receptor complex.

On occasion, **antibody-coated virions binding to Fc receptor molecules** on the surface of **monocytes** and other blood cells, which bind the "handle" or **Fc heavy chain dimer region** of antibody molecules, can result in virus uptake. This phenomenon has been shown to occur in a number of cases where **antibody-dependent enhancement** (**ADE**) of virus uptake occurs, in people or animals that have had prior exposure to the virus. Here, the presence of antivirus antibodies can occasionally result in **increased virus uptake by cells** and **increased pathogenicity** rather than **virus neutralization**, as would normally be expected. This is a severe problem with **dengue viruses**, where cross-reactive antibodies to one of the four strains of the virus can **increase uptake of another dengue virus strain**, which can result in dengue hemorrhagic fever (see Chapter 8). It has been suggested that this mechanism may also be important in the **uptake of HIV by macrophages and monocytes** and that this is a factor in the pathogenesis of acquired immune deficiency syndrome (AIDS).

In some cases for animal viruses, specific receptor binding can be side-stepped by **nonspecific** interactions between virions and cells—which is in fact the norm for plant viruses, as explained above. It is possible that virions can be "accidentally" taken up by cells via receptor nonspecific processes such as **pinocytosis** or **phagocytosis** (see Penetration below). However, in the absence of some form of physical interaction that holds the virion in close association with the cell surface, the frequency with which these accidental events happen is very low.

Penetration or entry

Penetration of the target cell normally occurs a very short time after attachment of the virus to its **receptor** in the cell membrane. Unlike attachment, cell penetration is often an **energy-dependent process**—that is, the **cell must be metabolically active** for this to occur. The type of entry employed by particular viruses into cells is largely governed by the type of cells being infected: as mentioned earlier, the barrier that cells have between them and the outside world may be relatively thick and rigid (nearly all prokaryotes, fungi, algae, microalgae and land plants), or essentially nonexistent (animal cells, protists). This is shown in Fig. 4.5.

Note that although prokaryotes have rigid cell walls, the thickness of these (up to 50 nm) is comparable to the diameter of small virions (e.g., ssRNA+ MS2, ssDNA microviruses), but much less thick than the length of larger viruses such as the tailed dsDNA-containing **sipho-** and **myoviruses** (>100 nm). This means that traversing these walls can be done by a variety of specific mechanisms, generally triggered by the **binding of the virion to the cell surface**. Other than the nonspecific process of cell entry by plant viruses, then, three main mechanisms are involved in penetration of or entry into cells:

1. **Penetration by physical and chemical means,** activated by the attached virion, of prokaryote and possibly also of microalgal and algal cell walls.

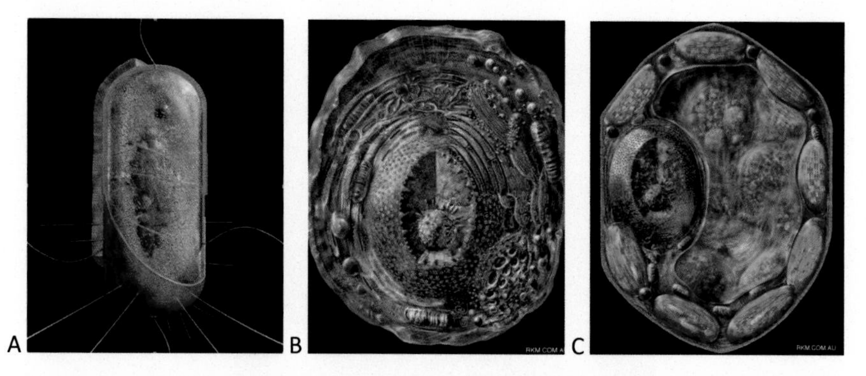

FIG. 4.5 Cell walls. (A) Idealized bacterial cell: undifferentiated cytoplasm inside a lipid bilayer within a cell 1 μm in length, relatively rigid cell wall ~50 nm thick. (B) Generic animal cell: nucleus, mitochondria, endoplasmic reticulum (ER) within lipid bilayer membrane. Typically around 10 μm in length. (C) Generalized diagram of a plant cell. Nucleus, vacuole, mitochondria, ER, and chloroplasts bounded by a lipid bilayer, within a rigid cell wall largely composed of the **carbohydrate polymer cellulose**. Cells about 20 μm in length, cell walls about 200 nm thick. Fungi may be regarded as being equivalent to plant cells, except their cell walls are largely composed of another carbohydrate named **chitin**. *Courtesy of Russell Kightley Media.*

2. **Fusion** of the **envelope** of enveloped virions with the eukaryotic cell membrane, at the cell surface (Fig. 4.6).
3. **Endocytosis** in animal cells of the virus into **intracellular vesicles** (Fig. 4.7). This is probably the most common mechanism of virus entry into animal cells.

Prokaryote cell entry

Bacterial cell walls are strong and relatively thick, to protect them from **osmotic lysis and predation**, and to give them shape. Gram-positive cells have a **single internal lipid bilayer**, and a **thick peptidoglycan** cell wall. Gram-negative cells have **an internal membrane**, a **thin peptidoglycan layer**, **an outer membrane**, and often a **polysaccharide-based capsule**.

Prokaryote viruses therefore have to have some means of breaching a quite formidable barrier if they are to enter the cell.

The very well-characterized **bacteriophage T4**—more properly *Enterobacteria phage T4*, genus *"T4-like Viruses,"* family *Myoviridae*—belongs to a family of viruses with **34–170 kb dsDNA genomes, isometric heads, and contractile tails**, and infects the gram-negative bacterium *Escherichia coli*. Myoviruses have one of the more complex entry mechanisms of prokaryote viruses, involving **an active injection process**: this effectively makes the virions the equivalent of nanoscale one-use hypodermic syringes, as explained below.

The **phage tail fibers** (see Fig. 2.15) are the virus attachment proteins; these individually bind the bacterial cell surface after random collision—specifically to certain **lipopolysaccharides** and to the surface outer membrane protein **OmpC**. This is **reversible binding**, and is probably due to **electrostatic interactions** as it is Mg^{2+} and Ca^{2+} concentration-dependent. After tail fiber binding has consolidated, the **baseplate** then settles down onto the surface and binds firmly to it as a result of conformational changes in the **short tail fibers** attached to it.

FIG. 4.6 Cell entry by direct fusion. This depicts the fusion of the envelope of a HIV-1 virion with a cell membrane. This results in envelope proteins being dissipated in the cell membrane, breakup of the virion matrix shell, and release of the intact capsid containing the virus RNA genome as a nucleoprotein complex, together with associated RT. *Courtesy of Russell Kightley Media.*

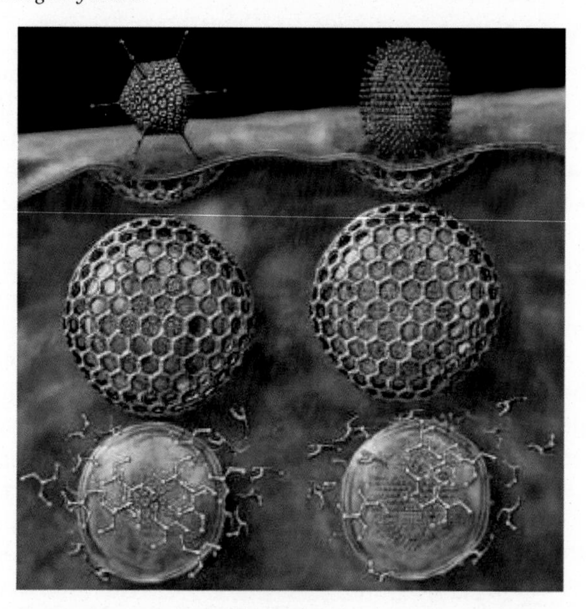

FIG. 4.7 Receptor-mediated endocytosis for virion entry into animal cells. The picture depicts binding of an adenovirus virion (*left*) and an influenza virion (*right*) in coated pits, their internalization in clathrin-coated vesicles, and release of clathrin prior to endosome fusion. *Courtesy of Russell Kightley Media.*

After this occurs, a **conformational change takes place in the tail sheath**, which then **contracts** in a wave **that propagates up the sheath from the baseplate**, pushing the **tail core** through the cell wall: this is aided by a **lysozyme activity** associated with the tip of the **tail tube**. This is an **irreversible process**. **DNA is then extruded** from the phage head into the bacterial cytoplasm, by means of a **molecular motor** at the base of the head. This is used for initial transcription and virus expression. This mode of genome entry to the cell is surprisingly conserved among certain large eukaryotic dsDNA viruses as well (see Maghsoodi et al., 2019).

Enterobacteria phage λ, genus "**λ-like viruses,**" family *Siphoviridae*, is a phage with a long rigid tail attached to an isometric head, with a **49 kb dsDNA genome**. It attaches via the **J protein in the tail tip** to **aporin**, the cell surface receptor that is responsible for transport of **maltose** across the outer membrane. Although the tail is **noncontractile**, a molecular motor-mediated DNA injection mechanism similar to that of T-even phages allows entry of DNA into the cell, via a **sugar transport protein (ptsG)** in the inner membrane, leaving the capsid behind.

The injection of DNA from bacterial virions into cells is usually enabled by a **molecular motor** (see also Chapter 2): this is housed at the **junction of the tail and head of the virion**, and is also responsible for **filling preassembled phage heads with DNA** in the infected cell as part of the virion assembly and maturation process. Thus, the motors are nanomachines which can run in both directions. This is generally an **ATP-driven process**—implying the presence of ATP in virion heads—and in this is similar to other cellular molecular motors such as **myosin, kinesin, and dynein** in eukaryotes (see below).

MS2 phage (*Enterobacteria phage MS2*, genus *Levivirus*, family *Leviviridae*)—the first organism ever sequenced—is an isometric ssRNA+ virus infecting *E coli*. Its virion is too simple to house the kind of machinery characteristic of the tailed dsDNA phages mentioned above. It attaches to the **pilin** (the building block of pili, or helical filaments on the surface of many bacteria) of the **F**(ertility) pili via its single attachment or **A protein,** located at a **5′ rotational axis of symmetry** in a ***T*=3 nucleocapsid composed 180 copies of the single coat protein**. The A protein is **covalently linked** to the 5′-end of the 3.6 kb genomic RNA; binding pilin **causes cleavage of the A protein** and releases it from the capsid after a conformational rearrangement. Thus, when the pilus is retracted into the cell, the **A protein and RNA are pulled with it**, leaving the empty capsid outside.

Plant and fungal cell entry

Plant cells are superficially similar to animal cells in basic construction, apart from having an extra organelle—**chloroplasts**—and often having extensive membrane-bounded **vacuoles** that are used as storage organelles. They do, however, have one large and fundamental difference to animal cells, which profoundly affects the way in which they are infected by viruses, and how viruses move between them. This is their possession of **thick, rigid, cellulose-based cell walls** (see Fig. 4.5). Every cell is separated from every other cell by thick cell walls (usually >**200 nm** thick), whose dimensions are far larger than the size of the average virion (e.g., isometric **bromoviruses** and **comoviruses,** ~**30** nm). This means that plant cells are effectively inaccessible to viruses, even given mechanisms of injection similar to the T-even phages—which no terrestrial plant viruses have. Cells in **vascular terrestrial plants** interconnect only via **specific discontinuities** in the cellulose walls: the most numerous of these are **plasmodesmata** (see Chapter 6).

Land plants' structure is possibly why plant virus virions are mostly **nonenveloped**: they do not appear to specifically interact with host cell membranes or cell walls, as do bacterial and animal viruses, and therefore do not need the envelope structures for cell entry. This is true even when the plant-infecting virus **is enveloped and also infects an insect** (e.g., plant **rhabdoviruses** and **bunyaviruses**): while the virus presumably behaves normally in the other host in terms of receptor recognition and cell entry, and even though apparently plant cells are capable of phagocytosis/endocytosis (see below), **this is not the mode of entry into plant cells**. The mechanisms employed to enter cells rather appear to be **passive carriage through breaches in the cell wall** in the first instance, followed by later cell-to-cell spread in a plant by means of specifically-evolved **"movement" functions** (see Chapter 3)—including for insect-infecting viruses—and perhaps also **spread via conductive tissue** such as phloem and xylem as whole virions.

The "**passive carriage**" referred to above could include the following mechanisms:

- a purely **mechanical injury** that breaches the cell wall and transiently breaches the plasma membrane of underlying cells;
- similar gross injury due to the **mouthparts of a herbivorous arthropod**, such as a beetle;
- **injection directly into cells** through the **piercing mouthparts** of sap-sucking insects or nematodes;
- carriage into plant tissue on or in association with **cells of a fungal parasite**;
- **vertical transmission** through infected seed or by **vegetative propagation**;
- transmission via **pollen**; and
- **grafting** of infected tissue onto healthy tissue.

There are certain superficial similarities between plants and fungi with respect to the cell wall; however, in **fungi, cell walls are composed of chitin**, a different **complex polysaccharide**. There is also a wide phylogenetic divide between single-celled fungi like yeasts, and filamentous fungi, which may be reflected in their viruses: filamentous fungal hyphae are often effectively tubes with no cross-walls, whereas yeasts and similar organisms have cell walls round every cell.

No fungal viruses appear to have any specific mechanisms for gaining entry to fungal cells; indeed, it is extremely difficult to demonstrate the infectivity of virus-like particles, and it is only since the advent of the gene gun or **biolistic transformation**, that many viruses have been shown to be infectious at all—by being "shot" into fungal cells adsorbed onto metal particles. It is probable that most fungal viruses—like the **plant-infecting cryptoviruses** in the mainly **fungus-infecting *Partitiviridae***—are only transmitted by **"grafting,"** or the physical connection of infected to healthy cells by ***anastomosis***. Thus, **fungal mating** is a good means of transmission, as it results in the mixing of cell contents of different hyphae. Otherwise, transmission would be **vertical**, or to progeny via **spore formation**.

Most fungal viruses appear to have no extracellular stage: indeed, narnaviruses (see Chapter 3) do not even make particles. They are also unusual in that viruses in one of the two genera of family *Narnaviridae*—**mitoviruses**—infect the **mitochondria** of certain filamentous fungi, making them the only viruses so far found to infect mitochondria. Viruses in the genus *Narnavirus* replicate in the cytoplasm of certain **oomycetes** and yeasts.

Animal cell entry

Animal cells do not have cell walls; they have only lipid bilayer cell membranes (=**plasmalemma or plasma membrane**) with associated embedded proteins. They also have a very highly developed **internal membrane complex** (see Fig. 4.5): this consists of the **double nuclear membrane**, with its specialized entry and exit ports; the **rough and smooth endoplasmic reticulum (ER)** which are continuous with the nuclear membrane; the multilayered **Golgi apparatus,** and various specialized vesicles such as **lysosomes and endosomes and peroxisomes**, which are involved in transport around the cell, and intracellular digestion of macromolecules (Box 4.3).

There are two types of **virus-driven membrane fusion**: one is **pH-dependent** and occurs in vesicles inside the cell (see below), and the other is **pH-independent** and is more characteristic of entry directly via the cell membrane. Direct fusion of an enveloped virion with the animal cell membrane after specific attachment to a receptor requires the presence of a specific **fusion protein** in the virus envelope—for example, **retrovirus TM glycoproteins** or **paramyxovirus fusion (F) proteins**. Typically, **"reversible attachment"** first occurs between virion and cell surface receptors after random collision events, with one or a few attachment proteins binding to the same number of receptors. **Consolidation** of receptor-attachment protein binding occurs when receptors migrate in the plasma membrane to bind the virion,

BOX 4.3

Eukaryote cell vesicle transport

Eukaryote cells have an **intricate system of vesicle transport,** centered on the **Golgi apparatus**: this involves export of protein(s) and vesicles from the endoplasmic reticulum (ER) to the Golgi; production of **export vesicles** containing proteins from this for fusion with the cell membrane (=**exocytosis**); production of **lysosomes** to fuse with **endosomes** for digestion of material internalized by **receptor-mediated** or nonspecific **endocytosis** (**phagocytosis** for particulates; **pinocytosis** for liquid). Endocytosis does not require any specific virus proteins (other than those already utilized for receptor binding) but relies on the normal **formation and internalization of coated pits** at the cell membrane. Receptor-mediated endocytosis is an efficient process for **specifically taking up and concentrating extracellular macromolecules** in the normal life of the cell. Roughly 2500 clathrin-coated vesicles leave the membrane of a cultured fibroblast every minute.

Other **cytoskeleton-directed** vesicle trafficking—**mediated mostly via molecular motors** such as **myosin, kinesin, and dynein,** which move along the cytoskeletal fibers—involves targeting of vesicles back to the Golgi and to the nuclear membrane. **Microtubules** are an important part of the cytoskeletal transport network for viruses (Ploubidou and Way, 2001): microtubules radiate from the cell centrosome located near the nucleus, to the cell periphery, with **"minus ends"** at the center of organization, and **"plus ends"** at the periphery. **"Retrograde transport"** from the point of entry into the cell involves recruitment of a **microtubule minus-ended directed molecular motor complex**, normally dynein, to move virions or other cargo toward the center of the cell. The **actin cytoskeleton** is also involved in transport of **baculovirus** virions.

attaching it more firmly to the cell and immobilizing it. Continued migration and binding results in **irreversible attachment**, where attempts to remove the virions would result in its disruption. At this point, the concerted action of several fusion protein molecules bound to receptors promote the joining or fusion of the cellular and virus membranes, which results in the **nucleocapsid** being deposited directly in the cytoplasm. Pox-, herpes-, SARS-CoV-2, and paramyxovirus virions typically enter cells by this process (Fig. 4.6).

As mentioned above, most virions—enveloped or naked—appear to enter animal cells via **intracellular vesicles** produced by receptor-mediated endocytosis. The process of endocytosis is **almost universal in animal cells** and deserves further consideration (Fig. 4.7). The formation of **coated pits** follows adventitious initial binding due to a chance encounter of phosphatidylinositol-2-bis phosphate (PIP2), the adaptor protein complex AP2 and clathrin **triskelia** (homotrimeric 3-armed protein formation). This forms a pit as clathrin triskelia induce inward curvature of the cell membrane. This very transient structure is stabilized if a virion that is bound to one or two receptor molecules finds itself in a pit, then recruits more receptors as these migrate in the fluid cell membrane and are then fixed in place by binding the virion (**consolidation of binding**). Pits deepen as a result of more clathrin molecules binding to the established structure forming the depression in the cell membrane. Eventually, a clathrin "cage" caused by accumulation of increasing numbers of triskelia surrounds receptor molecules bound to a virion in a nascent vesicle that is in close proximity to the internalized virion (see Fig. 4.7). Continuing cage assembly results formation of a coated vesicle joined only by a thin neck to the cell membrane. This is "pinched off," after action of the GTPase **dynamin** on the vesicle neck, by **budding**—with transport into the cytoplasm of the cell via the "**minus-end**" ***cytoskeletal motor protein myosin VI***. Clathrin cages are typically around **80 nm in diameter** (Paraan et al., 2020), varying between 60 and 200 nm; consequently, usually only single virions of an appropriate size can be internalized in one. The lifetime of these coated vesicles is very short: within seconds, most have lost their clathrin (recycled back to the cytoplasm) via action of several cytoplasmic proteins including one from the **heat shock protein 70 family** (Hsp70). These vesicles fuse with "**sorting endosomes,**" releasing their contents into these larger vesicles. These endosomes mature to **late endosomes**, then may fuse with **lysosomes**, depending on signaling molecules on their outside surfaces. The latter are specialized vesicles with a low internal pH (4.5–5.0) packed with a variety of **acid hydrolase enzymes** that can degrade cellular molecules such as proteins, nucleic acids, carbohydrates and lipids, as well as extraneous molecules including virions.

At this point, any virion contained within such structures and still bound to its receptors is cut off from the cytoplasm by a lipid bilayer, and therefore has not strictly entered the cell. Moreover, as endosomes fuse with **lysosomes**, the environment inside these vessels becomes increasingly hostile as the pH falls, while the concentration of degradative enzymes rises. This means that the virion must leave the vesicle and enter the cytoplasm before it is degraded, if it is to succeed in infecting the cell. There are a number of mechanisms by which this can occur, including **membrane fusion** for enveloped virions, and **direct interaction with membranes** for "naked" virions. This process is triggered by conformational changes in virion-associated proteins, including fusion proteins and virion surface and other proteins for naked virions such as adeno-, reo-, and picornaviruses. The release of virions from endosomes and their passage into the cytoplasm is intimately connected with (and often impossible to separate from) the process of **uncoating** (see below). The process is qualitatively identical to what occurs with direct virion envelope-host membrane fusion, in **that internal capsids or nucleoproteins are**

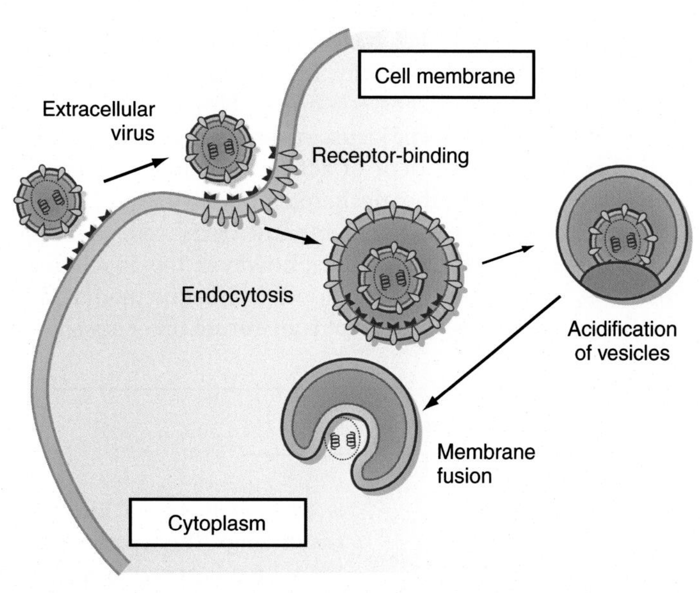

FIG. 4.8 Enveloped virion escape from endosomal vesicles. This process is dependent on the presence of a specific fusion protein on the surface of the virion which, under particular circumstances (e.g., acidification of the virus-containing vesicle; proteolytic cleavage), becomes activated, inducing fusion of the vesicle membrane and the virion envelope. The virion effectively turns inside-out, liberating its internal contents into the cell cytoplasm.

deposited into the host cell cytoplasm, with envelope proteins remaining in the vesicle membrane, and matrix shells being disrupted. A simple view of this process is shown in Fig. 4.8.

Caveolin-dependent endocytosis is probably an under-appreciated mode of entry into animal cells, because the mechanisms were obscure until recently. There are three mammalian caveolin proteins: all are 21 kDa in size, have cytoplasmic N- and C-termini with a transmembrane hairpin in between. The caveolins are transported to the Golgi as monomers, then associate with "**lipid rafts**": these are distinct **10–100 nm microdomains** of the plasma membrane that consist of combinations of **glycosphingolipids, cholesterol, and protein receptors,** which are more ordered and tightly packed than the rest of the membrane (see Box 4.5). The caveolins form oligomeric structures of 14–16 molecules, and these form **caveolae**: these are 50–100 nm **amphora-shaped invaginations** of the plasma membrane, and occur in many vertebrate cell types. They protect cells from mechanical stress, have a role in cell signaling and in lipid regulation, and act as **mechanosensors**. They can also be used for entry to the cell by some viruses such as SV40 polyomavirus and echovirus type 1 (*Picornaviridae*). Binding of virions to receptors on the inside of caveolae triggers them into budding inside cells by the action of the **GTPase dynamin II**, much as clathrin-coated vesicles do. These vesicles fuse with early endosomes to form **caveosomes**: these are endosomal compartments with neutral pH which do not fuse with lysosomes, which avoids the kind of virion degradation that can occur in these compartments.

It is worth mentioning that even structurally very similar viruses such as **picornaviruses** may use completely different means of entering cells. For example, while **polioviruses** and most other picornaviruses studied generally use the receptor-mediated **clathrin cage** mode of entry, the related **encephalomyocarditis virus (EMCV)**—which infects a wide variety of animals around the world—enters cells via **caveolae** in a dynamin and actin-dependent manner.

There is, more rarely, also **endocytosis based on other routes**, involving vesicles that contain neither clathrin nor caveolin (see Sobhy, 2017). However, like these pathways, they generally require **dynamin, cholesterol,** and/or **lipids**. However, **lymphocytic choriomeningitis virus**—an enveloped ssRNA– **arenavirus**—uses a **dynamin-, clathrin-,** and **caveolin-independent** route that is also independent of **actin, lipid rafts,** and the **pH.** Despite this being a very well-studied model virus, the mechanism of entry is still obscure.

Translocation of the entire virion across the membrane of the cell, resulting in its deposition in the cytoplasm in an unchanged state, is possible: however, this is relatively rare among viruses infecting animal cells, and is poorly understood. It must be mediated by proteins in the virus capsid and specific membrane receptors that **translocate** other specific proteins (Box 4.4).

BOX 4.4

Enveloped virus fusion proteins

Fusion proteins of enveloped viruses derive from host-to-virus or virus-to-virus acquisition.

These are common to all enveloped viruses, and there are **four types**.

Class I fusion proteins form virion "spikes" that are **homotrimers**, of three identical subunits. These are largely alpha-helical, with two subunits generated by proteolytic cleavage from a precursor. The original C-terminus is anchored to the viral membrane; the new N-terminus has a stretch of ~20 hydrophobic amino acids: this is the **fusion peptide**. Class I proteins all have a **trimeric helical coiled-coil** rod adjacent to the fusion peptide: this may act as a template for the refolding of protein segments during fusion, when the **"six helix bundle"** forms. They are found in **retroviruses** (HIVs, HTLVs), **orthomyxoviruses** (influenza), **coronaviruses** (SARSCoV), **paramyxoviruses** (mumps, measles), and **filoviruses** (Ebola, Marburg), among others (see fig. 2.13, Fig. 4.21B).

Class II fusion proteins predominantly have a **β-sheet-type structure** and are not cleaved; they are present as **closely-associated homodimers** on virions, where they generally provide a continuous isometric shell (see Fig. 8.8). The **"fusion peptide"** portion that inserts into the target membrane is thought to be an **internal hydrophobic fusion loop**. These proteins are typical of ssRNA+ **flaviviruses** such as **dengue, yellow fever** and **West Nile viruses, Semliki Forest** and **other alphaviruses**, and ssRNA– **bunyaviruses (Rift Valley fever, Crimean-Congo hemorrhagic fever viruses**).

Class III fusion proteins like Class I proteins have a central α-helical trimeric core; however, the fusion domains have two fusion loops at the tip of an elongated β-sheet, similar to Class II fusion proteins. These proteins are characteristic of ssRNA– **rhabdoviruses** like **rabies** and **vesicular stomatitis viruses**, and dsDNA **herpesvirus** gB attachment glycoproteins. Most of these virions are capable of pH-independent membrane or direct fusion at the cell surface.

There is a putative **Class IV** of fusion proteins: the E2 glycoproteins of **pesti-** and **hepaciviruses** (*Flaviviridae*) were found to have novel folds, unlike dengue, West Nile, etc. E1 and E2 proteins of pesti- and hepaciviruses define a new class of fusion machinery; structural data suggest that the fusion proteins evolved by **host-virus** or **virus-virus** transfer. Fusion proteins appear to evolve as **independent evolutionary units**, transferring from host to virus, or from virus to virus in coinfected hosts. Thus, fusion proteins potentially compete as **"selfish genes,"** often crossing the boundaries between virus species or between virus and host (Rey and Lok, 2018).

Uncoating of the viral genome

Uncoating is a general term for the events that occur after **penetration or entry**, in which the virus **capsid** is completely or partially removed and the virus **genome** is exposed, usually in the form of a **nucleoprotein complex**. For nearly all **prokaryote viruses**, uncoating is essentially complete after cell entry: the viral genomes that are **injected or translocated** into cells are seldom associated with significant amounts of proteins. The same is largely true for the eukaryotic **algae-infecting phycodnaviruses**: these are NCLDVs with evolutionary affinities to **mimiviruses**. However, while mimiviruses enter their protist host cells via a phagocytosis-like mechanism, ***Paramecium bursaria* chlorella virus (PBCV-1)** enters its host cells like a bacteriophage. Virions attach to host cells via a viral vertex, and **degrade the host chitin-based cell wall** at the site of attachment by means of chitinases and other enzymes. The 330 kb linear genome then enters the cell, leaving the capsid at the surface—presumably mediated by a **DNA motor**, as for phages.

The situation is generally more complex with viruses infecting animal cells. In one sense, the removal of a virus envelope that occurs during membrane fusion for viruses infecting animal cells is part of the uncoating process (see Fig. 4.8). The initial events in uncoating may occur inside endosomes, being triggered by the change in pH as the endosome is acidified, or directly in the cytoplasm. Endocytosis is potentially dangerous for viruses, because if they remain in the vesicle too long they will be irreversibly damaged by acidification or lysosomal enzymes. Some viruses can control this process: for example, the **influenza virus M2 protein**—a small transmembrane protein that makes one of the smallest **viroporins**, or **membrane ion channels** known—allows entry of hydrogen ions from the lysosome into the **nucleocapsid**, facilitating uncoating. The M2 protein is multifunctional and also has a role in influenza virus **maturation**.

With **picornaviruses**, penetration of the cytoplasm by exit of virus from endosomes is tightly linked to uncoating (Fig. 4.9). The acidic environment of the endosome causes a **conformational change** in the **capsid** which reveals hydrophobic domains not present on the surface of mature virions. The interaction of these hydrophobic patches with the endosomal membrane is believed to form pores through which the genome passes into the cytoplasm.

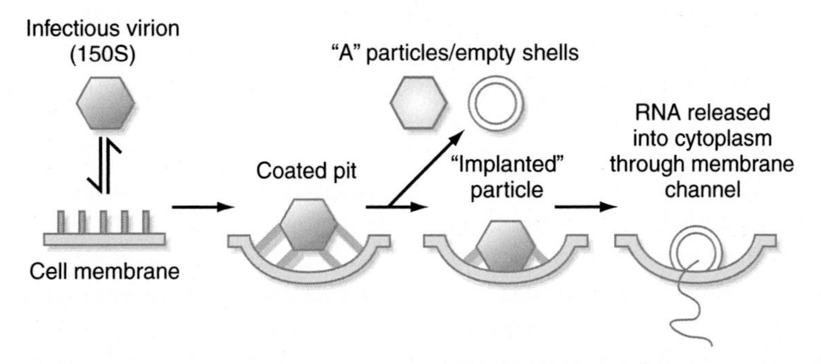

FIG. 4.9 Cell penetration and uncoating of polioviruses. Following receptor binding and consolidation, and generally inside an endosome, infectious poliovirus virions undergo a structural transition to an "implanted" or noninfectious "A particle" form with a lower sedimentation coefficient (135S). This interacts directly with the endosome membrane to form a pore so as to allow exit of the naked ssRNA− genome into the cytoplasm, leaving the empty capsid (80S) behind. The 150S–135S transition can also be triggered in vitro by heating or acidification.

Most nonenveloped viruses, such as the simple ss– and dsDNA viruses (**circoviruses, papillomaviruses**) and complex nonenveloped dsDNA **adenoviruses** and **dsRNA reoviruses**, also enter animal cells via endosomal vesicles after receptor consolidation. Adenoviruses enter via an endocytotic vesicle, then interact with the vesicle membrane—because of **structural alterations** in the faces of the icosahedral particle due to the pH shift—and expel a partially-uncoated viral core structure into the cytoplasm.

Some viruses may normally enter cells in more than one way: for example, both **poxviruses** and **iridoviruses** may have virions that are enveloped, or not (see Chapter 2). Enveloped virions **fuse directly with the cell membrane**; nonenveloped are **taken in via endocytotic vesicles**—indicating that both forms have appropriate attachment proteins for cell recognition and binding. **SARS-CoV-2** virions of the original Wuhan isolate, and variants alpha, beta, delta all enter cells mainly by direct fusion after binding either the **ACE2 protein receptor** (mainly), or **TMPRSS2**, a cell-surface serine protease that enhances membrane fusion by virions. Virions of the latest-emerging **Omicron variant**, however, bind more tightly to ACE2 than do the other variants—resulting in more efficient infection—and do not bind TMPRSS2. This results in virions being less able to cause syncytium formation, and entering mainly via the endocytic pathway, as well as being more efficient at infecting upper airway epithelial cells and less efficient at infecting lung epithelia. Omicron virions can also use ACE2 receptors from a wider range of host animal species than the earlier variants, which may increase their host range (see Jackson et al., 2022).

The product of uncoating depends on the structure of the virus nucleocapsid. In some cases, it might be relatively simple—picornavirus **virions** are the nucleocapsid, for example—or highly complex: **retrovirus capsids** (see Fig. 4.6) contain cores that are highly ordered nucleoprotein complexes containing the diploid RNA genome and the RT to convert the ssRNA genome into the DNA **provirus**. Naked **herpesvirus** and partially disassembled **adenovirus** capsids enter into the cytoplasm too, for later transport to the nucleus (see below). The **dsRNA reoviruses**, with their nonenveloped three-layer capsids, attach to cell-surface carbohydrate, and enter cells by receptor-mediated endocytosis. Inside lysosomes, the virion outer capsid undergoes **acid-dependent proteolysis** that exposes the viral membrane-penetration protein, $\mu 1$. Further proteolysis and conformational rearrangements of $\mu 1$ mediate disruption of the endosomal membrane, and delivery of reovirus **"core particles"** into the cytoplasm (see Danthi et al., 2010). These remain essentially intact and are transcriptionally active (see replication of Class III viruses). The same is true for **poxviruses** and **mimiviruses**: complete uncoating does not occur, and many of the reactions of genome replication are catalyzed by virus-encoded enzymes inside particles in the cell cytoplasm which still resemble the mature virions.

Nuclear localization

All eukaryotic viruses that require entry to the nucleus in order to replicate, must employ some means of **intracellular transport** to get their genomes to the nucleus. As discussed in Chapter 3, for example, retroviral reverse transcription can only occur inside an ordered **core particle** or **capsid**, and the resultant **dsDNA with LTRs** must find its way to the nucleus for integration—which is mediated by **integrase**, which is inside the capsid. **Herpesvirus, adenovirus, parvovirus,** and **polyomavirus capsids** undergo structural changes during and following **penetration**, but overall remain largely intact. All of these capsids contain sequences that are responsible for attachment to the **cytoskeleton**, and this interaction allows the transport of the entire capsid to the nucleus via **dynein motors** and the **microtubule network**. It is at the **nuclear pore complexes (NPCs)** (see Fig. 4.10) that uncoating of **DNA viruses** normally occurs, and **nucleocapsids** or **nucleoproteins** pass into the nucleus.

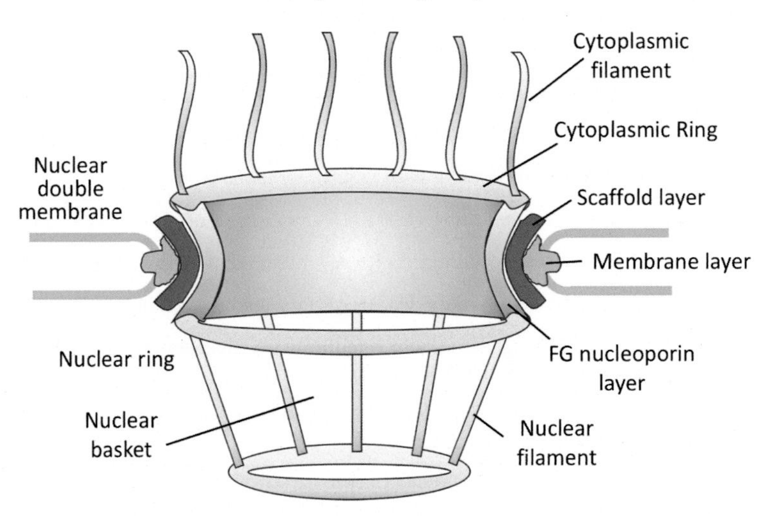

FIG. 4.10 Nuclear pore complex. The NPC in animal cells is a ~125 MDa molecular assembly of multiple copies of more than 30 **nucleoporins** (Nups), and is about 120 nm wide, with a channel diameter of around 5 nm. These are embedded in the nuclear membrane with eightfold radial symmetry, anchored to the nuclear envelope by a membrane layer that surrounds the scaffold layer. FG Nups are proteins which are rich in Phe-Gly repeats, which assist in both "gating" of the pores, and in shuttling proteins and larger assemblies into and out of the nucleus. *Modified from Azimi, M., Mofrad, M.R.K., 2013. Higher nucleoporin-importinb affinity at the nuclear basket increases nucleocytoplasmic import. PLoS One 8 (11), e81741. https://doi.org/10.1371/journal.pone.0081741.*

Herpes- and **adenovirus** cores bind to **importins** at the NPC; herpesvirions may pump out the DNA via the molecular motor into the nucleus through the pore, while adenovirions uncoat at the pore, and their DNA is transported in. For capsids/nucleoproteins of **HIV** and **influenza virions, nuclear localization signals** (NLS) in proteins associated with the viral genome are recognized by importins at the nuclear pore complex, and the nucleoprotein is imported intact along with associated enzymes (**RT, integrase** for HIV; **RdRp** and associated proteins for influenza virus).

Trafficking into the nucleus is generally dependent upon specific mechanisms; however, there are some nonspecific means. These include nonspecifically entering disrupted nuclei during mitosis (e.g., **murine leukemia virus** capsids); viruses that induce transient disruption of the nuclear membrane, enter through gaps, and uncoat (e.g., **parvovirus, SV40**); having virions that are small enough (e.g., **hepadnaviruses** and some **baculovirus** virions) to enter intact or partially intact via NPCs, followed by uncoating.

Genome replication and gene expression

What happens once a virus is uncoated or partially uncoated as a result of cell entry, depends largely upon what sort of virus it is. As explained earlier (Chapter 1), all **ribocells** have genomes consisting of dsDNA, which is replicated by similar principles across all cell types. Moreover, the principle of "**DNA makes RNA and RNA makes protein**"—often inaccurately referred to as "**The Central Dogma**" (see Cobb, 2017)—is true for all ribocells, with the exception of reverse-transcribing elements (see Chapter 3). By contrast, viral replication is far more complicated in terms of **information flow**: here, while DNA viruses behave according to the dogma, **RNA may make DNA** (retro- and pararetroviruses), **RNA can make RNA** (all other

TABLE 4.1 The Baltimore classification of viruses by replication strategy.

Baltimore class	Mode of replication
I	**dsDNA** genomes replicating via a **DNA** intermediate
II	**ssDNA** genomes replicating via a **DNA** intermediate
III	**dsRNA** genomes replicating via **ssRNA+** intermediate (conservative mode of replication)
IV	**ssRNA+** genomes replicating via **RNA−** intermediate (semiconservative)
V	**ssRNA−** genomes replicating via **RNA+** intermediate (semiconservative)
VI	"diploid" **ssRNA+** genomes which replicate via **reverse transcription** with a **greater-than-genome-length dsDNA** intermediate
VII	**dsDNA** genomes which replicate via **reverse transcription** with a **greater-than-genome-length ssRNA+** intermediate

RNA viruses), and while some RNAs can make protein, others have to be **transcribed** to do so. The **Baltimore Classification** of viruses by their **genome types and replication strategies** makes it fairly easy to predict the broad sort of strategy that a virus with a given genome will employ in order to get replicated. This classification was originally devised by David Baltimore in 1971; it originally only had six categories, but the discovery of "DNA retroviruses" or **pararetroviruses** in the 1980s (see Chapter 1) has necessitated a **Class VII** (Table 4.1).

The **information flow** for all types of viral genome can be depicted as shown in Fig. 4.11: here, the one-directional flow of genetic information due to the original central dogma paradigm—**DNA makes RNA makes protein**—is central to the diagram; however, all of the extra modalities employed by viruses complicate the picture considerably.

The size and nature of the virus genome has a major influence on the tactics used within a given replication strategy. For a virus with a small, very compact genome that only encodes the essential information for a few regulatory and structural proteins (e.g., most **RNA viruses, ssDNA microviruses, parvoviruses,** and **geminiviruses**), this would be one involving heavy reliance on the

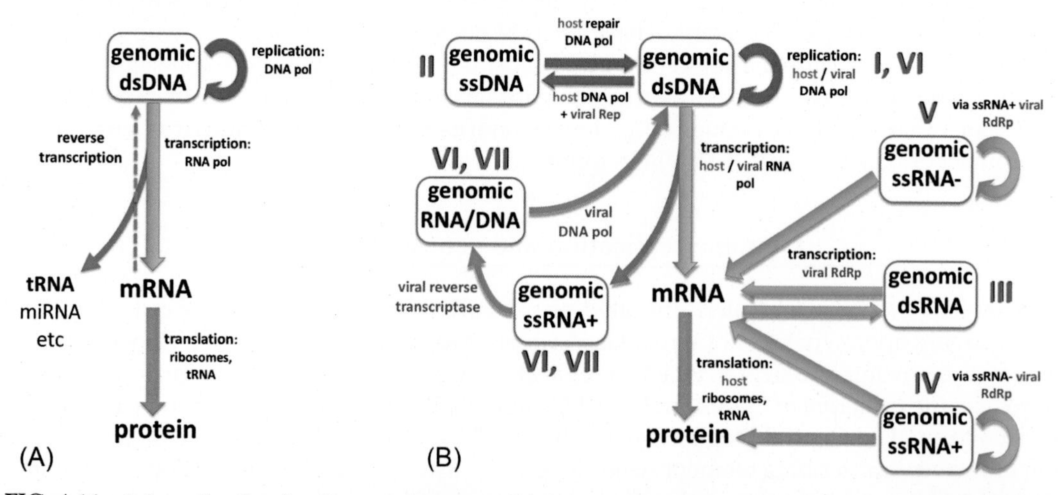

FIG. 4.11 Information flow for all types of genomes. (A) Genetic information flow in ribocells and (B) information flow in virocells. *Red* represents replication of the genome using cell enzymes, *blue* is cellular transcription, *green* is replication and transcription using RdRp or reverse transcriptase, and *orange* is translation. Baltimore class designations are shown for each class of virus.

host cell, whether it be prokaryote or eukaryote. Alternatively, large and complex virus genomes, such as those of **dsDNA poxviruses and other NCLDVs**, as well as large phages, encode most of the information necessary for replication, and the virus is only reliant on the cell for the provision of energy and the apparatus for macromolecular synthesis, such as ribosomes (see Chapter 1). Viruses with an RNA lifestyle (i.e., an RNA genome plus messenger RNAs) have no apparent need to enter the nucleus, although during the course of replication a few do (e.g., **influenza viruses**, family *Orthomyxoviridae*). DNA viruses of eukaryotes, as might be expected, mostly replicate in the nucleus, where host cell DNA is replicated and where the biochemical apparatus necessary for this process is located. However, some NCLDVs (e.g., **poxviruses, mimiviruses**) have evolved to have sufficient biochemical capacity to be able to **replicate in the cytoplasm**, with minimal requirement for either host cell **replication** or **transcriptional** machinery, which are nucleus restricted.

The **replication strategies** of the seven Baltimore groups are discussed below. For viruses with RNA genomes in particular, genome replication and the expression of genetic information are inextricably linked, therefore both of these criteria are taken into account. The control of gene expression determines the overall course of a virus infection (acute, chronic, persistent, or latent), and such is the emphasis placed on gene expression by molecular biologists that this subject is discussed in detail in Chapter 5 (Box 4.5).

BOX 4.5

The problem with DNA chromosome "Ends"

While most **bacterial and archaeal** genomes are **circular**, and thus replication can be initiated anywhere or at multiple locations, **eukaryote chromosomes are linear**. This introduces a problem at the molecular level when it comes to DNA replication, as because **new strand synthesis initiation** is dependent upon the presence of a **free 3′-OH on a short (10 bases) random oligoribonucleotide primer base-paired to the template strand**, there is no guarantee that this will happen at **exact ends of a chromosome**—which means **these would get progressively shorter with every replication cycle**.

Eukaryotes solve this problem by having ***telomeres*** at the ends of each chromosome: in vertebrates, these consist of around **2500 repeats of the sequence TTAGGG**, with a 3′ single-stranded-DNA overhang, which are lengthened regularly in a template-independent manner by a **reverse-transcriptase-containing** enzyme called telomerase (see Chapter 3), which contains its own RNA to act as template.

Virus DNA genomes do not have telomeres. However, most depend on the same mechanism to initiate DNA synthesis: namely, **oligoribonucleotide-priming of new DNA synthesis**. This accounts for why most viral dsDNA genomes are either **circular**, or **circularize during replication**, or have **covalently-closed ends**: these are all mechanisms by which to **avoid having free ends** which will become **abraded or shortened during replication**.

The few bacteria with **linear chromosomes** avoid the same problem by using either **proteins covalently bound to the 5′ ends** of each DNA strand, or having **hairpin loops of single-stranded DNA** at the 3′ ends. Interestingly, two groups of DNA viruses that may be evolutionarily related—**eukaryote-infecting adenoviruses and bacteria-infecting tectiviruses**—also have **5′ genome-linked proteins**, which serve to prime DNA synthesis on the opposite DNA strand (see adenoviruses, Chapter 5).

Class I: Double-stranded DNA

These viruses all essentially replicate their genomes as cellular DNAs are replicated, or **semiconservatively**: that is, double-helical dsDNA templates are unwound by **helicases,** and the **DNA-dependent DNA polymerase complex** (DNA pol) synthesizes two new DNA strands using the two virion DNA strands as templates, that remain base-paired to the template strands. This class contains most prokaryote viruses—which replicate in the cytoplasm of their host cells as these have no nuclei—and two groups of eukaryote viruses. For most dsDNA viruses of eukaryotes replication is exclusively **nuclear** (Fig. 4.13): replication of these viruses is relatively dependent on cellular factors such as **DNA-dependent DNA and RNA polymerases** used for **replication and transcription**, as well as of certain cellular transcription factors and other machinery. For some **NCLDVs,** as mentioned above, replication occurs in the **cytoplasm** (e.g., the **poxviruses** and **mimiviruses**), in which case the viruses have evolved, or acquired by transfer from the host, all the necessary factors for transcription and replication of their genomes, and are therefore largely independent of the cellular machinery. This also means that they can have control sequences such as **promoters and transcription factor binding sites** that are very unlike those of their host cells, which is a factor in expression (see Chapter 5).

It has long been known that many prokaryote dsDNA viruses may have both **lysogenic** and **lytic** phases in their life cycles—that is, their genomes can insert into host genomes like **DNA transposons** (see Chapter 5) and remain dormant until some stimulus triggers their excision, subsequent replication and expression, and exit by **lysis** of the cells. It has become apparent recently that this is also a factor in the life cycles of some eukaryotic dsDNA viruses, and especially **NCLDVs** and their **large satellite viruses ("virophages")** infecting protists and algae. However, this is presently a very under-studied area of virology (Fig. 4.12).

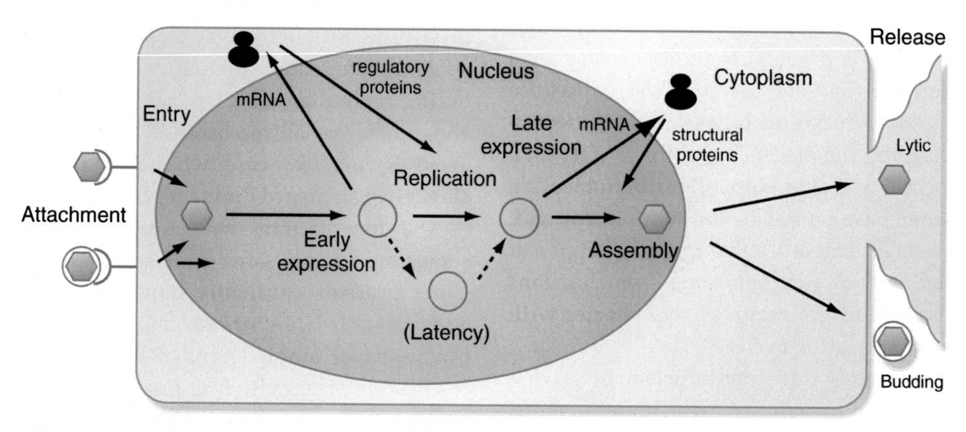

FIG. 4.12 Schematic representation of the replication of most Class I viruses. Details of the events that occur for genomes of this type are given in the text. Note that all prokaryote viruses and certain large dsDNA viruses of eukaryotes (e.g., poxviruses, mimiviruses) replicate in the cytoplasm of infected cells. Enveloped viruses will bud from host cells, while prokaryotic and some eukaryotic viruses will cause cell lysis or will be released upon death of the cell.

Class II: Single-stranded DNA

Single-stranded DNA virus genomes are essentially "fixed" by host **repair DNA pol** complexes, given that ssDNA is essentially defective or a transient state inside cells, to dsDNA intermediate forms. For prokaryotic viruses, replication occurs in the cytoplasm, as all the necessary machinery is there. For eukaryotic viruses, replication occurs in the nucleus. In both cases this involves the formation of a **dsDNA replication intermediate** initially by repair DNA pol, and then by mechanisms involving **rolling circle replication** (see Chapter 3). This results in the synthesis of **single-stranded progeny DNA**, which is rapidly converted to dsDNA unless it is sequestered by **ssDNA binding protein** (in prokaryotes), or accumulating **coat protein** (all cell types) (Fig. 4.13). There is now a rapidly-expanding catalogue of Class II viruses, as metaviromic sequence trawls through marine and even soil environments are turning up a huge diversity of these genomes, linked to both prokaryotes and eukaryotes.

Class III: Double-stranded RNA

These viruses mostly have segmented genomes (e.g., prokaryote-infecting **cystoviruses**, and eukaryote-infecting **reoviruses** and **partitiviruses**), although some important groups

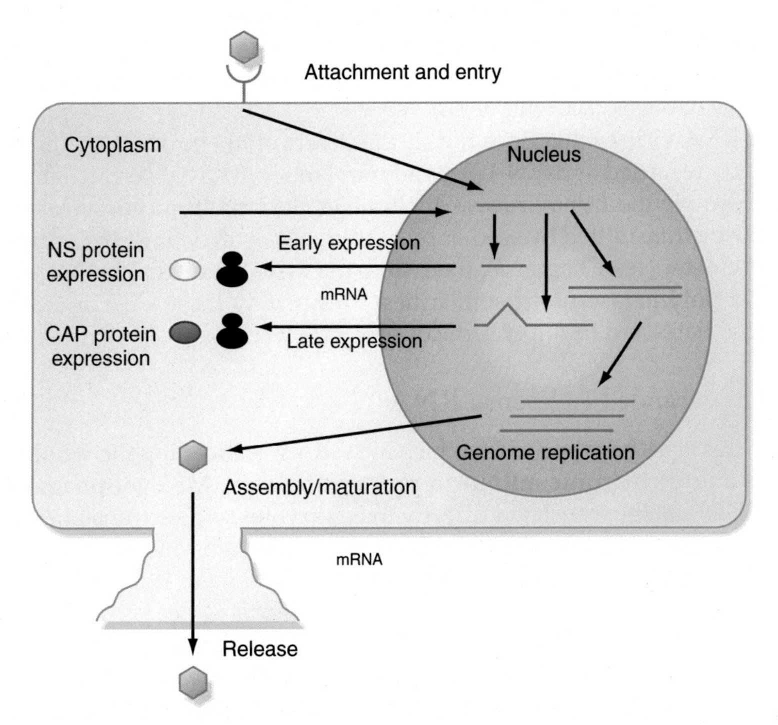

FIG. 4.13 Schematic representation of the replication of Class II viruses. All Class II viruses have relatively small genomes, that are mainly circular. Prokaryote viruses replicate in the cytoplasm; all eukaryote viruses replicate in the nucleus. All of these, and linear genomes such as those of parvoviruses, replicate by a rolling circle model that involves a replication-associated protein encoded by the viruses. Assembly will occur in the nucleus for eukaryotic viruses. None of the viruses have envelopes, so all are released from host cells by lysis or after cell degradation.

(e.g., **totiviruses**) have single-component genomes. Each segment of the multipartite genomes of reoviruses, cystoviruses and partitiviruses is transcribed separately: for eukaryotic viruses, this produces individual **monocistronic** or single-ORF mRNAs, whereas for cystoviruses, **polycistronic** mRNAs (having more than one ORF for translation) are produced. This reflects the differing requirements for translation in pro- and eukaryotic cells. All of the viruses take a **RNA-template-dependent RNA polymerase (RdRp) complex** into cells with them, as **transcription** is the first step in their replication cycle—and their host cells do not have the enzyme.

The viruses are **polyphyletic in origin** (see Chapter 3), and there is almost certainly a wide variety of mechanisms used for expression and replication. However, many of the viruses have not been well studied, so details are lacking. Most of the viruses, however, are known to have **conservative replication mechanisms**: this means that only one of the genomic strands is transcribed into a full-length (+) sense RNA that is used as a mRNA, then used as template to synthesize a new (−)sense RNA. A feature of many of the viruses is that the genomic dsRNA and the transcriptional complex are sequestered away from the cell by remaining inside "**core particles**" in the cell cytoplasm: this is true for **cysto-** and **reo-** and **partiti-** and **totiviruses**, for example. Viral translation products also accumulate as **viroplasms**: these are associations of viral structural and polymerase proteins and mRNAs that form **immature (provirion) particles**, inside which ssRNA+ molecules are transcribed to give RNA(−) molecules, with which they become base-paired. **This is the best-characterized example of conservative replication for any organism**. Part of the reason for this sequestration in eukaryotes is that **intracellular dsRNA-binding Toll-like receptors** (TLRs) and other defense molecules bind dsRNAs, and trigger defense mechanisms (see Chapter 6).

A group of dsRNA viruses that does not fit the description above are the **endornaviruses**: these were formerly regarded as dsRNA plasmids of plants, but have been recognized as viruses recently. They resemble the **hypoviruses**—themselves derived from **potyviruses**—in lacking particles, but may be transmitted by seed or by grafting. They may have their origin within the **alpha-like virus cluster** (see Chapter 3): their 10 kb dsRNAs have a single ORF with recognizable **helicase and polymerase motif** similarities. Presumably these exist as RdRp-associated replicative intermediates and multiply semiconservatively, like ssRNA(+) viruses (Fig. 4.14).

Class IV: Single-stranded (+) sense RNA

All viruses in this class have genomes which are ssRNA+, meaning the **whole genome can be translated as a multicistronic mRNA** in prokaryotes (e.g., **MS2 coliphage**), and **at least the 5′-proximal ORF** can be translated directly in eukaryotes, where translation is differently regulated. The most important product of this translation of genomic RNA is always the **replicase**, RdRp, as host cells cannot provide this. The first product of RdRp interaction with the genome will be full-length **(−) strand genomic RNA**, as this is necessary for production of **(+) sense genomic RNA** as well as for mRNAs should these be made. The (−) strand RNA is only found complexed with (+) sense RNA in essentially **double-stranded replication complexes**; over-production of (+) sense relative to (−) sense supplies the genome for encapsidation—which in some cases (e.g., **comoviruses**) is tied to replication, meaning virions cannot be assembled in vitro. The eukaryotic viruses can be essentially subdivided into three groups.

1. Viruses where the genomic RNA or RNAs have **only a single ORF**, and these are individually translated to form a **polyprotein** product. Polyproteins are subsequently cleaved, often sequentially and by means of a virus-encoded protease, to form the mature proteins. **Picornaviruses** have single ORF genomes; plant-infecting **comoviruses**

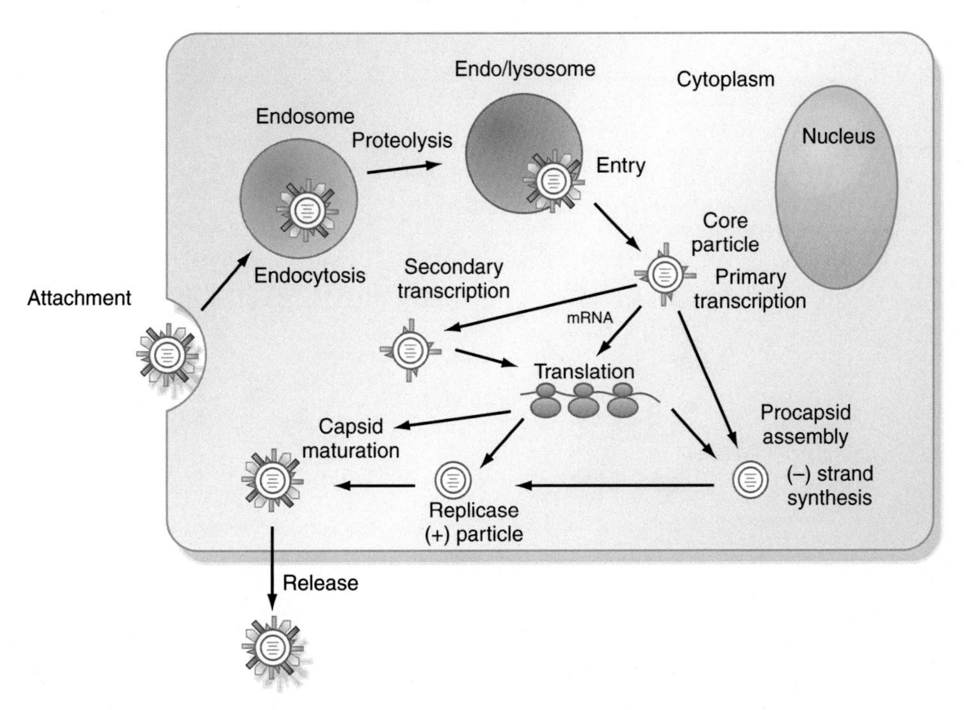

FIG. 4.14 Schematic representation of the replication of Class III viruses. Replication of these viruses takes place in the cytoplasm of infected cells, whether these are pro- or eukaryotic—although there are very few prokaryotic dsRNA viruses. Endosome entry only occurs in eukaryotes. Replication is conservative: one transcribed strand from an infecting genome within a core particle containing RdRp is used as template for a new dsRNA in a new "provirion" or immature particle. Virions do not bud but may be released from cells by lysis or after cell death, among other mechanisms.

(e.g., **cowpea mosaic virus**, in the picornavirus-like superfamily) have **two-component genomes**, each with a single ORF.

2. Viruses whose genomes—single- or multicomponent—are expressed partly by **direct translation** (5′-proximal ORF(s) only), and partly via **transcription**. For these viruses, genome replication is required before transcription of sequences downstream of the 5′-proximal ORF—encoding the **RdRp**—can occur, as this has to have a **(–) strand RNA template** to allow **subgenomic mRNA transcription** from **internal RNA promoter sequences** on the (–) strand. **Tobamoviruses** (e.g., **TMV**) have single-component genomes with two downstream ORFs expressed via individual 3′-coterminal mRNAs; **bromoviruses** (e.g., **brome mosaic, cowpea chlorotic mottle viruses**) have 3-component genomes, two of which have single ORFs, and the smallest has two ORFs with one (the **CP gene**) expressed via mRNA (see Chapter 5).
3. Viruses whose genomes have both a **polyprotein-encoding ORF** that is translated and then processed to provide regulatory proteins (e.g., **RdRp** and accessory proteins), and downstream ORFs **expressed via mRNAs** (Fig. 4.15). The **SARS-CoV-2 genome** is a good example here (see Fig. 1.26). Most of the 5′ end of the genomic RNA is a single ORF, which produces a polyprotein that is proteolytically processed into more than **15 individual proteins**, included the RdRp-containing **replicase complex**. After replication, (–) strand RNA is used to transcribe **8 mRNAs**, which can be used to produce another **13 or so proteins** (Fig. 4.15).

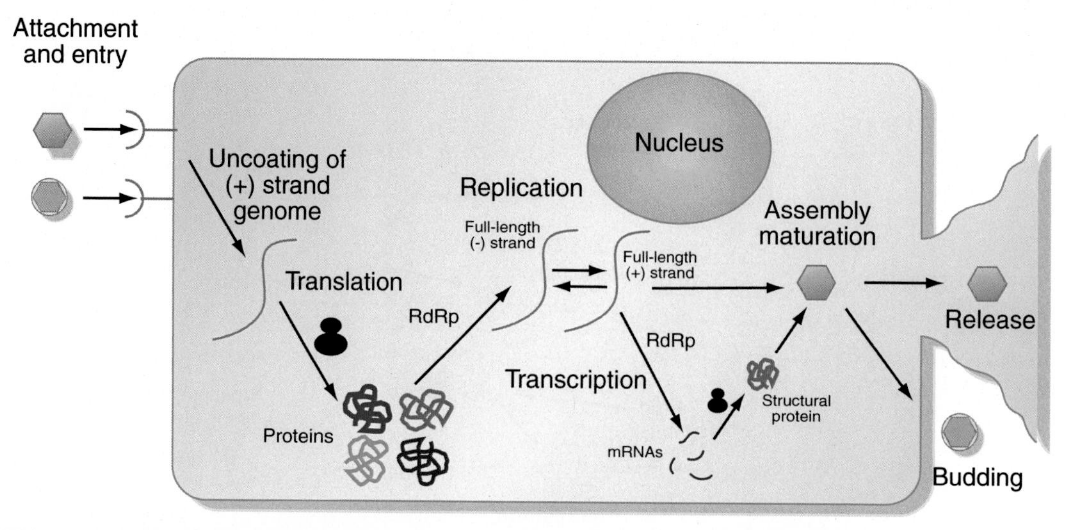

FIG. 4.15 Schematic representation of the replication of Class IV viruses. All replication of Class IV viruses takes place in the cytoplasm—for both pro- and eukaryotic cells. Prokaryote viruses (e.g., MS2, *Leviviridae*) may have polycistronic or multiple genes on a single RNA, all of which can be individually translated. With eukaryotic viruses, only the ORFs nearest the genome 5′ terminus are translated. Enveloped virions bud from cells, while naked virions may exit the cell by lysis or after cell death and degradation.

Class V: Single-stranded (−)sense RNA

These viruses only infect eukaryotes, and only a subset of organisms developing from marine invertebrates and vertebrates, and terrestrial insects, vertebrates, and plants. They are therefore all more closely related in an evolutionary sense than are the very diverse ssRNA+ viruses. As detailed in Chapter 3, ssRNA− viruses arose as an offshoot of the flavivirus-like supergroup of ssRNA+ viruses, meaning there are distinct similarities in the mode of replication. Essentially, ssRNA- genomes enter the cell together with the **genome-associated replicase complex**, which transcribes **mRNAs** in order to provide proteins necessary for replication and assembly. Following this, **full-length ssRNA+** is transcribed from the genomic RNA, and provides a template in replication complexes for **transcription/replication of the genomic ssRNA−**. The RNA+ strand is only found in **replication complexes**; ssRNA− accumulates in excess due to interactions of viral proteins with the complexes. The genomes of these viruses can be divided into **three basic types**, all of which replicate similarly, although expression of the genomes may differ (see Fig. 4.16).

1. **Nonsegmented** or single-component genomes (all viruses in order *Mononegavirales*; e.g., **rhabdo, paramyxo-,** and **filoviruses**). These viruses are all quite closely related phylogenetically (see Chapter 3), and share a common genetic structure and gene order. For these viruses the first step in replication is **transcription of the ssRNA− genome,** by the **RdRp** brought into cells in the virion, to produce **monocistronic (single gene) mRNAs**.
2. **Segmented genomes** (*Orthomyxoviridae*; **e.g., influenza viruses**), for which replication **occurs in the nucleus**, with monocistronic mRNAs for each of the virus genes produced

by the virus **transcriptase** in the nucleoprotein complex taken in the nucleus, from the full-length virus genome segments (see Chapter 5).

3. Segmented genomes with some of the viruses having an **ambisense** organization (Box 4.6): this means that one or more genome segments may have a ORF that is **directly readable by ribosomes**, while the rest of the genome can only be expressed via mRNA transcription (e.g., certain **bunyaviruses, arenaviruses**). These viruses all still take a RdRp complex into the cell in virions (Fig. 4.16).

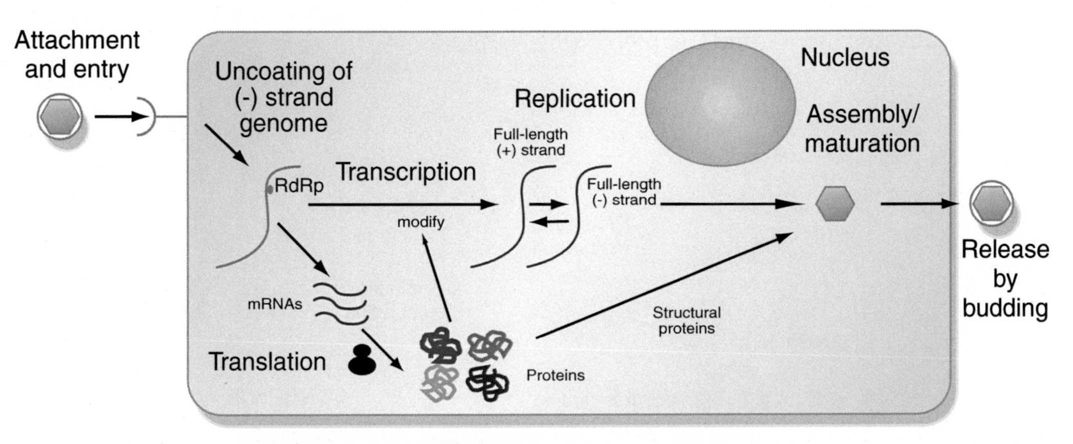

FIG. 4.16 Schematic representation of the replication of Class V viruses. Class V viruses are only found in eukaryotes—in marine invertebrates and vertebrates, and terrestrial insects and plants and vertebrates (see Chapter 3). Most viruses replicate in the cytoplasm; only orthomyxoviruses (e.g., influenza virus) may replicate in the nucleus, with enzymes taken into that location in a nucleoprotein complex—and imported from the cytoplasm via NLS transport through NPCs (see Fig. 4.11). Nearly all ssRNA– viruses have enveloped virions; only the plant-infecting ophioviruses do not—so nearly all exit the host cell by budding (Box 4.6).

BOX 4.6

Can't make your mind up? Do both!

Ambisense virus genomes contain at least **one RNA segment which is part (+)sense and part (−)sense—in the same molecule**. In spite of this, genetically they have more in common with ssRNA- viruses than ssRNA+ RNA viruses. But why on Earth would any virus bother with such a complicated gene expression strategy? In general, it is more difficult for RNA viruses to control gene expression that it is for DNA viruses to upregulate and downregulate individual gene products. Most ambisense viruses can replicate in a range of hosts, such as mammals and insects or insects and plants. In their vector or reservoir host, infection is usually asymptomatic. However, in another host, multiplication of the virus can be lethal. Having two different strategies for gene expression may help them to successfully span this divide—as well as providing an **extra gene for movement** in plants, as the plant-infecting **tospoviruses** (family *Bunyaviridae*) do.

Class VI: Single-stranded (+)sense RNA with LTR+ DNA intermediate

All of these viruses **only infect or are found in eukaryotes**, and share the attribute of having integrated copies of the genome with added **long terminal repeats** (LTRs) in cell nuclei as a necessary part of the replication cycle. **Retro-** and **meta-** and **pseudovirus** genomes found in particles are **ssRNA+** (see Chapter 3), but are unique in that they are **diploid**—each particle contains two copies of the genome—and they do not serve directly as mRNA. Rather, the RNAs as a template for **reverse transcription into DNA with LTRs**, using **RT inside the virion capsid**. **Integration** of the dsDNA into the nuclear genome occurs by means of the viral **integrase**, also present in the **nucleoprotein complex** that is transported into the nucleus (see Nuclear localization above and Chapter 3) (Fig. 4.17).

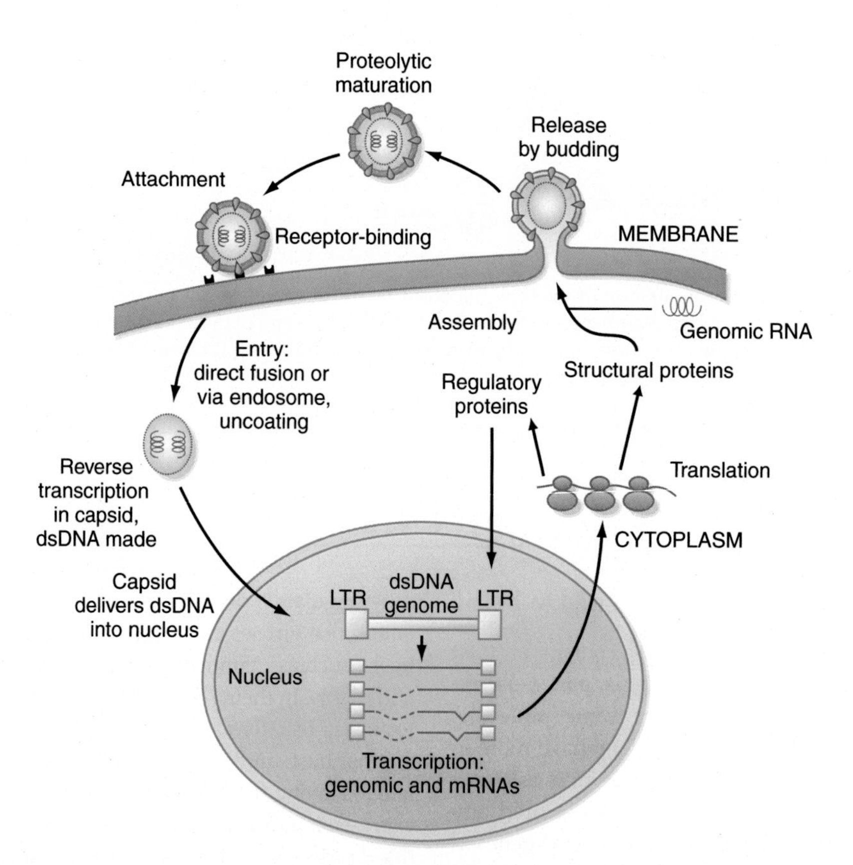

FIG. 4.17 Schematic representation of the replication cycle of Class VI viruses. Class VI viruses are all retroviruses, or endogenous LTR+ retrotransposons similar to retroviruses, found only in eukaryotes (see Chapter 3). Retroviruses only occur in vertebrates; meta- and pseudoviruses are also found in fungi and plants. Metaviruses make complete enveloped virions and may be infectious; pseudoviruses make only capsids (they do not have Env proteins, so do not bud), and are not infectious.

Class VII: Double-stranded DNA with LTR+ RNA intermediate

This group of viruses also relies on reverse transcription, but unlike the retroviruses (Class VI), where this occurs in capsids after infection and before integration and expression, this process occurs inside a **provirion** during maturation, which follows infection and expression by the dsDNA genome. On infection of a new cell, the first event to occur is **repair** of the gapped circular dsDNA genome, followed by transcription of mRNAs for translation and accumulation of the structural and other proteins necessary to form the provirion in the cytoplasm (Fig. 4.18). Full-length RNA transcripts that include LTRs—created by transcription that over-runs a transcriptional stop site (see Chapter 5)—associate with the structural proteins and RT that accumulate in the cytoplasm, and provirions form. These mature to form virions, with accompanying **RT-dependent synthesis** of gapped open circular dsDNA forms of the genome (Fig. 4.18).

Interference with replication

Understanding how particular types of genomes are replicated allows one to formulate specific strategies for interfering with the process (see Box 4.1). For example, while Class I viruses at first sight would seem to replicate and transcribe their genomes as the host does—which is true for the simplest viruses in this class—viruses with large genomes that encode their own DNA and RNA polymerases, and transcription factors and other replication-modifying proteins, will have functions specific to their proteins that are not shared by cellular proteins, and can be selectively interfered with. For example, replication of Class VI agents—typically **retroviruses**, but also **retrotransposons**—can be controlled to a large extent by one or more of

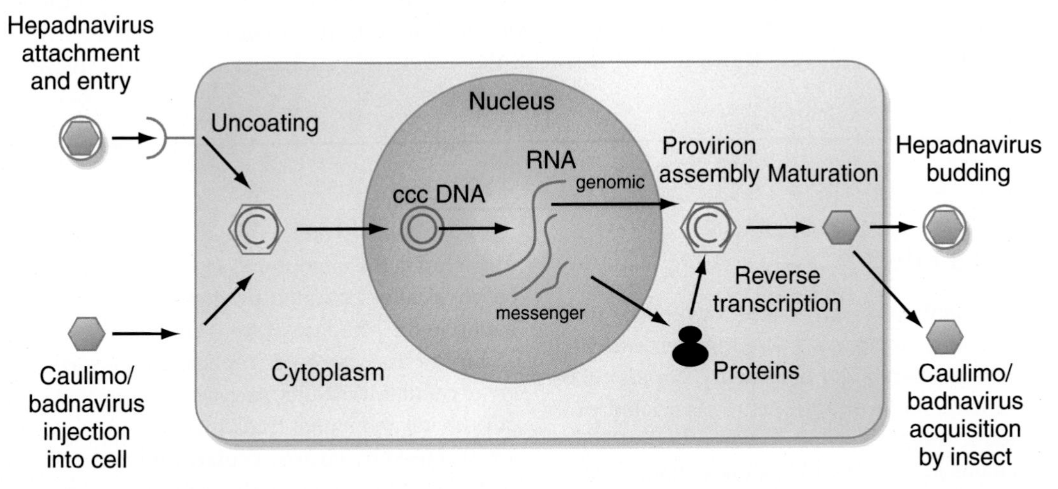

FIG. 4.18 Schematic representation of the replication of Class VII viruses. The two major classes of these viruses are very different: hepadnaviruses are enveloped and infect vertebrates; caulimoviruses and badnaviruses are non-enveloped and infect only plants via injection by insect vectors. Both make provirions—immature particles—in the cytoplasm from LTR+ RNA and translation products, that include structural proteins and RT. The RT makes partially double-stranded or "gapped" circular dsDNA genomes from RNA templates inside the provirions or maturing virions.

the growing arsenal of **antiretrovirals**. These include **nucleoside** and **nonnucleoside reverse transcriptase inhibitors** (NRTIs and NNRTIs), which interfere with actual replication, or production of DNA from an RNA template. **Protease** and **fusion inhibitors** (PIs, FI,) inhibit the kinds of processing necessary for maturation of proteins and virions, and entry of virions into cells by membrane fusion, respectively. **Integrase inhibitors** stop the action of **integrase** that allows the insertion of viral DNA into the healthy cell's DNA. These and other strategies that specifically combat other types of viruses—for example, **thymidine kinase inhibitors** as chemotherapeutics against certain **herpesviruses; remdesivir** as an anti-**Ebola RdRp** drug—will be discussed in more detail in Chapter 6.

Assembly of virions

The **assembly** process involves the collection of all the components necessary for the formation of the mature **virion** at a particular site in the cell. During assembly, the basic structure of the virion is formed. The site of assembly depends on the site of replication within the cell and on the mechanism by which the virus is eventually released from the cell—and this process differs for different viruses. For example, with **picornaviruses, poxviruses,** and **reoviruses**, assembly occurs in the cytoplasm; with **adenoviruses, polyomaviruses**, and **parvoviruses**, it occurs in the nucleus (Box 4.7).

As with the early stages of replication, it is not always possible to identify the **assembly**, **maturation**, and **release** of virions as distinct and separate phases. The site of assembly has a profound influence on all these processes. In the majority of cases involving **enveloped** virions, cellular membranes are used to anchor virus proteins, and this initiates the process of assembly. In the case of **nonenveloped** virions, assembly takes place in the nucleus of eukaryotes, or in the cytoplasm of all prokaryote and many eukaryote cells, with the genomes localized there.

In spite of considerable study, the control of virus assembly is generally not well understood. In general, it is thought that intracellular levels of virus **structural proteins** and

BOX 4.7

Lipid rafts

As mentioned above (see **Caveolin**), lipid rafts are membrane microdomains enriched with glycosphingolipids (or glycolipids), cholesterol, and a specific set of associated proteins. A high level of saturated hydrocarbon chains in sphingolipids allows cholesterol to be tightly interleaved in these rafts. The lipids in these domains differ from other membrane lipids in having **relatively limited lateral diffusion** in the membrane, and they can also be physically separated by density centrifugation in the presence of some detergents.

Lipid rafts have been implicated in a variety of cellular functions, such as apical sorting of proteins and signal transduction, but they are also used by viruses as **platforms for cell entry** (e.g., HIV, SV40, and rotavirus), and as **sites for particle assembly, budding, and release** from the cell membrane (e.g., influenza virus, HIV, measles virus, and rotavirus).

genomic nucleic acids reach a critical concentration, and that this triggers the process. Many viruses achieve high levels of newly synthesized structural components by concentrating these into **subcellular compartments**, visible in light microscopes, which are known as **inclusion bodies,** or **virus factories**. These are a common feature of the late stages of infection of prokaryotic or eukaryotic cells by many different viruses. The size and location of inclusion bodies in infected cells is often highly characteristic of particular viruses: for example, **rabies virus** infection results in large perinuclear "**Negri bodies**," first observed using an optical microscope by Adelchi Negri (see Chapter 1). Alternatively, local concentrations of virus structural components can be boosted by lateral interactions between membrane-associated proteins. This mechanism is particularly important in enveloped viruses released from the cell by **budding**. Another important and recently-recognized mode of forming inclusion bodies is **liquid-liquid phase separation** (see Box 4.8).

BOX 4.8

Worlds within worlds

We think of eukaryotic cells as simply compartmentalized into nucleus and cytoplasm, but the true situation is much more complicated than that (see Fig. 4.5): the cell interior is a **mesh of cytoskeletal filaments**, with mitochondria, nucleus, vesicles and much-ramified ER taking up space, so that the interior is really rather crowded. There are also other biochemical rather than physical compartments within a cell. One is the **lipid/aqueous division**: proteins with **hydrophobic** (water-fearing) domains don't like to be in a soluble form within the cytoplasm. They only start to act when they're in the natural environment of a membrane.

But it's not that simple: there are **different domains within membranes** where different processes occur, such as **lipid rafts** (Box 4.5). Viruses have used these lipid rafts for particular functions, such as entering or leaving the cell, and forming tiny factories where new particles are assembled.

There's also **liquid-liquid phase separation**, a recent area of research attracting much attention: replication complexes or "**viroplasms**" of ssRNA+ viruses tend to be compartments surrounded by membranes; however, ssRNA− paramyxovirus viroplasms are nonmembrane bounded **condensates** that form by phase separation due to interaction of nucleoprotein (N) and phosphoprotein (P), resulting in concentration and sequestration of the proteins. Virus RNA colocalizes to the condensates, triggering N to assemble into nucleoprotein helices (see Guseva et al., 2020). The SARS-CoV-2 NP also undergoes phase separation with viral RNA, which also has the effect of suppressing the innate antiviral defense (see Chapter 6 and Wang et al., 2021). The value of knowing this is that **there are drugs that can inhibit phase separation**, which could result in broad-spectrum interference with virus replication.

And then there's **time**: the processes of virus replication do not happen in a random order—they are carefully sequenced to optimize the process. This control is directed by the biochemistry of the components involved, which may only start to function as their **concentration** within an infected cell reaches a critical level, which is most noticeable during virion assembly. And all of this goes on within the minute world of an infected cell, too small to see with the eye alone.

As discussed in Chapter 2, the formation of virions may be a relatively simple process which is driven only by interactions between the subunits of the **capsid and genome**, and controlled by the rules of symmetry. The encapsidation of the virus genome may occur early in the assembly of the particle: for example, assembly for many viruses with **helical** symmetry is nucleated on the genome (see Fig. 1.17). For small isometric virions MS2 coliphage or $T=3$ plant virus virions, high affinity association of a few CPs at low CP concentration nucleates assembly that is completed at higher CP concentration. **Papillomavirus** virions (e.g., HPV types) have 360 copies of a major (L1) and up to 72 copies of a minor (L2) capsid protein. Virions assemble in the **nucleus** by an initial rapid assembly of **pentamers of L1**, usually with inclusion of 1 L2 molecule, then a slow process of assembly of 72 pentamers with L2 at an internal position near 5′ rotational axes of symmetry, together with the circular dsDNA genome. In fact, particles assemble in vivo or in vitro around any DNA of between 5 and 8 kbp, meaning they are highly convenient for making **pseudovirions**, or properly-assembled capsids containing nonviral DNA.

In other cases, assembly is a **highly complex, multistep process** involving not only virus structural proteins and possibly proteolysis, but also virus-encoded and cellular **scaffolding proteins** that act as templates to guide the assembly of virions. It may also occur at a late stage, when the genome is packed into an almost completed complex protein shell by the action of a **molecular motor**. Examples of some of these sorts of assembly are shown below.

Picornavirus capsids contain four structural proteins. Sixty copies each of the three major proteins VP1–3 (see fig. 2.9) constitute the visible capsid. There are also 60 copies of a small fourth protein, VP4: this is located on the inside of the capsid. The whole picornavirus genome is translated as one long ORF; however, "**autocatalytic cleavage**"—by a protease inside the sequence—cuts the polyprotein up as it is still being translated. VP4 is formed from cleavage of the **structural protein precursor VP0** into VP2 and VP4 late in assembly (see Fig. 4.19),

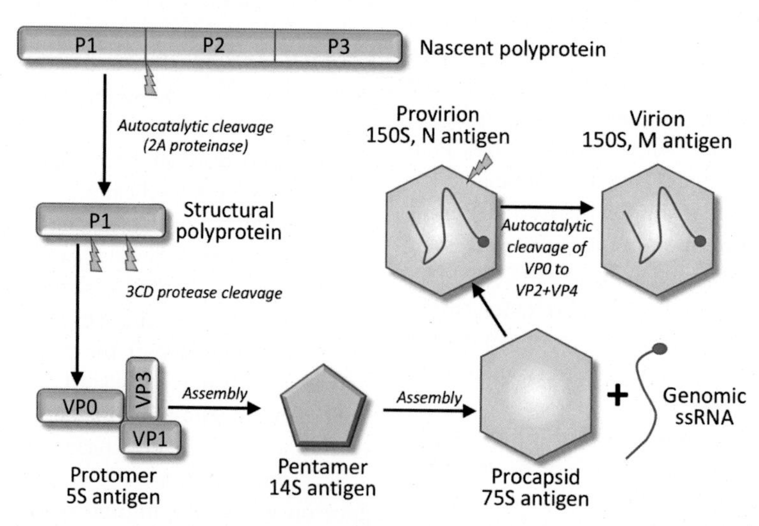

FIG. 4.19 Proteolytic processing and self-assembly of picornavirus capsids. Picornavirus proteins, and those of some other evolutionarily related ssRNA+ viruses, are produced by cleavage of a long polyprotein into the final products needed for replication and capsid formation (see Chapter 5).

and is myristoylated at its aminoterminus (i.e., it is modified after translation by the covalent attachment of myristic acid, a 14-carbon unsaturated fatty acid). Five VP4 monomers form a hydrophobic micelle, driving the assembly of a pentameric subassembly. There is biochemical evidence that these pentamers, which form the vertices of the mature capsid, are a major precursor in the assembly of the particle; hence, the chemistry, structure, and symmetry of the proteins that make up the picornavirus capsid reveal how the assembly is driven.

Viruses with tailed virions in the *Caudovirales* (see Chapter 3), in the families *Myoviridae*, *Siphoviridae*, and *Podoviridae*, have been extensively studied for several reasons, among them their highly complex processes of virion self-assembly. These viruses are easy to propagate in bacterial cells, can be obtained in high titers and are easily purified, making biochemical and structural studies comparatively straightforward. **T4-type coliphages** are especially interesting because of their **contractile tail structures**. The heads of these particles consist of an **icosahedral shell** with $T=7$ symmetry—some with an "expansion ring" of hexamers that allows elongated heads—attached by a **collar** to the **contractile, helical tail**. At the end of the tail is a **base plate** that functions in attachment to the bacterial host, and also in **penetration** of the bacterial host cell wall by means of lysozyme-like enzymes associated with the plate (see above). In addition, thin protein fibers attached to the tail plate are involved in **binding to receptor molecules** in the wall of the host cell. Inside this compound structure there are also internal proteins and polyamines associated with the genomic DNA in the head, and an internal tube structure inside the outer sheath of the helical tail. The sections of the particle are put together by **separate assembly pathways for the head and tail sections** inside infected cells: these come together at a late stage to make up the infectious virion (Fig. 4.20).

A major problem viruses must overcome is how to **specifically select and encapsidate the virus genome** from the background noise of cellular nucleic acids. In most cases, by the late stages of virus infection when assembly of virions occurs (see Chapter 5), transcription of cellular genes has been reduced and a **large pool of virus genomes** has accumulated. Overproduction of virus nucleic acids eases but does not eliminate the problem of specific genome packaging. A virus-encoded **capsid or nucleocapsid protein** is required to achieve specificity, and many viruses, including **retroviruses** (see below), SARS-CoV-2 and **rhabdoviruses**, encode this type of protein.

Certain viruses with **segmented genomes**, such as **reo- and orthomyxoviruses** (see Chapter 3) face further problems. Not only must they encapsidate only virus nucleic acid and exclude host cell molecules, but they must also package **at least one of each** of the required genome segments **in the same particle**. During particle assembly, errors frequently occur. These can be measured by **particle:infectivity** ratios: these are the ratio of the **total number of particles** in a virus preparation (counted by electron microscopy) to the number of particles **able to give rise to infectious progeny** (measured by **plaque** or limiting dilution assays). This value is in some cases found to be **several thousand particles** to each **infectious unit**—presumed to be a virion—and only rarely approaches a ratio of 1:1. This is also a factor for viruses with nonsegmented genomes such as MS2 phage, which indicates how much worse it could be for segmented genomes.

However, calculations show that viruses such as **influenza** have far lower **particle:infectivity ratios** than could be achieved by **random packaging** of eight different genome segments. Until recently, the process by which this was achieved was poorly understood, but a clearer picture has emerged of a mechanism for specifically packaging a full genome, mediated by

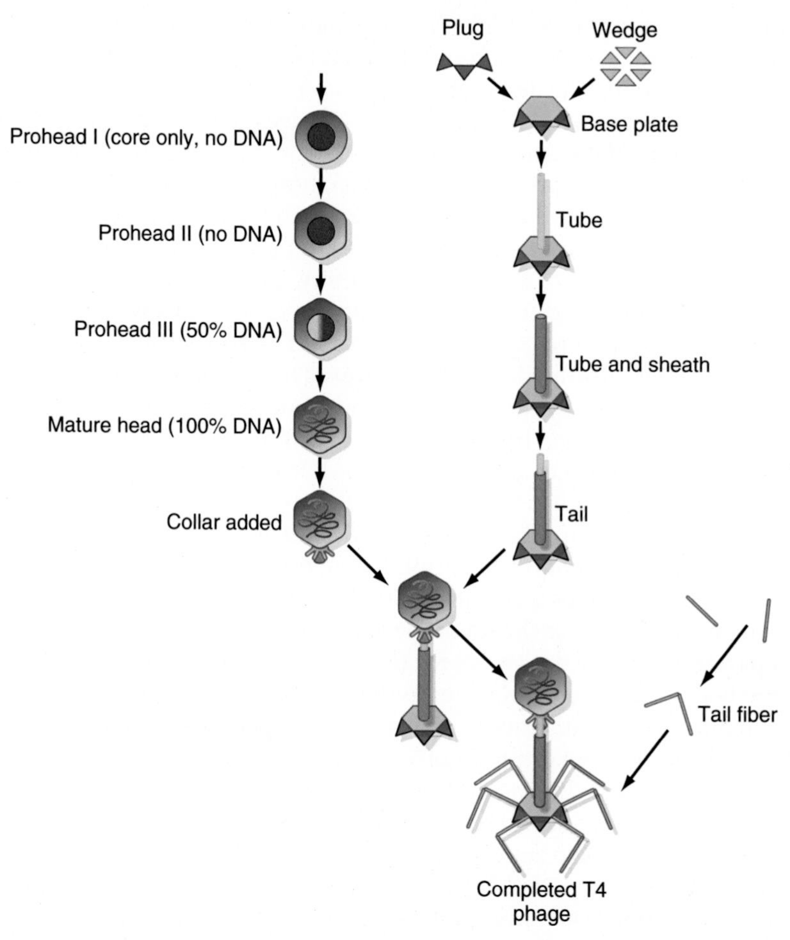

FIG. 4.20 Assembly pathway of bacteriophage T4. Simplified version of the assembly of bacteriophage T4 particles (*Myoviridae*). The head and tail sections are assembled separately and are brought together at a relatively late stage. This complex process was painstakingly worked out by the isolation of phage mutants in each of the virus genes involved. In addition to the major structural proteins, a number of minor "scaffolding" proteins are involved in guiding the formation of the complex particle.

***cis*-acting packaging signals** in the **genomic or vRNAs** (see Li et al., 2021). Consequently, genome packaging, previously thought to be a random process in influenza viruses, turns out to be quite highly ordered. **Rotaviruses**, with their **11 genome segments**, face an even more complex problem, and it is not clear how they have solved this issue. Current evidence indicates a "**concerted packaging model,**" with the 11 ssRNA+ molecules transcribed from the genomic segments interact with one other via *cis*-acting sequences and RNA-encoded structural elements, prior to encapsidation within a precore structure in which second strand synthesis takes place (see McDonald and Patton, 2011).

While many ssRNA+ exclusively plant-infecting viruses have segmented genomes—viruses in family ***Bromoviridae*** have three genome components, for example, and **comoviruses** have two—there does not seem to be the same pressure to have all genome segments in the same particle, and in fact only **one genome segment is found per virion**. This is probably because, unlike orthomyxoviruses like influenza A viruses or reoviruses in animal hosts, the virions reach very high concentrations in infected plants and often aggregate in **ordered quasicrystalline arrays**, and each infection is probably initiated by many virions. By contrast, animal-infecting virions rarely reach the same concentrations, and infection is likely to be initiated by far fewer particles than is the case with plants—meaning that all genome components need to be in the same particle.

Although **retrovirus** genomes are not segmented, retrovirions are only infectious if they contain **two complete copies of the genome**, because of the need for two copies during reverse transcription (see Chapter 5). Different retroviruses solve the problem of how to package not one but two genomes in different ways. There are specific nucleotide sequences in the genome (the **packaging signal**—which forms a complex **secondary structure**, much as TMV RNA does) which permit the **virus structural protein** to select **genomic nucleic acids** from the cellular background. The packaging signal from a number of virus genomes has been identified. Examples are the ψ (**Psi**) signal in retrovirus genomes which has been used to package synthetic "**retrovirus vector**" genomes into a **pseudovirion**, and the sequences responsible for packaging the genomes of several **DNA virus** genomes (e.g., some **adenoviruses** and **herpesviruses**) which have been clearly and unambiguously defined. Efficient genome packaging requires information not only from the linear nucleotide sequence of the genome but also from regions of secondary structure formed by the folding of the genomic nucleic acid into complex forms. In many cases, attempts to find a **unique, linear packaging signal** in virus genomes have failed. The probable reason for this is that the key to the specificity of genome packaging in most viruses lies in the **secondary structure of the genome**.

Like many other aspects of virus assembly, the way in which packaging is controlled is, in many cases, not well understood. However, in some cases detailed knowledge about the mechanism and specificity of genome encapsidation is now available. These include both viruses with **helical symmetry** and some with **icosahedral symmetry**.

Undoubtedly the best understood packaging mechanism is that of the **ssRNA+ helical plant virus, TMV**. This is due to the relative simplicity of this virus, which has only a single major coat protein and will spontaneously assemble from its purified RNA and protein components in vitro, due to the presence of an "**origin of assembly**" sequence near the genome 3′ end (see Fig. 1.17 and Chapter 2).

Bacteriophage M13 is another helical virus where protein-nucleic acid interactions in the virion are relatively simple to understand (fig. 2.4). The primary sequence of the **g8p** molecule determines the orientation of the protein in the capsid. The inner surface of the rod-like phage capsid is positively charged and interacts with the negatively charged genome, while the outer surface of the cylindrical capsid is negatively charged. However, the way in which the capsid protein and genome are brought together is a little more complex. During replication, the genomic DNA is associated with a **ssDNA-binding protein, g5p**. This is the most abundant of all virus proteins in an M13-infected *E. coli* cell, and it **coats the newly replicated single-stranded phage DNA** in a concentration-dependent process, forming an intracellular rod-like structure similar to the mature phage particle but somewhat longer and thicker (1100×16nm). The function of this protein is to protect the genome from host cell nucleases and incidentally to **interrupt genome replication**, sequestering newly-formed **ssDNA**

strands as substrates for encapsidation as the concentration of g5p increases. Newly synthesized coat protein monomers (**g8p**) are associated with the **inner (cytoplasmic) membrane** of the cell, and it is at this site that assembly of the virion occurs. The g5p coating is stripped off as the particle passes out through the membrane and is exchanged for the mature g8p coat, plus accessory proteins. The forces that drive this process are not fully understood, but the **protein-nucleic acid interactions** that occur appear to be rather simple and involve **opposing electrostatic charges** and the stacking of the DNA bases between the planar side-chains of the proteins. This is confirmed by the variable length of the M13 genome and its ability to freely encapsidate extra genetic material.

Protein-nucleic acid interactions in other helical viruses, such as **rhabdoviruses** (see Fig. 2.5), are rather more complex. In most **enveloped** virions with **helical nucleoproteins**, the latter forms first as the **N or NP** accumulates in the cell, and is then coated by **matrix proteins**, and then the **envelope and its associated glycoproteins** during budding (see Fig. 2.12 for the budding process). Assembly of rhabdoviruses follows a well-orchestrated program. It begins with the **RNA-containing nucleocapsid** stretched out as a ribbon. The ribbon curls into a tight ring and then is physically forced to curl into larger rings that eventually form the **helical trunk** at the center of the particle. **Matrix** (M) protein subunits bind on the outside of the nucleocapsid, rigidifying the bullet-shaped tip and then the trunk of the particle, and create a **triangularly-packed platform** for binding **glycoprotein** (G) trimers and the **envelope membrane**, all in a coherent operation during budding.

Rather less is known about the arrangement of the genome inside most particles with **icosahedral symmetry**. Exceptions to this statement, however, are the ***T*=3 icosahedral RNA viruses** whose subunits consist largely of the "eight-strand antiparallel β-barrel" structural motif (see Chapter 2, Figs. 2.8 and 2.9). In these virions, positively charged inward-projecting arms of the capsid proteins interact with the RNA in the center of the particle. In **bean pod mottle virus** (BPMV), a $T=3$ **comovirus** with a bipartite genome (see Chapter 3), X-ray crystallography has shown that the extensively structured RNA is folded in such a way that it **assumes icosahedral symmetry**, corresponding to that of the capsid surrounding it. The regions that contact the capsid proteins are **single stranded** and appear to interact by **electrostatic forces**. The atomic structure of **ϕX174** microvirus also shows that a portion of the DNA genome **interacts with +ve-charged arginine residues** exposed on the inner surface of the capsid in a manner similar to BPMV.

A consensus about the physical state of nucleic acids within icosahedral virus capsids appears to be emerging. Just as the icosahedral capsids of many genetically unrelated viruses are based on monomers with a common **SJR structural motif** (see Chapters 2 and 3), the genomes inside also appear to display icosahedral symmetry, the vertices of which interact with **basic amino acid residues on the inner surface** of the capsid. These common structural motifs may explain how viruses selectively package the required genomic nucleic acids and may even offer opportunities to design specific drugs to inhibit these vital interactions.

Maturation is the stage of the replication cycle at which the virus becomes infectious. This process usually involves structural changes in the virion that may result from specific cleavages of **capsid** proteins to form the mature products or conformational changes which occur in proteins during assembly. Such events frequently lead to substantial structural changes in the capsid that may be detectable by measures such as differences in the antigenicity of incomplete and mature virions, which in some cases (e.g., picornaviruses) alters radically. Alternatively, **internal structural alterations**—for example, the **condensation of nucleoproteins** with the virus genome—often result in such changes. As already stated, for some viruses virion assembly and maturation occur inside the cell and are inseparable, whereas for others maturation events may occur **only after release** of the virion from the cell. In all cases, the process of maturation prepares the particle for the infection of subsequent cells.

Virus-encoded proteases are frequently involved in maturation, although cellular enzymes or a mixture of virus and cellular enzymes are used in some cases. Clearly there is a danger in relying on cellular proteolytic enzymes in that their relative lack of substrate specificity could easily completely degrade the capsid proteins. In contrast, virus-encoded proteases are usually **highly specific** for particular amino acid sequences and structures, frequently only **cutting one particular peptide bond** in a large and complex virus capsid. Moreover, they are often further controlled by being **packaged into virions** during assembly and are only activated when brought into close contact with their target sequence by the conformation of the capsid (e.g., by being placed in a local **hydrophobic** environment or by **changes of pH** or **metal ion concentration** inside the capsid).

Retrovirus proteases are good examples of enzymes involved in maturation which are under this tight control. The retrovirus **core particle** is composed of proteins expressed from the *gag* gene, and the protease is packaged into the core before its release from the cell by **budding (see fig. 2.12)**. Gag polyprotein precursors accumulate inside the cell membrane due to myristylated residues at their N-termini. There is an encapsidation signal (**Psi**, a highly structured sequence near the 5′ end of **genomic RNA**) that promotes specific binding of dimers of the ssRNA+ molecules to near the C-terminus of Gag. At some stage of the budding process (the exact timing varies for different retroviruses), the polyprotein is cleaved in situ into **matrix (MA), capsid (CA),** and **nucleocapsid (NC)** proteins by the protease (**Pro**) sequence that is included in **1/20th of Gag molecules** by *gag-pol* readthrough during translation. The protease cleaves the **Gag** protein precursors into the mature products—the **capsid (CA), nucleocapsid (NC),** and **matrix (MA) proteins** of the mature virion (Fig. 2.13 and Fig. 4.21). The **Pol protein—RT** and **integrase**—is also provided by this process, and is associated with the genome inside the assembling capsid.

Not all protease cleavage events involved in maturation are this tightly regulated. Native **influenza virus HA** undergoes posttranslational modification (PTM; in this case **glycosylation** in the Golgi apparatus) and at this stage exhibits receptor-binding activity. However, the protein must be **cleaved into two fragments** (HA_1 and HA_2) to be able to produce **membrane fusion** during infection. Cellular trypsin-like enzymes are responsible for this process, which occurs in **secretory vesicles** as the virus buds into them prior to **release** at the cell surface. **Amantadine and rimantadine** are two drugs that are active against influenza A viruses (Chapter 6). The action of these closely related compounds is complex, but they block **cellular membrane ion channels**. The target for both drugs is the **influenza matrix protein 2 (M2,** a "viroporin"), but resistance to the drug may also map to the HA gene. The replication of some strains of influenza virus is inhibited at the **penetration** stage and that of others at **maturation**. The biphasic action of these drugs results from the inability of drug-treated cells to **lower the pH of the endosomal compartment** (a function normally controlled by M2), and hence to cleave HA during maturation. Similarly, retrovirus **envelope** glycoproteins require cleavage into the **surface (SU)** and **transmembrane (TM)** proteins for activity: in HIV these **are gp41 (TM)** and **gp120 (SU)**, derived from a **gp160 Env** precursor. This process is also carried out by cellular enzymes but is in general poorly understood; however, it is a target for **inhibitors** which may act as **antiviral drugs**.

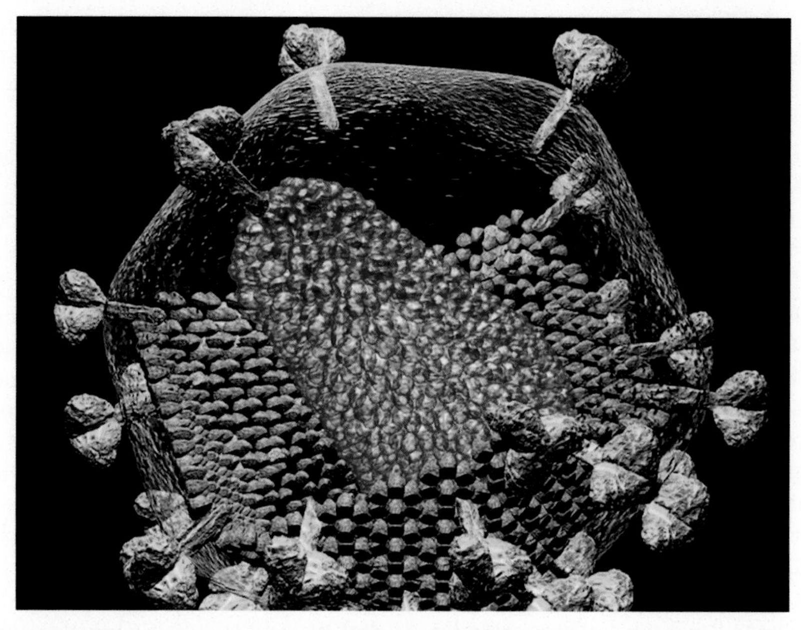

FIG. 4.21 Processing of retrovirus virion structural proteins after budding. Fully-processed mature HIV virion. The Gag polyprotein has been processed by the protease function of the Pol polyprotein into matrix (MA, *blue*), capsid (CA, *purple*), and nucleocapsid (NC, *orange*), with the nucleoprotein complex inside the capsid within the matrix and membrane. *Courtesy of Russell Kightley Media.*

Release of virions from cells is a highly varied process, depending on the virus and on the host cell. As described earlier, plant viruses face particular difficulties caused by the **structure of plant cell walls** when it comes to leaving cells and infecting others. In response, they have evolved particular strategies to overcome this problem: these are discussed in detail in Chapter 6. All other viruses escape the cell by **one of three mechanisms**—the first of which is also true for plant viruses.

Many plant virus virions simply **accumulate** in their infected host cells until concentrations are high enough that virions form **crystalline** or **quasicrystalline arrays**: these are released by death and degradation or by mechanical disruption of cells, for onward **mechanical transmission** to cells in other plants. The same is basically true for **occluded baculoviruses** in insect tissues (see Chapter 2, Fig. 2.20). For **lytic** viruses—such as most **prokaryote viruses** and most **nonenveloped animal viruses, release** is a simple process—the infected cell breaks open, due either to virus-specific processes (e.g., **lysins** for phages) or due to the **host immune system killing the cell**, and releases the virus. **Enveloped viruses** acquire their lipid membrane as the virus buds out of the cell through the cell membrane or into an intracellular vesicle prior to subsequent release. **Virion** envelope proteins are picked up during this process as the virion is extruded. This process is known as **budding** (see Chapter 2 and Fig. 2.12). Release of virions in this way may be highly damaging to the cell (e.g., **paramyxoviruses, rhabdoviruses,** and **togaviruses**), or in other cases, may appear not to be (e.g., **retroviruses**), but in either case the process is controlled by the virus. That is, the physical interaction of the **capsid** proteins on the **inner surface of the cell membrane** forces the particle out through the membrane. As mentioned earlier, assembly, maturation, and release are usually **simultaneous processes** for virions formed by budding. The type of membrane from which the virus buds depends on the virus concerned. In most cases, budding involves **cytoplasmic** or **cell outer membranes** (retroviruses, togaviruses, orthomyxoviruses, paramyxoviruses, bunyaviruses, coronaviruses, rhabdoviruses, and hepadnaviruses). In other cases, budding can involve the **nuclear membrane** (herpesviruses), or the **endoplasmic reticulum** (pox-, irido-, asfarviruses; see Chapter 2).

In a few cases, notably in **human retroviruses** such as HIV and human T-cell leukemia virus, viruses utilize **direct cell-to-cell spread** rather than release into the external environment and reuptake by another cell. This process requires intimate contact between cells and can occur at **tight junctions** between cells or in **neurological synapses**. These structures have been subverted by human retroviruses, which engineer a novel structure in infected cells known as a **virological synapse** to promote more efficient spread within the host organism (Fig. 4.22).

The release of mature virions from susceptible host cells by budding presents a problem in that these particles are designed to enter, rather than leave, cells—and the **receptors** by which they entered are present on the cell they bud from. How do these particles manage to leave the cell surface? The details are not known for many viruses, but there are clues as to how the process is achieved. Certain **virus envelope proteins** are involved in the **release** phase of replication as well as in the initiating steps. A good example of this is the **neuraminidase (NA)** protein of **influenza virus**. In addition to being able to reverse the attachment of virions to cells via HA, NA is also believed to be important in preventing the aggregation of influenza virions and may well have a role in virus release. The attachment protein **hemagglutinin/neuraminidase (HN)** in **paramyxoviruses** has the same function. The

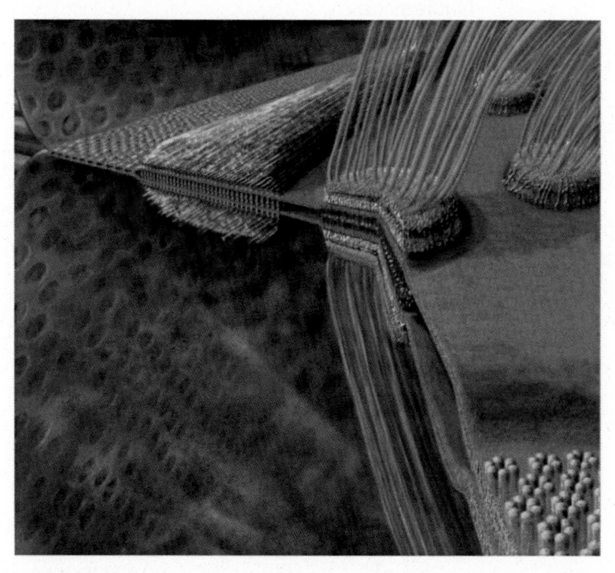

FIG. 4.22 Mammalian cell tight junctions. Depiction of zonula occludens (tight junction; provide a link between the membrane proteins and the cytoskeleton), zonula adherens (adherens junction, *top left*; a belt-like junction that maintains the physical integrity of the epithelium), macula adherens (desmosome, *middle*; cell structure specialized for cell-to-cell adhesion) and gap junctions (*bottom right*; aggregates of intercellular channels allowing cell-to-cell transfer of ions and small molecules) between cells in solid tissues. *Courtesy of Russell Kightley Media.*

influenza NA receptor **glycan cleavage process** is targeted by newer drugs such as **oseltamivir** (trade name "Tamiflu") and **zanamivir** ("Relenza") (Chapter 6).

In recent years, a group of proteins known as **viroporins** has been discovered in a range of different viruses. These are small **hydrophobic pore-forming** proteins that modify the permeability of cellular membranes and customize host cells for efficient virus propagation. For example, they promote the release of viral particles from infected cells. These proteins are usually not essential for the replication of viruses, but their presence often enhances virus growth. The **M2 protein in influenza viruses** is the best characterized of these: it is the smallest known hydrogen ion pump pore-forming protein (four copies to form a pore), and because it is so conserved among influenza viruses, is the target for so-called "universal vaccine" strategies (see Chapter 6).

In addition to using specific proteins, enveloped viruses that bud from cells (see also Chapter 2) have also solved the problem of release by the **careful timing of the assembly-maturation-release pathway**. Although it may not be possible to separate these stages by means of biochemical analysis, this does not mean that careful spatial separation of these processes has not evolved as a means to solve this problem. Similarly, although we may not understand all the subtleties of the many conformation changes that occur in virus capsids and envelopes during these late stages of replication, virus replication clearly works, despite our lack of knowledge.

Blocking virus entry, uncoating, expression, replication, and release was touched earlier in the chapter in the appropriate sections. However, it is as well to condense this information into a diagram that allows one, at a glance, to understand exactly what can be blocked, at which stage of the virus life cycle (Fig. 4.23).

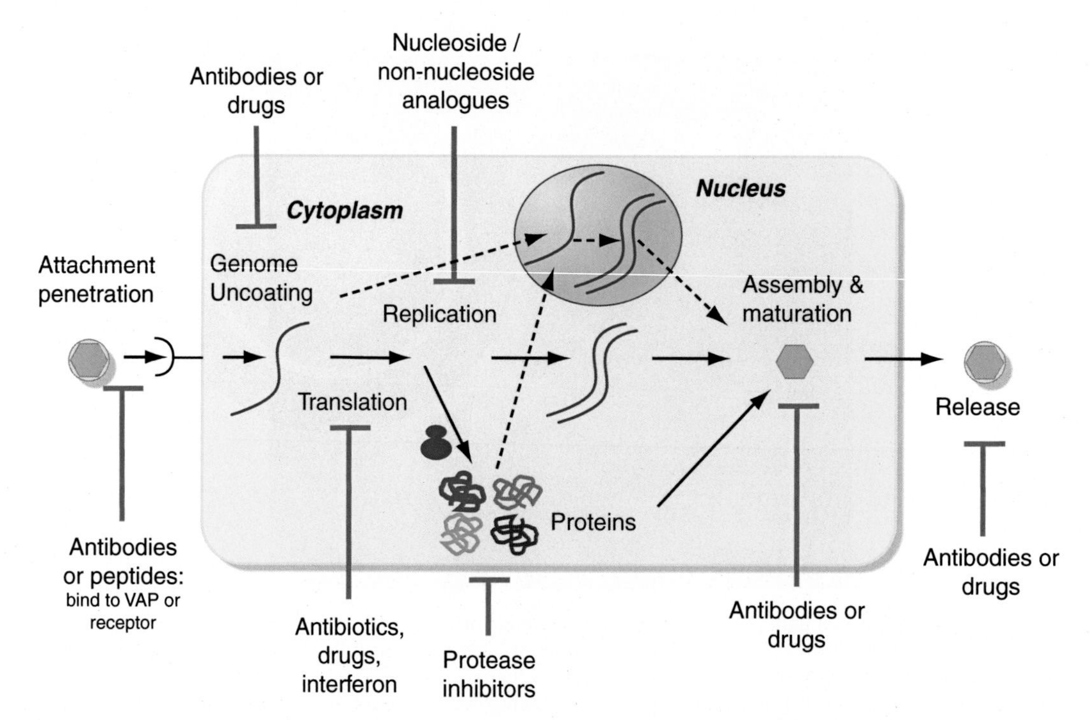

FIG. 4.23 Stages of a typical virus replication cycle. The diagram illustrates where and how the process can be blocked.

As always in virology, every year brings completely unexpected discoveries—and one recent one was that hepatitis A virus (*Picornaviridae*) is released from infected hepatocytes as single or multiple virions encased in **exosomes** derived from **late multivesicular bodies** (MVB), so as to form "**quasi-enveloped**" particles. This renders them invisible to circulating antibodies as the form found in the circulatory fluids and blood is almost exclusively enveloped, which facilitates their dissemination within hosts. Quasienveloped particles are infectious, although their mode of entry into cells is obscure. Given that the main route of infection among people is fecal-oral, however, it is interesting that virions in feces are nonenveloped: they are apparently stripped of membranes by bile acids present in the pathway of secretion leading to the gut. The same is true for **hepatitis E virus**, a ssRNA+ **norovirus** related to the diarrhea-causing Norwalk virus—which indicates that this is probably a much more common phenomenon than was hitherto thought (see Feng et al., 2014).

Summary

In general terms, virus replication involves three broad stages carried out by all types of virus. These the **initiation** of infection, **replication**, and **expression** of the **genome**, and, finally, **release** of **virions** from the infected cell. At a detailed level, there are many differences in the replication processes of different viruses which are imposed by the biology of the host cell and the nature of the virus genome. Nevertheless, it is possible to derive an overview of virus replication and the common stages which, in one form or another, are followed by all viruses.

References

Cobb, M., 2017. 60 years ago, Francis crick changed the logic of biology. PLoS Biol. 15 (9), e2003243. https://doi.org/10.1371/journal.pbio.2003243.

Danthi, P., Guglielmi, K.M., Kirchner, E., Mainou, B., Stehle, T., Dermody, T.S., 2010. From touchdown to transcription: the reovirus cell entry pathway. Curr. Top. Microbiol. Immunol. 343, 91–119. https://doi.org/10.1007/82_2010_32.

Feng, Z., Hirai-Yuki, A., McKnight, K.L., Lemon, S.M., 2014. Naked viruses that Aren't always naked: quasi-enveloped agents of acute hepatitis. Annu. Rev. Virol. 1 (1), 539–560. https://doi.org/10.1146/annurev-virology-031413-085359.

Guseva, S., Milles, S., Ringkjøbing Jensen, M., Salvi, N., Kleman, J.-P., Maurin, D., Ruigrok, R., Blackledge, M., 2020. Measles virus nucleo- and phosphoproteins form liquid-like phase-separated compartments that promote nucleocapsid assembly. Sci. Adv. 6 (14), eaaz7095. https://doi.org/10.1126/sciadv.aaz7095.

Jackson, C.B., Farzan, M., Chen, B., Hyeryun, C., 2022. Mechanisms of SARS-CoV-2 entry into cells. Nat. Rev. Mol. Cell Biol. 23, 3–20. https://doi.org/10.1038/s41580-021-00418-x.

Li, X., Gu, M., Zheng, Q., Gao, R., 2021. Packaging signal of influenza a virus. Virol. J. 18, 36. https://doi.org/10.1186/s12985-021-01504-4.

Maghsoodi, A., Chatterjee, A., Andricioaei, I., Perkins, N.C., 2019. How the phage T4 injection machinery works including energetics, forces, and dynamic pathway. Proc. Natl. Acad. Sci. U. S. A. 116 (50), 25097–25105. https://doi.org/10.1073/pnas.1909298116.

McDonald, S.M., Patton, J.T., 2011. Assortment and packaging of the segmented rotavirus genome. Trends Microbiol. 19 (3), 136–144. https://doi.org/10.1016/j.tim.2010.12.002.

Paraan, M., Mendez, J., Sharum, S., Kurtin, D., He, H., Stagg, S.M., 2020. The structures of natively assembled clathrin-coated vesicles. Sci. Adv. 6 (30), eaba8397. https://doi.org/10.1126/sciadv.aba8397.

Ploubidou, A., Way, M., 2001. Viral transport and the cytoskeleton. Curr. Opin. Cell Biol. 13 (1), 97–105. https://doi.org/10.1016/S0955-0674(00)00180-0.

Rey, F.A., Lok, S.M., 2018. Common features of enveloped viruses and implications for immunogen design for next-generation vaccines. Cell 172 (6), 1319–1334. https://doi.org/10.1016/j.cell.2018.02.054. 29522750.

Sobhy, H., 2017. A comparative review of viral entry and attachment during large and giant dsDNA virus infections. Arch. Virol. 162, 3567–3585. https://doi.org/10.1007/s00705-017-3497-8.

Wang, S., Dai, T., Qin, Z., Pan, T., Chu, F., Lou, L., Zhang, L., Yang, B., Huang, H., Lu, H., Zhou, F., 2021. Targeting liquid–liquid phase separation of SARS-CoV-2 nucleocapsid protein promotes innate antiviral immunity by elevating MAVS activity. Nat. Cell Biol. 23, 718–732. https://doi.org/10.1038/s41556-021-00710-0.

Recommended reading

Anti-retrovirals., 2022. https://www.webmd.com/hiv-aids/aids-hiv-medication.

Fay, N., Panté, N., 2015. Nuclear entry of DNA viruses. Front. Microbiol. 6, 467. https://www.frontiersin.org/article/10.3389/fmicb.2015.00467. https://doi.org/10.3389/fmicb.2015.00467.

CHAPTER

5

Expression of virus genomes

INTENDED LEARNING OUTCOMES

On completing this chapter, you should be able to:

- Discuss the mechanisms by which cells and viruses **express** the information stored in their genes.
- Describe various **genome coding strategies**.
- Explain how viruses control gene expression via **transcription** and **posttranscriptional** mechanisms.

Expression of genetic information

As described in Chapter 1, no virus yet discovered has the genetic information that encodes all of the genes necessary for the **generation of metabolic energy**, or for **protein synthesis**—even though every new giant virus discovery widens the spectrum of such genes that viruses do encode. Thus, all viruses are dependent on their host cells for these functions—but the way in which viruses persuade their hosts to express their genetic information for them varies considerably. Patterns of virus replication are determined by tight controls on virus gene expression. There are fundamental differences in the control mechanisms of these processes in prokaryotic and eukaryotic ribocells, and these differences inevitably affect the viruses that utilize them as hosts. In addition, the **relative simplicity and compact size** of most virus genomes compared with those of most cells, creates further limits. Cells have evolved varied and complex mechanisms for controlling gene expression by using their often extensive genetic capacity, while viruses have had to achieve **highly specific quantitative, temporal, and spatial control of expression** with generally much more limited genetic resources. Viruses have counteracted their genetic limitations by the evolution of a range of solutions to these problems. These mechanisms include:

- Powerful **positive and negative** signals that promote or repress gene expression.
- **Highly compressed genomes** in which **overlapping reading frames** are common.
- **Control signals** that are frequently **nested within other genes**.
- Strategies which allow **multiple polypeptides** to be created from a **single mRNA.**

Cann's Principles of Molecular Virology
https://doi.org/10.1016/B978-0-12-822784-8.00005-2

Gene expression involves **regulatory loops** mediated by signals that act either in *cis* (affecting the activity of neighboring genetic regions) or in *trans* (giving rise to diffusible products that act on regulatory sites anywhere in the genome). For example, **transcription promoters** are ***cis*-acting** sequences that are located upstream in terms of transcriptional direction of the genes whose transcription they control, while proteins such as "**transcription factors**" which bind to specific sequences present on any stretch of nucleic acid present in the cell, are examples of ***trans*-acting** factors. The relative simplicity of virus genomes and the elegance of their control mechanisms were important historical models that informed our current understanding of cellular genetic regulation—and continue to do so in this **"-omics age."** This chapter assumes that you are familiar with the mechanisms involved in cellular control of gene expression. However, we will start with a brief reminder of some important aspects (Box 5.1).

Control of prokaryote gene expression

Prokaryote ribocells are second only to viruses in the specificity and economy of their genetic control mechanisms. In bacteria and archaea, genetic control operates both at the **level of transcription** and at subsequent or **posttranscriptional** stages of gene expression.

The **initiation of transcription** is regulated primarily in a **negative** way by the synthesis of *trans*-acting **repressor proteins**, which bind to ***operator sequences*** upstream of protein-coding sequences. Collections of metabolically related genes are grouped together and coordinately controlled as "***operons***." Transcription of these operons typically produces a **polycistronic mRNA** that encodes several different proteins in separate **open reading frames (ORFs)**. During subsequent stages of expression, transcription is also regulated by a number of mechanisms that act as "**genetic switches**," turning on or off the transcription of different genes. Such mechanisms include **antitermination**, which is controlled by *trans*-acting factors that promote the synthesis of longer transcripts encoding additional genetic information, and by various modifications of RNA polymerase. Bacterial σ **(sigma) factors**—which are

BOX 5.1

It's all about the gene

Even before Richard Dawkins wrote "The Selfish Gene" in the 1970s, the molecular biology revolution in the 1960s had ensured that the gene became the biological object that our thinking revolved around. In Dawkins' view, **genes are only concerned with their own survival**. This is simplistic if you consider that **transposases** are some of the most ubiquitous "selfish" genes in nature (see Chapter 3), but must offer some selective advantage(s) to the genomes and ecosystems they inhabit, or they would be strongly selected against. When we look at how malleable virus genomes are, and how easily genes flow from host to virus and from one virus to another, you can understand, though, why this view has become common. In many ways, this chapter on gene expression is at the very heart of this book. Understanding the mechanisms of transmission and expression which act on genes—and in this case, viral genes—is central to understanding modern biology.

homologous to **archaeal transcription factor B** and to **eukaryotic transcription factor IIB**—are proteins that affect the specificity of the **DNA-dependent RNA polymerase holoenzyme** (RNA pol; active form) for different **promoters**. Several bacteriophages (e.g., phage **SP01** of *Bacillus subtilis*) encode proteins that function as alternative σ factors, sequestering RNA polymerase and altering the rate at which phage genes are transcribed. Phage T4 of *Escherichia coli* encodes an enzyme that carries out a covalent modification (**adenosine diphosphate [ADP]-ribosylation**) of the host cell RNA pol. This is believed to eliminate the requirement of the polymerase holoenzyme for σ factor, and to achieve an effect similar to the production of modified σ factors by other bacteriophages.

At a posttranscriptional level, gene expression in bacteria is also regulated by **control of translation**. The best-known virus examples of this phenomenon come from the study of small bacteriophages of the family *Leviviridae*, such as **R17, MS2,** and **Qβ**. In these ssRNA+ phages, the **secondary structure** of the genome not only regulates the quantities of different phage proteins that are translated, but also operates **temporal (timed) control** of a switch in the ratios between the different proteins produced in infected cells.

Control of expression in bacteriophage λ

The genome of **phage** λ has been studied in great detail and illustrates several of the mechanisms described above, including the action of repressor proteins in regulating **lysogeny** versus **lytic** replication, and antitermination of transcription by phage-encoded *trans*-acting factors. Such has been the impact of these discoveries that no discussion of the types of control of virus gene expression is complete without detailed examination of this phage.

Phage λ was discovered by Esther Lederberg in 1949. Experiments at the Pasteur Institute by André Lwoff in 1950 showed that some strains of *Bacillus megaterium*, when irradiated with ultraviolet light, stopped growing and subsequently lysed, releasing a crop of virions. Together with Francois Jacob and Jacques Monod, Lwoff subsequently showed that the cells of some bacterial strains carried a phage in a dormant form, known as a **prophage**, and that the phage could be made to alternate between the **lysogenic** (nonproductive) and **lytic** (productive) growth cycles. Subsequently, the virus was shown to have a linear dsDNA genome 46–54 kb in size. After many years of study, our understanding of phage λ has been refined into a picture that represents one of the best understood and most elegant genetic control systems yet to be investigated. A simplified genetic map of λ is shown in Fig. 5.1.

The components involved in genetic control are as follows:

1. P_L is the **promoter** responsible for transcription of the **left-hand side** of the λ genome, including *N* and *cIII*.
2. O_L is a short noncoding region (NCR) of the phage genome (approximately 50 bp) which lies between the *cI* and *N* genes next to P_L.
3. P_R is the **promoter** responsible for transcription of the **right-hand side** of the λ genome, including *cro*, *cII*, and the genes encoding the structural proteins.
4. O_R is a short NCR of the phage genome (approximately 50 bp) which lies between the *cI* and *cro* genes next to P_R.
5. *cI* is transcribed from its own promoter and encodes a **repressor protein** of 236 aa which binds to O_R, **preventing transcription** of *cro* but allowing transcription of *cI*, and to O_L, preventing transcription of *N* and the other genes at the left-hand end of the genome.

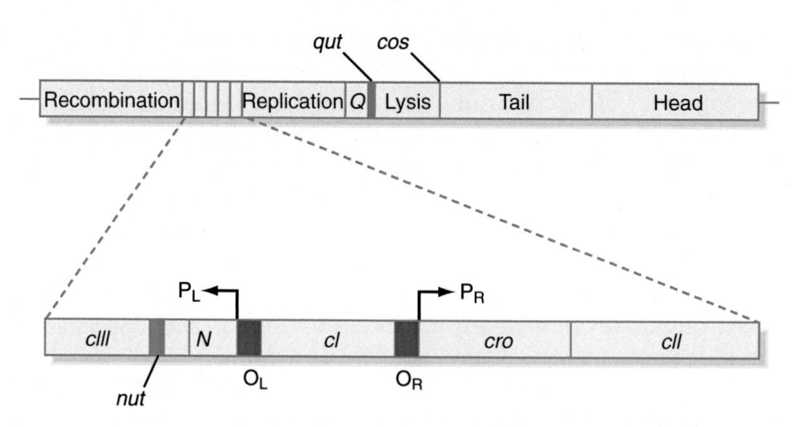

FIG. 5.1 Simplified genetic map of bacteriophage λ. The *top part* of this figure shows the main genetic regions of the phage genome and the *bottom part* is an expanded view of the main control elements described in the text.

6. *cII* and *cIII* encode **activator proteins** that bind to the genome, **enhancing the transcription** of the *cI* gene.
7. *cro* encodes a 66-amino acid protein that binds to O_R, **blocking binding** of the repressor to this site.
8. *N* encodes an **antiterminator protein** that acts as an **alternative ρ (rho) factor** for host-cell RNA polymerase, modifying its activity and permitting extensive transcription from P_L and P_R.
9. *Q* is an antiterminator similar to *N*, but it only permits extended transcription from P_R.

In a newly infected cell, *N* and *cro* are transcribed from P_L and P_R, respectively (Fig. 5.2). The N protein allows RNA polymerase to transcribe a number of phage genes, including those responsible for **DNA recombination** and **integration of the *prophage*** to establish **lysogeny** (see Fig. 5.3), as well as cII and cIII. The N protein acts as a **positive transcription regulator**. In the absence of the N protein, the RNA pol stops at certain sequences located at the end of the *N* and *Q* genes, known as the *nut* and *qut* sites, respectively. However, RNA pol-N protein complexes are able to overcome this restriction and **permit full transcription** from P_L and P_R. The RNA pol-Q protein complex results in **extended transcription from P_R only**. As levels of the cII and cIII proteins in the cell build up, transcription of the *cI* repressor gene from its own promoter is turned on.

At this point, the critical event that determines the outcome of the infection occurs. The cII protein is constantly degraded by host-cell proteases. If levels of cII remain below a critical level, transcription from P_R and P_L continues, and the phage undergoes a productive replication cycle that culminates in **lysis of the cell** and the release of phage particles (see Fig. 1.7A). This is the sequence of events that occurs in the vast majority of infected cells. However, in a few rare instances, the concentration of cII protein builds up, transcription of *cI* is enhanced, and intracellular levels of the cI repressor protein rise. The repressor binds to O_R and O_L which **prevents transcription of all phage genes** (particularly *cro*; see below) except itself. The level of cI protein is maintained automatically by a **negative feedback mechanism**, as

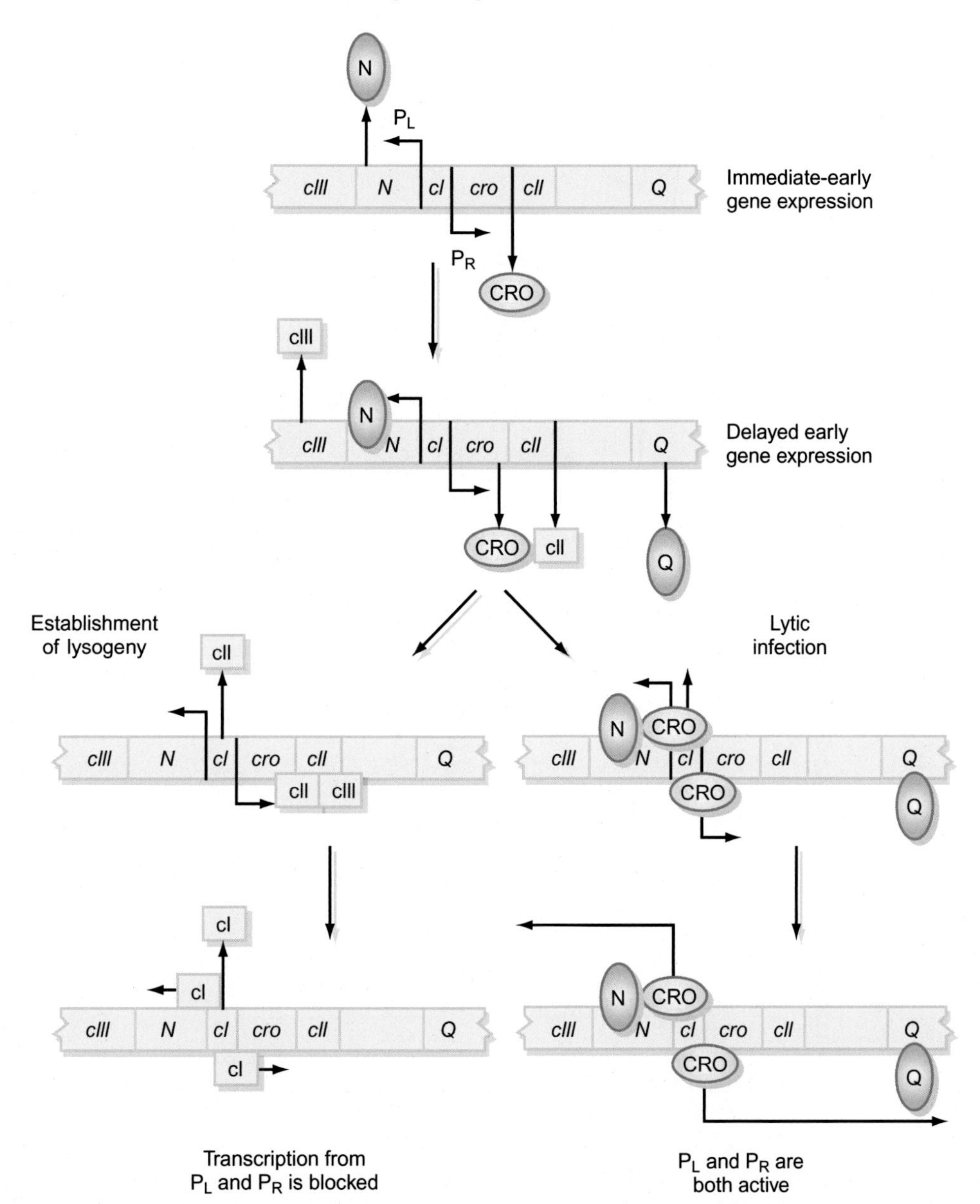

FIG. 5.2 Control of expression of the bacteriophage λ genome. See text for a detailed description of the events that occur in a newly infected cell and during lytic infection or lysogeny.

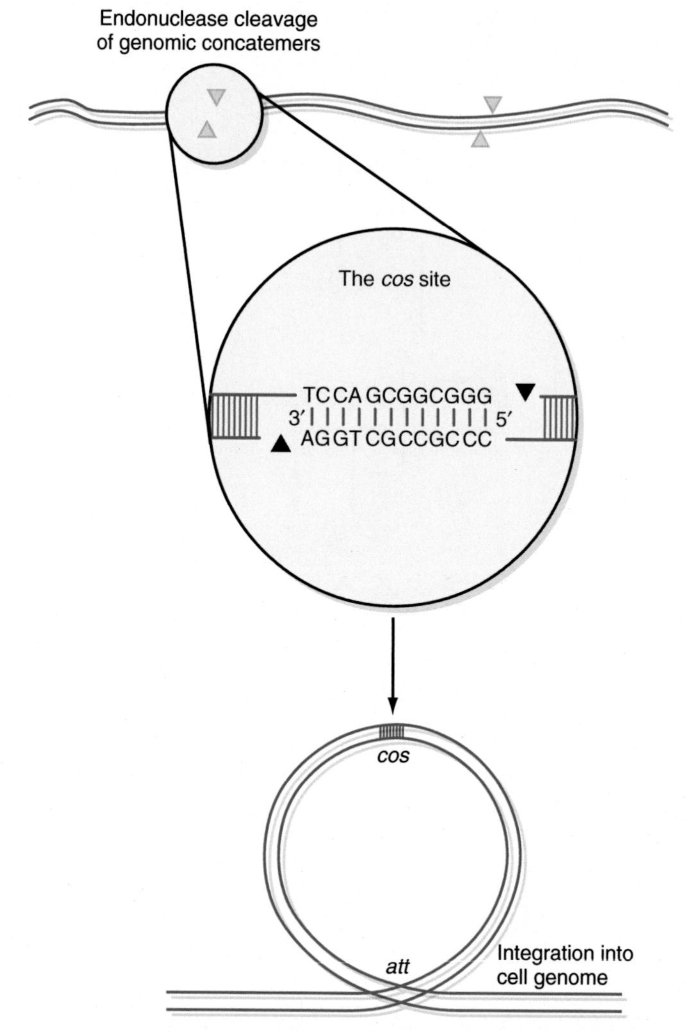

FIG. 5.3 Integration of the bacteriophage λ genome. The cohesive sticky ends of the cos site in the λ genome are ligated together in newly infected cells to form a circular molecule. Integration of this circular form into the *E. coli* chromosome occurs by specific recognition and cleavage of the *att* site in the phage genome.

at high concentrations the repressor also binds to the left-hand end of O_R and prevents transcription of cI (Fig. 5.2). This autoregulation of cI synthesis keeps the cell in a stable state of **lysogeny**.

If this is the case, how do such cells ever leave this state and enter a productive, lytic replication cycle? **Physiological stress** and particularly **ultraviolet irradiation** of cells result in the induction of a host cell protein, **RecA**. This protein's normal function is to induce the expression of cellular genes that permit the cell to adapt to and survive in altered environmental conditions: here, it **cleaves the cI repressor protein**. This would not normally be sufficient to prevent the cell from reentering the lysogenic state; however, when repressor protein is

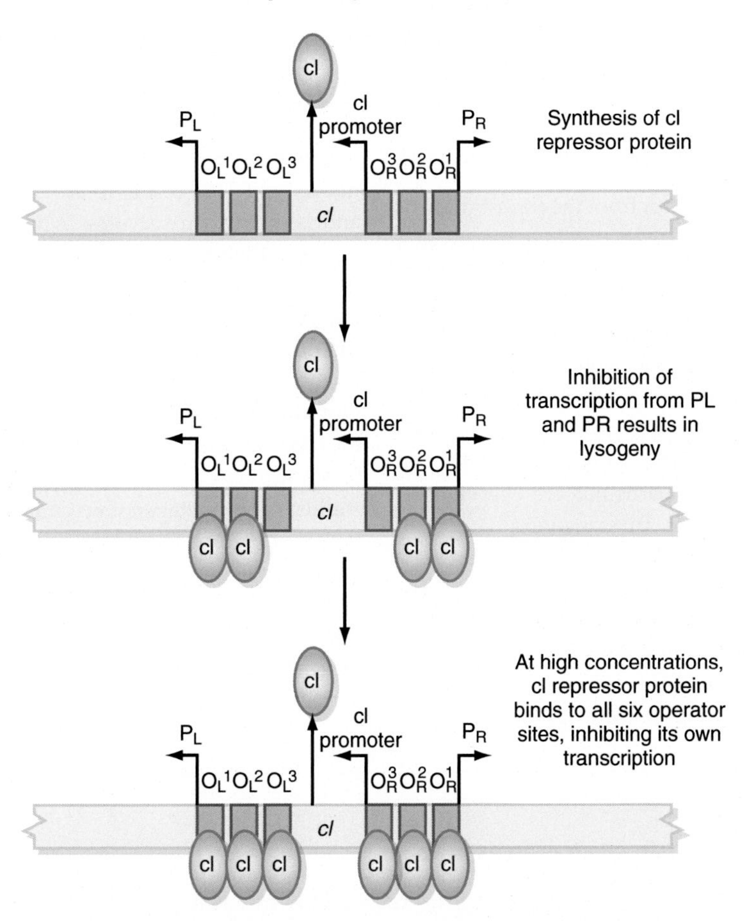

FIG. 5.4 Control of lysogeny in bacteriophage λ. See text for a detailed description of the events that occur in the establishment and maintenance of lysogeny.

not bound to O_R, *cro* is transcribed from P_R. Cro also binds to O_R but, unlike cI, which preferentially binds the right-hand end of O_R, the Cro protein binds preferentially to the left-hand end of O_R, preventing the transcription of *cI* and **enhancing its own transcription in a positive-feedback loop**. The phage is then locked into a lytic cycle and cannot return to the lysogenic state (Fig. 5.4).

The molecular details of λ gene have contributed greatly to our understanding of genetic regulation in prokaryotic and eukaryotic ribocells. Determination of the structures of the proteins involved has allowed us to identify the fundamental principles behind the observation that many proteins from unrelated organisms can **recognize and bind to specific sequences** in DNA molecules. The concepts of proteins with independent **DNA-binding and dimerization** domains, **protein cooperativity** in DNA binding, and **DNA looping** allowing proteins bound at distant sites to interact with one another, have all risen from the study of λ. See References for papers that explain more fully the nuances of gene expression in this complex virus (Box 5.2).

BOX 5.2

Bacteriophages are so old-fashioned!

No, they're not. Apart from the contribution of bacteriophages to understanding viruses as a whole—and there's no better example of that than λ—some of the most exciting work in virology over the last couple of decades has been about phages. When we finally raised our eyes from the glassware in our laboratories and went out hunting for viruses in the natural environment, we were staggered at what we found. Phages in particular are everywhere, and in everything, in staggering quantities and variety. It has been estimated that every second on Earth, **1×10^{25} bacteriophage infections** occur—which is a powerful number, even considering that there are probably 10^{30} bacterial cells available. Moreover, marine viruses are estimated to **kill 20% of the marine microorganism biomass every day**—and this makes up ~70% of the total marine biomass and is mainly prokaryotes. While such studies are in their infancy, it's found that the same is probably true for soils as well. This means that **phages control the turnover of such large quantities of organic material that this has a major impact on nutrient cycling and the global climate**—and that this must be taken into account for ***any*** accurate modeling of carbon and other nutrient cycling. Increasingly sophisticated metagenomic and metaviromic trawls of natural and disturbed environments are throwing up such incredible diversity of prokaryotes and their viruses that it is clear that we are only still scratching the surface of understanding their ecology.

Old fashioned? Wrong.

Control of eukaryote gene expression

Control of gene expression in **eukaryotic ribocells** is much more complex than in prokaryotes, and involves a multilayered approach in which diverse control mechanisms exert their effects at different levels. It is important to realize, for the DNA viruses that infect eukaryotic cells, that the requisite machinery for replication must either be **immediately available**, or must be able to be **switched on**, for most DNA viruses to be able to replicate. While cells are mostly transcriptionally active to some extent all the time, cell DNA synthesis and replication only occurs in **S phase** (synthesis phase) of **interphase** during the cell cycle (see Fig. 5.5). Therefore, for DNA virus replication to succeed, cells must either be in S phase, or the virus must be able to **induce a transition** or partial transition to S phase—usually by means of expression of one or more **transcription factors** that act upon the promoters of the requisite cellular genes. This requirement may govern exactly which cells may be infectable by viruses, as regardless of whether or not they bear the appropriate receptors, cells which are in interphase may not support replication of viruses that cannot induce S phase or the necessary DNA pol and accessory proteins.

The **first level of control** of eukaryotic gene expression occurs prior to transcription, and depends on the local configuration of the DNA. DNA in eukaryotic cells has an elaborate structure, forming complicated and dynamic but far from random complexes with numerous proteins to form the **nucleosomes** that are the basis for ***chromatin structure***. Although the contents of eukaryotic cell nuclei appear amorphous in electron micrographs (at least in

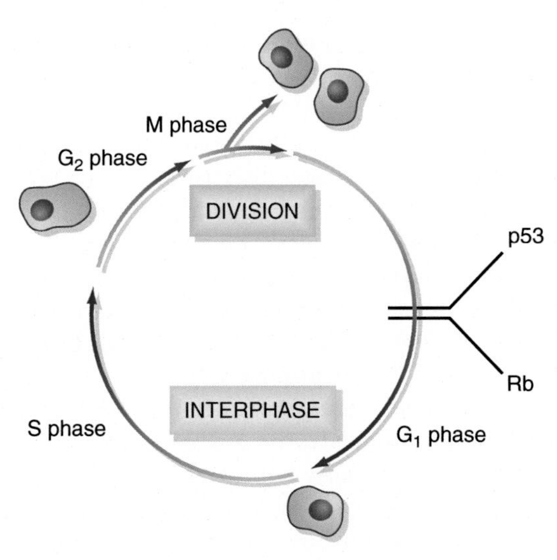

FIG. 5.5 The eukaryotic cell cycle. p53 and Rb are cell cycle regulatory proteins often interfered with by virus-encoded proteins to induce S phase.

interphase), they are actually **highly ordered**. Chromatin interacts with the structural backbone of the nucleus—the ***nuclear matrix***—and these interactions are thought to be important in controlling gene expression. Locally, nucleosome configuration and DNA conformation, particularly the formation of **left-handed helical "Z-DNA,"** are also important. **DNAse I** digestion of chromatin does not give an even, uniform digestion pattern but reveals a pattern of **DNAse-hypersensitive sites** believed to indicate differences in the function of various regions of the chromatin. It is likely, for example, that **retroviruses are more likely to integrate** into the host-cell genome at these sites than elsewhere. **Transcriptionally active** DNA is also **hypomethylated**—that is, there is a relative scarcity of nucleotides modified by the covalent attachment of methyl groups in these regions compared with the frequency of methylation in **transcriptionally quiescent** regions of the genome. The methylation of integrated retrovirus genomes (**proviruses**) has been shown to suppress the transcription of the provirus genome.

The **second level of control** rests in the **process of transcription itself**, which again is much more complex than in prokaryotes. There are **three forms of RNA pol** in eukaryotic cells that are involved in the expression of different classes of genes (Table 5.1). The rate at which transcription is initiated is a **key control point** in eukaryotic gene expression. Initiation is influenced dramatically by sequences **upstream of the transcription start site** which function

TABLE 5.1 Forms of RNA polymerase in eukaryotic cells.

RNA polymerase	Sensitivity to α-amanitin	Cellular genes transcribed	Virus genes transcribed
I	Unaffected	Ribosomal RNAs	–
II	Highly sensitive	Most single-copy genes	Most DNA virus genomes
III	Moderately sensitive	5S rRNA, tRNAs	Adenovirus VA RNAs

by acting as recognition sites for families of highly specific DNA-binding proteins, known as "**transcription factors**." Immediately upstream of the transcription start site is a relatively short region known as the **promoter**. It is at this site that transcription complexes, consisting of RNA pol plus accessory proteins, bind to the DNA and transcription begins. However, sequences further upstream from the promoter also influence the efficiency with which transcription complexes form. The rate of initiation depends on the combination of transcription factors bound to these **transcription enhancers**. The properties of these **enhancer sequences** are remarkable in that they can be inverted and/or moved around relative to the position of the transcription start site without losing their activity, and can exert their influence even from a distance of several kb away. This emphasizes the **flexibility of DNA**, which allows **proteins bound at distant sites to interact with one another**, as also shown by the protein-protein interactions seen in regulation of phage λ gene expression (above). Transcription of eukaryotic genes results in the production of **monocistronic mRNAs**, each of which is transcribed from its **own individual promoter**, unlike the case with prokaryotes.

At the **next level of control**, gene expression is influenced by the **structure of the mRNA produced**. The stability of eukaryotic mRNAs varies considerably, some having comparatively long half-lives in the cell (e.g., many hours). The half-lives of others, typically those that encode regulatory proteins, may be very short (e.g., a few minutes). The stability of eukaryotic mRNAs depends on the speed with which they are degraded. This is determined by such factors as its **terminal sequences**, which consist of a **methylated cap structure at the 5′ end** (7-methyl-guanosine; disguises the 5′ terminus), and **polyadenylic acid (polyA)** sequence **at the 3′ end**, as well as on the **overall secondary structure** of the message. However, gene expression is also regulated by differential **splicing** of heterogeneous (heavy) nuclear RNA (**hnRNA**) precursors in the nucleus, which can alter the genetic meaning of different mRNAs transcribed from the same gene. In eukaryotic cells, control is also exercised during **export of RNA** from the nucleus to the cytoplasm via the **nuclear pore complexes** (NPCs, see Fig. 4.10).

Finally, the **process of translation** offers further opportunities for control of expression. The efficiency with which different mRNAs are translated varies greatly. These differences result largely from the **efficiency with which ribosomes bind** to different mRNAs and recognize AUG translation initiation codons in different sequence contexts, as well as the speed at which different sequences are converted into protein. Certain sequences act as **translation enhancers**, performing a function analogous to that of transcription enhancers: virus-derived versions are increasingly being recognized as valuable tools in biotechnology.

MicroRNAs (miRNAs) are small (approximately 22 nt) RNAs that play important roles in the regulation of gene expression, typically in **silencing gene expression** by directing repressive protein complexes to the **3′ untranslated region** of specific target mRNA transcripts (a similar mode of action to siRNAs Chapter 6). First discovered to play important roles in **posttranscriptional gene regulation** in eukaryotic cells, many virus-encoded miRNAs are now known from diverse virus families, most with DNA genomes but also some RNA viruses, including **HIV** and **Ebola virus**. Some of the best characterized virus miRNA functions include **prolonging the longevity of infected cells** (several different herpesviruses), **evading the immune response** (in polyomaviruses), and **regulating the switch to lytic infection** (herpesviruses). This enables gene expression to continue in infected cells in circumstances where in the absence of the miRNAs it would be shut down.

The point of this extensive list of eukaryotic gene expression mechanisms is that they are **all utilized by viruses to control gene expression**. Examples of each type are given in the sections below. If this seems remarkable, remember that the control of gene expression in eukaryotic cells was unraveled largely by **utilizing viruses as model systems**, therefore finding examples of these mechanisms in viruses is really only a self-fulfilling prophecy.

In recent years, the importance of ***epigenetics*** in shaping virus gene expression has been recognized. Epigenetics refers to control of gene expression independent of changes in the DNA sequence of the gene affected. Some of the processes involved include **DNA methylation, histone modifications, chromatin-remodeling proteins,** and **DNA silencing**. In viruses, examples of epigenetic mechanisms controlling gene expression include **influenza virus NS1 protein mimicking histone H3**, chromatin assembly and histone and DNA modifications in the **control of latent infections of herpesviruses**, and the importance of epigenetics in controlling **virus-associated (VA) tumor formation**.

As discussed in Chapter 1, advances in biotechnology such as next generation sequencing methods have had a major impact on the understanding of virus replication. In understanding gene expression, **rapid direct RNA sequencing methods** (**"RNAseq"**) has been particularly important in recent years. **Microarray technology** evolved from Southern blotting: tens of thousands of DNA probes, usually **chemically synthesized oligonucleotides**, can be attached to a glass slide, and the genes they represent can all be analyzed in a single experiment. To achieve the very high density of DNA probes in the array, **spotting** or **printing** of the slides is carried out by specialized robots. DNA microarrays are most commonly used to detect mRNAs and this is referred to as **expression analysis** or **expression profiling** (see References). Because of the power of this technique, microarrays have become the preeminent technology for the investigation of **functional genomics**—the area of modern biology which focuses on dynamic aspects such as gene transcription, translation, and protein-protein interactions.

Virus genome coding strategies

In Chapter 4, genome structure was one element of a classification scheme used to divide viruses into seven groups. The other part of this scheme is the way in which the genetic information of each class of virus genomes is expressed. The **replication and expression of virus genomes are inseparably linked**, and this is particularly true in the case of RNA viruses. Here, the seven classes of virus genome described in Chapter 4 are reviewed again, this time examining the way in which the genetic information of each class is expressed (Box 5.3).

Class I: Double-stranded DNA

In Chapter 4 we said that this class of virus genomes in eukaryotes can be subdivided into two further groups: those in which genome replication is exclusively **nuclear** (e.g., *Adenoviridae, Polyomaviridae, Herpesviridae*) and those in which replication occurs in the **cytoplasm** (*Poxviridae, Mimiviridae*). In one sense, all of these viruses can be considered to be similar—because their **genomes all resemble double-stranded cellular DNA**, and they are essentially transcribed by the same mechanisms as cellular genes. However, there are profound differences between them relating to the degree to which each family is reliant on the host-cell machinery.

BOX 5.3

So many viruses!

Do I have to remember them ALL?! Good question- and the answer is that you *don't* need to remember all the details about every virus—even people who write virology textbooks cannot do that. What you do need to do is to have a framework which allows you to think "Yes, I've seen something like this before, so I can guess what's likely to happen." And that's where the seven classes of virus genomes described in the previous chapter come in. Add on to that an understanding of how gene expression works for each type, and you're pretty much there.

There's one small catch, however. Even for viruses with very similar genome structures, there are often surprising differences in mechanisms of gene expression. Biology is all about variation: if you wanted everything to be predictable, you should have signed up for the physics class!

Polyomaviruses and papillomaviruses

These two distantly-related small nonenveloped DNA viruses have covalently closed circular (ccc) dsDNA genomes of around 5 kbp (polyomaviruses) and 8 kbp (papillomaviruses) in size. The genomes of both types of viruses are condensed by **host histone binding**, to form **chromatin-like complexes**. Polyomaviruses have **bidirectional** transcription from the origin of replication (**ori**) (see Fig. 5.6), while papillomavirus genes are all transcribed in the same direction so that only one strand of the dsDNA genome is transcribed. They are heavily dependent on cellular machinery for both replication and gene expression. Both viruses express early and late proteins. The two pre-mRNAs transcribed for polyomaviruses are processed by alternative splicing to yield 5–9 distinct mRNAs, depending on the virus. Papillomavirus early genes are transcribed similarly, with alternative splicing producing up to seven distinct proteins. Later transcription of the two structural proteins occurs only in **terminally-differentiating keratinocytes**, and the two proteins (L1 and L2) are similarly produced by splicing. All genes are transcribed by host RNA pol II.

Polyomaviruses encode ***trans*-acting** factors (**T-antigens**) which stimulate transcription and genome replication. The equivalent in papillomaviruses is the **E1 and E2 proteins**, which bind cooperatively to the **ori** to form an E1–E2-ori complex which initiates DNA replication. **Coat protein mRNA transcription** occurs after replication for both virus types. Papillomaviruses in particular are dependent on the cell for replication, which only occurs in terminally differentiated keratinocytes and not in other cell types, although they do encode several *trans*-regulatory proteins (Chapter 7). HPV transcription is **largely controlled by the E2 protein**, with host cells **transcription factors** binding to sequences within the **viral Long Control Region (LCR),** with these being controlled by which stage of differentiation the host cell is in.

Adenoviruses

Adenovirus genomes consist of **linear dsDNA of 30–38 kbp**, depending on the specific viruses. The genomes contain **30–40 genes** (Fig. 5.7). The terminal sequence of each DNA strand

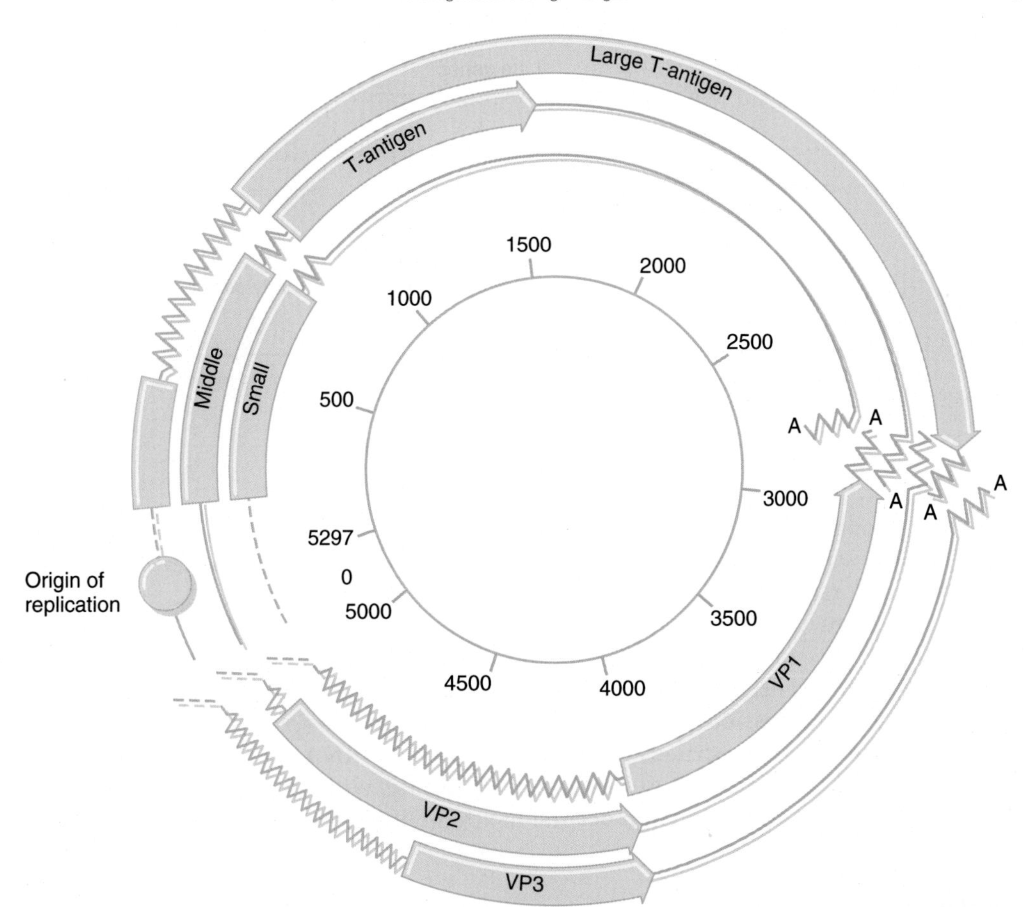

FIG. 5.6 The polyomavirus genome. The complex organization of the polyomavirus genome results in the compression of much genetic information into a relatively short sequence. Differential splicing of pre-mRNAs transcribed from common promoter in both directions (shown by zig-zag lines) from the origin of replication results in several different products for each direction. Early products are the different T regulatory antigens, while late products are the structural proteins VP1–3.

is an **inverted repeat** of 100–140bp, and the denatured single strands can form "**panhandle**" secondary structures. These structures are important in DNA replication, as is a 55-kDa protein known as the **terminal protein** that is covalently attached to the 5′ end of each genome strand. During genome replication, this protein acts as a **primer**: as explained in Chapter 3 (see table 3.2), **adenoviruses** and other small dsDNA viruses of prokaryotes share the property of using **pPolB**—this is a **protein-primed variant** of the hallmark gene encoding **DNA pol B** found in all large dsDNA viruses.

The expression of adenovirus genomes is quite complex. Clusters of genes are expressed from a limited number of **shared promoters**. **Multiply-spliced mRNAs** and **alternative splicing patterns** are used to express a variety of polypeptides from each promoter. Splicing of pre-mRNAs was in fact discovered using adenoviruses, in work reported in 1977: coding

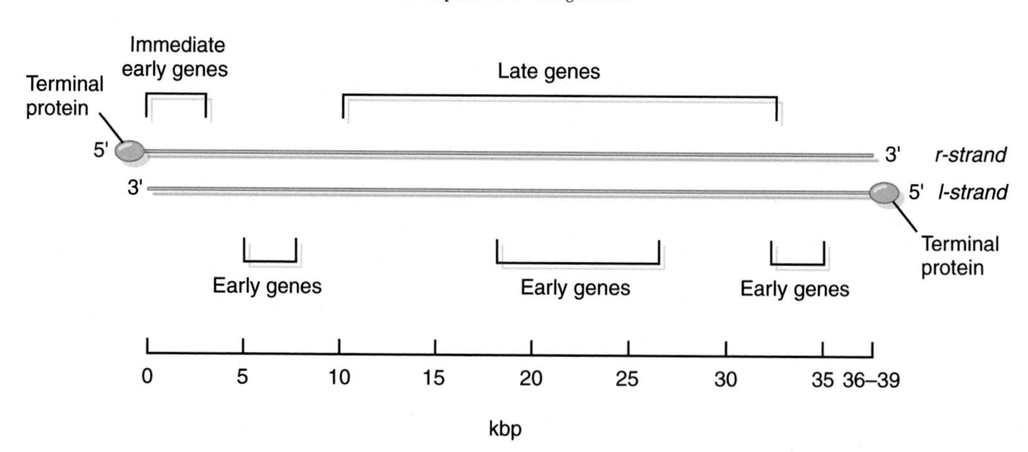

FIG. 5.7 Adenovirus genomes.

regions of adenoviral genes (now known as **exons**) were shown to be separated by **noncoding DNA** which was not involved in protein expression. These sequences—**introns**—are cut out from the precursor mRNAs (see Chapter 7 and Fig. 5.6).

Adenoviruses are also heavily dependent on the cellular apparatus for transcription, but they possess various mechanisms that specifically regulate virus gene expression. These include *trans*-acting transcriptional activators such as the **E1A protein**, and posttranscriptional regulation of expression, which is achieved by alternative splicing of mRNAs and the virus-encoded **VA** (viral associated) **noncoding regulatory RNAs** (Chapter 7). Adenovirus infection of cells is divided into two stages, **early and late**, with the late phase commencing **at the time when genome replication occurs**. In adenoviruses, these phases are less distinct than in herpesviruses (below).

Herpesviruses

The *Herpesviridae* is a large family containing more than 100 different members, at least one for most animal species that have been examined to date—including marine coral polyps. There are members of the NCLDV group (see Chapter 3), and viruses replicate in the nucleus. There are eight **human herpesviruses (HHVs 1–8)**, all of which share a common overall genome structure but which differ in the fine details of genome organization and at the level of nucleotide sequence. Herpesviruses have large genomes composed of up to **235 kbp of linear, double-stranded DNA** and correspondingly large and complex virus particles containing about **35 virion polypeptides** (see Fig. 2.18). All encode a variety of enzymes involved in **nucleic acid metabolism, DNA synthesis,** and **protein processing** (e.g., protein kinases). The different members of the family are widely separated in terms of genomic sequence and proteins, but all are similar in terms of structure and genome organization (Fig. 5.8A). Some but not all herpesvirus genomes consist of two covalently joined sections, a unique long (UL) and a unique short (US) region, each bounded by inverted repeats. The repeats allow structural rearrangements of the unique region. This arrangement allows these genomes to exist as a **mixture of four structural isomers**, all of which are functionally equivalent (Fig. 5.8B).

Herpesvirus genomes also contain multiple repeated sequences and, depending on the number of these, the genome size of different isolates of a particular virus can vary by up to

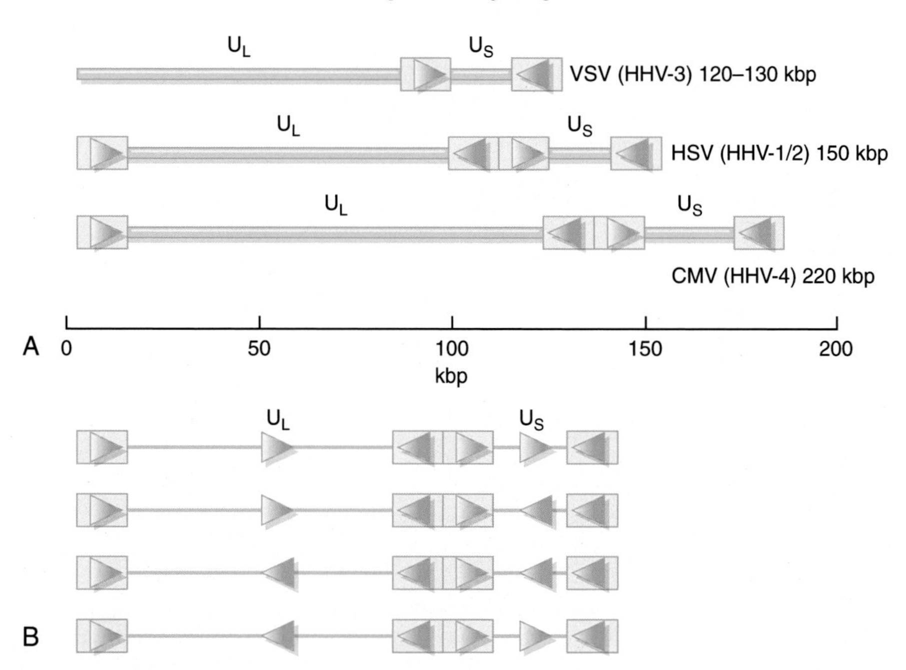

FIG. 5.8 Herpesvirus genomes. (A) Some herpesvirus genomes consist of two covalently joined sections, UL and US, each bounded by inverted repeats (*triangular arrows*). (B) This organization permits the formation of four different isomeric forms of the genome.

10 kbp. The prototype member of the family is **herpes simplex virus type 1 (HSV-1)**, whose genome consists of approximately 152 kbp of double-stranded DNA, the complete nucleotide sequences of a number of isolates have now been determined. This virus contains about **80 genes**, densely packed and with **overlapping open reading frames**. Each gene is expressed from its own promoter (see adenovirus discussion above).

These viruses are less reliant on cellular enzymes than the previous groups. They encode many enzymes involved in **DNA metabolism** (e.g., **thymidine kinase**) and a number of *trans*-acting factors that regulate the **temporal expression of virus genes**, controlling the phases of infection. Transcription of the large, complex genome is **sequentially regulated in a cascade fashion** (Fig. 5.9). At least 50 virus-encoded proteins are produced after transcription of the genome by host-cell RNA polymerase II.

Three distinct classes of mRNAs are made:

- **α: Immediate-early (IE)** mRNAs encode *trans*-acting regulators of virus transcription.
- **β: (Delayed) early** mRNAs encode further nonstructural regulatory proteins and some minor structural proteins.
- **γ: Late** mRNAs encode the major structural proteins.

Gene expression in herpesviruses is tightly and coordinately regulated, as indicated (Fig. 5.9). If translation is blocked shortly after infection (e.g., by treating cells with **cycloheximide**), the **production of late mRNAs is blocked**. Synthesis of the **early gene product** turns off the IE products and initiates genome replication. Some of the **late structural proteins** (γ1)

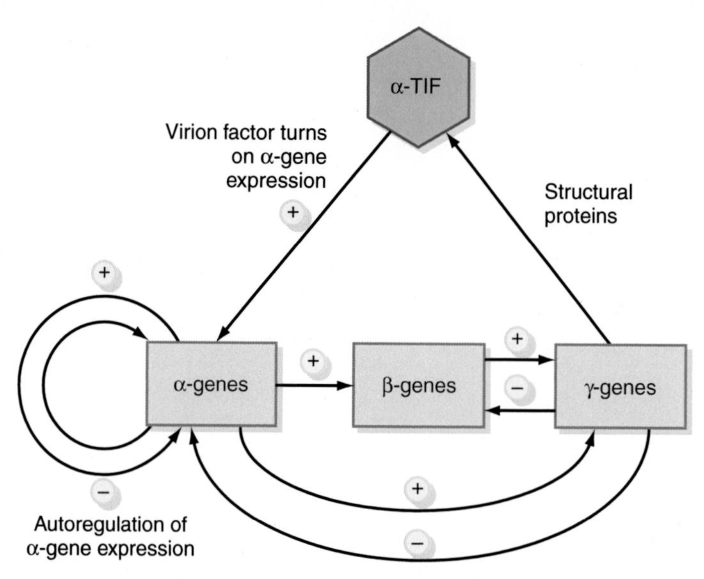

FIG. 5.9 Control of expression of the herpes simplex virus (HSV) genome. HSV particles contain a protein called α-gene transcription initiation factor (α-TIF) which turns on α-gene expression in newly infected cells, beginning a cascade of closely regulated events which control the expression of the entire complement of the 70 or so genes in the virus genome.

are produced independently of genome replication; others (γ2) are only produced after replication. Both the IE and early proteins are required to initiate genome replication. A **virus-encoded RNA-primed DNA pol** (PolB, see table 3.2) and a **DNA-binding protein** are involved in genome replication, together with a number of enzymes (e.g., thymidine kinase) that alter cellular biochemistry. The production of all of these proteins is closely controlled.

Poxviruses

Poxvirus genomes are linear structures ranging in size from **140 to 290 kbp**. As with the herpesviruses, **each gene tends to be expressed from its own promoter**. Characteristically, the central regions of poxvirus genomes tend to be highly conserved and to contain essential genes which are **essential for replication in culture**, while the outer regions of the genome are more variable in sequence and **at least some of the genes located here are dispensable** (Fig. 5.10). In contrast, the noncoding nucleic acid structures at the ends of the genome are **highly conserved and vital for replication**. There are no free ends to the linear genome because these are **covalently closed by "hairpin" arrangements**. Adjacent to the ends of the genome are other noncoding sequences which play vital roles in replication (see Chapter 4).

While poxviruses are NCLDVs, they—like mimiviruses—**replicate in the cytoplasm**, and not the nucleus. Consequently, and because all the eukaryotic cell machinery for both functions is in the nucleus, genome replication and gene expression in poxviruses are almost **independent of cellular mechanisms.** As mentioned earlier (Chapter 4), because of this poxvirus **promoters, transcription terminators, transcription factors** and their **binding sites**, are mostly very unlike their cellular counterparts, enabling **differentiation of virus- vs cell-specific control** of replication and expression. Poxvirus genomes encode numerous enzymes involved

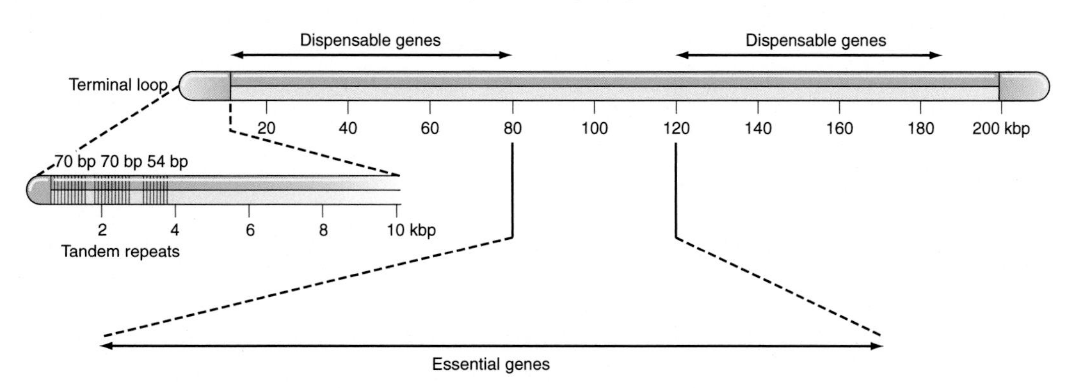

FIG. 5.10 Poxvirus genome organization. In these large and complex genomes, essential genes are located in the central region of the genome. Genes that are dispensable for replication in culture are located closer to the ends of the genome while sequences at the end of the strand contain many sequence repeats important for genome replication.

in **DNA metabolism, virus gene transcription,** and **posttranscriptional modification** of mRNAs. Many of these enzymes are packaged within the virion (which contains >100 proteins), enabling transcription and replication of the genome to occur in the cytoplasm almost totally under the control of the virus—in so-called "**virus factories.**" Gene expression is carried out by virus enzymes associated with the **core of the particle**—which remains **undegraded** in the cytoplasm—and is divided into two rather indistinct phases.

- **Early genes**: These comprise about 50% of the poxvirus genome and are expressed before genome replication inside a partially uncoated core particle (Chapters 2 and 4), resulting in the production of 5′ capped, 3′ polyadenylated but unspliced mRNAs.
- **Late genes**: These are expressed after genome replication in the cytoplasm, but their expression is also dependent on virus-encoded rather than on cellular transcription proteins. Like herpesviruses, late gene promoters are dependent on prior DNA replication for activity. Products of late gene expression form virus factories, where large aggregates of viral products and assembling virions are found.

The giant viruses: Mimivirus, megavirus, pandoravirus, pithovirus

These NCLDVs—like pox- but unlike herpesviruses—replicate in the **cytoplasm**, forming dense vesicles which act as virus factories. Some of these virus genomes contain **large numbers of noncoding repeated elements** in a similar way to eukaryotic genomes. Of the 2500 predicted proteins of *Pandoravirus*, **most are unique** and are not found in other viruses or cellular genomes: this is true for many putative giant virus genomes found by metaviromics, meaning even viruses with similar-looking virions **may share only a small proportion of their genomes with other viruses**—the so-called "**core assembly.**" Typically, the coding genes are expressed as monocistronic mRNAs with very short noncoding regions (NCRs) at each end, shorter than those found in eukaryotic transcripts, leaving little room for regulatory signals in the transcript. Many of the giant genomes contain entire pathways worth of genes involved in cellular metabolism, and some even have a **subset of ribosomal proteins encoded**—indicating that the viruses may profoundly affect host cell metabolism and even gene expression. This type of discovery is typically heralded as "blurring the distinction between viruses and cells," but of course, it does not: all known viruses are still dependent on host cells for the complete pathway of protein synthesis. This is a very young field, however, and significant surprises can be expected in the near future.

Class II: Single-stranded DNA

Single-stranded DNA viruses fall into two groups: these are those with circular genomes, and those with linear genomes (families *Parvoviridae* and *Bidnaviridae*). All of the former group of diverse viruses replicate via a "rolling circle" strategy, whether they infect prokaryotes (**micro-, inoviruses**), animals (**anello-, circoviruses**) or plants (**nano-, geminiviruses**). The two groups of linear ssDNA viruses have very similar genetic organization and very similar "**rolling hairpin**" replication strategies (see below), and infect only animals.

Parvovirus genomes are linear, nonsegmented, ssDNA of **about 5kb** in length. The genomes are (−)sense, but some parvoviruses package equal amounts of (+) and (−) strands into virions. Even the replication-competent parvovirus genomes contain **only two genes**: these are ***rep***, which encodes proteins involved in transcription and replication, and ***cap***, which encodes the **coat proteins**. The expression of these genes is rather complex, somewhat resembling adenoviruses, with **multiple splicing patterns** seen for each gene. Parvoviruses avoid the "**ends**" problem for linear viral DNA genomes (see box 4.5) by having **palindromic sequences of about 115nt** at 5′ and 3′ ends of the genome, which form "**hairpins**" (Fig. 5.11). These structures are **essential for the initiation of genome replication**, again emphasizing the importance of the sequences at the ends of the genome. They also determine which strand is packaged into virus particles.

The members of the **replication-defective** *Dependovirus* genus of the *Parvoviridae* (e.g., **adeno-associated viruses**) are entirely dependent on **adenovirus** or **herpesvirus superinfection** for the provision of further helper functions essential for their replication beyond those present in normal cells. The adenovirus genes required as helpers are the **early, transcriptional regulatory genes such as E1A** rather than late structural genes, but it has been shown that

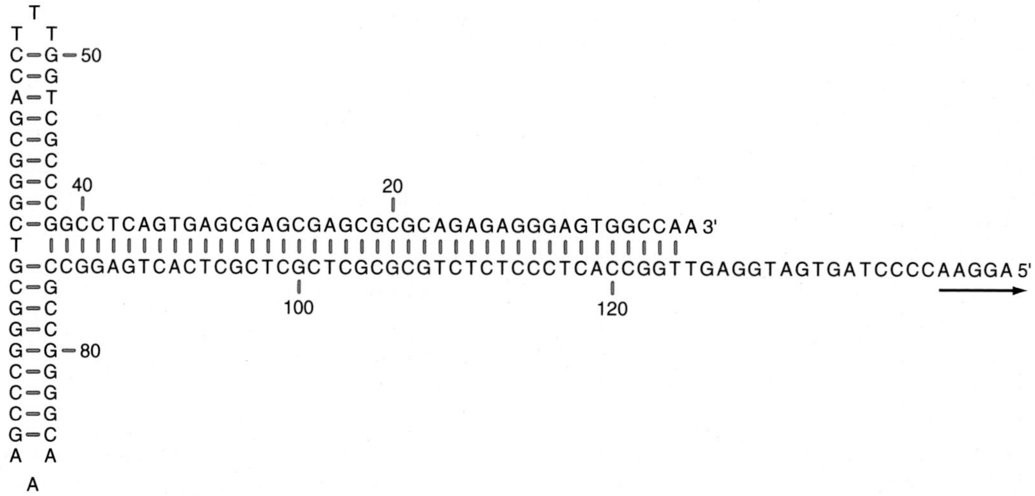

```
   T
 T   T
 C-G-50
 C-G
 A-T
 G-C
 C-G
 G-C
 G-C 40                   20
 G-C  |                    |
 C-GGCCTCAGTGAGCGAGCGAGCGCGCAGAGAGGGAGTGGCCAA 3'
 T  ||||||||||||||||||||||||||||||||||||||||||
 G-CCGGAGTCACTCGCTCGCTCGCGCGTCTCTCCCTCACCGGTTGAGGTAGTGATCCCCAAGGA 5'
 C-G             |                   |
 G-C            100                 120                    ---->
 G-C
 G-G-80
 C-G
 C-G
 C-G
 G-C
 A   A
   A
```

FIG. 5.11 Parvovirus genomic hairpin sequences. Palindromic sequences at the ends of parvovirus genomes result in the formation of hairpin structures involved in the initiation of replication.

treatment of cells with **ultraviolet light, cycloheximide,** or some **carcinogens** can replace the requirement for helper viruses. Therefore, the help required appears to be for a modification of the cellular environment—possibly to activate **DNA repair** processes—rather than for a specific virus protein. Both the independent or **autonomous** and the **helper virus-dependent** parvoviruses are highly reliant on host-cell assistance for gene expression and genome replication because of their very small genome sizes. These viruses show an extreme form of parasitism, utilizing the normal functions present in the nucleus of their host cells for **both expression and replication** (Fig. 5.12).

Viruses in the plant-infecting family *Geminiviridae* also have ssDNA genomes, but these are **circular**. The expression of their genomes is quite different from that of parvoviruses but nevertheless still relies heavily on host-cell functions. There are open reading frames in both orientations in the virus DNA, which means that both (+) and (−)sense strands are transcribed during infection, in either direction from a "**long intergenic region**" (LIR). The mechanisms involved in control of gene expression have not been fully investigated, but at least some geminiviruses (in genus *Mastrevirus*) use **RNA splicing** for expression of one or more genes (Fig. 5.13).

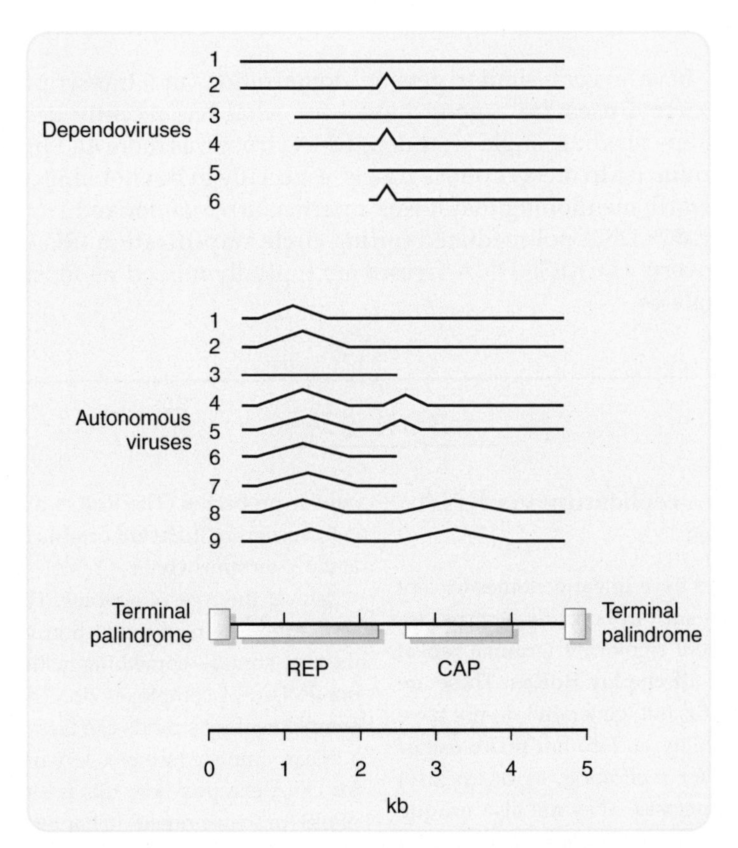

FIG. 5.12 Transcription of parvovirus genomes. Transcription of parvovirus genomes is heavily dependent on host cell factors and results in the synthesis of a series of spliced, subgenomic mRNAs that encode two types of proteins: Rep, which is involved in genome replication, and Cap, the capsid protein (see text).

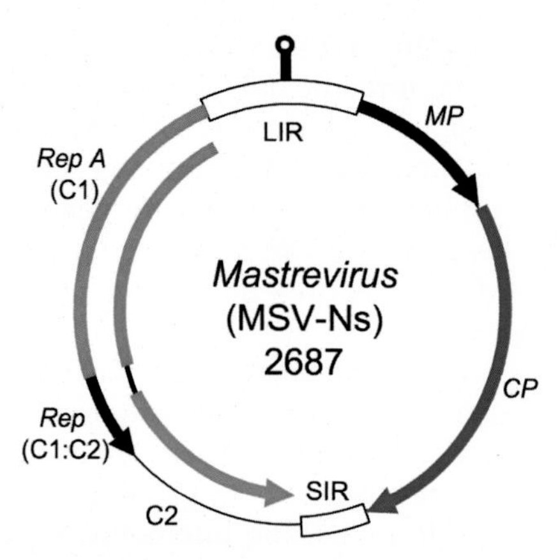

FIG. 5.13 Maize streak virus genome. The genome organization of the Nigerian isolate first sequenced (MSV-Ns) is shown (genus *Mastrevirus*, family *Geminiviridae*). *LIR*, long intergenic region and bidirectional promoter; *MP*, *CP*, movement and coat protein genes. Single pre-mRNAs are transcribed for (+)sense ORFs, and spliced to give MP- and CP-expressing mRNAs. C1 and C2 exons are differentially spliced for expression of Rep A and Rep proteins, which constitute an operational nicking-ligating complex for genome replication. *Picture courtesy of Darren Martin, PhD thesis, 2001.*

Geminiviruses have a very similar genetic organization and transcription strategy to animal-infecting **circoviruses** are a good model for what has recently become known as **CRESS** (circular Rep-encoding single-stranded) DNA viruses, as more and more unclassified genomes keep turning up in metaviromics trawls, especially in but not limited to, marine environments. It is worth mentioning that it was a particular virus-derived DNA amplification technique—phage Φ29 DNA pol-mediated **rolling circle amplification** (RCA)—that enabled much of the discovery, as CRESS DNA viruses are typically missed by most environmental metagenomics analyses.

BOX 5.4

RNA viruses replicating via RNA intermediates

All RNA viruses have **linear** genomes (except for deltaviruses such as hepatitis D virus, see table 3.1), without significant terminal repeat sequences, and **all employ RdRps**. These are **template-specific**, but (generally) do not have proofreading ability, and **do not make use of RNA primers** for replication, as do (nearly) **all DNA polymerases**. They are also **unique to RNA viruses**: there is no host enzyme with similar properties. The RdRps also all specifically recognize **different origins of replication** at the 3′-termini of both (+) and (−)sense RNAs, whatever the type of genome. They also transcribe the origin of replication sequence into the new strand—something neither DNA pool nor RNA pol complexes do. Thus, complete complementary strands can be synthesized on a linear template (see box 4.5), unlike the case for DNA except where this is synthesized by pPolB (protein-primed) (Chapter 3).

BOX 5.4 (cont'd)

Once virions are in the cytoplasm, they are generally uncoated to some extent by a variety of processes, including simple dissociation and/or **enzyme-mediated partial degradation** of the particles, to release the viral genome as a **naked RNA** or as a **nucleoprotein** complex or **core particle**. Thereafter, all events occur in the cytoplasm except for a few ssRNA- viruses like **influenza** and related **orthomyxoviruses**. The majority of **ssRNA+ viruses** cause massive rearrangements of ER membranes by interaction of their **nonstructural regulatory proteins**: this creates a partially or even completely enclosed (see Box 5.5) microenvironment where replication occurs, also known as "**replication factories (RF).**" These structures segregate the essentially **dsRNA viral replication complex** from the host cell, and probably from dsRNA-sensing proteins in the cell (see Mazeaud et al., 2018).

The **replication** of these viruses is intimately involved with the **expression** of their genomes: all of the viruses must produce all or most of the components of an **RdRp**, and often other proteins as well, in order to transcribe full-length complementary RNA molecules from RNA templates. RNA genomes tend to have **higher mutation rates** than those composed of DNA because they are copied less accurately, and only few RdRps have proofreading or repair mechanisms. These reasons tend to drive RNA viruses toward smaller genome sizes: ssRNA+ genomes vary in size from those of **coronaviruses** at approximately 30 kb—which do have **proof-reading** (see Chapter 3)—to those of bacteriophages such as **MS2** and **Qβ**, at about 3.5 kb, and do not have proofreading.

Class III: Double-stranded RNA

These viruses are a diverse collection, with a **polyphyletic** origin (see Chapter 3): while eukaryote-infecting **reoviruses** and prokaryote-infecting **cystoviruses** may share an origin, the other dsRNA viruses (all of eukaryotes) derive from various branches of the ssRNA+ virus tree as a result of convergent evolution—and consequently, may well have different modes of replication as a result (see Chapter 4). However, all of the viruses that have been studied in any detail share the characteristics of **taking a RdRp complex into cells** with the genome, **protecting the dsRNA genome** from intracell sensors, and **expressing ssRNA+s** and possibly **mRNAs** from the protected genome.

Reoviruses have **multipartite genomes** (see Chapters 3 and 4) and replicate in the cytoplasm of the host cell. Characteristically for viruses with segmented RNA genomes, a separate **monocistronic** mRNA is produced from each segment (Fig. 5.14). Early in infection, transcription of the **dsRNA** genome segments by virus-specific transcriptase (RdRp) activity occurs inside partially uncoated **core particles**. At least **five enzymatic activities** are present in reovirus particles to carry out this process, although these are not necessarily all separate proteins (Table 5.2). This primary transcription results in **capped transcripts** (m7Gppp added at 5′ end) that are not polyadenylated. These leave the virus core via **pores in vertex spikes** at the 5-fold rotational axes of symmetry, at the **base** of each of which is the transcriptional apparatus, to be **translated** in the cytoplasm. The various genome segments are transcribed/translated at **different frequencies**, which is perhaps the main advantage of a segmented

TABLE 5.2 Enzymes in reovirus particles.

Activity	Virus protein	Encoded by genome segment
RNA-dependent RNA polymerase (RdRp)	λ3	L3
RNA triphosphatase	μ2	M1
Guanyltransferase (capping enzyme)	λ2	L2
Methyltransferase	λ2	L2
Helicase (Hel)	λ1	L1

genome. Genome RNA is transcribed **conservatively**: that is, only **(−)sense strands** are used as templates, resulting in synthesis of **(+)sense mRNAs only**, which are capped inside the core. **Secondary transcription** occurs later in infection, inside new **provirions**—partially-assembled particles made from mRNAs and newly-translated and accumulating proteins—and results in **uncapped, nonpolyadenylated** transcripts. The genome is replicated in a **conservative fashion**, which is unique among cell or viral genomes. New (−)sense strands are transcribed to many (+)sense, thus excess (+)sense strands are produced which serve as late mRNAs—and as templates for (−)sense strand synthesis (Fig. 5.14).

Class IV: Single-stranded (+)sense RNA

This type of genome occurs in many animal viruses and plant viruses. In terms of both the number of different families and the number of individual viruses, this is the **largest single class of virus genomes**—and contains the oldest ssRNA lineages (see Chapter 3). Essentially, these virus genomes act as **mRNAs** that are translated **immediately after infection** of the host cell (Chapter 4). The virus genomes usually have a cell mRNA-like **5′-cap structure** (7-methyl guanosine triphosphate, **m7Gppp**)—although picorna-like viruses have a **genome-linked protein** instead—and while many genomes have **polyA** tails like eukaryotic mRNAs, a significant number of alphavirus-like superfamily members have a **complex 3′-terminal structure**, often resembling a **tRNA** (and ***aminoacylatable***), and otherwise a series of ***pseudoknots***. This serves both to protect the 3′-end from exonucleases and as a **specific RdRp recognition site.**

As stated earlier (Chapter 4), **prokaryote** ssRNA+ virus genomes are essentially **polycistronic mRNAs**, from which all ORFs can be translated, while eukaryotic ssRNA+ virus genomes are **monocistronic**—only the **ORF nearest the 5′ end** can normally be translated. Not surprisingly with so many representatives, this class of genomes displays a very diverse range of strategies for controlling gene expression and genome replication. However, in very broad terms, the viruses in this class can be subdivided into three groups, as previously shown in Chapter 4:

1. Viruses where the genomic RNA or RNAs each have **only a single ORF**, and these are individually translated to form a **polyprotein** product that is processed further by proteolytic cleavage (see Fig. 5.15).
2. Viruses whose genomes—**single- or multicomponent**—are expressed partly by **direct translation** (5′-proximal ORF(s) only), and partly via **transcription**. For these viruses, **genome replication** is required before transcription of sequences downstream of the 5′-proximal ORF—encoding the **RdRp**—can occur.

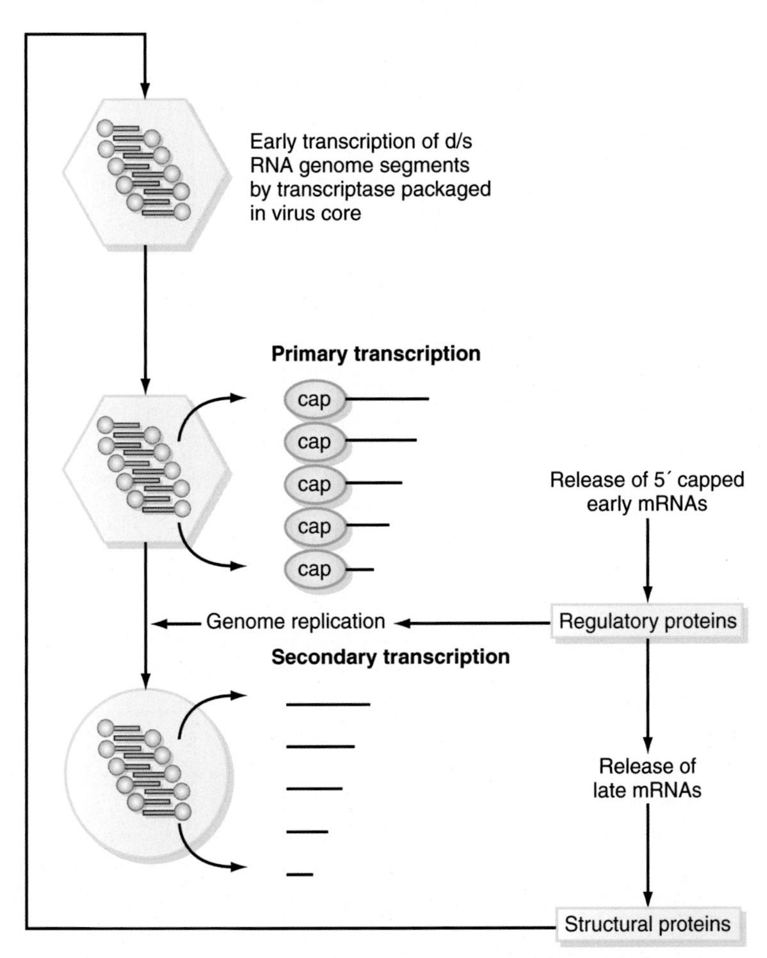

FIG. 5.14 Expression of reovirus genomes. Expression of reovirus genomes is initiated by a RdRp complex packaged inside every virus particle. Subsequent events occur in a tightly regulated pattern, the expression of late mRNAs encoding the structural proteins being dependent on prior genome replication.

3. Viruses whose genomes have both a **polyprotein-encoding ORF** that is translated and then processed to provide regulatory proteins (e.g., **RdRp** and accessory proteins), and downstream ORFs **expressed via mRNAs** (Fig. 5.16).

All the viruses in this class have evolved mechanisms that allow them to regulate their gene expression in terms of the **concentrations** of **individual virus-encoded proteins,** the **ratios of different proteins,** and the **stage of the replication cycle** when they are produced. Compared with the two classes of DNA virus genomes described above, these mechanisms operate largely independently of those of the host cell. The power and flexibility of these strategies are reflected very clearly in the overall success of the viruses in this class, as determined by the number of different representatives known and the number of different hosts they infect.

Production of proteins via processing of a single ORF essentially results in **equimolar amounts** of each, all at the same time—which can be a problem if proteins are toxic, or

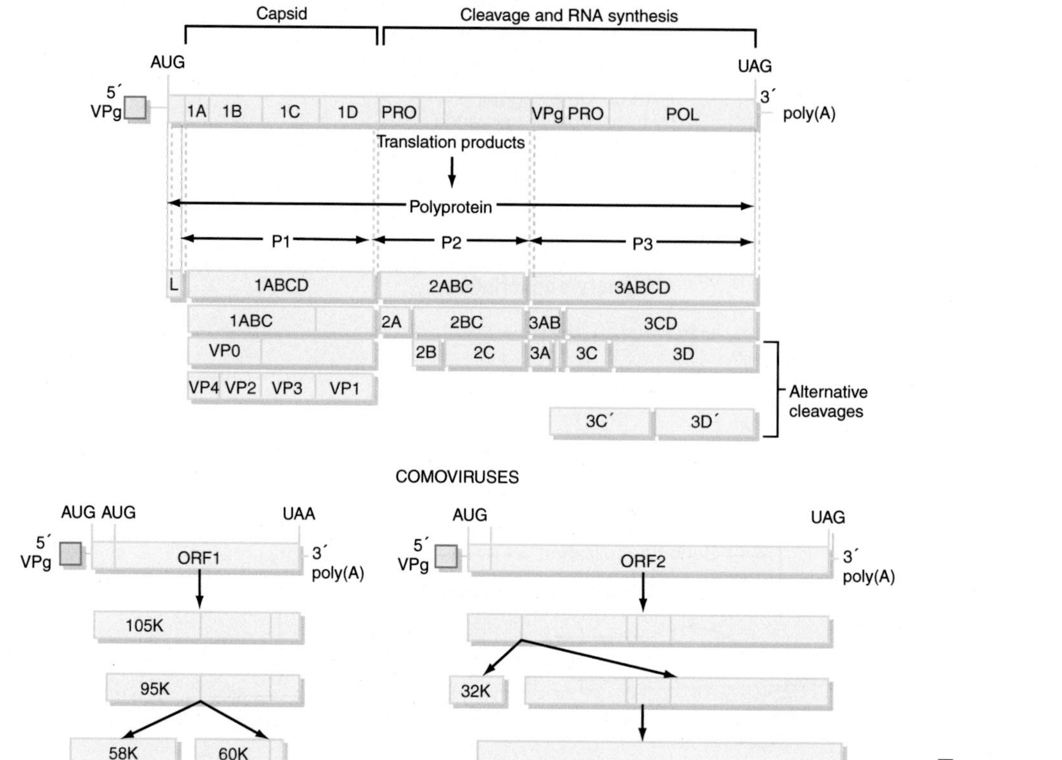

FIG. 5.15 Gene expression in picorna-like viruses. Many positive-sense RNA virus genomes or segments are translated to form a long polyprotein, which is subsequently cleaved by highly specific virus-encoded protease(s) to form the mature polypeptides. Picornaviruses and comoviruses and potyviruses are examples of this mechanism of gene expression: comoviruses have two-component genomes that resemble a split picornavirus genome. Comoviruses and nodaviruses resemble picornaviruses that have been cut into two segments; the bymoviruses in family *Potyviridae* similarly resemble a cleaved potyvirus genome.

only a little is needed (e.g., of **RdRp**). **Picornaviruses** and similar viruses in the superfamily have the strategy of synthesizing **"one-use" RdRps**: the replication complex initiates (−) sense RNA transcription at the genomic 3′ end, using a component of itself (**virus protein genome-linked**, or **Vpg** in picornaviruses) as a **primer**: the first nucleotide of the new sequence is attached to the Vpg portion of the replicase, and synthesis commences from the free 3′-OH of the nucleotide after cleavage of Vpg from the rest of the RdRp. Thus, each new strand of RNA has a Vpg covalently attached at the 5′ end, and the RdRp complex cannot re-initiate transcription. Similarly, various of the **potyvirus** genome products such as the **NIa**

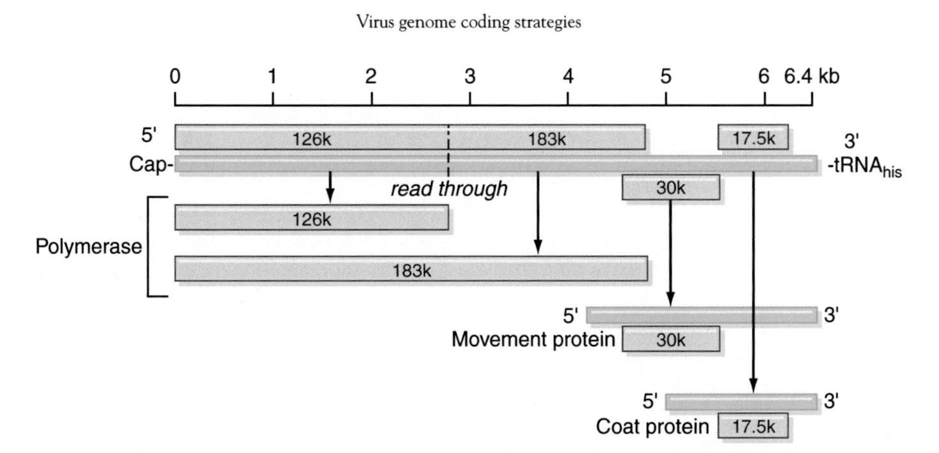

FIG. 5.16 Organization and mRNA expression of the TMV genome. The TMV genome makes use of a number of strategies to get around the limitation of eukaryotes requiring mRNAs to be monocistronic. First, the replication-associated genes are expressed as two polypeptides with a common 126 kDa N-terminal sequence by translational readthrough. Second, movement and coat proteins are expressed off mRNAs transcribed from different RNA promoters in the full-length (−)sense RNA that is synthesized after production of RdRp: these start at different places in the genome, but have the same 3′ termini. The 3′ end of the genome and all mRNAs is also a sequence that can act as a $tRNA_{His}$: this protects the 3′ end from exonuclease digestion, as does the 5′ m7Gppp cap structure at the 5′ end.

protease and **NIb protein** are sequestered in **plant cell vacuoles** or **nuclei** to get them out of the cytoplasm, where they may form quasicrystalline arrays or characteristic "**pinwheel**" inclusions. Viruses in the picorna-like virus supergroup tend to have 3′-polyadenylated (polyA) genome segments, with the polyA sequences **being part of the genome** and copied into polyU in (−)sense RNA.

Production of proteins via mRNAs, on the other hand, is more controllable in terms of both quantity and timing: transcription of mRNAs from (−)sense RNAs depends upon **"RNA subgenomic promoter"** sequences, which may be of different efficiencies if there are more than one, leading to the possibility of differential expression of mRNAs. These "promoters" differ from origins of replication at genome and complementary RNA 3′ ends in that they are **not copied into mRNA transcripts**, which are therefore not **replicatable**. This expression allows temporal separation of **early (=regulatory)** and **late (=structural)** genes. The context of translation of the mRNAs is also important: different **5′ untranslated sequence regions (UTRs)**, or even **3′ UTRs**, can significantly affect translation efficiency—again, meaning **differential translation** is possible for different genes, allowing slower or less accumulation of different products. This strategy is typical of the **alpha-like supergroup** of viruses, and is exemplified by the generic **tobamoviruses**.

Temporal regulation of expression from ssRNA+ genomes is not possible in the same way as it is for larger DNA genomes, simply due to a far lower capacity to produce regulatory proteins (see Box 5.4). Accordingly, the replication/expression process is regulated by differential accumulation of different products—such as RdRp—and **concentration-driven processes** such as **nucleation** then **completion** of virion assembly by coat proteins.

Genomes in the third group of ssRNA+ viruses have a **polyprotein** processing strategy for products of ORFs **nearest the 5′ end** of the genome, and **mRNA expression** for ORFs **nearer the 3′ end**. This possibly indicates an ancient recombinational event between a picorna-like virus and an alpha-like virus (see Chapter 3). Such viruses include **alphaviruses** and **coronaviruses** (see Fig. 1.26 for the SARS-CoV-2 genome and its ORFs, and below for discussion of expression of the virus). **Togaviruses** (e.g., *Togaviridae*, including **Sindbis virus, Chikungunya virus**) are an important collection of mainly **arthropod-borne (arbo-) viruses** that replicate in insects and in vertebrates. They have enveloped virions and a ~12 kb genome, that is capped and polA tailed (see Fig. 5.17). This has two ORFs, with the 3′-proximal one expressed via mRNA synthesis from the (−)sense complementary strand after initial

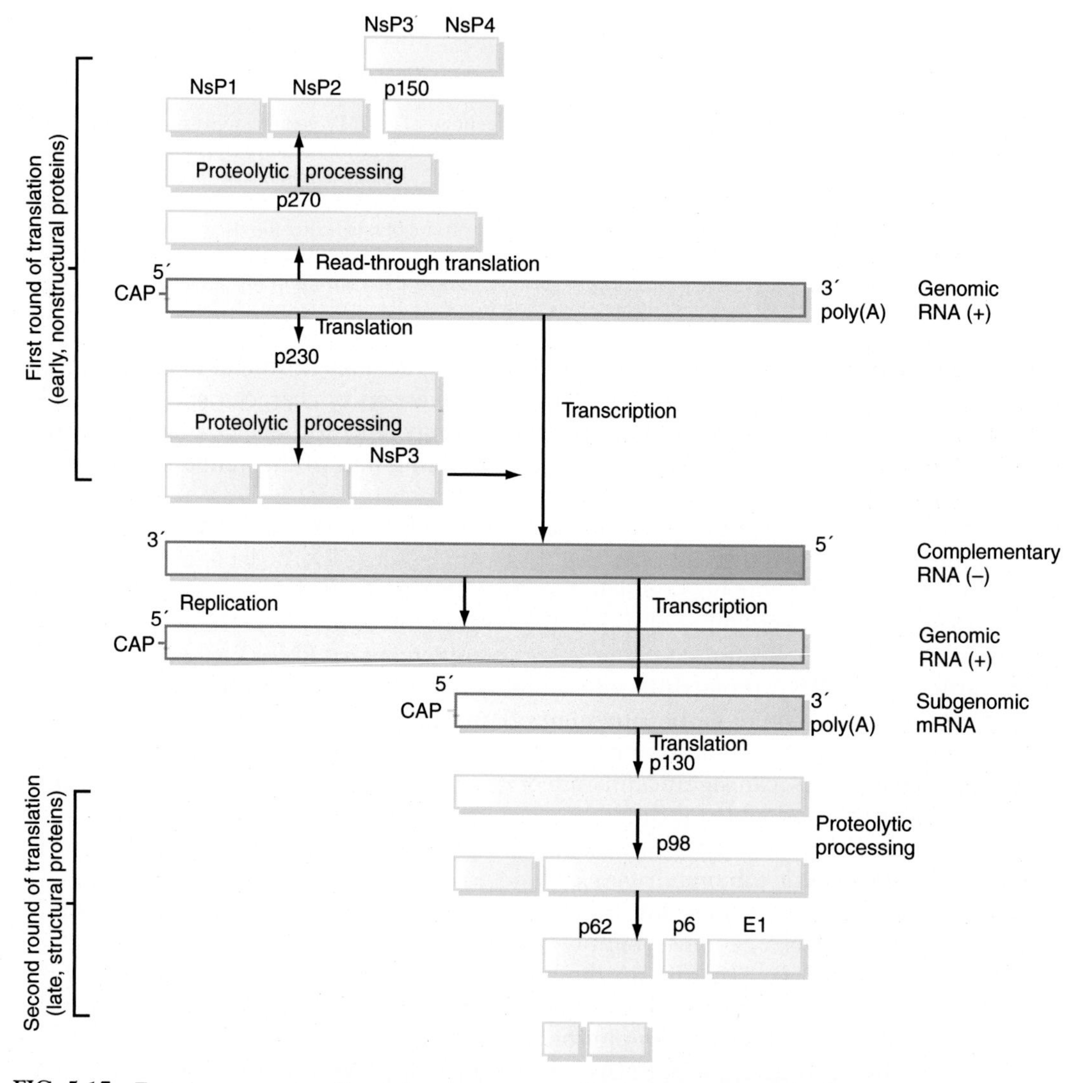

FIG. 5.17 Togavirus expression. These viruses are expressed by two separate rounds of translation, involving the production of regulatory proteins from a 5′-proximal ORF by immediate translation, and production of a monocistronic subgenomic mRNA at a later stage of replication. Translation of both 5′- and 3′-proximal ORFs produces polyproteins, that are proteolytically processed for regulatory proteins (5′ ORF) and structural proteins (3′ ORF).

replication. Both ORFs encode polyproteins, which are processed by cleavage by a variety of virus-specified proteases.

Togaviruses have an interesting strategy for regulating replication and transcription, in that the 5′-proximal ORF-encoded **P1234 polyprotein** is proteolytically cleaved by **nsP2** to form the **(−)strand** replicase complex, consisting of **P123** and **nsP4.** This initiates synthesis of the **full-length (−)strand RNA** on the (+)strand genome template. If the P123 polyprotein is further processed—to give **nsP1** and **P23**—this results in a switch from (−)strand to **(+)strand synthesis**, resulting in the production of both full-length **(+)strand genomic** and **subgenomic RNAs**. **P23 has a half-life of a few seconds**, so needs to be constantly produced if (+)strand synthesis is to continue—which is another way of regulating both protein accumulation as a result of polyprotein processing, and of (+)strand RNA synthesis. Further cleavage of P23 into **nsP2 and nsP3** produces a **fully cleaved** viral replicase complex, which strongly favors production of the mRNA, while still being capable of full-length (+)strand RNA synthesis (Rupp et al., 2015).

The strange case of coronaviruses

Coronaviruses have of course achieved recent notoriety because of the pandemic caused by **severe acute respiratory coronavirus type 2, or SARS-CoV-2**—but let us not forget its older cousins **SARS-CoV** and **Middle East respiratory syndrome coronavirus (MERS-CoV)**, which also burst out of obscurity in this century to cause disease and panic in people (see Chapter 8). While these viruses are similar to **flaviviruses** and others in their general expression strategy—a mixture of **large polyprotein processing** and **multiple monocistronic mRNAs**—it turns out that they have a most interesting transcription strategy.

In the example shown in Fig. 5.18—for **mouse hepatitis virus**—you will see that seven mRNAs are produced (9 for SARS-CoV-2). One might expect that these would be synthesized from a full-length (−)strand template, but are in fact synthesized from (−)strand templates of varying length—and each mRNA has a **~60 base 5′ leader sequence** corresponding to the 5′ end of the genomic (+)sense RNA. The model for this process is that a **membrane-bound replication-transcription complex (RTC)** initiates (−)strand synthesis at the 3′ end of genomic RNA, and proceeds until it meets a short (~10 nt) motif named the transcription regulating sequence, or TRS: these sequences are found at the 5′ side of all ORFs that end up in mRNAs, and at the 3′ end of the 5′ leader. Upon encountering the first TRS, the RTC either stalls after transcribing it, or continues—to meet further TRSs, and stall or continue there. If it stalls, the RTC relocates by **"hopping" to the genome 5′-proximal TRS**, assisted by the complementarity of it to the 3′ end of the newly-synthesized (−)strand and completes transcription by including this sequence. Thus, a collection of (−)sense strands accumulates, all with the same 3′ terminal sequence. This is recognized by the RTC, which transcribes (+) sense strands of varying length—the subgenomic mRNAs—from each of the subgenomic (−) sense strands. Full-length (−)sense strands—resulting from a continuous RTC traverse of the RTC—are transcribed to genomic RNA.

Class V: Single-stranded (−)sense RNA

As discussed in Chapters 3 and 4, the genomes of these viruses may be either **segmented** or **nonsegmented**: because of the difficulties of gene expression and genome replication inherent in having a nontranslatable genomic RNA, they tend to have larger genomes encoding more

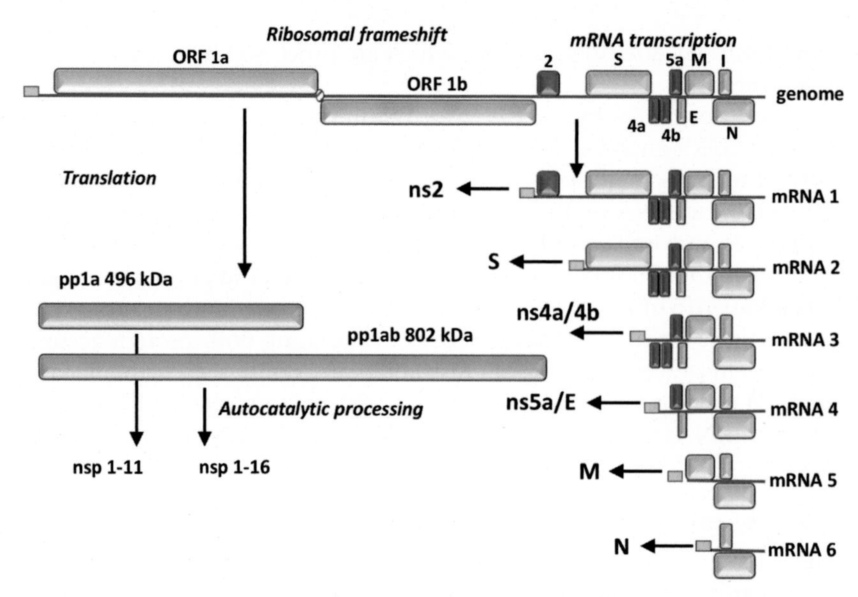

FIG. 5.18 Coronavirus expression. The genome and expression strategy of mouse hepatitis virus (MHV-A59) are shown. The 31 kb genome and mRNAs are shown as *purple lines*, translation products of ORFs 1a and 1b in *blue*. *Blue rectangles* depict nonstructural proteins (nsps) that are expressed from mRNAs; *green rectangles* are the structural protein ORFs. The common 5′ leader sequence is shown as an *orange box*. *The diagram was adapted from Sawicki, S.G., Sawicki, D.L., Siddell, S.G., 2007. A contemporary view of coronavirus transcription. J. Virol. 81(1):20–29. https://doi.org/10.1128/JVI.01358-06.*

genetic information than Class IV ssRNA+ viruses. Because of this, segmentation is a common, although not universal, feature of such viruses (Fig. 5.19). **None of these genomes is infectious** as purified RNA as all have to carry in their own RdRp complex into new cells inside their virions, much as dsRNA viruses do. None are capped at the 5′, or polyadenylated at the 3′ end.

Using influenza viruses (family *Orthomyxoviridae*) as an example, the first step in the replication of these eight-component segmented genomes is **transcription of the (−)sense genomic or virion RNA (vRNA)** by the **virion-associated RdRp** to produce (predominantly) **monocistronic mRNAs**, which also serve as the template for subsequent genome replication (Fig. 5.19). Atypically for RNA viruses, influenza and its close relatives (*Orthomyxoviridae*) **replicate in the nucleus of the host cell** rather than the cytoplasm. This allows these viruses to use the **splicing apparatus** present in this compartment to express subgenomic messages from some, although not all, of its genome segments. All of the eight segments have **common nucleotide sequences** at the 5′ and 3′ ends respectively (Fig. 5.20). These sequences are complementary to one another, and, inside the particle, the **ends of the genome segments** are held together by base-pairing and form a **panhandle structure** that is believed to be involved in replication, and initiation of transcription—which starts with "**cap snatching**" by the RdRp of a short 5′ capped sequence from cellular mRNAs (see below) (Table 5.3).

In the other families that have **nonsegmented** genomes—such as all the evolutionarily-related members of the *Mononegavirales* order, such as **rhabdo-** and **paramyxoviruses—monocistronic mRNAs** are also produced. However, these messages must be produced from a single, long (−)sense vRNA molecule. While details of how this is achieved for all of these viruses is not completely clear,

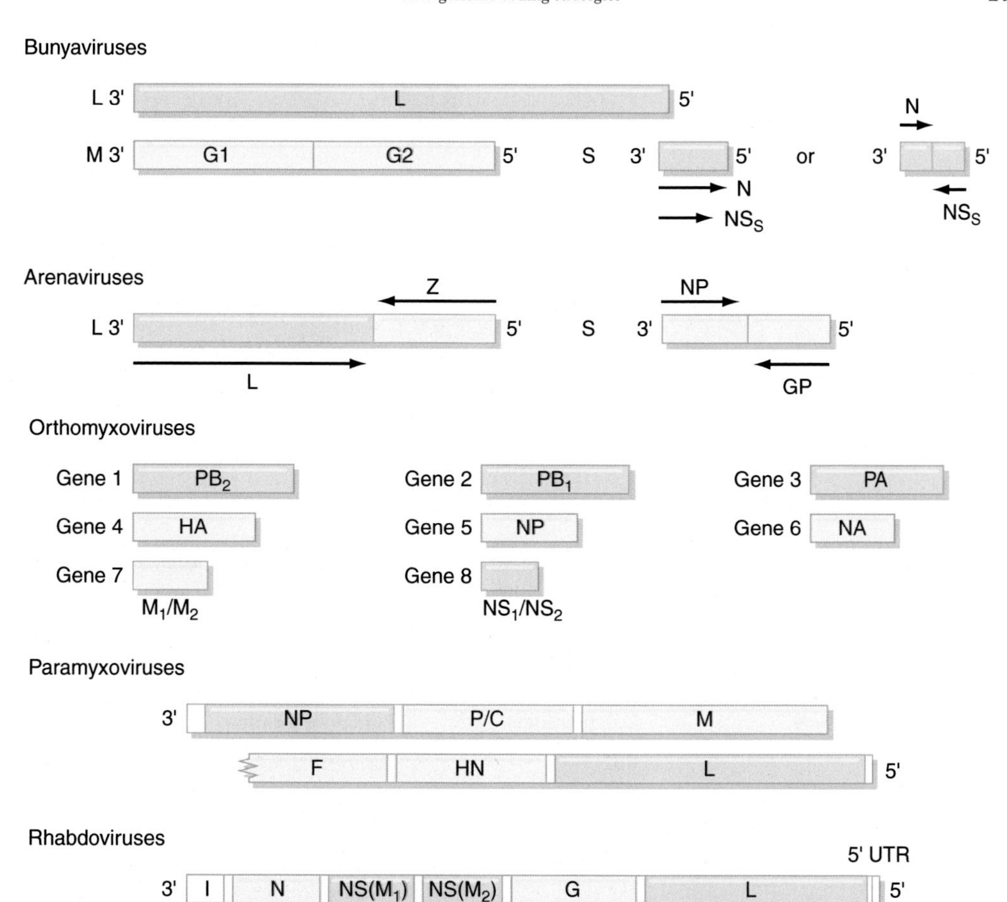

FIG. 5.19 Genome organization of negative-stranded RNA viruses. The fundamental distinction in the negative-strand RNA viruses is between those viruses with segmented genomes and those with nonsegmented genomes: otherwise, most mechanisms of transcription, replication, and expression are similar. In the nonsegmented genomes, there are conserved sequence regions between each of the genes that are relevant for transcription.

transcription apparently usually initiates **at the genomic 3′-end** with synthesis of a 50-base leader. The complex then moves to **internal promoter sites** for individual gene transcription. Transcripts are **capped and polyA** tailed; the RdRp complex (consisting of L, N/NP, and other proteins) adds caps, while tails are apparently added by RdRp stuttering at short polyU repeats at the end of genes. The **stop-and-start** mechanism of transcription is regulated by the **conserved intergenic sequences** present between each of the virus genes (see Fig. 5.21). **Splicing** mechanisms cannot be used because these viruses replicate in the cytoplasm.

Such a scheme of gene expression might appear to offer few opportunities for regulation of the relative amounts of different virus proteins. If this were true, it would be a major disadvantage, as all viruses require far more copies of the structural proteins (e.g., **nucleocapsid** protein, produced in the highest amounts of all virion proteins) than of the nonstructural proteins (e.g., **polymerase**) for each **virion** produced. In practice, the **ratio** of different proteins is regulated both during transcription and afterward. In **paramyxoviruses**, for example, there is

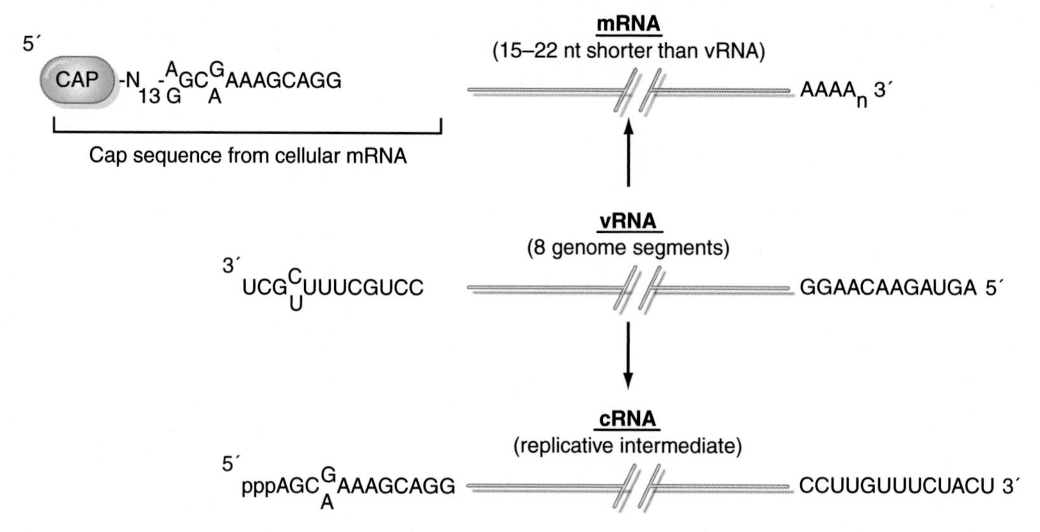

FIG. 5.20 Common terminal sequences of influenza RNAs. Influenza virus genome segments are a classic example of how sequences at the ends of linear virus genomes are crucial for gene expression and for replication.

TABLE 5.3 Influenza virus genome segments.

Segment	Size (bases)	Polypeptide(s)	Function, location
1	2341	PB2	Transcriptase: cap binding
2	2341	PB1	Transcriptase: elongation
3	2233	PA	Transcriptase: (?)
4	1778	HA	Hemagglutinin
5	1565	NP	Nucleoprotein: RNA binding; part of transcriptase complex
6	1413	NA	Neuraminidase
7	1027	M1	Matrix protein: major component of virion M2 Integral membrane protein ion channel
8	890	NS1	Nonstructural-nucleus: function unknown
		NS2	Nonstructural-nucleus + cytoplasm: function unknown

a clear **polarity of transcription** from the 3′ end of the virus genome to the 5′ end: this results in the synthesis of far more mRNAs for the **structural proteins** encoded in the 3′ end of the genome than for the **nonstructural proteins** located at the 5′ end (Fig. 5.22). Similarly, the advantage of producing monocistronic mRNAs is that the **translational efficiency** of each message can be varied with respect to the others, as each will have different **5′ UTR sequences** (see the section "Posttranscriptional control of expression").

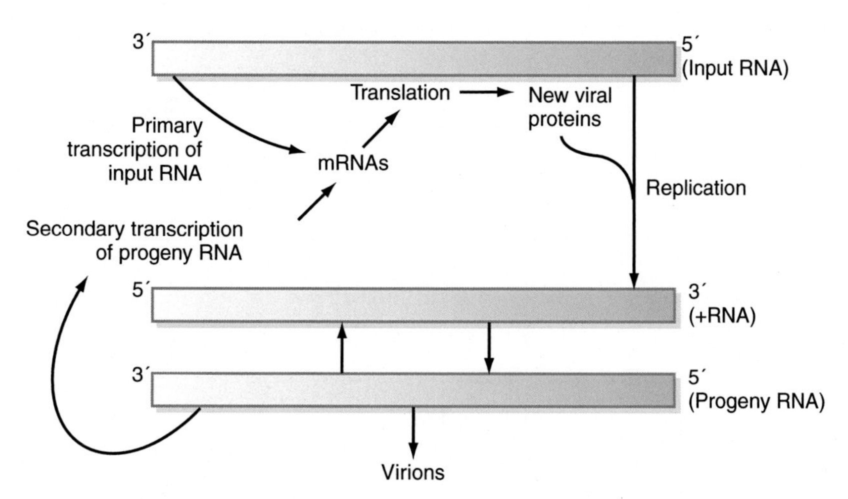

FIG. 5.21 General scheme for the expression of negative-sense RNA virus genomes. All negative-sense RNA viruses face the problem that the information stored in their genome cannot be interpreted directly by the host cell. They must therefore include mechanisms for transcribing the genome into mRNAs within the nucleocapsid or partially disassembled virus particle.

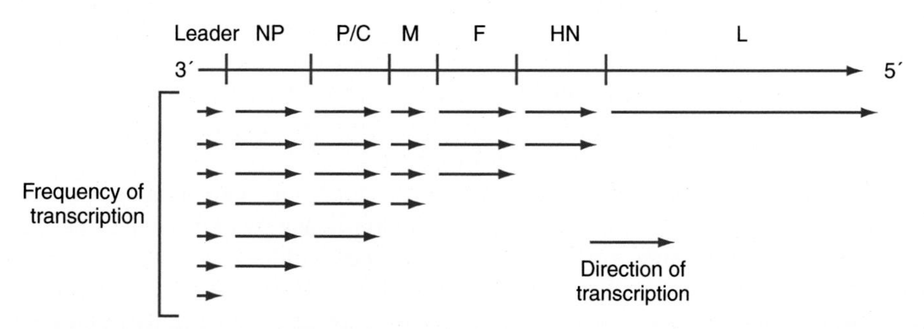

FIG. 5.22 Transcriptional expression of paramyxovirus genomes. Paramyxovirus genomes exhibit "transcriptional polarity." Transcripts of genes at the 30 end of the virus genome are more abundant than those of genes at the 50 end of the genome, permitting regulation of the relative amounts of structural (30 genes) and nonstructural (50 genes) proteins produced.

Production of **full-length (+)sense RNA** rather than of mRNAs is triggered by binding mainly of **newly-synthesized N** (nucleoprotein) but also of **P** (a RdRp minor subunit) proteins to the 5′-leader sequence, causing the RdRp **to ignore all termination and polyadenylation signals**. The RNA+ is then used as template for vRNA-transcription: this also has a 5′-leader, which is also recognized as an **assembly origin** by N protein. Thus, concomitant genome or antigenome synthesis and nucleoprotein assembly occur, with a **bias for (−) strand synthesis**, possibly due to preferential recognition of the (+) strand 3′-origin, and also associated with **nucleation of assembly** by N protein.

Viruses with **ambisense** genome organization—where genetic information that can be **translated** is encoded in both the **(−)sense vRNA** and the **(+)sense complementary** orientation on the **same strand of RNA** (see Chapter 4)—must express their genes in **two rounds**

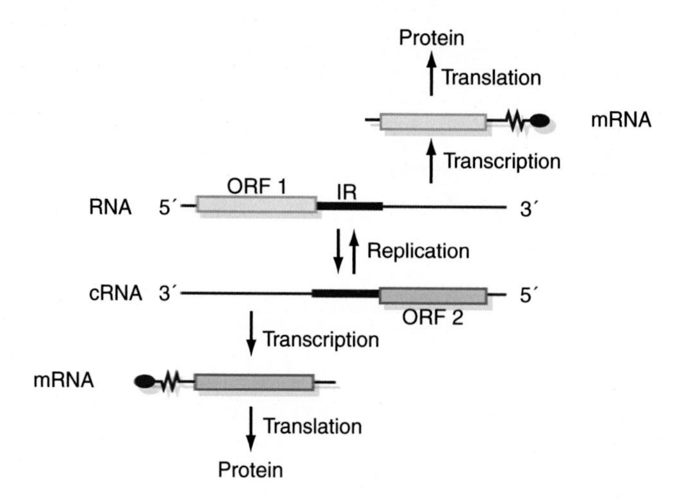

FIG. 5.23 Expression of ambisense virus genomes. Ambisense virus genomes require two rounds of gene expression so that information encoded in both strands of the genome is turned into decodable mRNA at some point in the replication cycle.

of expression so that both are turned into decodable mRNA at some point (Fig. 5.23). There is almost certainly some advantage to this mode of expression for it to have been developed and then conserved—and it probably lies in being able to express viral proteins immediately on entry into the cell, for some **regulatory purpose**. With **bunyaviruses**, all three RNA species are linear, but in the virion they appear **circular** and **panhandle-like** because the ends are held together by base-pairing, as with influenza viruses.

The trisegmented **bunyaviruses** (Fig. 5.19) have an interesting "**cap stealing**" strategy, also common to **orthomyxoviruses**: virion-associated **L or replicase protein** cleaves **cellular mRNAs 12–18nt from their 5′-ends**, and uses these "**capped leaders**" to **prime transcription** of nonpolyadenylated mRNA on the three virion L, M, and S (−) RNAs. These mRNAs are **shorter than the genome segments**, as transcription is apparently terminated by **hairpin loops**. It is not certain how bunyaviruses switch from mRNA to full-length RNA(+) transcription; however, the N protein may act similarly to the way it does in mononegaviruses such as paramyxoviruses. Viruses in two genera of the *Bunyaviridae* also have at least one ambisense RNA: **phleboviruses** and **tospoviruses** transcribe a mRNA from the 3′-end of the **S segment RNA(+)**; the plant-infecting **tospoviruses** in addition have an ambisense **M RNA**, with the extra gene (5′-end of M RNA(−)) being involved in **movement functions** in plants. Arenaviruses (e.g., Lassa virus; see Fig. 5.19) have two ambisense genome segments: the L segment ORF translates the matrix-equivalent **Z protein**, and the S segment expresses the **GP envelope glycoprotein**.

Class VI: Single-stranded (+)sense RNA genomes with DNA replication intermediates

The **retroviruses** and newly-renamed retrotransposons (**meta-, pseudo-, belpaoviruses**; see Chapter 3) are the ultimate case of reliance on the host-cell transcription machinery, although like all RNA viruses, they encode their own replicase—that is, **RNA-dependent RNA**

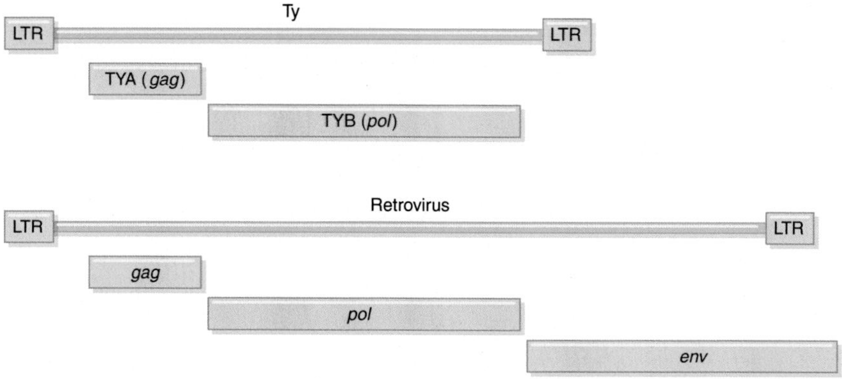

FIG. 5.24 Retrotransposon and retrovirus genomes. The genetic organization of retrotransposons such as Ty (pseudovirus; *above*) and retrovirus genomes (*below*) shows a number of similarities, including the presence of direct LTRs at either end. Metavirus genomes resemble retroviruses in having an env gene; belpaovirus genomes resemble pseudovirus genomes in lacking this gene (see Chapter 3).

polymerase, or reverse transcriptase. The RNA genomes form a template for reverse transcription to DNA: these are the only ssRNA+ viruses whose genomes **do not serve as mRNA** on entering the host cell (Chapters 3 and 4) (Fig. 5.24).

Retrovirus-like genomes have four unique features:

- They are the only viruses that are truly **diploid**: there are two identical copies of the ssRNA+ version of the genome included in each virion.
- They are the only RNA viruses whose genome is produced by **cellular transcriptional machinery** without any participation by a virus-encoded polymerase.
- They are one of only two groups of viruses (with members of the *Caulimoviridae*) that **require a specific cellular tRNA** for initiation of replication.

Once integrated into the host cell genome (see Chapter 3; figs. 3.8 and 3.9), the DNA **provirus** is under the control of the host cell, and is transcribed **exactly as are other cellular genes**. Some retroviruses, however, have evolved a number of transcriptional and posttranscriptional mechanisms that allow them to control the expression of their genetic information, and these are discussed in detail later in this chapter.

Class VII: Double-stranded DNA genomes with (+)sense RNA replication intermediates

As discussed in Chapter 3, the so-called "**DNA retroviruses**" appear at first sight to be very different to their Class VI counterparts—however, they are joined in the "**retroid cycle**" (Fig. 3.7), and the only effective difference is in **which genome gets encapsidated**, or where the **long terminal repeats or LTRs** are. Expression of the genomes of these viruses is complex and relatively poorly understood.

Hepadnavirus genomes (see Fig. 5.25) contain a number of overlapping reading frames clearly designed to squeeze as much coding information as possible into a compact genome. **Hepatitis B virus (HBV)** is the type member of the family *Hepadnaviridae* and the

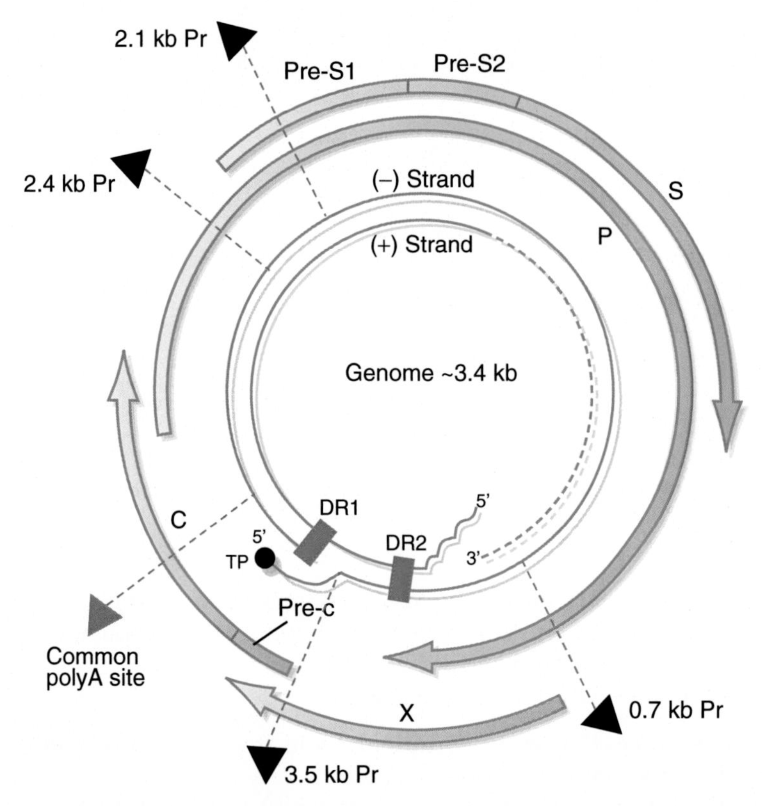

FIG. 5.25 The HBV genome and its transcription. Structure and organization of the HBV genome, with promoter positions for the four different mRNAs marked. The common polyadenylation site (polyA) for all four transcripts is shown. *P*, polymerase; *pre-C*, pre-core; *c*, core, Pre-S1 and -S2 and S are the envelope proteins. DR1 and DR2 are short direct repeat sequences. Note the (−)strand overlap, and the variable length (*dotted line*) of the (+)strand: this is caused by intraparticle DNA synthesis of this strand failing at unpredictable places. *TP*, domain of P protein that covalently attaches as primer to 5′ end of (−)strand DNA.

best studied. HBV virions are spherical, lipid-containing particles, 42–47 nm diameter, and contain a partially dsDNA genome—termed **"gapped,"** with a **(−)sense strand** of 3.0–3.3 kb (varies between virus types) that has a **5′-linked protein** that acted as primer for DNA synthesis, and a variable length (+)sense strand (1.7–2.8 kb) that has a 5′-linked short length of RNA—and an **RT complex** (Fig. 5.25). The (−)sense DNA has a short **direct repeat sequence** as shown. Genomes enter cell nuclei inside virions, and uncoat. The genome is "repaired" with excision of the (−)sense and (+)sense protein and RNA primers and truncation of the **(−) strand overlap**, extension of the (+)strand, and ligation of both so as to be a **covalently closed circular dsDNA** by host repair DNA pol. Three major genome transcripts are then produced in infected cells by means of **RNA Pol II**: these are 3.5-, 2.4-, and 2.1-kb mRNAs. All are transcribed from the **same strand** of the virus genome and have the same 3′ ends, but are **initiated at different start sites**. There are four known genes in the virus:

- **C** encodes the core protein.
- **P** encodes the **polymerase** complex (RT).

- **S** encodes the **three versions** of the **surface antigen**: pre-S1, pre-S2, and S. These are derived from **alternative transcriptional start sites** of the two RNAs encoding S, and a choice of translational start in the 2.4 kb mRNA.
- **X** encodes a **transactivator** of virus transcription and possibly also of cellular genes.

The X transcriptional *trans*-activator is believed to be analogous to the **human T-cell leukemia virus (HTLV) tax** protein (see later). The **largest (3.5 kb) mRNA**—which is **longer than the genome**, thanks to RNA Pol II overshooting a **transcriptional stop site** adjacent to the promoter on the first time around the circular genome—is also the **template for reverse transcription**, which occurs during the formation of the **provirion** after transcription and accumulation of translation products. This forms as a result of binding of **P (RT)** protein to an **encapsidation signal called** ε, which is a **short stem-loop** structure near the 5′ end of the 3.5 kb RNA—also known as pgRNA—which is not present on a longer version of the mRNA that includes the preC domain, which is accordingly not encapsidated. This interaction leads to recruitment of **C (core)** protein dimers, which **initiates packaging by C** of the RNA-P complex. At this point, reverse transcription begins on the 3.5 kb RNA as template, with a nucleotide covalently linked to the **terminal protein (TP)** domain of P as a primer (see Beck and Nassal, 2007). Hepadnaviruses are unique among reverse-transcribing elements in their use of a protein rather than of a tRNA to prime reverse transcription: this is reminiscent of the DNA **pPolB** used by **adeno**- and other small prokaryote **dsDNA** viruses with linear genomes.

The genome structure and replication of **cauliflower mosaic virus (CaMV)**, the prototype member of the *Caulimoviridae* family, is reminiscent of that of **hepadnaviruses**, although there are important differences. The CaMV genome (Fig. 5.26) consists of a gapped, circular dsDNA molecule of about 8 kbp, one strand of which is known as the α-strand and contains a single gap, and a complementary strand, which contains two gaps (Fig. 3.23). There are eight genes encoded in this genome, although not all eight products have been detected in infected cells:

- ORF I—P1: **movement** protein
- ORF II—P2: aphid/insect **transmission factor**
- ORF III—P3: **virion-associated** protein (VAP): structural protein, DNA-binding
- ORF IV—P4: **capsid protein** (CP)
- ORF V—P5: pro-pol: **protease, RT and RNaseH**
- ORF VI—P6: **transactivator**/viroplasmin: inclusion body formation/trafficking; possibly other functions
- ORF VII/VIII—unknown function

Replication of the CaMV genome is similar to that of HBV. The first stage is the migration of the gapped virus DNA to the **nucleus** of the infected cell where it is repaired to form a covalently closed circle. This DNA is transcribed to produce **two polyadenylated transcripts**, one—the so-called **35S RNA**—that is **longer than the genome**, made by a similar mechanism as happens with HBV, and one that is shorter (**19S**), transcribed from different promoters. It is a matter of history that the CaMV 35S promoter is now one of the most used elements of **plant expression vectors**, given that it is one of the **strongest plant promoters known**. In the cytoplasm, the 19S mRNA is translated to produce a protein that forms **large inclusion bodies** in the cytoplasm of infected cells, and it is in these sites that the **second phase of replication**

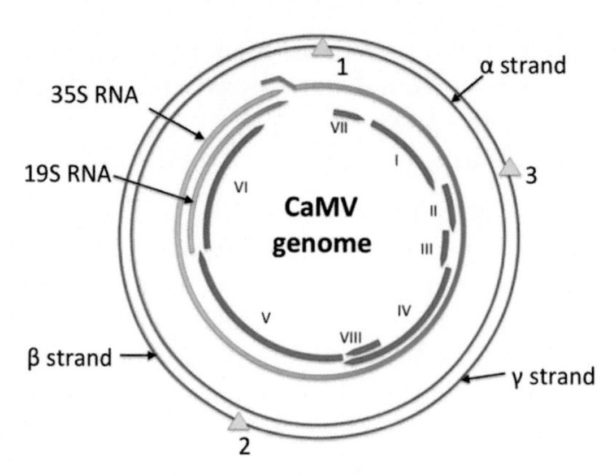

FIG. 5.26 The cauliflower mosaic virus genome and its transcription. The genome encodes eight proteins. It is transcribed into only two mRNAs—named 35S and 19S, on the basis of their sedimentation properties—and contains one discontinuity in the alpha or coding strand, and two in the noncoding strand (beta and gamma).

occurs. In these **replication complexes**, some copies of the 35S mRNA are translated while others are **reverse transcribed** using a $tRNA_{fmet}$ as a primer, and packaged into virions as they form.

The 19S transcript is **monocistronic**, whereas the 35S is **multicistronic**—which is supposedly anathema for a eukaryotic mRNA. This is in fact translated by a so-called "**relay race**" model, most unusual for eukaryotes, which involves an association of the **transactivator (TAV)** product of **gene VI**, and **elongation initiation factor III (eiFIII)**: this prevents dissociation of eiFIII from **ribosomes**, and enhances resumption of ribosome "**scanning**" and **reinitiation** at a AUG start codon downstream, following termination of translation. This maintains the ribosome in a competent state to commence translation of downstream ORFs without dissociating, as is normally the case.

Transcriptional control of expression

Having looked at general strategies used by different groups of viruses to regulate gene expression, the rest of this chapter concentrates on more detailed explanations of specific examples from some of the viruses mentioned earlier, beginning with control of transcription in **SV40**, a member of the *Polyomaviridae*. Few other genomes, virus or cellular, have been studied in as much detail as SV40, which has been a model for the study of eukaryotic transcription mechanisms for many years. In this sense, SV40 provides a eukaryotic parallel with the **bacteriophage λ** genome. In vitro systems exist for both **transcription** and **replication** of the **5.2 kbp dsDNA** SV40 genome, and it is believed that all the virus and cellular DNA-binding proteins involved in both of these processes are known. The SV40 genome encodes two T-antigens ("**tumor antigens**") known as **large T-antigen** and **small T-antigen** after the sizes of the proteins (Fig. 5.27A). Replication of the genome of SV40 occurs in the nucleus of the host cell. Transcription of the genome is carried out by host RNA pol II, and large T-antigen plays a vital role in **regulating transcription** of the virus genome. Small T-antigen is not essential for virus replication but does allow virus DNA to **accumulate in the nucleus**.

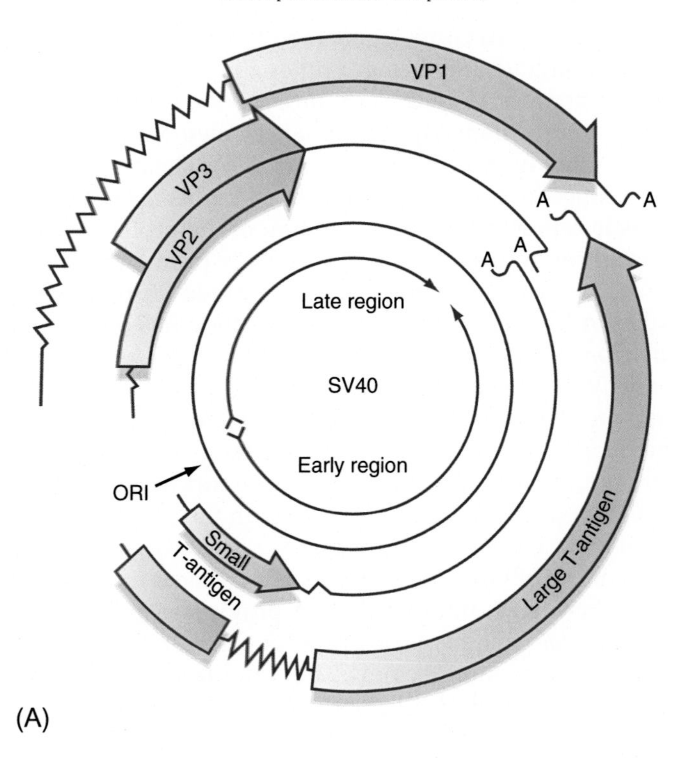

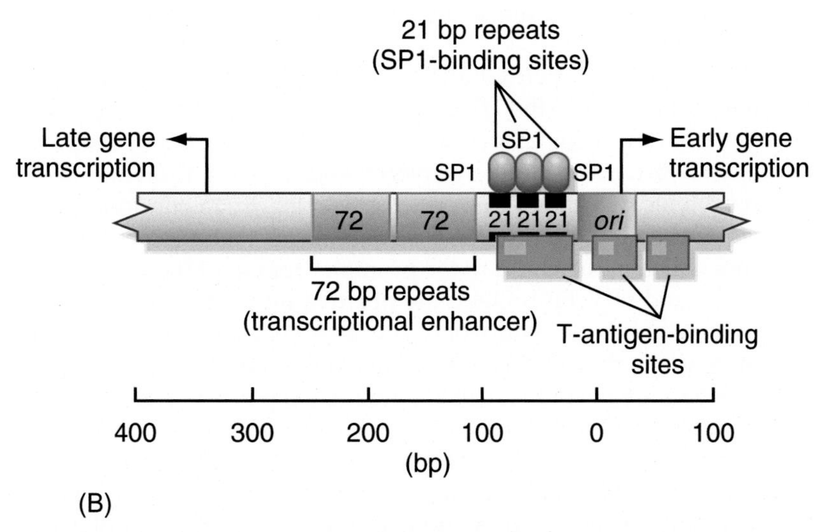

FIG. 5.27 Organization and transcriptional control of the SV40 genome. (A) The highly compact SV40 genome includes genetic information encoded in overlapping reading frames on both strands of the DNA genome and mRNA splicing to produce alternative polypeptides from one open reading frame. Contrast this genetic organization with that of the mouse polyomavirus in Fig. 5.5: the latter has more ORFs for T antigens and is in the opposite orientation. (B) Multiple virus-encoded (T-antigen) and cellular proteins bind to the ori region of the SV40 genome to control gene expression during different phases of replication (see text for details).

Both proteins contain **nuclear localization signals (NLSs)** which result in their accumulation in the nucleus, where they migrate after being synthesized in the cytoplasm.

Soon after infection of permissive cells, early mRNAs are expressed from the **early promoter**, which contains a strong **transcription enhancer element** (72-bp sequence repeats), allowing it to be active in **newly infected cells** (Fig. 5.27B). The early proteins made are the **two T-antigens**. As the concentration of large T-antigen builds up in the nucleus, **transcription of the early genes is repressed** by direct binding of the protein to the origin region of the virus genome, **preventing transcription** from the early promoter and causing the switch to the late phase of infection. Large T-antigen is also required for replication of the genome: this is discussed further in Chapter 7. After DNA replication has occurred, **transcription of the late genes** occurs from the late promoter and results in the synthesis of the **structural proteins VP1, VP2, and VP3**. This process illustrates two classic features of control of virus gene expression. First, the definition of the "**early**" and "**late**" phases of replication, when **different sets of genes** tend to be expressed, is **before/after** genome replication. Second, there is usually a crucial protein, in this case **T-antigen**, whose function is comparable to that of a "**switch**": compare the pattern of replication of SV40 with the description of bacteriophage λ gene expression control given earlier in this chapter.

Another area where control of virus transcription has received much attention is in the **human retroviruses, HTLV** and **HIV**. Integrated DNA **proviruses** are formed by reverse transcription of the RNA retrovirus genome, as described in Chapter 3. The presence of numerous binding sites for cellular transcription factors in the **long terminal repeats (LTRs)** of these viruses have been established (Fig. 5.28). Together, the "distal" elements (such as NF-κB and SP1-binding sites) and "proximal" elements (such as the TATA box) make up a transcription **promoter** in the U3 region of the LTR (Chapter 3). However, the basal activity of this promoter on its own is relatively weak, and results in only limited transcription of the provirus genome by RNA pol II.

Both HTLV and HIV encode proteins that are *trans*-acting **positive regulators** of transcription: the **Tax** protein of **HTLV** and the **HIV Tat** protein (Fig. 5.29). These proteins act to **increase transcription** from the virus LTR by a factor of at least **50–100 times** that of the basal rate from the "unaided" promoter. Unlike T-antigen and the early promoter of SV40, neither the Tax nor the Tat proteins (which have no structural similarity to one another) bind directly to its respective LTR. Instead, these proteins function indirectly by **interacting with cellular transcription factors** which in turn bind to the promoter region of the virus LTR.

Thus, the HTLV Tax and HIV Tat proteins are **positive regulators** of the basal promoter in the provirus LTR and are under the control of the virus, as synthesis of these proteins is dependent on the promoters which they themselves activate (Fig. 5.30). On its own, this would be an unsustainable system because it would result in unregulated positive feedback, which might be acceptable in a lytic replication cycle but would not be appropriate for a retrovirus integrated into the genome of the host cell. Therefore, each of these viruses encodes an additional protein (the Rex and Rev. proteins in HTLV and HIV, respectively), which further regulates gene expression at a posttranscriptional level (see the section "Posttranscriptional control of expression").

Control of transcription is a critical step in virus replication and in all cases is closely regulated. Even some of the simplest virus genomes, such as **SV40**, encode proteins that regulate their transcription. Many virus genomes encode *trans*-acting factors that modify and/or direct the cellular transcription apparatus. Examples of this include HTLV and HIV, as described earlier, but also the **X protein** of **hepadnaviruses**, **Rep protein** of **parvoviruses**,

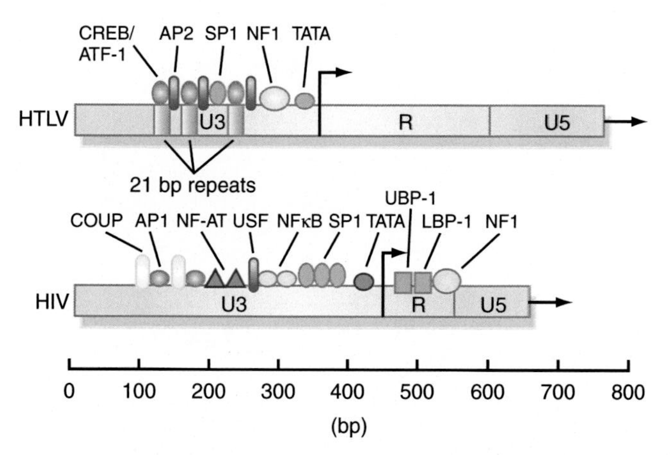

FIG. 5.28 Cellular transcription factors that interact with retrovirus LTRs. Many cellular DNA-binding proteins are involved in regulating both the basal and trans-activated levels of transcription from the promoter in the U3 region of the retrovirus LTR.

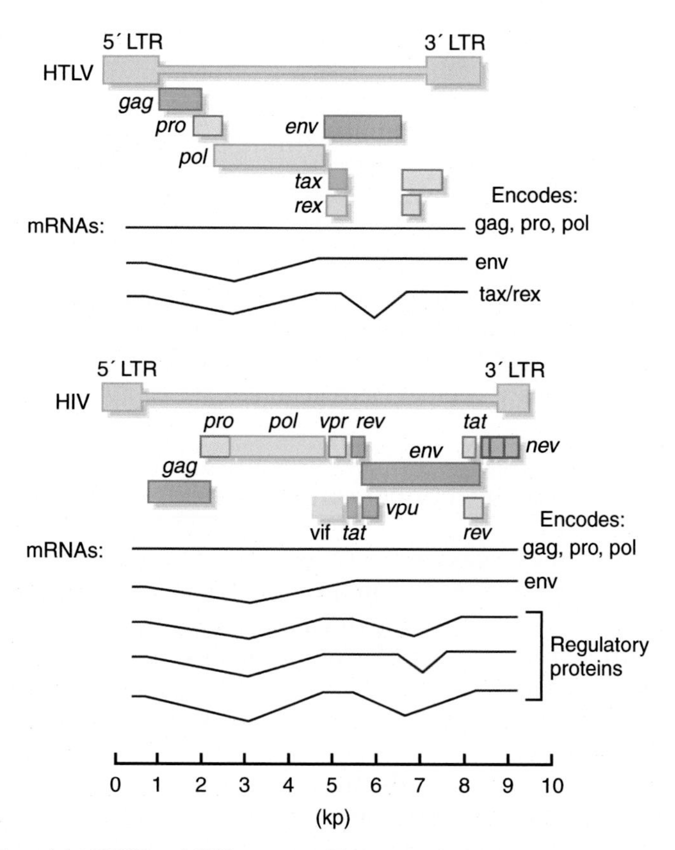

FIG. 5.29 Expression of the HTLV and HIV genomes. These complex retroviruses contain additional genes to the usual retrovirus pattern of *gag*, *pol*, and *env*, and these are expressed via a complicated mixture of spliced mRNAs and *trans*-acting proteins.

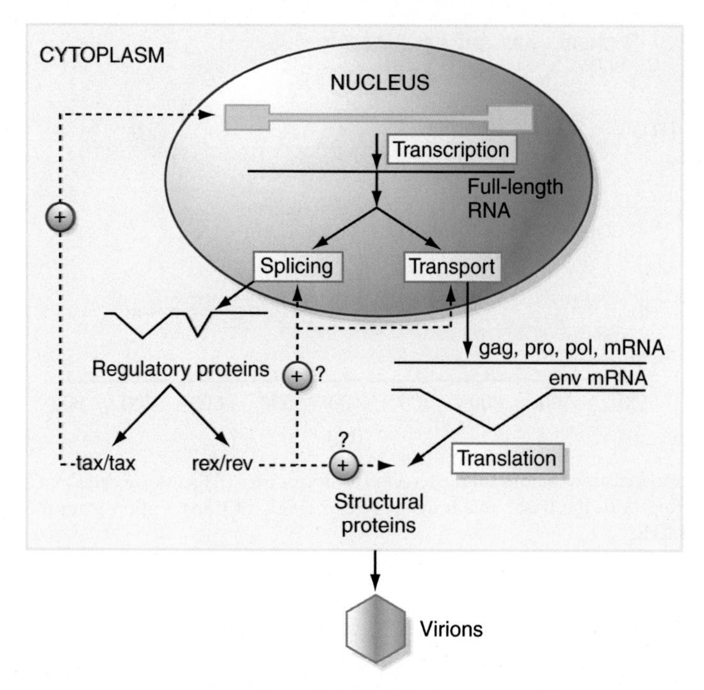

FIG. 5.30 *Trans*-acting regulation of HTLV and HIV gene expression by virus proteins. The Tax (HTLV) and Tat (HIV) proteins act at a transcriptional level and stimulate the expression of all virus genes. The Rex (HTLV) and Rev (HIV) proteins act posttranscriptionally and regulate the balance of expression between virion proteins and regulatory proteins.

E1A protein of **adenoviruses** (see below), and the **IE proteins** of **herpesviruses**. The expression of RNA virus genomes is similarly tightly controlled, but this process is carried out by virus-encoded transcriptases and has been less intensively studied, and is generally much less well understood than transcription of DNA genomes.

Posttranscriptional control of expression

In addition to control of the process of transcription, the expression of virus genetic information is also governed at a number of additional stages between the **formation of the primary RNA transcript** and completion of the **finished polypeptide**. Many generalized subtle controls, such as the **differential stability of various mRNAs** and variability in the **efficiency of translation** are employed by viruses to regulate the flow of genetic information from their genomes into proteins. This section describes only a few well-researched, specific examples of posttranscriptional regulation.

Many DNA viruses that replicate in the nucleus encode mRNAs that must be spliced by cellular mechanisms to remove intervening sequences—or **introns**—before being translated. This type of modification applies only to viruses that replicate in the **nucleus** (and not, e.g., to poxviruses), as it requires the **processing of mRNAs by nuclear machinery** before

BOX 5.5

Viruses—Making more from less

If viruses were as wasteful with their genomes as cells are, they would struggle to exist. For most cellular genes, it's one gene, one protein—and some of those genes are very large indeed, with many introns. While some of the big DNA viruses do that, most viruses have to work harder and get more than one protein out of a gene. There are lots of ways they do this. Splicing, alternative start codons, ribosomal frameshifting, alternate protease cleavages—all these are used by different viruses to squeeze the maximum amount of information into the minimum space. Of course, the trick is to ensure that you can get the information, in the form of proteins, out again. The host cell is generally tricked into doing this by virus-specific signals (e.g., **ribosomal frameshift** or **slippery sequences** in mRNAs), so in addition to the range of virus genomes, the range of host organisms also means that many different gene expression mechanisms are used.

they are transported into the cytoplasm for translation. However, several virus families have taken advantage of this capacity of their host cells to compress more genetic information into their genomes. A good example of such a reliance on splicing are the **parvoviruses**, transcription of which results in **multiple spliced, polyadenylated transcripts** in the cytoplasm of infected cells, enabling them to produce multiple proteins from their 5 kb genomes (see Fig. 5.12), and, similarly, polyomaviruses such as **SV40** (Fig. 5.6 and 5.27). In contrast, the large genetic capacity of **herpesviruses** makes it possible for these viruses to produce mostly unspliced monocistronic mRNAs, each of which is expressed from its own promoter, thereby rendering unnecessary the requirement for extensive splicing to produce the required range of proteins.

One of the best-studied examples of the splicing of virus mRNAs is the expression of the **adenovirus genome** (Fig. 5.31). Several "families" of adenovirus genes are expressed via differential splicing of precursor **hnRNA** transcripts. This is particularly true for the early genes that encode *trans*-acting regulatory proteins expressed immediately after infection. The first proteins to be expressed, **E1A** and **E1B**, are encoded by a transcriptional unit on the *r*-strand at the extreme left-hand end of the adenovirus genome (Fig. 5.31). These proteins are primarily transcriptional *trans*-regulatory proteins comparable to the Tax and Tat proteins described above (Fig. 5.30), but are also involved in **transformation** of adenovirus-infected cells (Chapter 6).

Five polyadenylated, spliced mRNAs are produced (13S, 12S, 11S, 10S, and 9S) which encode five related **E1A polypeptides** (containing 289, 243, 217, 171, and 55 amino acids, respectively) (Fig. 5.32). **All of these proteins are translated from the same reading frame** and have the same amino and carboxy termini. The differences between them are a consequence of **differential splicing of the E1A transcriptional unit** and result in major differences in their functions. The 289 and 243 amino acid peptides are transcriptional activators. Although these proteins activate transcription from all the early adenovirus promoters, it has been `discovered that they also seem to be "promiscuous," activating most RNA polymerase II-responsive promoters that contain a ***TATA box***. There are no obvious common sequences

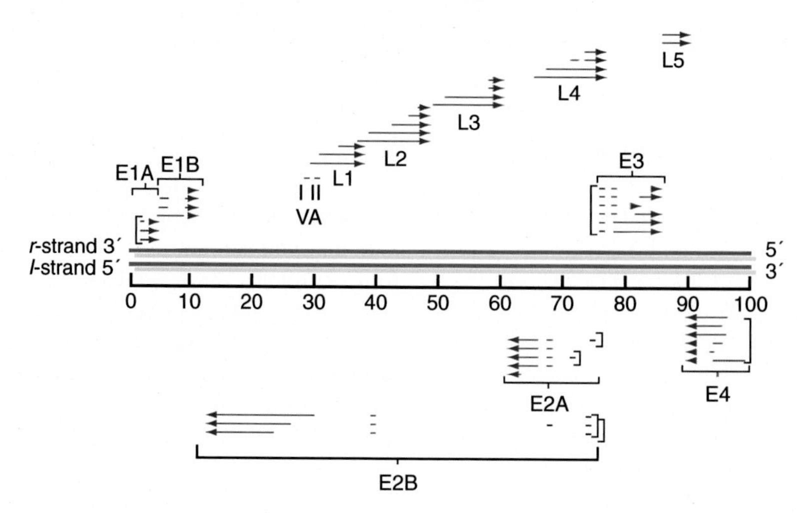

FIG. 5.31 Transcription of the adenovirus genome. The *arrows* in this figure show the position of exons in the virus genome which are joined by splicing to produce families of related but unique virus proteins.

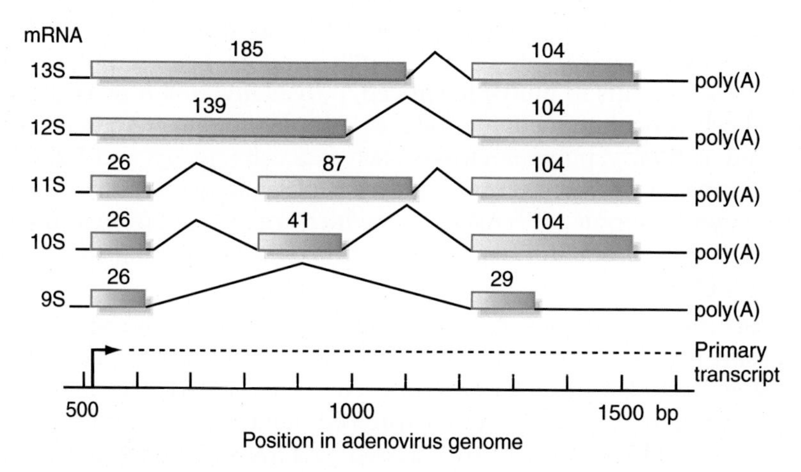

FIG. 5.32 Expression of the adenovirus E1A proteins. The *numbers shown above each box* are the number of amino acids encoded by each exon.

present in all of these promoters, and there is no evidence that the E1A proteins bind directly to DNA. E1A proteins from different adenovirus serotypes contain three conserved domains: **CR1, CR2, and CR3**. The E1A proteins interact with many other cellular proteins, primarily through binding to the three conserved domains. By binding to components of the basal transcription machinery—activating proteins that bind to upstream promoter and enhancer sequences and regulatory proteins that control the activity of DNA-binding factors—**E1A can both activate and repress transcription.**

Synthesis of the adenovirus E1A starts a **cascade of transcriptional activation** by turning on transcription of the other adenovirus early genes: E1B, E2, E3, and E4 (Fig. 5.33). After the virus genome has been replicated, this cascade eventually results in **transcription of the**

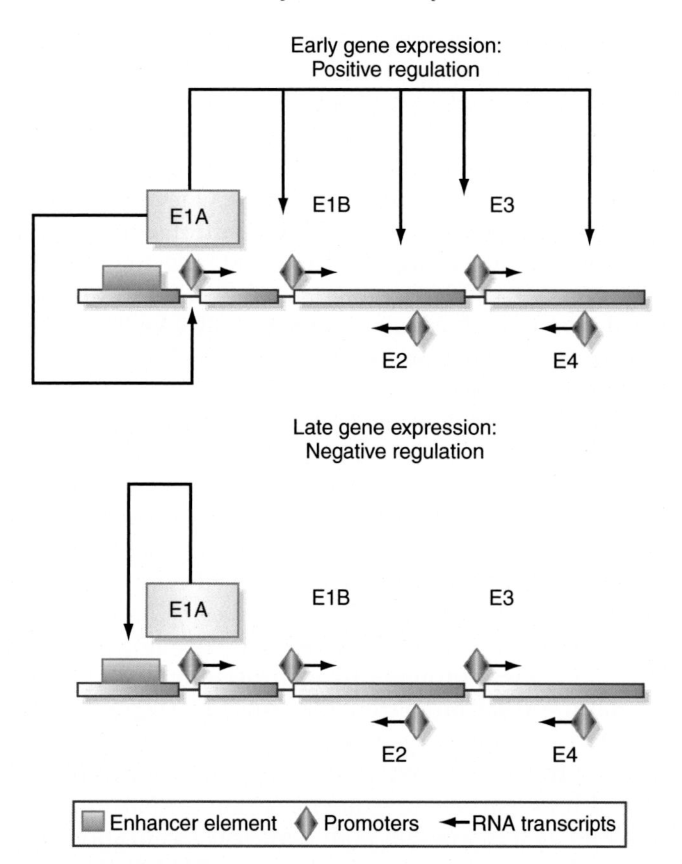

FIG. 5.33 Regulation of adenovirus gene expression. Adenovirus early proteins are involved in complex positive and negative regulation of gene expression.

late genes encoding the structural proteins. The transcription of the E1A itself is a **balanced, self-regulating system**. The IE genes of DNA viruses typically have strong **enhancer elements** upstream of their promoters. This is because in a newly infected cell there are no virus proteins present and the enhancer is required to "kick-start" expression of the virus genome. The IE proteins synthesized are transcriptional activators that turn on expression of other virus genes, and E1A functions in exactly this way. However, although E1A *trans*-activates its own promoter, the protein represses the function of the upstream enhancer element so, at high concentrations, it also **downregulates its own expression** (Fig. 5.33).

The next stage at which expression can be regulated is during **export of mRNA from the nucleus** and preferential translation in the cytoplasm. Again, the best-studied example of this phenomenon comes from adenoviruses. The **VA (viral associated) genes** (see earlier) encode two small (~160 nt) RNAs, unusually transcribed during the late phase of virus replication from the *r*-strand of the genome by RNA pol III: the normal function of this enzyme is to transcribe similar small RNAs such as 5S ribosomal RNA and tRNAs (Fig. 5.30). Both VA RNA I and VA RNA II have a **high degree of secondary structure**, and neither molecule encodes any polypeptide—in these two respects they are **similar to tRNAs**—and they accumulate to high levels in the cytoplasm of adenovirus-infected cells. The way in which these two RNAs act is not completely understood, but their net effect is to **boost the synthesis** of adenovirus **late proteins**. The VA RNAs are processed by the host cell to form virus-encoded **miRNAs**. These operate through **RNA interference** (see Chapter 6) to **downregulate** a large number of cellular genes involved in RNA binding, splicing, and translation. In addition, virus infection of cells stimulates the production of ***interferons*** (Chapter 6). One of the actions of interferons is to activate a cellular **protein kinase** known as **PKR** that **inhibits the initiation of translation**. VA RNA I binds to this kinase, preventing its activity and relieving inhibition on translation. The effects of interferons on the cell are generalized (discussed in Chapter 6), and result in **inhibition of the translation** of both cellular and virus mRNAs. The effect of the VA RNAs is to **promote selectively the translation of adenovirus mRNAs** at the expense of cellular mRNAs whose translation is inhibited.

The HTLV Rex and HIV Rev. proteins mentioned earlier also act to **promote the selective translation** of specific virus mRNAs. These proteins regulate the differential expression of the virus genome but do not substantially alter the expression of cellular mRNAs. Both of these proteins appear to function in a similar way, and, although not related to one another in terms of their amino acid sequences, the HTLV Rex protein can substitute functionally for the HIV Rev. protein. **Negative-regulatory sequences** in the HIV and HTLV genomes **cause the retention of virus mRNAs in the nucleus** of the infected cell. These sequences are located in the **intron** regions that are removed from spliced mRNAs encoding the Tax/Tat and Rex/Rev proteins (Fig. 5.29), therefore, **these proteins are expressed immediately after infection**. Tax and Tat stimulate **enhanced transcription** from the virus LTR (Fig. 5.30). However, unspliced or singly spliced mRNAs encoding the *gag*, *pol*, and *env* gene products are only expressed when sufficient Rex/Rev. protein is present in the cell. Both proteins bind to a region of secondary structure formed by a particular sequence in the mRNA and shuttle between the nucleus and the cytoplasm as they contain both a nuclear localization signal and a nuclear export signal, **increasing the export of unspliced virus mRNA** to the cytoplasm where it is **translated** and **acts as the virus genome** during particle formation.

The efficiency with which different mRNAs are translated varies considerably and is determined by a number of factors, including the **stability and secondary structure** of the RNA. However, the main factor appears to be the particular nucleotide sequence surrounding the **AUG translation initiation codon** that is recognized by ribosomes. The most favorable sequence for initiation is GCC(A/G)CCAUGGG, although there can be considerable variation within this sequence. A number of viruses use variations of this sequence to regulate the amounts of protein synthesized from a single mRNA. Examples include the **Tax and Rex** proteins of **HTLV** which are encoded by overlapping reading frames in the same doubly spliced 2.1 kb mRNA (Fig. 5.30). The AUG initiation codon for the Rex protein is upstream of that for Tax but provides a **less favorable context** for initiation of translation than the sequence surrounding the Tax AUG codon. This is known as the "**leaky scanning**" mechanism because it is believed that the **ribosomes scan along the mRNA** before initiating translation. Therefore, the relative abundance of Rex protein in HTLV-infected cells is considerably less than that of the Tax protein, even though **both are encoded by the same mRNA**.

Picornavirus genomes illustrate an alternative mechanism for controlling the initiation of translation. Although these genomes are genetically economical (i.e., have discarded most **cis-acting** control elements and express their entire coding capacity as a single polyprotein), they have **retained long noncoding regions (NCRs)** at their 5′ ends, comprising approximately 10% of the entire genome. These sequences are involved in the replication and possibly **packaging** of the virus genome. Translation of most cellular mRNAs is initiated when ribosomes recognize the 5′ end of the mRNA and scan along the nucleotide sequence until they reach an **AUG initiation codon**. Picornavirus and similar genomes (e.g., **comoviruses**) are not translated in this way: the 5′ end of the RNA is not capped and thus is not recognized by ribosomes in the same way as other mRNAs, but it is **modified by the addition of the VPg** (see also Chapters 3 and 6). There are also **multiple AUG codons** in the 5′ NCR upstream of the start of the polyprotein coding sequences which are not recognized by ribosomes. In picornavirus-infected cells, a virus protease cleaves the **220-kDa "cap-binding complex"** involved in binding the m7Gppp cap structure at the 5′ end of the mRNA during initiation of translation. Translation of artificially mutated picornavirus mRNAs in vitro and the construction of bicistronic picornavirus genomes bearing additional 5′ NCR signals in the middle of the polyprotein have resulted in the concept of the **ribosome "landing pad,"** or **internal ribosomal entry site (IRES)** (Fig. 5.34). Rather than scanning along the RNA from the 5′ end, ribosomes **bind to the RNA via the IRES** and begin translation internally. This is a precise method for controlling the translation of virus proteins. Very few cellular mRNAs utilize this mechanism but it has been shown to be used by a variety of viruses, including **picornaviruses, hepatitis C virus, coronaviruses**, and **flaviviruses**.

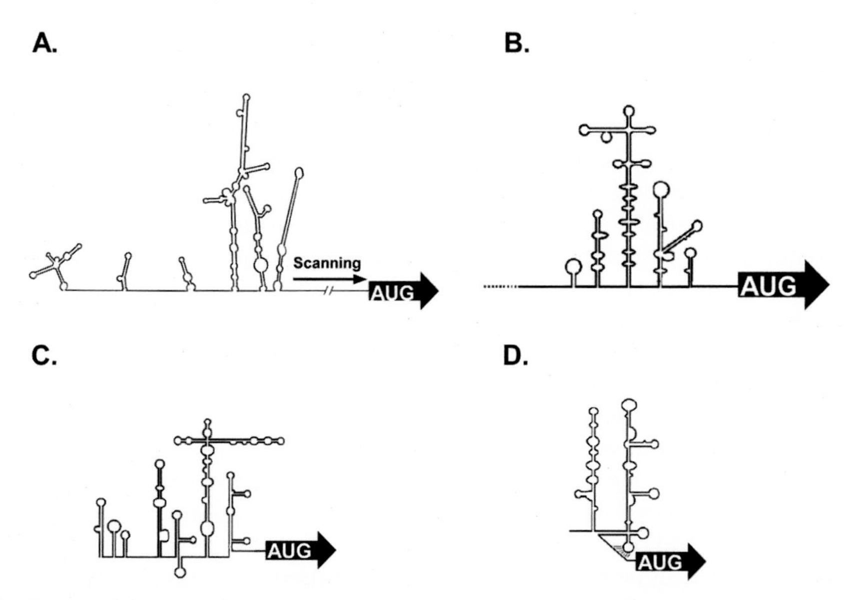

FIG. 5.34 Structural features of types I, II, III, and IV picornavirus IRES elements. RNA secondary structures of four types based on M-fold software. (A) Enteroviruses (PV, CVB3 and EV71) and rhinoviruses contain type I IRES elements. (B) Aphthoviruses (FMDV) and cardioviruses (EMCV and TMEV) contain type II IRES. (C) HAV (hepatoviruses) contains a type III IRES. (D) Porcine reschovirus serotype 1 (PTV-1) contains a type IV IRES. *From Lin, J.Y., Chen, T.C., Weng, K.F., et al., 2009. Viral and host proteins involved in picornavirus life cycle. J. Biomed. Sci. 16, 103. https://doi.org/10.1186/1423-0127-16-103, CC BY 2.0 licence.*

Many viruses belonging to different families compress their genetic information by encoding different polypeptides in overlapping reading frames. The problem with this strategy lies in decoding the information. If each polypeptide is expressed from a **monocistronic** mRNA transcribed from its own **promoter**, the additional ***cis*-acting** sequences required to control and coordinate expression might cancel out any genetic advantage gained. More importantly, there is the problem of **coordinately regulating the transcription and translation** of multiple different messages. Therefore, it is highly desirable to **express several polypeptides from a single RNA transcript**, and the examples described above illustrate several mechanisms by which this can be achieved, e.g., differential **splicing** and control of RNA export from the nucleus, or initiation of translation. The **CaMV "relay race model"** is also a good example (see earlier).

An additional mechanism known as "**ribosomal frameshifting**" is used by several groups of viruses to achieve the same effect. The best-studied examples of this phenomenon come from **retrovirus** genomes, but many viruses use a similar mechanism. Such frameshifting was first discovered in viruses but is now known to occur also in **prokaryotic** and **eukaryotic** cells. Retrovirus genomes are transcribed to produce at least two 5′ capped, 3′ polyadenylated mRNAs. Spliced mRNAs encode the **envelope** proteins, as well as, in more complex retroviruses such as HTLV and HIV, additional proteins such as the Tax/Tat and Rex/Rev proteins (Fig. 5.30). A **long, unspliced transcript** encodes the *gag*, *pro*, and ***pol* genes** and also forms the **genomic RNA** packaged into virions. The problem faced by retroviruses is how to express three different proteins from one long transcript—and the arrangement of the three genes varies in different viruses. In some cases (e.g., **HTLV**), they occupy **three different reading frames**, while in others (e.g., **HIV**), the **protease (*pro*) gene** forms an **extension at the 5′ end of the *pol* gene** (Fig. 5.35). In the latter case, the **protease** and **polymerase** (i.e., **reverse transcriptase**) are expressed as a **polyprotein** that is autocatalytically cleaved into the mature proteins in a process that is similar to the cleavage of picornavirus polyproteins.

At the boundary between each of the three genes is a particular sequence that usually consists of a **tract of reiterated nucleotides**, such as UUUAAAC (Fig. 5.36). This sequence is rarely found in protein-coding sequences and therefore appears to be specifically used for this type of regulation. Most ribosomes encountering this sequence will translate it without difficulty and continue on along the transcript until a translation stop codon is reached. However, a proportion of the ribosomes that attempt to translate this sequence will **slip back by one nucleotide** before continuing to translate the message, but now in a different (i.e., −1) reading frame. Because of this, the UUUAAAC sequence has been termed the "**slippery sequence**," and the result of this frameshifting is the **translation of a polyprotein containing alternative information from a different reading frame**. This mechanism also allows the virus to control the **ratios** of the proteins produced. Because only a proportion of ribosomes undergoes frameshifting at each slippery sequence, there is a **gradient of translation** from the reading frames at the 5′ end of the mRNA to those at the 3′ end.

The slippery sequence alone results in only a **low frequency of frameshifting**, which appears to be inadequate to produce the amount of protease and reverse transcriptase protein required by the virus. Therefore, there are additional sequences that further regulate this system and increase the frequency of frameshift events. A short distance downstream of the slippery sequence is an **inverted repeat** that allows the formation of a **stem-loop structure** in the mRNA (Fig. 5.36). A little further on is an additional sequence complementary to the nucle-

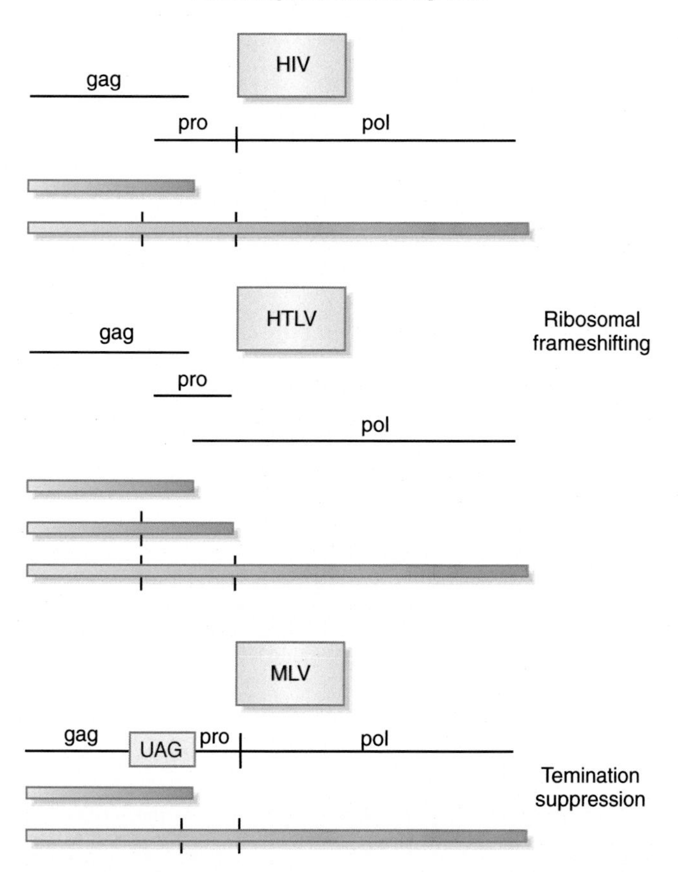

FIG. 5.35 Ribosomal frameshifting and termination suppression in retroviruses. Ribosomal frameshifting and termination suppression are posttranscriptional methods used to extend the range of proteins produced by retrovirus genomes.

otides in the loop that allows **base-pairing between these two regions of the RNA**. The net result of this combination of sequences is the formation of what is known as an **RNA pseudoknot**. This secondary structure in the mRNA causes ribosomes translating the message to **pause at the position of the slippery sequence upstream**, and this slowing or pausing of the ribosome during translation **increases the frequency at which frameshifting occurs**, thus boosting the relative amounts of the proteins encoded by the downstream reading frames. It is easy to imagine how this system can be fine-tuned by subtle mutations that alter the stability of the pseudoknot structure and thus the relative expression of the different genes.

Yet another method of translational control is **termination suppression**. This is a mechanism similar in many respects to frameshifting that permits **multiple polypeptides to be expressed from individual reading frames in a single mRNA**. In some retroviruses, such as **murine leukemia virus (MLV)**, the *pro* gene is separated from the *gag* gene by a UAG termination codon rather than a slippery sequence and pseudoknot (Fig. 5.36). In the majority of cases, translation of MLV mRNA **terminates** at this sequence, giving rise to the Gag proteins. However in a few instances, the **UAG stop codon is suppressed and translation continues**, producing a

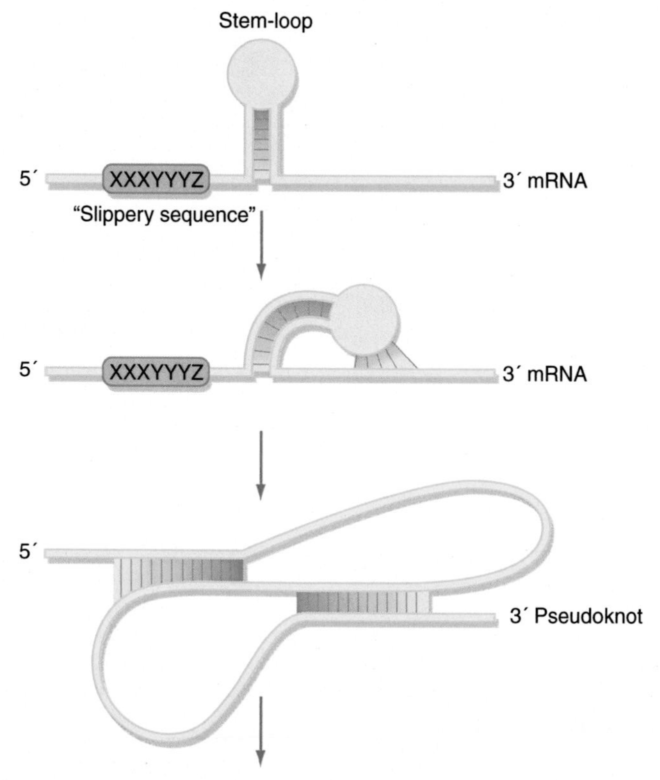

FIG. 5.36 RNA pseudoknot formation. RNA pseudoknot formation is the mechanism by which ribosomal frameshifting occurs in a number of different viruses and a few cellular genes (see text for details).

Gag-Pro-Pol polyprotein (see Chapter 3), which subsequently **cleaves itself** (=autocatalytic cleavage) to produce the mature proteins. The overall effect of this system is much the same as ribosomal frameshifting, with the **relative ratios of Gag and Pro/Pol proteins** being controlled by the frequency with which ribosomes traverse or terminate at the UAG stop codon.

A recent study that collected the coding sequences of over 500 human-infecting viruses, and then analyzed these against a database of **tRNA abundances** in 23 different human tissues, allowed researchers to determine that **viruses with a distinct tissue tropism are adapted to the codon usage of that tissue** (see Hernandez-Alias et al., 2021). For example, they showed that **SARS-CoV-2** is specifically adapted to infecting cells in the **upper respiratory tract and lung alveoli**. There was also an indication that **early regulatory proteins** are generally better adapted to the codon usage in the preferred target tissue, compared to **structural proteins**. Similar observations had previously been made for **human papillomaviruses**, where codon usage by virus genes is very similar to that preferred specifically by **differentiating keratinocytes** (see earlier).

Summary

Control of gene expression is a vital element of virus replication. Coordinated expression of groups of virus genes results in successive phases of gene expression. Typically, "**immediate early** (IE) genes encode "**activator**" proteins, **early genes** encode further **regulatory**

proteins, and **late genes** encode **virus structural proteins**. Viruses make use of the biochemical apparatus of their host cells to express their genetic information as proteins and, consequently, utilize the appropriate biochemical language recognized by the cell. Thus viruses of **prokaryotes** produce **polycistronic** mRNAs, while viruses with **eukaryotic** hosts produce more **monocistronic** mRNAs. Some viruses of eukaryotes do however produce polycistronic mRNA to assist with the coordinate regulation of multiple genes.

References

Beck, J., Nassal, M., 2007. Hepatitis B virus replication. World J. Gastroenterol. 13 (1), 48–64. https://doi.org/10.3748/wjg.v13.i1.48. 17206754. PMC4065876.

Hernandez-Alias, X., Benisty, H., Schaefer, M.H., Serrano, L., 2021. Translational adaptation of human viruses to the tissues they infect. Cell Rep. 34 (11), 108872. https://doi.org/https://doi.org/10.1016/j.celrep.2021.108872.

Mazeaud, C., Freppel, W., Chatel-Chaix, L., 2018. The multiples fates of the flavivirus RNA genome during pathogenesis. Front. Genet. 9, 595. https://www.frontiersin.org/article/10.3389/fgene.2018.00595. https://doi.org/10.3389/fgene.2018.00595.

Rupp, J.C., Sokoloski, K.J., Gebhart, N.N., Hardy, R.W., 2015. Alphavirus RNA synthesis and non-structural protein functions. J. Gen. Virol. 96 (9), 2483–2500. https://doi.org/10.1099/jgv.0.000249.

Recommended reading

Alberts, B., 2014. Molecular Biology of the Cell. Garland Science, New York.

Brandes, N., Linial, M., 2019. Giant viruses-big surprises. Viruses 11 (5), 404. https://doi.org/10.3390/v11050404.

Firth, A.E., Brierley, I., 2012. Non-canonical translation in RNA viruses. J. Gen. Virol. 93 (7), 1385–1409.

Gómez-Díaz, E., Jordà, M., Peinado, M.A., Rivero, A., 2012. Epigenetics of host–pathogen interactions: the road ahead and the road behind. PLoS Pathog. 8 (11), e1003007.

Kannian, P., Green, P.L., 2010. Human T lymphotropic virus type 1 (HTLV-1): molecular biology and oncogenesis. Viruses 2 (9), 2037–2077.

Karn, J., Stoltzfus, C.M., 2012. Transcriptional and posttranscriptional regulation of HIV-1 gene expression. Cold Spring Harb. Perspect. Med. 2 (2), a006916.

Kincaid, R.P., Sullivan, C.S., 2012. Virus-encoded microRNAs: an overview and a look to the future. PLoS Pathog. 8 (12), e1003018.

Lin, J.Y., Chen, T.C., Weng, K.F., Chang, S.C., Chen, L.L., Shih, S.R., 2009. Viral and host proteins involved in picornavirus life cycle. J. Biomed. Sci. 16, 103 (2009) https://doi.org/10.1186/1423-0127-16-103.

López-Lastra, M., 2010. Translation initiation of viral mRNAs. Rev. Med. Virol. 20 (3), 177–195.

Resa-Infante, P., Jorba, N., Coloma, R., Ortín, J., 2011. The influenza RNA synthesis machine. RNA Biol. 8 (2), 207–215.

Santos, F., Martinez-Garcia, M., Parro, V., Antón, J., 2014. Microarray tools to unveil viral-microbe interactions in nature. Front. Ecol. Evol. 2, 31.

Skalsky, R.L., Cullen, B.R., 2010. Viruses, microRNAs, and host interactions. Annu. Rev. Microbiol. 64, 123–141.

Zhao, H., Dahlö, M., Isaksson, A., Syvänen, A.C., Pettersson, U., 2012. The transcriptome of the adenovirus infected cell. Virology 424 (2), 115–128.

CHAPTER

6

Infection and immunity

INTENDED LEARNING OUTCOMES

On completing this chapter, you should be able to:

- Discuss the similarities and the differences between virus infections of prokaryotes, plants and animals.
- Explain how the immune responses to viruses enables organisms to resist infection, and how viruses respond to this pressure.
- Describe and understand how virus infections are prevented and treated.

Virus infection of prokaryotes

Prokaryotes are the most diverse ribocells on our planet, and the viruses that infect them are the most diverse organisms. The sheer antiquity of prokaryotes and their viruses—both have probably been around for **at least 3.8 billion years** at current estimates, compared to 2 billion for eukaryotes—means that the "**arms race**" between them has had that long to evolve. By this we mean the evolution of measures in prokaryotes to defend against virus infection, and the evolution of strategies in viruses to counter those defenses. The fact that both sets of organisms are still with us indicates that the race is even: neither prokaryotic ribocells nor their viruses have the upper hand, and the ancient war continues. This is evidenced by the estimate discussed earlier (Box 5.2) that up to **20% of all marine prokaryotes are killed per day** by viruses.

While a common view of prokaryotes is that they are "**primitive**" organisms, the opposite is in fact true: while they may be genetically and structurally simpler than most eukaryotes, what we see with us today is the end result of 3.8 billion years of evolution of ribocells that have shorter generation times than most eukaryotes, and hence **much more chance to be naturally selected** for fitness—and therefore to evolve. The same is true for their defense systems: these have been continuously honed by natural selection—meaning predation by viruses—to the point that they are **sophisticated** to an extent many of you will find surprising. Accordingly, and because their antiviral defenses have already found significant application in biotechnology in applications including human medicine, we will briefly discuss some of the mechanisms prokaryotes use to defend themselves against viruses.

https://doi.org/10.1016/B978-0-12-822784-8.00006-4

Clustered regularly interspaced short palindromic repeats (CRISPRs)

In 1987, in the investigation of a particular enzyme in *E. coli*, five 29-base identical repeat sequences were found, spaced out by 32-base unique sequences: this pattern was repeated 12 times in the *E. coli* genome. Subsequent work found similar clusters of sequences in the related enterobacteria *Shigella dysenteriae* and *Salmonella enterica*, and multiple 36-bp repeats with 35–41-base unique spacers in *Mycobacterium tuberculosis*. The first evidence for similar clusters of sequences in archaea was found in 1993—and it is now estimated that there are such clusters in around 45% of bacteria and 84% of archaea. The term **CRISPR** was coined in 2002, and at the same time it was reported that the CRISPR loci were closely associated with a group of genes that were well conserved in CRISPR-containing cells, but were not present in those lacking such clusters. These became known as **CRISPR-associated (*cas*) genes**, and by 2006 **more than 50 families of Cas proteins** had been identified.

By 2005 it had been shown that CRISPR loci in two different archaea (*Archaeoglobus fulgidus* and *Sulfolobus solfataricus*) were **transcribed into large RNAs** that were subsequently processed into **smaller RNAs** whose sizes corresponded to **the size of a single "repeat-spacer unit"** of 67–77 bases. In the next 2 years it was shown that the sources of spacer sequences were "**extrachromosomal**"—i.e., derived from **phages** or **conjugative plasmids**—and by 2007, CRISPR was definitively linked to an "**acquired resistance**" in prokaryotes to viral infection, that was suggested to work via an antisense RNA interference mechanism. It was proved in 2007 that cells could acquire **new phage-derived sequences** by challenging nonresistant *Sulfolobus thermophilus* with two different viruses, and showing that surviving cells had **new spacers in their CRISPR loci derived from these phages**, adjacent to older spacers—and that they were resistant to subsequent challenge by those same viruses.

Linkage of Cas proteins to the acquired resistance phenomenon came when it was shown that *E. coli* K12 **CRISPR-derived small RNAs (crRNAs)** contained the last 8 bases of the upstream repeat sequence at their 5′ ends, the full sequence of each spacer, and the 5′-sequence of the downstream repeat sequence at their 3′ ends, and formed complexes in vitro with recombinant **Cas6** protein. Cas6 also **specifically recognized and cleaved within the repeat region** of the pre-crRNA transcript, meaning it was **responsible for generating mature crRNAs**. The next step in determining just how CRISP/Cas worked as an antiviral mechanism was the demonstration that in vivo activity **required the Cas3 nuclease** in addition to the crRNA containing a spacer complementary to viral DNA, which indicated **an anti-DNA activity** and not mRNA antisense or "**small interfering RNA**" (**siRNA**; see later) activity. This was tested in detail by a number of research groups, and in a short time it was shown that there are **three main types of CRISPR/Cas systems—I, II,** and **III**, linked to **Cas3, -9,** and **-10 as signature proteins**, respectively—and that types I-III act against infecting **viral DNA** for which there are cognate spacer regions, by **cleaving the DNA** that is identified by complementary binding, and type III can act against transcribed **RNA** as well (see Rath et al., 2015). Type I and III systems are found in both bacteria and archaea, while Type II are only found in bacteria.

All Type I systems encode a "**Cascade-like**" complex, or **CRISPR-associated complex for antiviral defense**. This binds crRNA and uses it to locate the target DNA; most cascades also **process the crRNA** from longer transcripts. Cascade can also enhance **spacer acquisition**. Type II systems encode **Cas1** and **Cas2**—both nucleases—and **Cas9**, and sometimes also either Csn2 or Cas4. Cas9 is involved in acquisition of new spacer DNA (also known as **adaptation**), and in crRNA processing, and cleaves the target DNA together with crRNA and **tracrRNA**, which is derived from the 3′ repeat region. Type III systems contain **Cas10**, whose function is unclear. This is included in complexes in different cell types that are similar to cascades.

A **Type IV system** has also been proposed: this has several cascade genes but **no CRISPR sequences,** or ***cas1*** or ***cas2***, that are almost universal in CRISPR systems. This sort of complex would presumably be guided by **protein-DNA interaction** rather than by crRNA, and effectively constitutes a prokaryote **innate immune system** that can attack certain sequences but is otherwise not adaptable. It has been proposed that the **adaptive immune systems** evolved from an ancestral Type IV that associated with a **transposon** (specifically a **casposon**; see Chapter 3) containing *cas1* and *cas2*: the idea is that this lost genes, the inverted terminal repeat sequences evolved into a CRISPR cluster, and the assembly eventually became **Type I** and **Type III** CRISPR/Cas systems. **Type II** systems evolved when a **mobile genetic element** containing ***cas9***—which strongly resembles a **transposase** gene—replaced the cascade genes (see Rath et al., 2015).

CRISPR/Cas complexes function to **degrade incoming viral DNA or potentially RNA** through the **crRNA** bound to **Cas protein(s)** locating the corresponding "**protospacer**"—target sequence in the viral genome—to trigger **endonuclease-type activity and degradation** of the target by specific Cas nucleases. Spacer selection in the first place is guided by sequence elements in the target DNA: there is a short motif called the **protospacer adjacent motif (PAM)** next to the target sequence that is vital for **self/nonself-discrimination**—because it is not included in the spacers in the prokaryote genome. Thus, the PAM functions both in spacer acquisition and in interference in phage infection. In **Type I and II systems** interference with infecting dsDNA genomes requires a PAM sequence in the viral genome, and exact protospacer-crRNA complementarity in the sequence located adjacent to the PAM. There are no obvious PAM sequences for **Type III** systems: **self/nonself-discrimination** occurs as a result of an extension of crRNA complementary base pairing into the repeat region of the prokaryote genomic DNA CRISPR locus, which results in "**self-inactivation**" (Fig. 6.1).

Given that these defense systems have been evolving for a couple of billion years, it is not surprising that bacterial and archaeal viruses have evolved **counter-measures** to CRISPR/Cas systems—and have even, in some cases, **adapted them for their own use**. A number of **anti-CRISPR (*acr*) genes** are known to be encoded by a variety of mobile genetic elements (MGEs): the proteins encoded by these interfere with CRISPR/Cas action inside cells, and negate the acquired immunity. These genes are also found in phages, notably myo- and siphoviruses, and may be gathered into operons. In 2020 it was found in Jennifer Doudna's lab that "**huge bacteriophages**"—10 clades of which were found by metavirome trawling, with genomes up to 735,000 bp in size—in the "**Biggiephage**" clade encoded a highly novel single-component Cas system. This has been termed **CasΦ**, and is the smallest and simplest system yet found. The probable reason for the presence of this, and the accompanying CRISPR array, is to **augment host CRISPR systems of host bacteria**, in order to target other, competing viruses.

A most unexpected recent discovery was that **mimiviruses** may use a defense system against **virophages** that strongly resembles the CRISPR/Cas system hitherto found only in prokaryotes. So-called "lineage A" strains of mimiviruses are resistant to a virophage dubbed

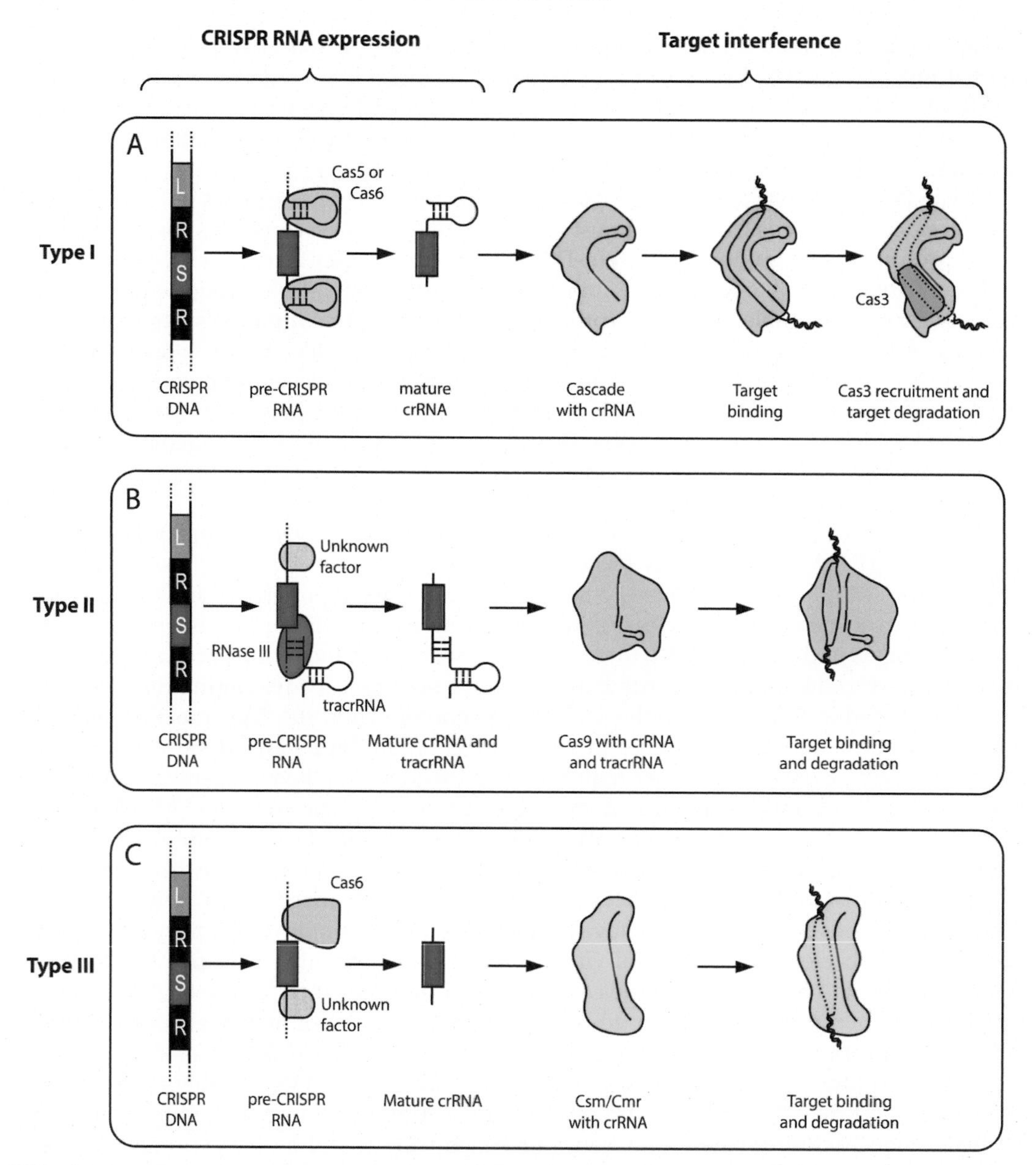

FIG. 6.1 CRISPR and Cas defense systems in prokaryotes. (A) In Type I systems, the pre-crRNA is processed by Cas5 or Cas6. DNA target interference requires Cas3 in addition to cascade and crRNA. (B) Type II systems use RNase III and tracrRNA for crRNA processing together with an unknown additional factor that perform 5′ end trimming. Cas9 targets DNA in a crRNA-guided manner. (C) The Type III systems also use Cas6 for crRNA processing, but in addition an unknown factor performs 3′ end trimming. Here, the Type III Csm/Cmr complex is drawn as targeting DNA, but RNA may also be targeted. *Adapted from Rath, D., Amlinger, L., Rath, A., Lundgren, M., 2015. The CRISPR-Cas immune system: biology, mechanisms and applications. Biochimie 117, 119–128. https://www.sciencedirect.com/science/article/pii/S0300908415001042.*

"Zamilon," and contain in their genomes four 15-base repeated sequences homologous to Zamilon, in what resembles an operon region that also contains a helicase and an endonuclease, that are involved in virophage immunity. The CRISPR-like array has been dubbed "**MIMIVIRE**".

Although CRISPRs originate in prokaryotes, they also work in eukaryotic cells if introduced by recombinant DNA technology. This provides a convenient way of targeting genes in cells, including human cells—which has led to an explosion in research directed at doing just this, for everything from fruit trees to human embryos. Recent work suggests that CRISPRs might also be involved in control of bacterial gene expression as well as in immunity. We will undoubtedly see much more widespread use of CRISPRs in biotechnology over the next few years.

Tailocins

We have already mentioned "**tailocins**" in passing in Chapter 1: these are a specialized **bacteriocin**. Bacteria often make use of bacteriocins, or proteins that mediate a variety of **bacteriotoxic** effects frequently directed at even close relatives of the strains producing them. A particular subset of bacteriocins have recently been labeled tailocins, or **bacteriophage tail-like bacteriocins**: these are produced by both Gram-positive and -negative bacteria, and one species of bacteria may produce more than one kind of tailocin. For example, the human pathogen *Pseudomonas aeruginosa* produces tailocins with either a **flexible (F) or rigid (R) appearance**, which are denoted as **F- and R-types**: these have served as models for similar tailocins found widely among bacteria. Tailocins are devices that **penetrate the cell envelope** of target bacteria and cause lysis. Similar assemblies, related to R-type tailocins and called **type VI secretion systems (T6SS)**, are used by certain bacteria to **inject toxic proteins into other bacterial and eukaryotic cells**. Notably, tailocins have to be released from bacterial cells by **lysis**—meaning the cell is killed by the process. T6SS assemblies, however, operate from **inside** the parent cell and are not released—and are much longer than tailocins, although they share the contractile tail sheath. Cells are immune to their own tailocins: these generally target related bacteria, and kill the cells by making multiple holes in their cell walls (Fig. 6.2).

Among *Pseudomonas* spp., the two kinds of tailocins appear to derive from two lineages of **myoviruses** (rigid tailed dsDNA viruses, family *Myoviridae*; R-type) and several lineages of **siphoviruses** (flexible tailed dsDNA viruses, family *Siphoviridae*; F-type). The bacteria encode the several genes necessary for specifying the individual tailocins in distinct clusters—and it appears as though recombination between various components of the clusters can produce tailocins with **altered target surface specificities** (Fig. 6.3). This can potentially be engineered to produce **novel specificities** such as for the human enteric pathogens **enterotoxigenic *Escherichia coli*** (**ETEC**), and ***Clostridium difficile***. A recent review commented that:

> *"Tailocins illustrate the daedalian capacity of bacteria to accommodate exogenous genetic elements and domesticate them for their own benefit. The stinging device used by tailed (bacterio)phages against bacteria has been cunningly converted into tools to manipulate eukaryotic cells and into precision weapons for interbacterial warfare."*
> Ghequire and de Mot (2015)

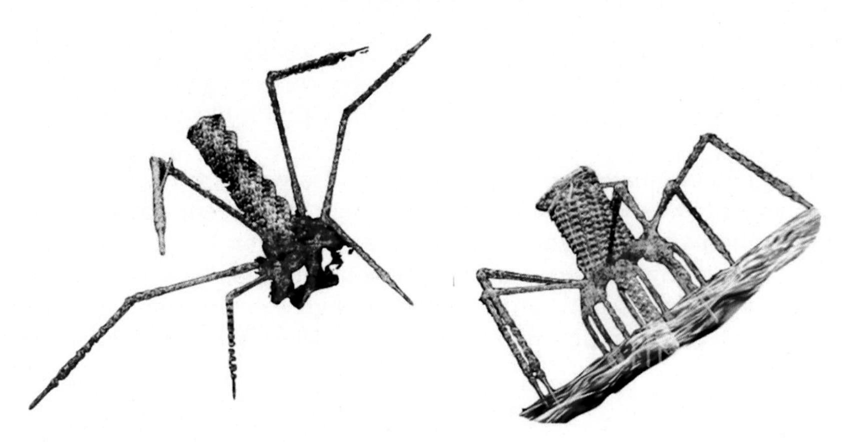

FIG. 6.2 R-type tailocin. These are produced by bacteria from tail structure genes derived from bacterial viruses, to kill other bacteria. The picture shows two R-type tailocins, one free and one attached via tail fibers and baseplate to a bacterial cell, where the binding-induced tail core penetration of cells can cause cell damage and lysis. *Courtesy of Russell Kightley Media.*

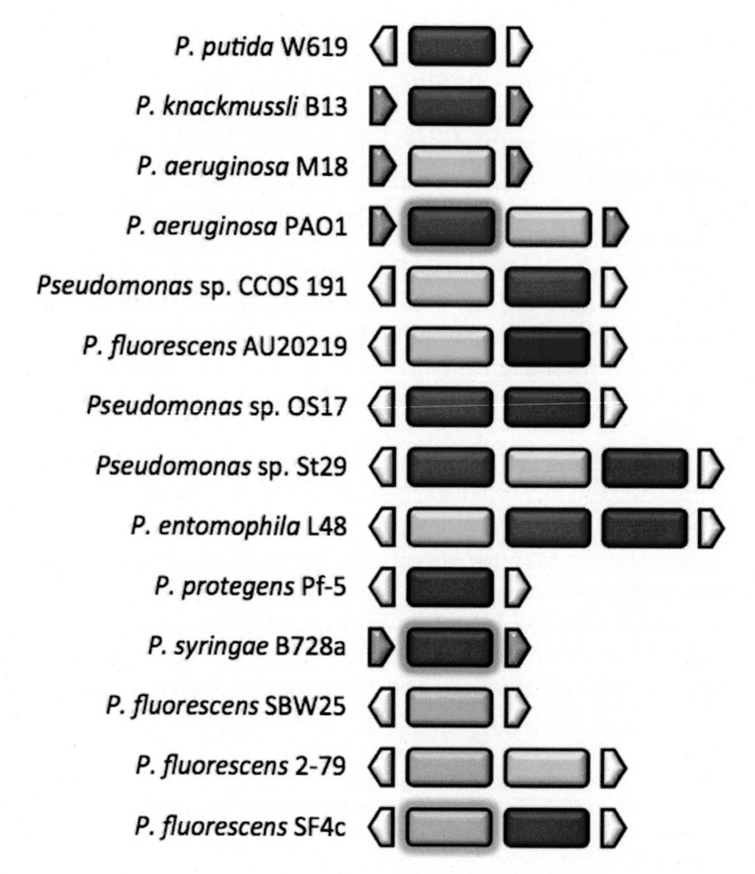

FIG. 6.3 Tailocin genes in Pseudomonas spp. Tailocin-encoding regions in pseudomonads located between *trpE* and *trpG* (*gray arrows*) or between *mutS* and *cinA* (*white arrows*). Tailocin gene clusters occur individually, or as pairs or triplets. There are four different subtypes: three myovirus variants (*red*, *purple*, and *orange*) and an F-type with similarity to siphovirus phage tails (*pale blue*). *Adapted from Ghequire, M.G.K., De Mot, R., 2015. The Tailocin tale: peeling off phage tails. Trends Microbiol. 23 (10), 587–590. doi:10.1016/j.tim.2015.07.011.*

Virus infections of plants

While it is estimated that **prokaryotes constitute about 15%** of the total biomass on our planet, **plants**—and mainly **terrestrial plants—constitute 80%** or so of the estimated total of 550 billion tons. Terrestrial life on Earth depends on the **primary productivity of plants**—the production of organic molecules from inorganic molecules such as CO_2—with some an additional contribution from some bacteria, primarily photosynthetic marine organisms. From the smallest single-celled alga in the ocean to the largest forest giant tree, and all of our crop plants, they are vitally important. **Photosynthetic algae in the oceans** play a major role in controlling the atmosphere and the climate, and interaction with viruses is one of the major mechanisms which in turn control the algae. All higher animals depend on the primary productivity of plants for their food. Thus, plants are a big deal, and anything which affects plant growth is of great importance to us.

In purely economic terms, viruses are only of importance if it is likely that they will affect crops during their commercial lifetime, a likelihood that varies greatly between very short extremes in horticultural production (a few weeks for fresh green-leaf vegetables) and very long extremes in forestry (decades for certain large trees). Some estimates have put the total worldwide cost of plant virus infections as high as **US$ 6×10^{10}—$60 billion**—per year. The mechanisms by which plant viruses are transmitted between hosts, and how they cause disease, are therefore of great importance. There are a number of routes by which plant viruses may be transmitted:

- **Seeds:** These may transmit virus infection either by **external contamination** of the seed with virions during its development, or by **infection of the living tissues** of the seed itself. Transmission by this route leads to **early outbreaks of disease** in new crops which are usually initially **focal in distribution**—affect one plant or small groups of plants—but may subsequently be transmitted to the remainder of the crop by other mechanisms.
- **Vegetative propagation/grafting:** These techniques are inexpensive and easy methods of plant propagation, and are in fact necessary for cultivation of bananas, grapevines and many other fruit trees—but also provide the ideal opportunity for viruses to spread to new plants from infected **rootstocks** or "**scions**" (transplanted material).
- **Vectors:** Many different groups of living organisms can act as vectors and spread viruses from one plant to another:
  - ➢ **Bacteria**: while natural transmission of plant viruses by bacteria has not been characterized, recombinant *Agrobacterium tumefaciens* has been used **experimentally** to transmit virus genomes to plants via cloned DNA or cDNA forms of their genomes.
  - ➢ **Fungi**: 30 **soil-borne** viruses are transmitted by five species of fungal vectors. Nine isometric **tombusviruses** are carried essentially nonspecifically on the surfaces of spores of their *Olpidium* spp. vectors, while 18 rod-shaped **furo-** and **bymoviruses** are found **within the resting spores** of their various vectors.
  - ➢ **Nematodes**: these are tiny nonsegmented worms—~50 μm in diameter and 1 mm in length—that are highly abundant in soils and even in sea-floor ooze. Most species are not plant parasites, but the few that are can transmit viruses, mainly to plant roots, by feeding damage to plant cells after picking up viruses from cell sap by feeding. **Nepoviruses**—nematode-transmitted polyhedral ssRNA+ viruses—are the best example.

- ➢ **Arthropods**: these include **insects** (e.g., aphids, leafhoppers, planthoppers, beetles, thrips), and **arachnids** (e.g., mites). Carriage may be **persistent propagative** (i.e., virus **infects** vector; e.g., **tomato spotted wilt virus** in **thrips**), **persistent nonpropagative** (e.g., geminiviruses in whiteflies or leafhoppers); **semipersistent** (i.e., does not infect, but **interacts** with internal organs and proteins of the vector; e.g., **barley yellow dwarf viruses** in **aphids**), or **nonpersistent** (i.e., is carried by **surface contamination** of, e.g., aphid or beetle mouthparts; e.g., **cucumber mosaic, brome mosaic viruses** in some **aphids**).
- **Mechanical:** Mechanical transmission of viruses is the most widely used method for experimental infection of plants, and is usually achieved by **inadvertently** (in nature) or **deliberately** (by humans) **rubbing virus-containing preparations into the leaves**, which in most plant species are the organs most susceptible to infection. This is an important natural method of transmission: virions may contaminate **soil** for long periods and be transmitted to the leaves of new host plants as **wind-blown dust or as rain-splashed mud**, or leaves of neighboring plants may simply rub together. In agriculture, the most common means of mechanical transmission is via **farm implements**, or **manual pruning or cleaning**.

The problems plant viruses face in initiating infections of host cells have already been described (Chapter 4), as has the fact that **no known plant virus employs a specific cellular receptor** of the types that animal and bacterial viruses use, to attach to and enter plant cells. Transmission of plant viruses by **insects** is of particular agricultural importance. Huge areas of monoculture and the inappropriate use of pesticides that kill natural predators can result in **massive population booms** of pest insects such as **aphids** and **whiteflies**. As described in Chapter 4, plant viruses rely on a **mechanical breach** of the integrity of a cell wall to directly introduce a virion into a cell, which can then heal over the breach. Introduction or transfer by insect vectors is a particularly efficient means of virus transmission. Insects that **bite into** (beetles) or **suck from** plant tissues (aphids, whiteflies, leafhoppers, thrips) are an ideal means of transmitting viruses to new hosts (see above).

In some instances, viruses are only transmitted **mechanically** from one plant to the next by the vector and the insect is only a means of distribution, through flying or being carried on the wind for long distances (sometimes hundreds of kilometers).This is known as **nonpersistent, nonpropagative transmission,** and can generally occur **almost immediately** after the vector has fed on an infected plant, as the virus is associated with the **stylets** of piercing mouthparts, or contaminates the **mandibles** of beetles.

For the type of transmission known as **semipersistent, nonpropagative,** the plant viruses do not multiply in their insect or arthropod hosts, but after entry into the vector body via the **mouthparts** and **digestive tract**, may specifically bind to the chitin lining the gut and slowly elute from here. **Cauliflower mosaic virus**, for example, synthesizes an **aphid acquisition factor** in infected cells that aids in this process. Similarly, the **helper component-proteinase (HC-Pro)** of potyviruses is also essential for aphid transmission, as it binds the CP to the inside of aphid stylets. This type of transmission is slower than the previous mode, as it requires some specific intermolecular interactions to occur.

For **persistent nonpropagative** (circulative) transmission, the virus may actually be trafficked specifically across the **gut wall** by **exocytosis**, into the **hemolymph**, and then into the **salivary glands** across their cell membranes, via **specific interactions** with a number of **cell membrane-associated** and other proteins (e.g., **luteoviruses** such as barley yellow dwarf and poleroviruses; **geminiviruses**). The **virus CPs** are specifically implicated in recognition of host proteins and trafficking across membranes, but virion-associated plant host proteins may also be involved. With **whitefly-transmitted geminiviruses** (e.g., begomoviruses), bacterial endosymbionts may be involved too: **heat shock protein 70** (HSP70) and **GroEL chaperone** proteins are implicated, respectively, in transport of virions to, and stabilization in, the insect hemolymph.

In the case of **persistent, propagative** transmission—also known as **circulative, propagative**—the virus may also **infect and multiply** in the tissues of the insect (e.g., plant **rhabdoviruses**, some **reoviruses**) as well as those of host plants. In these cases, the vector serves as a means not only of distributing the virus, but also of **amplifying** the infection—and acting as a **reservoir host**, much as insects do for human- and animal-infecting arboviruses (see below). Here, the viruses typically enter the insect host by uptake the via piercing mouthparts into the gut, and then infect cells of the gut via means such as **receptor-mediated endocytosis** after **specific recognition of cell receptors by capsid or envelope proteins** (see Chapter 4). Here is where such viruses have to bridge two worlds: they use **animal cell-type** infection and propagation mechanisms to multiply in their vectors, and then **plant cell-type mechanisms** (i.e., cell wall breach by injection via vector mouthparts) to get directly into the cytoplasm of plant cells. Interestingly, **tenuiviruses**—which resemble **bunyavirus** nucleoproteins as they have no envelope—nevertheless have distinct tissue tropism in their planthopper and leafhopper vectors, doubtless mediated by **N-protein-mediated** cell receptor binding. Viruses spread from their initial localization in **midgut epithelial cells** via a number of mechanisms, including entry to the **hemocoel**, to infect a variety of other tissue types. This eventually includes infecting cells in the **salivary glands**, from where they are **injected back into plant cells** by the initial ejection of saliva that begins the feeding process (see Dietzgen et al., 2016).

Initially, most plant viruses **multiply at the site of infection** when introduced into plants, giving rise to **localized symptoms** such as **necrotic spots on the leaves**. The virus may subsequently be distributed to all parts of the plant either by **direct cell-to-cell spread** or by the **vascular system**, resulting in a **systemic infection**, or one that involves the whole plant. However, the problem these viruses face in reinfection and recruitment of new cells is the same as the one they faced initially—**how to cross the barrier of the plant cell wall**. Plant cell walls necessarily contain channels called "**plasmodesmata**" which allow plant cells to communicate with each other and to pass metabolites between them. However, these channels are not just breaks in the thick cell wall, but complex channels **lined by plasma membrane**, with a central **rod-like membrane**-based structure with an external diameter of around 15 nm called the **desmotubule,** which is **continuous with the endoplasmic reticulum**. The desmotubule's normal function is to transport lipid molecules between cells. Molecules pass through plasmodesmata either via the internal lumen of the desmotubule by **simple diffusion**—if **less than 1 kDa** in size for mesophyll cells—or via a **specific transport**

FIG. 6.4 Plasmodesmata. The picture depicts a number of plasmodesmata piercing the wall between two plant cells. The channels are closely lined by membranes derived from the cells' plasma membranes, and contain a central tube-like structure—the desmotubule—formed by and continuous with the endoplasmic reticulum of each cell, and whose inner diameter can be varied by specific protein-protein interactions between chaperonins and other transport proteins, proteins associated with the plasma membrane lining and the desmotubule, and the actin and myosin that form a helical structure in the tubule. *Courtesy of Russell Kightley Media.*

mechanism mediated by plasmodesmata-associated proteins: this may occur internally in the desmotubule, or in the cytoplasmic "sleeve" surrounding it. Small ssRNAs involved in gene silencing (**siRNAs**; see later) are known to move systemically: this is in association with chaperonins which interact with the plasmosdesmatal machinery so as to **increase the "pore" size** (see Fig. 6.4), and may engage with **actin and/or mysosin fibers**, which extend through the cytoplasmic sleeve.

As plasmodesmata do not admit particles as large as virions, plant viruses have therefore evolved **specific movement functions**, mediated by **one or more virus-specified proteins**. All plant-infecting viruses possess one or more **movement-related protein (MP) genes**: these are very varied, although there are **distinct groups of them**, and they appear to largely derive from host plant genes for **chaperonins and plasmodesmata-associated proteins**, which interact with the plasmosdesmatal machinery to allow specific transport of **viral nucleoprotein complexes** rather than of assembled virions. One of the best known examples of this is the **30kDa** protein of **TMV**. This is expressed from a subgenomic mRNA (Fig. 3.12), and its function is to modify plasmodesmata and allow **genomic RNA coated with 30kDa protein** to be transported from the infected cell to neighboring cells (Fig. 6.5). Other viruses such as the

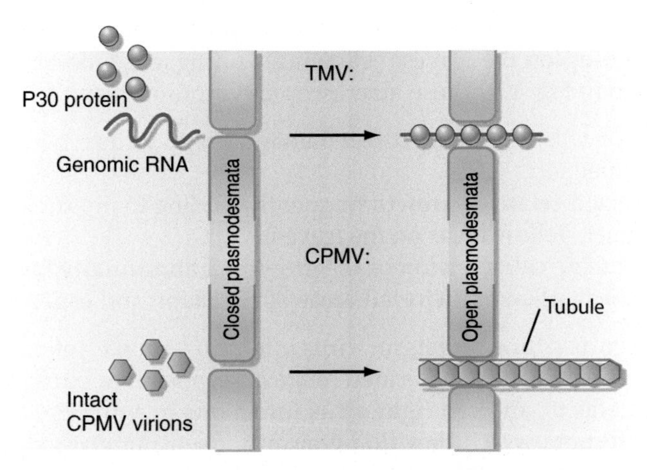

FIG. 6.5 Plant virus movement proteins. Plant virus movement-related proteins allow plant viruses to infect new cells without having to penetrate the cell wall from the outside for each new cell: this can happen via specific protein-mediated plasmodesmatal transport of nucleoproteins, or via virus-specific tubule formation, which allows passage of small virions.

isometric **cowpea mosaic virus** (CPMV; *Comoviridae*) employ a different molecular mechanism. In CPMV, the 58/48 kDa movement-associated proteins form **tubular structures** allowing the passage of **intact virions** to pass from one cell to another (Fig. 6.5). The ssDNA **geminiviruses** have a movement protein that forms a **loose nucleoprotein structure** with the genome for plasmodesmatal movement, but may also make tubules, in the case of the two-component viruses (**begomoviruses**). The latter also have a protein encoded by the B component genome (if present) that is a **nuclear localization protein**, to move the genome into the nucleus.

It could be that plasmodesmata are the reason that most types of plant viruses only have genomes that have a **single-stranded phase** in their replication cycles. This includes viruses in Classes II-VII, or **ssDNA** (gemini-, nanoviruses), **dsRNA** (e.g., reoviruses), **ssRNA+** (most plant viruses), **ssRNA−** (rhabdo-, bunyaviruses) and the dsDNA **pararetroviruses** (e.g., caulimo- badnaviruses), but not **Class I dsDNA** viruses. This could be because **double-stranded genomes** are probably too inflexible and possibly too wide to fit through plasmodesmata, which can apparently only admit **nucleocapsid complexes** with **single-stranded nucleic acids**.

While plasmodesmatal transport allows virus genomes to move cell-to-cell, **longer distance transport** is necessary for systemic infection. This is generally facilitated by the **virus CP**, although not necessarily, so it is that **virions** are often responsible for infection at a distance from the inoculation site, although less structured nucleoproteins may also be. For example, the two-component ssDNA begomovirus **African cassava mosaic virus** can spread systemically even if the genome **lacks the CP gene**. The avenue for movement is generally the **phloem tissue**, and other proteins are frequently involved: these include virus-encoded **movement proteins**, and potentially host factors such as chaperonins. Virion or nucleoprotein movement via phloem is regulated at the level both of entry into phloem, but also exit from phloem in distant leaves (see Solovyev and Savenkov, 2014).

Typically, virus infections of plants might result in effects such as growth retardation, distortion, mosaic patterning on the leaves, yellowing, wilting, etc. Good examples of striking symptoms can be seen in Fig. 1.5. These macroscopic symptoms result from:

- **Necrosis** or death of cells, caused by **direct damage** due to virus replication or to the host plant response to infection.
- **Hypoplasia**—localized **retarded growth** frequently leading to **mosaicism** (the appearance of thinner, yellow areas on the leaves).
- **Hyperplasia**—excessive **cell division** or the growth of **abnormally large cells**, resulting in the production of swollen or distorted areas of the plant, and especially of leaves.

Plants might be seen as sitting targets for virus infection—unlike animals, they cannot run away. However, plants exhibit a sophisticated range of responses to virus infections designed to minimize harmful effects, and can **fight virus infections** in a number of ways. First, they need to **detect** the infection, which they do by means of sensing **virus signature molecules** (so-called **pathogen-associated molecular patterns or PAMPs**, e.g., particular viral proteins) via dedicated receptors. This is an example of an **innate immune response**, similar in concept to the prokaryote response discussed earlier. When this happens, the production of **resistance proteins** that activate highly specific resistance mechanisms is triggered. In response, plant viruses attempt to **evade** these defense mechanisms by **altering protein structures** where possible—by mutation, similar to **antigenic drift** in influenza viruses (see later)—and by producing proteins which bind to and hide **small RNAs** which could otherwise trigger **RNA silencing**—a phenomenon first described in plants, and formerly simply known as "**post-transcriptional gene silencing.**" Infection can result in a "**hypersensitive response**," manifested as:

- The synthesis of a range of new proteins, the **pathogenesis-related (PR) proteins**.
- An increase in the production of **cell wall phenolic substances**.
- The release of active oxygen species.
- The production of **phytoalexins**.
- The accumulation of **salicylic acid**—amazingly, plants can even warn each other that viruses are coming by **airborne signaling** with volatile compounds such as methyl salicylate.

The hypersensitive response involves synthesis of a wide range of different molecules. Some of these PR proteins are **proteases**, which presumably destroy virus proteins while the response is activated, limiting the spread of the infection. There is some similarity here between the design of this response and the **production of interferons** (IFNs) by animals. The visible manifestation of the response is often the formation of **local lesions**, or small areas of dead tissue that "pock" the inoculated or initially infected leaves: this sort of response severely limits or even completely negates **spread** of the virus out of the infected areas (see below). This phenomenon has been studied since the early 20th century (see Chapter 1, Culturing viruses, and Fig. 1.8), as it was the basis of the first **accurate assay method** for plant viruses, akin to the plaque assay for bacteriophages.

Systemic resistance to virus infection is a naturally occurring phenomenon in some strains of plant. This is clearly a highly desirable characteristic that is prized by plant breeders, who try to spread this attribute to economically valuable crop strains. There are probably many different mechanisms involved in systemic resistance, but in general terms there is a tendency of these processes to **increase local necrosis** when substances such as proteases and **peroxidases** and **polyphenoloxidase** are produced by the plant to destroy the virus or create an environment inimical to its survival, and thus to prevent its spread and subsequent **systemic infection**. An example of this is the **tobacco *N* gene**, which encodes a cytoplasmic protein with a nucleotide-binding site which interferes with the TMV replicase. When present in plants, this gene causes TMV to produce a localized, necrotic infection (**local lesion**) rather than the systemic mosaic symptoms normally seen. A personal testimony to the effectiveness of polyphenoloxidase comes from attempts to purify the multicomponent ssRNA+ **bromovirus** broad bean mottle virus from infected broad bean plants in the 1970s: disruption of plant tissues caused instant release of so much polyphenoloxidase that liquid extracts turned black as a result of **polyphenol** formation within a few minutes, all proteins bound the molecules and also turned black, and no infectious virions could be purified.

Virus-resistant plants have been created by the production of transgenic plants expressing **recombinant virus proteins or nucleic acids** which interfere with virus replication without producing the pathogenic consequences of infection. Examples of strategies:

- **Virus coat protein (CP) genes**, which have a variety of complex effects, including inhibition of virus **uncoating** by virus CP, and interference of **expression** of the virus at the level of **RNA** ("**gene silencing**" by RNAI; see later).
- Intact or partial **virus replicases** which interfere with genome **replication**.
- Antisense RNAs.
- **Defective** virus genomes.
- **Satellite** sequences (see Chapter 8).
- Catalytic RNA sequences (**ribozymes**).
- Modified movement proteins.
- **CRISPR/Cas**-based antivirus strategies.

These are very promising strategies that offer the possibility of substantial increases in agricultural production without the use of **expensive**, **toxic**, and **ecologically damaging** chemicals (fertilizers, herbicides, or pesticides), and the engineering of **resistance** or agronomically desirable traits (e.g., **drought tolerance**) that are not available in the natural gene pool of the host plants. In some countries, notably in Europe, public resistance to genetically engineered plants has so far prevented the widespread adoption of new varieties produced by genetic manipulation, without considering the environmental cost of not utilizing these new approaches to plant breeding. This will become increasingly harder to justify with application of the **new gene-editing technologies**—largely based on CRISPR/Cas (see earlier)—where it may be impossible to genetically distinguish engineered from natural plants, and where **no exogenous DNA** may have been used.

Immune responses to virus infections in animals

It could be argued that complex multicellular organisms owe their immune systems to the very processes that enable multicellularity—and that is the **cell-to-cell recognition** that allows **differentiation** of cellular functions and the development of "***tissues***," or groups of **structurally** and **functionally similar** cells. In simple multicellular organisms such as **sponges** this is fairly rudimentary—that is, cells recognize one another by means of **membrane protein** interactions so as to be able to organize themselves back into a whole organism if it is carefully broken up by sieving, for example. There is also "**nonself**" recognition in terms of not accepting transplants from distantly related sponges. In vertebrates this is very much more complex, however, and formidable **defense systems** have evolved from cell-cell recognition mechanisms, with **nonself-recognition** resulting in both **innate** and **adaptive** immune responses.

Innate immune responses

We have introduced the concept of innate immune responses in prokaryotes earlier in this chapter: this was in the context of these organisms having mechanisms that are "**hard wired**" into their genetics for the **recognition by genome-encoded proteins** of specific virus **DNA** or potentially **protein** sequences. This could be termed an **ancestral** response, where the capacity for the response results from repeated exposure to a specific set of DNA and/or proteins sequences, and the evolution of responses to target these. These responses are essentially **nonadaptable,** in that the genetic hard wiring is conceptually similar to the recognition of a **silhouette of a predator eagle** by chicks of another bird species: no previous exposure by that particular instance of that organism is required for a response to occur. An example closer to the current discussion is that of **PAMPs** in plants, discussed above: these are **pathogen-associated molecular patterns** that are recognized by **specific receptors** in or on the membranes of plant cells—and those receptor genes are the hard-wired part of the genome that has evolved over geological time (many millions of years) to deal with a **specific type or class of pathogen**.

Invertebrate innate defenses

This sort of response is exemplified by innate defenses in **invertebrates**, which have cellular receptors that bind to "**foreign**" elements, mainly **proteins**, in order to differentiate self from nonself—something that developed over **550 million years** ago, in the Cambrian era. In multicellular animals this ability is associated with specialized **phagocytes**—cells that are highly active in **phagocytosis**—that have names like **amoebocytes, hemocytes,** and **coelomocytes**, depending upon their host organism. The spectrum of animals that have these cells includes **sponges, worms, cnidarians, molluscs, crustaceans, chelicerates, insects,** and **echinoderms (sea stars and urchins)**. The phagocytic cells look like macrophages from vertebrates, and have well-conserved pathogen recognition receptors—also known as **pattern-recognition receptors (PRRs)**—for certain **viruses, bacteria, fungi, protozoans,** and **helminths**. These recognize specific molecular signatures such as PAMPs (see earlier) and **damage-associated molecules patterns** (**DAMPs**). Receptors include molecules also expressed in vertebrates, such as **scavenger receptors, Toll-like receptors** (**TLRs**), and **Nod-like receptors** (**NLRs**), all of which recognize evolutionarily-conserved signal sequences. Once bound by a virion or other pathogen,

signal transduction initiates, and a complex cascade of cellular reactions follows. This leads to production, via stimulation of expression of nuclear genes, of **effector molecules**, which participate in **elimination** or **inactivation** of the intruder organism. These include **reactive oxygen (ROS)** and **nitric oxide (NOS)** species of molecules for oxidative killing, **fibrinogen-related peptides (FREPS)** for agglutination and coagulation, and the production of **prophenoloxidases** for production of **melanin** and **cytotoxic quinones** for encapsulation and inactivation of pathogens (see plants, above). Molecules produced by the vertebrate lineage-basal **jawless fish**, which mostly lack **adaptive immune systems**, include **antimicrobial peptides (AMPs)**, **lysozymes** for bacterial cell wall degradation, **hemolysins**, **transferrins,** and **lectins**.

Vertebrate innate defenses

Passive defenses

It is not perhaps appreciated that the **two major organs** in vertebrates that are involved in defense against virus and other pathogen infection are the **skin** and the **gut**, or **intestines**, and that much of this defense is **passive**, or happens because of **physical** or **chemical** barriers. The main physical barriers are **cornified cells** (i.e., squamous epithelial cells converted into tough protective layers), and **mucus** (slimy, carbohydrate-rich substance secreted by the mucous membranes and glands for lubrication and protection). **Chemical barriers** include **saliva** (pH, enzymes), **mucus** (enzymes), **low pH in the stomach**, and **secreted enzymes** in the gut. You will recall (Chapter 4) that **lysosomes** also are a **low pH** environment, and one that is full of enzymes such as proteases, lipases and nucleases: these are also a good natural and **nonspecific barrier** to virus entry into cells.

Another natural barrier that is live cell-based is **dendritic cells**: these are present in all **lymph nodes** (see below) as well as in the **skin** (Langerhan's cells), and **passively trap** antigens—including virions—and **present** them to T-cells and B-cells (see below) for initiation of the **adaptive immune response.**

Active defenses

There are a wide variety of specialized cells in the tissues and circulatory fluids of vertebrates that are involved in defense against pathogens. We will deal with the most important in relation to innate responses here, and with cells such as T- and B-lymphocytes that are involved in the adaptive response, below. The various **phagocytic cells** discussed above in relation to invertebrates have their counterparts in vertebrates: these are so-called **granulocytes** or **agranulocytes** (possess or do not possess **granules** under the microscope), and include the cells and functions shown in Table 6.1.

Macrophages are a heterogeneous population of **phagocytic effector cells**, found in tissues all over the mammalian body. They differentiate from blood **monocytes** that have left the circulatory systems for the various tissues they are found in. The different types of macrophages (e.g., **alveolar** in lungs, **Kupffer** cells in the liver, **microglia** in the brain and central nervous system) can recognize different types of pathogens as well as different types of **dead cells**, as one of their main functions is elimination of these. They recognize bacteria and their

TABLE 6.1 Granulo- and agranulocytic cells in vertebrates.

Type	Granule color	Life span and concentration	Functions
Neutrophils[a]	Red or dark pink	1–4 days, 50%–70%[b]	Phagocytose bacteria
Eosinophils	Red or dark pink	1–2 weeks, 1%–4%	Kill helminths, other parasites; modulate local inflammation
Basophils	Dark blue or purple	Months; ~1%	Modulate inflammation, release histamine during allergic responses
Monocytes	None	Hours to years; 2%–8%	Macrophage precursors, also other "mononuclear phagocytes"
Macrophages (many versions)	None	Months to years	Detection, phagocytosis and destruction of bacteria and other cells and virions; cytokine release; antigen presentation

[a] *These cells are historically named according to how they stain with histological stains used by pathologists.*
[b] *Proportion of peripheral blood mononuclear cells (PBMCs) determined by microscopy.*
Adapted from Mescher, A.L., 2018. Junqueira's Basic Histology: Text and Atlas. 15th ed. ISBN 978-1-260-02617-7.

products and certain viruses by means of **Toll-like receptors (TLRs)**, that bind things like **lipopolysaccharides** (LPSs) and **flagellin** from bacteria, as well as **RNA, DNA** and certain viral or bacterial **proteins**. They quite strongly resemble the invertebrate phagocytic cells.

"**Natural killer**" (**NK**) cells are a class of **lymphocytes** carry out cell lysis independently of conventional immunological specificity: that is, they do not depend on **clonal antigen recognition** for their action, as B- and T-cells do. In other works, NK cells are able to recognize virus-infected cells **without being presented with a specific antigen** by a macromolecular complex consisting of MHC antigens plus the T-cell receptor/CD3 complex. The advantage of this is that NK cells have **broad specificity** (many antigens rather than a single epitope) and are also active without the requirement for **sensitizing antibodies**. They are therefore the first line of defense against virus infection. NK cells are most active in the early stages of infection (i.e., in the first few days), and their activity is stimulated by **IFN-α/β** (see the section "Interferons"). NK cells are not directly induced by virus infection—they exist even in immunologically naive individuals and are "revealed" in the presence of IFN-α/β. They are thus part of the "innate" rather than the "**adaptive**" immune response. Their function is complementary to and is later taken over by cytotoxic T-lymphocytes (CTLs) which are part of the adaptive immune response. Not all of the targets for NK cells on the surface of infected cells are known, but they are **inhibited by MHC class I antigens** (which are present on all nucleated cells), allowing recognition of "**self**" (i.e., uninfected cells) and preventing total destruction of the body. It is well known that some virus infections **disturb normal cellular MHC-I expression** and this is one mechanism by which NK cells recognize virus-infected cells. NK cell cytotoxicity is activated by IFN-α/β, directly linking NK cell activity to virus infection (Fig. 6.6).

Viruses and apoptosis

Apoptosis, or "**programmed cell death**," is a critical mechanism in animal tissue remodeling during development and in cell killing by the immune system—and is a process central to

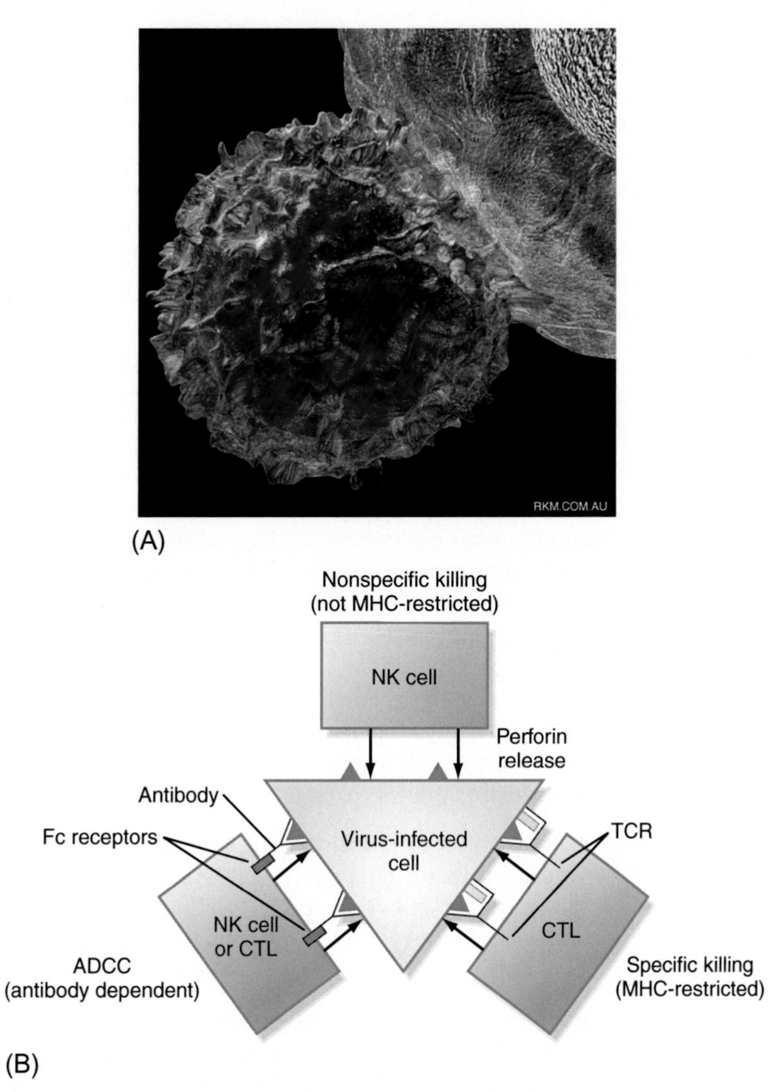

FIG. 6.6 (A) Natural killer cells. A NK cell is shown as a purple cell attached to a green target cell. These cells can act nonspecifically to destroy transformed (i.e., cancer) cells or cells that have been infected by a virus. Cytotoxic granules are shown as blue spheres at the edge of the killer cell where it touches the target cell. *Courtesy of Russell Kightley Media.* (B) Mechanisms of cell-mediated immunity. The diagram illustrates the three main mechanisms by which cell-mediated immunity kills virus-infected cells.

plant development as well, triggered by **reactive oxygen species (ROSs)**. There are basically two ways in which a cell can die apart from massive direct damage by outside agencies: these are **necrosis** or **apoptosis**.

- **Necrosis** is the normal response of cells to injury caused by **toxins** or **environmental stress**. Necrosis is marked by **nonspecific** changes such as disruption of the plasma membrane and nuclear envelope, **rupture** of membrane-bounded organelles such as

mitochondria and lysosomes, cell swelling, **random fragmentation** of DNA/RNA, **influx of calcium ions** into the cell, and **loss of membrane electrical potential**. The release of cellular components from the dying cell causes a **localized inflammatory response** by the cells of the animal immune system. This frequently leads to damage to adjacent cells/tissue—"**bystander**" cell damage.
- **Apoptosis** is, in contrast, a **tightly regulated process** that relies on complex molecular cascades for its control. It is marked by **cell shrinkage**, **condensation**, and **clumping of chromatin**, a regular pattern of **DNA fragmentation**, and "**bubbling off**" of cellular contents into **small membrane-bounded vesicles** ("blebbing") which are subsequently **phagocytosed by macrophages**, preventing inflammation.

When triggered by the appropriate signals, **immune effector cells** such as CTLs and NK cells release previously manufactured **lytic granules** stored in their cytoplasm. These act on the target cell and induce apoptosis by two mechanisms:

- Release of **cytotoxins** such as: (1) **perforin** (aka cytolysin), a peptide related to complement component C9 which, on release, polymerizes to form polyperforin, which forms **transmembrane channels**, resulting in permeability of the target cell membrane and (2) **granzymes**, which are **serine proteases** related to trypsin. These two effectors act collaboratively, the membrane pores allowing the entry of granzymes into the target cell. The membrane channels also allow the release of **intracellular calcium** from the target cell, which also acts to trigger apoptotic pathways.
- In addition, **CTLs** (but not NK cells) express **Fas ligand** on their surface which binds to Fas on the surface of the target cell, triggering apoptosis. Binding of Fas ligand on the effector cell to Fas (**CD95**) on the target cell results in activation of cellular proteases known as "**caspases**," which in turn trigger a cascade of events leading to apoptosis.

The process of induction and repression of apoptosis during virus infection has received much attention during the last few years. It is now recognized that this is an important innate response to **virus infection**. The regulation of apoptosis is a complex issue that cannot be described fully here (see References and Fig. 6.7 for a summary), but **virus infections disturb normal cellular biochemistry** and frequently trigger an apoptotic response, by the following mechanisms:

- **Receptor signaling:** Binding of virions to cellular receptors may also trigger signaling mechanisms resulting in apoptosis (e.g., HIV [see Chapter 7], reovirus).
- **PKR activation:** The IFN effector PKR (RNA-activated protein kinase) may be activated by some viruses (e.g., HIV, reovirus).
- **p53 activation:** Viruses that interact with p53 (Chapter 7) may cause either growth arrest or apoptosis (e.g., adenoviruses, SV40, papillomaviruses; see Fig. 5.5).
- **Transcriptional dysregulation:** Viruses that encode transcriptional regulatory proteins may trigger an apoptotic response (e.g., HTLV Tax).
- **Foreign protein expression:** Overexpression of virus proteins at late stages of the replication cycle can also cause apoptosis by a variety of mechanisms.

In response to this cellular alarm system, many if not most viruses have evolved mechanisms to **counteract** this effect and **repress apoptosis**:

- **Bcl-2 homologues:** A number of viruses encode Bcl-2 (a negative regulator of apoptosis) homologues (e.g., **adenovirus** E1B-19k, **human herpesvirus 8** (HHV-8) KSbcl-2).

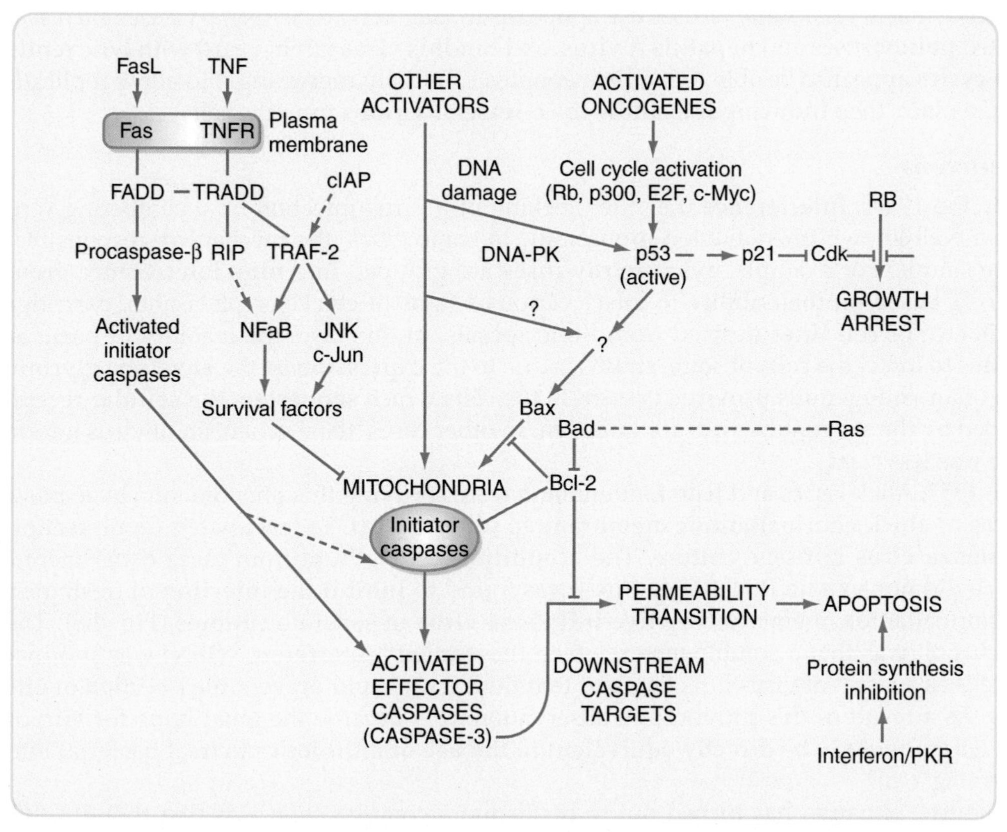

FIG. 6.7 Overview of apoptosis. The pathways controlling apoptosis are very complex. This diagram represents only a simple summary of some of the mechanisms of major significance in virus infections.

- **Caspase inhibition:** Caspases are a family of **cysteine proteases** that are important inducers of apoptosis. Inhibiting these enzymes is an effective way of preventing apoptosis (e.g., **baculovirus** p35, serpins, vIAPs—"**inhibitors of apoptosis**").
- **Fas/TNF inhibition:** Viruses have evolved several mechanisms to block the effects of Fas/TNF, including blocking signaling through the plasma membrane (e.g., adenovirus E3), **tumor necrosis factor receptor** (TNFR) mimics (e.g., **poxvirus** crmA), mimics of **death signaling factors** (vFLIPs), and interactions with signaling factors such as **Fas-associated death domain** (FADD) and **TNFR-associated death domain** (TRADD) (e.g., **HHV-4** (Epstein-Barr virus, EBV) LMP-1).
- **p53 inhibition:** A number of viruses that interact with p53 have evolved proteins to counteract possible triggering of apoptosis (e.g., **adenovirus** E1B-55k and E4, **SV40** T-antigen, **papillomavirus** E6).
- **Miscellaneous:** Many other apoptosis-avoidance mechanisms have been described in a wide variety of viruses (see References).

Without such inhibitory mechanisms, most viruses would simply not be able to replicate due to the **death of the host cell before the replication cycle was complete**. However, there is evidence that at least some viruses use apoptosis to their benefit. SsRNA+ viruses such as the related **polioviruses** and **hepatitis A virus**, and **Sindbis virus** (alphavirus) with **lytic replication cycles** appear to be able to regulate apoptosis, initially repressing it to allow replication to take place, then **inducing it** to **allow the release of virions** from the cell.

Interferons

By the 1950s, **interference** (i.e., the blocking of a virus infection by a competing virus) was a well-known phenomenon in virology. In some cases, the mechanism responsible is quite simple. For example, **avian retroviruses** are grouped into **nine interference groups** (A to I), based on their ability to infect various strains of chickens, pheasants, partridges, quail, etc., or **cell lines** derived from these species. In this case, the inability of particular viruses to infect the cells of some strains is due to the **expression of the envelope glycoprotein** of an **endogenous provirus** present in the cells which **sequesters the cellular receptor** needed by the exogenous virus for infection. In other cases, the mechanism of virus interference was less clear.

In 1957, Alick Issacs and Jean Lindenmann were studying this phenomenon by exposing pieces of **chick chorioallantoic membrane** to ultraviolet (UV)-inactivated (noninfectious) **influenza virus** in tissue culture. The "**conditioned**" medium from these experiments—which did not contain infectious virus—was found to **inhibit the infection** of fresh pieces of chorioallantoic membrane by **live influenza virus** in separate cultures (Fig. 6.6). Their conclusion was that a soluble factor, which they called "**interferon,**" (**IFN**) was produced by cells as a result of virus infection and that this factor could prevent the infection of other cells. As a result of this provocative observation, IFN became the great hope for virology and was thought to be **directly equivalent to the use of antibiotics** to treat bacterial infections (Fig. 6.8).

The true situation has turned out to be far more complex than was first thought. IFNs do have **antiviral properties**, but by and large their effects are exerted indirectly via their major function as **cellular regulatory proteins**. IFNs are immensely potent; fewer than **50 molecules per cell** show evidence of antiviral activity. Hence, following the initial discovery, many fairly fruitless years were spent trying to purify minute amounts of naturally produced IFN. This situation changed with the development of molecular biology and the cloning and

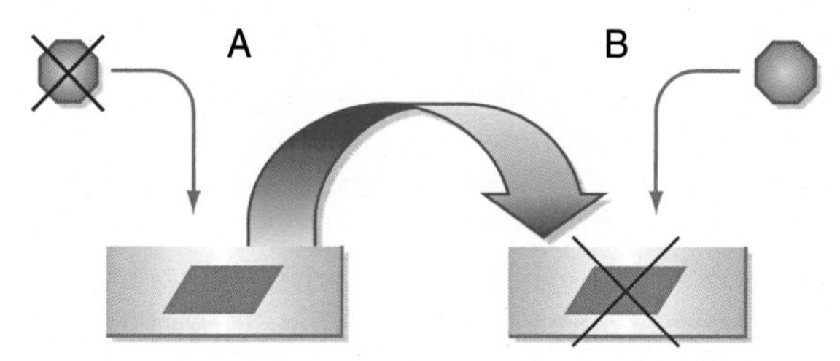

FIG. 6.8 Discovery of interferons. Interferons were discovered by the following experiment. (A) Pieces of chick chorioallantoic membrane were exposed to UV-inactivated (noninfectious) influenza virus in tissue culture. (B) The "conditioned" medium from these experiments (which did not contain infectious virus) was found to inhibit the infection of fresh pieces of chick chorioallantoic membrane by infectious influenza virus in separate cultures. The inhibitory substance in the conditioned medium was named "interferon."

recombinant expression of IFN genes, which has led to rapid advances in our understanding over the last 15 years. There are a number of different types of IFNs:

- **IFN-α**: There are at least **15 molecular species of IFN-α**, all of which are closely related; some species differ by only one amino acid. They are synthesized predominantly by **lymphocytes**. The mature proteins contain 143 amino acids, with a minimum homology of 77% between the different types. All the genes encoding IFN-α are located on human chromosome 9, and **gene duplication** is thought to be responsible for this proliferation of genes.
- **IFN-β**: The single gene for IFN-β is also located on human chromosome 9. The mature protein contains 145 amino acids and, unlike IFN-α, is **glycosylated**, with approximately 30% homology to other IFNs. It is synthesized predominantly by **fibroblasts**, or muscle cell precursors.
- **Other IFNs**: The single gene for **IFN-γ** is located on human chromosome 12. The mature protein contains 146 amino acids, is **glycosylated**, and has very low sequence homology to other IFNs. It is synthesized predominantly by **lymphocytes**. IFNs such as **IFN-γ, -δ, -k, -τ**, etc., play a variety of roles in **cellular regulation** by acting as **signaling molecules** or **cytokines**, but are not directly involved in controlling virus infection.

Because there are clear biological differences between the two main types of IFN, IFN-α and -β are known as **type I IFN**, and IFN-γ as **type II IFN**. Induction of IFN synthesis results from upregulation of transcription from the IFN gene **promoters**. There are three main mechanisms involved:

- **Virus infection:** This mechanism is thought to act by the **inhibition of cellular protein synthesis** that occurs during many virus infections, resulting in a reduction in the concentration of **intracellular repressor proteins,** and hence in increased **IFN gene transcription**. In general, **RNA viruses are potent inducers** of IFN—largely because of **dsRNAs** involved in replication complexes (see Chapter 4 and below)—while DNA viruses are relatively poor inducers; however, there are exceptions to this rule (e.g., **poxviruses** are very potent inducers). The molecular events in the induction of IFN synthesis by virus infection are not clear. In some cases (e.g., influenza virus), **UV-inactivated virus** is a potent inducer; therefore, virus replication is not necessarily required. Induction by viruses might also involve **perturbation of the normal cellular environment** in addition to production of small amounts of double-stranded RNA.
- **Double-stranded (ds) RNA:** All naturally occurring dsRNAs (e.g., **reovirus genomes; replication complexes** of ssRNA viruses; **highly base-paired** ssRNAs) are potent inducers of IFN, as are **synthetic molecules** (e.g., poly I:C); therefore, this process is independent of nucleotide sequence. Linear ssRNA and dsDNA are not inducers. This mechanism of induction is thought to depend on the **secondary structure of the RNA** rather than any particular nucleotide sequence.
- **Metabolic inhibitors:** Compounds that inhibit **transcription** (e.g., **actinomycin D**) or **translation** (e.g., **cycloheximide**) result in induction of IFN. **Tumorigenic** chemicals such as tetradecanoyl phorbol acetate or dimethyl sulfoxide are also inducers. Their mechanism of action remains unknown but they almost certainly act at the level of transcription.

The effects of IFNs are exerted via specific receptors that are ubiquitous on nearly all cell types (therefore, **nearly all cells are potentially IFN responsive**). There are distinct receptors for type I and type II IFN, each of which consists of two polypeptide chains. Binding of IFN to the type I receptor activates a **specific cytoplasmic tyrosine kinase** (Janus kinase, or Jak1), which phosphorylates another cellular protein, **signal transducer and activator of transcription 2** (STAT2). This is transported to the nucleus and turns on **transcriptional activation of IFN-responsive genes** (including IFN, resulting in **amplification** of the original signal). Binding of IFN to the **type II receptor** activates a different cytoplasmic tyrosine kinase (**Jak2**), which **phosphorylates the cellular protein STAT1**, leading to transcriptional activation of a different set of genes.

The main action of IFNs is on **cellular regulatory activities** and is rather complex. IFN affects both cellular proliferation and immunomodulation. These effects result from the induction of transcription of a wide variety of cellular genes, including other cytokines. The net result is complex regulation of the ability of a cell to **proliferate, differentiate,** and **communicate**. This cell regulatory activity itself has indirect effects on virus replication. **Type I IFN** is the **major antiviral mechanism**—other IFNs act as potent cellular regulators, which may have indirect antiviral effects in some circumstances.

The effect of IFNs on virus infections in vivo is extremely important. Animals experimentally infected with viruses and injected with **anti-IFN antibodies** experience much more severe infections than control animals infected with the same virus. This is because **IFNs protect cells from damage and death**. However, they do not appear to play a major role in the **clearance** of virus infections—the other parts of the immune response are necessary for this. IFN is a "**firebreak**" that inhibits virus replication in its earliest stages by several mechanisms. Two of these are understood in some detail, but a number of others (in some cases specific to certain viruses) are less well understood.

IFNs induce transcription of a cellular gene for the **enzyme 2′,5′-oligo A synthetase** (Fig. 6.9). There are at least four molecular species of 2′,5′-oligo A, induced by **different forms of IFN**. This compound activates an **RNA-digesting enzyme, RNAse L**, which **digests virus genomic RNAs, virus and cellular mRNAs**, and **cellular ribosomal RNAs**. The end result of this mechanism is a **reduction in protein synthesis** (due to the degradation of mRNAs and rRNAs)—therefore the cell is protected from virus damage. The second method relies on the **activation** of a 68-kDa protein called **PKR**.

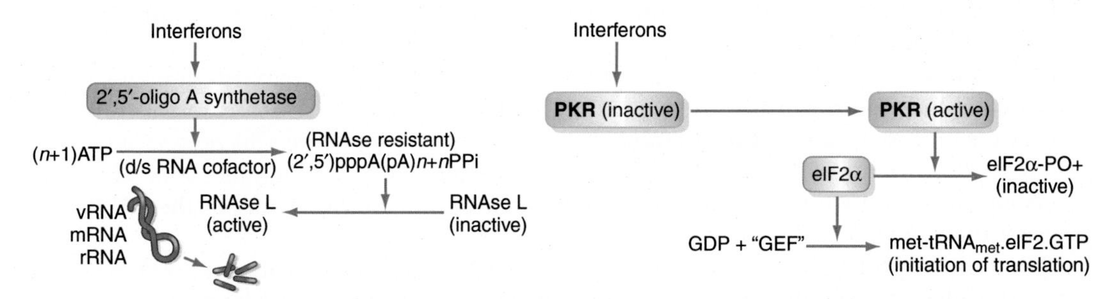

FIG. 6.9 Induction of 2′,5′-oligo A synthetase and PKR by IFNs. The modified nucleic acid 2′,5′-oligo A is involved in one of the major mechanism by which IFNs counteract virus infections. The protein kinase PKR is another major mechanism by which IFNs counteract virus infections.

TABLE 6.2 Therapeutic uses of IFNs.

Condition	Virus
Chronic active hepatitis	HBV, HCV
Condylomata accuminata (genital warts)	Papillomaviruses
Tumors	
Hairy cell leukemia	–
Kaposi's sarcoma (in AIDS patients)	Human herpesvirus 8 (HHV-8) (?)
Congenital diseases	
Chronic granulomatous disease (IFN-γ reduces bacterial infections)	–

PKR phosphorylates a **cellular factor, elongation initiation factor 2α (eIF2α)**, which is required by ribosomes for the **initiation of translation**. The net result of this mechanism is also the **inhibition of protein synthesis** and this reinforces the 2′,5′-oligo A mechanism. A third, well-established mechanism depends on the **M_x gene**, a single-copy gene located on human chromosome 21, the **transcription** of which is induced by **type I IFN**. The product of this gene inhibits the **primary transcription of influenza virus** but not of other viruses. Its method of action is unknown. In addition to these three mechanisms, there are many additional recorded effects of IFNs. They inhibit the **penetration** and **uncoating** of SV40 and some other viruses, possibly by altering the composition or structure of the cell membrane; they inhibit the **primary transcription of many virus genomes** (e.g., SV40, HSV) and also **cell transformation** by retroviruses. None of the molecular mechanisms by which these effects are mediated has been fully explained.

IFNs are a powerful weapon against virus infection, but they act as a blunderbuss rather than a "**magic bullet**." The **severe side effects** (fever, nausea, malaise) that result from the powerful cell-regulatory action of IFNs means that they will never be widely used for the treatment of trivial virus infections—they are not the cure for the common cold. However, as the cell-regulatory potential of IFNs is becoming better understood, they are finding increasing use as a **treatment for certain cancers** (e.g., the use of IFN-α in the treatment of **hairy cell leukemia**). Current therapeutic uses of IFNs are summarized in Table 6.2. The long-term prospects for their use as antiviral compounds are less certain, except for possibly in life-threatening infections where there is no alternative therapy (e.g., chronic viral hepatitis).

Adaptive immune responses in vertebrates

The evolution of more advanced vertebrates—starting with **cartilaginous and bony fish**—marked the development of a greater degree of immunological sophistication, or **adaptive immunity**. These animals developed the **major histocompatibility complex (MHC) proteins, T-cell receptors** and **immunoglobulins** as highly tuned, specific pathogen responsive regulatory systems. Thus, the most significant response to virus infection in vertebrates is activation of both the **cellular and humoral** parts of the **adaptive immune system**. A complete description of all the events involved in the immune response to the presence of foreign antigens is beyond the scope of this book, so you should refer to material in the References at the end of this chapter to ensure that you are familiar with all the immune mechanisms (and jargon!) described below. A brief summary of some of the more important aspects is worth considering however, beginning with the **humoral immune response**, which results in the production of antibodies (Fig. 6.10).

The major impact of the humoral immune response—responses primarily involving **antibodies (Abs)**, or **immunoglobulins (Igs)** produced by **activated B lymphocytes**—is the eventual clearance of virus from the **circulatory fluids** in the body. All mammalian antibodies—**IgA, -D, -E, -G, and -M**—are basically **four-chain molecules**, which start out as **B cell membrane-attached dimers** of a **light chain-heavy chain dimer**. **IgD** is only ever found as a **membrane receptor** on B-cells. Antibodies found in circulation are secreted forms lacking a membrane attachment domain. **Serum neutralization** of virions blocks their infectivity and stops the spread of virus to uninfected cells. This allows other defense mechanisms—**NK cells, macrophages, cytotoxic T-lymphocytes**—to mop up the infection, partly by **killing infected cells**. Fig. 6.10 shows a very simplified version of the basic mammalian Ig molecule, and Fig. 6.11 a synopsis of the humoral response to infection.

Virus infection induces at least **three classes of antibody**: **immunoglobulin G (IgG), IgM,** and **IgA**. IgM is a large, **multivalent** (5 copies of the basic Ig) molecule that is most effective

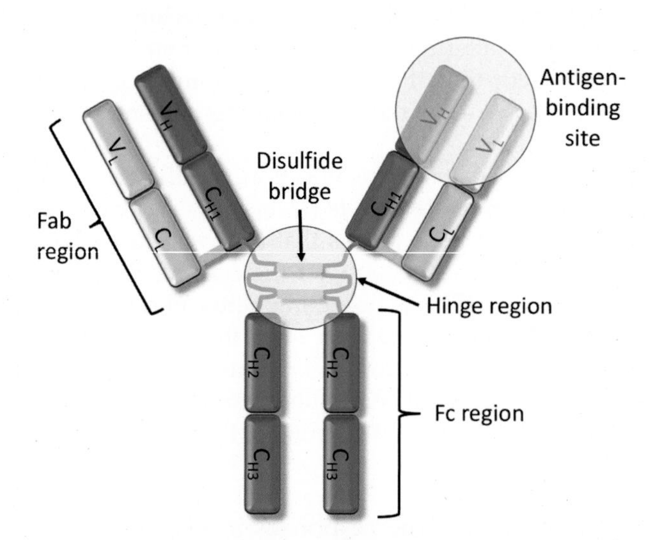

FIG. 6.10 Basic immunoglobulin structure. The basic structure of a "universal" mammalian antibody molecule. All of these molecules have two copies each of a light (L) and a heavy (H) chain, disulfide bridged to one another via cysteine residues, each with a variable N-terminal domain of around 12 kDa in size (V_L, V_H). The light chain has another constant domain (~12 kDa), while heavy chains have 3 (IgD, IgE, IgG) or 4 (IgA, IgM) constant domains of the same size. The light-heavy chain heterodimer is dimerized by disulfide bridges between heavy chains to give a four-chain Ig with two identical light and two identical heavy chains. Variable and constant domains are stabilized by internal –S–S– bridges in the flexible "hinge region" of the H chains.

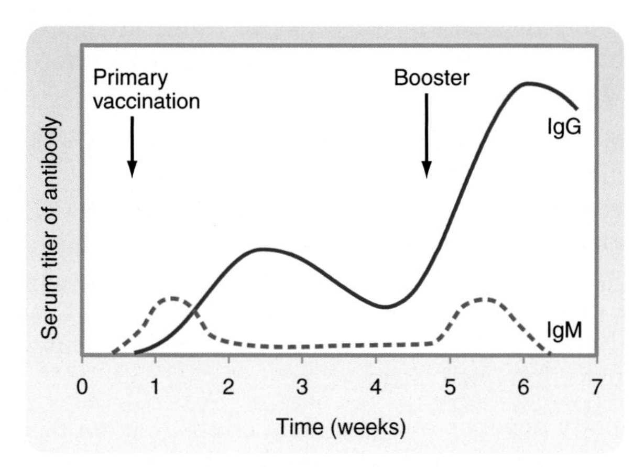

FIG. 6.11 Kinetics of the immune response. Simplified version of the kinetics of the mammalian humoral response to a "typical" foreign virus (or other) antigen, measured as titers of specific serum antibodies.

at **cross-linking large targets** (e.g., bacterial cell walls or flagella) but is probably less important in combating virus infections. In contrast, the production of **IgA** is very important for initial protection from virus infection. **Secretory IgA**—which is generally a **dimeric molecule** with attached **secretory component**—is produced at **mucosal surfaces** and results in "**mucosal immunity**," an important factor in preventing infection from occurring. Induction of mucosal immunity depends to a large extent on the way in which antigens are presented to and recognized by the immune system. Similar antigens incorporated into different vaccine delivery systems (see the section "Prevention and therapy of virus infection") can lead to very different results in this respect, and mucosal immunity is such an important factor that similar vaccines may vary considerably in their efficacy. **IgG,** which is always **monomeric, is probably the most important class of antibody** for direct neutralization of virions in serum and other body fluids into which it diffuses.

Direct virus **neutralization** by antibodies results from a number of mechanisms, including **conformational changes** in the virus capsid caused by antibody binding, or **blocking of the function** of the virus target molecule (e.g., **receptor binding**) by steric hindrance. A secondary consequence of antibody binding is **phagocytosis of antibody-coated ("opsonized") target molecules** by mononuclear cells or polymorphonuclear leukocytes (see earlier). This results from the presence of the **Fc receptor** on the surface of these cells, which binds the Fc or constant region of the Ig heavy chain. As has already been noted in Chapter 4, in some cases, **opsonization** or coating of virions by the binding of **nonneutralizing antibodies** can result in enhanced virus uptake. This has been shown to occur with **rabies virus**, and in the case of B may **promote uptake of the virus by macrophages**. Nonphagocytic cells can also destroy antibody-coated viruses via an **intracellular pathway involving the TRIM21 protein**. Antibody binding also leads to the activation of the ***complement cascade***, which assists in the neutralization of virions. **Structural alteration of virions** by complement binding can sometimes be visualized directly by electron microscopy. Complement is particularly important early in virus infection when limited amounts of **low-affinity antibodies** are made—complement enhances the action of these early responses to infection.

Despite all the above mechanisms, in overall terms **cell-mediated immunity** is probably more important than humoral immunity in the **control of virus infections**. This is demonstrated by the following observations:

- Congenital defects in cell-mediated immunity tend to result in **predisposition** to virus (and parasitic) infections, rather than to **bacterial** infections.
- The functional defect in **acquired immune deficiency syndrome (AIDS)** is a **reduction in the ratio of T-helper ($CD4^+$):T-suppressor ($CD8^+$) cells** from the normal value of about 1.2–0.2. AIDS patients commonly suffer many opportunistic virus infections—e.g., **herpes simplex virus** (HSV, now **HHV-1**), **cytomegalovirus** (CMV), and **Epstein-Barr virus** (EBV)—which may have been present before the onset of AIDS, but were previously suppressed by the intact immune system.

Cell-mediated immunity depends on three main effects (Fig. 6.12). These all act via molecular mechanisms that were explained earlier in this chapter (see the section "Viruses and apoptosis"):

- **Nonspecific cell killing**, mediated by "**natural killer**" (NK) cells (see above).
- **Specific cell killing**, mediated by **cytotoxic T-lymphocytes (CTLs)**.
- **Antibody-dependent** cellular cytotoxicity (**ADCC**).

CTLs are usually of **$CD8^+$** (suppressor) phenotype: that is, they express **CD8 molecules** as membrane proteins on their surface. CTLs are the **major cell-mediated immune response** to virus infections and are ***major histocompatibility complex*** **(MHC) restricted**—clones of cells recognize a specific antigen only when **a peptide from it is presented by MHC-I antigen** on the target cell to the **T-cell receptor/CD3 complex** on the surface of the CTL. MHC-I antigens are expressed on all nucleated cells in the body; MHC class II antigens are expressed only on the surface of the antigen-presenting cells of the immune system—**T-cells, B-cells,** and **macrophages**. CTL activity requires "help" (i.e., **cytokine production**) from **T-helper cells ($CD4^+$)**. The CTLs themselves recognize foreign antigens through the T-cell receptor/CD3 complex, which "docks" with antigen presented by MHC-I on the surface of the target cell (Fig. 6.12). The **mechanism of cell killing by CTLs** is similar to that of NK cells (explained above). The induction of a CTL response also results in the release of many different **cytokines**—important **cell signaling proteins**—from **T-helper cells (T_H)**, some of which result in **clonal proliferation of antigen-specific CTLs**, and others that have **direct antiviral effects**—for example, IFNs. The kinetics of the CTL response (peaking at about 7 days after infection) is somewhat slower than the NK response (e.g., 3–7 days, cf. 0.5–3 days)—so **NK cells and CTLs are complementary systems**.

The induction of a CTL response is dependent on recognition of **specific T-cell epitopes** by the immune system. These are distinct from the **B-cell epitopes** recognized by the humoral arm of the immune system. While B-cell epitopes (bound by antibodies acting as cell receptors) are generally either **short linear sequences** (5–7 aa) or **conformational epitopes** contributed to by several parts of a protein sequence, T-cell epitopes are **all linear sequences** that are bound into the constrained groove of **MHC Class I receptors for CD8+ T-cell recognition** (9–11 aa), or the "open" groove of **MHC Class II receptors for CD4+ T-cells** (9–22 aa). T-cell epitopes are more highly **conserved** (less variable) than B-cell epitopes, which are more able to mutate quickly to escape immune pressure. These

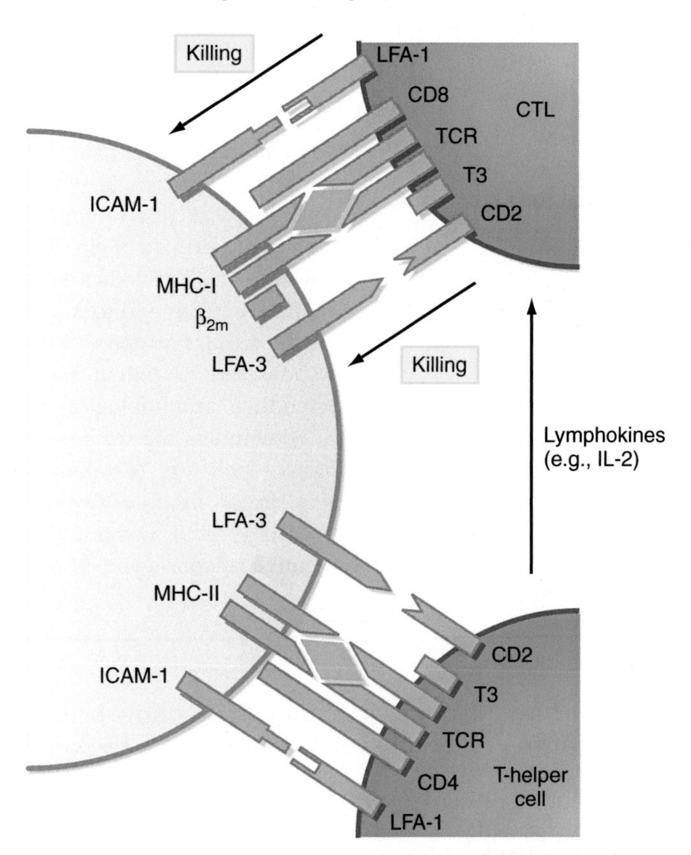

FIG. 6.12 Cell-surface proteins involved in immune recognition. Close contact between cells results in cell-to-cell signaling which regulates the immune response. The connections shown constitute "immune synapses," which regulated interactions between cells of the immune system much as neuronal synapses do between neurons.

are important considerations in the design of antiviral vaccines. The specificity of cell killing by CTLs is not absolute: although they are better "behaved" than NK cells, diffusion of **perforin** and local **cytokine production** frequently results in inflammation and **bystander cell damage**. This is a contributory cause of the pathology of many virus diseases (see Chapter 7), but the less attractive alternative is to allow virus replication to proceed unchecked (Fig. 6.12).

ADCC is less well understood than either of the two mechanisms mentioned above. ADCC can be carried out by **NK cells** or by **CTLs**. The mechanism of cell killing is the same as that described in the previous section, although **complement** may also be involved in ADCC. The distinguishing feature of ADCC is that this mechanism is dependent on the **recognition of antigen** on the **surface of the target cell** by means of an **antibody** bound on the surface of the **effector cell**. The antibody involved is usually **IgG**, which is bound to **Fc receptors** on the surface of the cytotoxic T-cell. ADCC therefore requires a preexisting antibody response and hence **does not occur early** during primary virus infections—it is part of the adaptive immune response. The overall contribution of ADCC

BOX 6.1

Collateral damage

We all walk around with a time bomb inside us: it's called your immune system. When it ticks away quietly in the background, we don't notice it, but when things go wrong ... it's very bad news. Your immune system has to keep working with Goldilocks precision—not too strong, not too weak—for decade after decade. As soon as a virus turns up and starts to take over your cells, your immune system has to show up right away (leave it a few days and it's probably too late), and it has to get it right every time. Fighting viruses is warfare and people get hurt—mostly you. Fever, muscle pain, headaches, vomiting, dead neurons in your brain or spinal cord: that's all due to your immune system working to rid you of circulating virions and infected cells. Antibodies work to mop up circulating antigen and to mark cells as targets for the cellular response, which is ultimately responsible for killing infected cells—and providing "immunological memory."

Sometimes the response is TOO vigorous—look up "cytokine storm," which is a factor in both severe influenza and COVID-19—but without the machinery to mount a response, you'd be dead long ago.

to the control of virus infections is not clear, although it is now believed that it plays a significant part in their control—especially where **neutralizing responses** are not elicited by the pathogen (Box 6.1).

Evasion of immune responses by viruses

In total, the many innate and adaptive components of the immune system present a powerful barrier to virus replication. Simply by virtue of their continued existence, however, it is obvious that viruses have, over millennia, evolved effective "counter-mechanisms" to the innate and adaptive immune responses in this molecular arms race.

Inhibition of MHC-I-restricted antigen presentation

As described above, CTLs can only respond to foreign peptides presented by **MHC-I** complexes on the target cell. A number of viruses interfere with MHC-I expression or function to disrupt this process and evade the CTL response. Such mechanisms include **downregulation of MHC-I** expression by **adenoviruses** and interference with the **antigen processing** required to form an MHC-I-antigen complex by **herpesviruses**.

Inhibition of MHC-II-restricted antigen presentation

The **MHC-II** receptors are essential in the adaptive immune response in order to stimulate the development of **antigen-responsive clones of effector cells**. Again, **herpesviruses** and **papillomaviruses** interfere with the processing and surface expression of MHC-II-foreign peptide complexes, inhibiting CD4+ cell (T_H cell) binding, and consequently both the CTL and humoral antibody response due to lack of T_H cell activation.

Inhibition of NK cell lysis

The **poxvirus** that causes the skin disease *Molluscum contagiosum* encodes a **homologue of MHC-I** that is expressed on the surface of infected cells but is **unable to bind an antigenic peptide**, thus avoiding killing by NK cells that would be triggered by the absence of MHC-I on the cell surface. Similar proteins are made by other viruses, such as **HHV-5 (CMV)**. **Herpesviruses** in general appear to have a number of sophisticated mechanisms to avoid NK cell killing.

Interference with apoptosis

See the section "**Viruses and apoptosis**" earlier in this chapter.

Inhibition of cytokine action

Cytokines are **secreted polypeptides** that coordinate important aspects of the immune response, including **inflammation, cellular activation, proliferation, differentiation,** and **chemotaxis**. Some viruses are able to inhibit the expression of certain chemokines directly. Alternatively, **herpesviruses** and **poxviruses** encode "**viroceptors**"—virus homologues of host **cytokine receptors** that compete with cellular receptors for cytokine binding but fail to give transmembrane signals. High-affinity binding molecules may also **neutralize cytokines directly**, and molecules known as "**virokines**" block cytokine receptors, again without activating the intracellular signaling cascade.

IFNs (see earlier) are cytokines which act as an effective means of curbing the worst effects of virus infections. Part of their wide-ranging efficacy results from their **generalized, nonspecific effects** (e.g., the inhibition of **protein synthesis** in virus-infected cells). This lack of specificity means that it is very difficult for viruses to evolve strategies to counteract their effects; nevertheless, there are instances where this has happened. The anti-IFN effect of **adenovirus VA RNAs** has already been described in Chapter 5. Other mechanisms of virus resistance to IFNs include:

- **EBV EBER RNAs** are similar in structure and function to the **adenovirus** VA RNAs. The EBNA-2 protein also blocks IFN-induced signal transduction.
- **Vaccinia virus (VV)** is known to show resistance to the antiviral effects of IFNs. One of the early genes of this virus, **K3L**, encodes a protein that is **homologous to eIF-2α**, which inhibits the action of **PKR**. In addition, the **E3L** protein also binds dsRNA and inhibits PKR activation.
- **Poliovirus** infection activates a **cellular inhibitor of PKR** in virus-infected cells.
- **Reovirus capsid protein σ3** is believed to **sequester dsRNA** and therefore prevent activation of PKR.
- **Influenza virus NS1 protein** suppresses IFN induction by blocking signaling through the Jak/STAT system.

Evasion of humoral immunity

Although direct humoral immunity is less significant than cell-mediated immunity, the **antiviral action of ADCC and complement** make this a worthwhile target to inhibit. The most frequent means of subverting the humoral response is by **high-frequency genetic variation** of the B-cell epitopes on antigens to which antibodies bind. This is only possible for viruses that are genetically highly variable (e.g., **influenza virus** and **HIV**). **Herpesviruses** use alternative strategies such as **encoding viral Fc receptors** to prevent Fc-dependent immune activation.

Evasion of the complement cascade

Poxviruses, herpesviruses, and some **retroviruses** encode mimics of normal regulators of **complement activation proteins** (e.g., secreted proteins that block **C3 convertase assembly** and accelerate its decay). **Poxviruses** can also inhibit **C9** polymerization, preventing membrane permeabilization.

RNA interference

RNA interference (**RNAi**) is a **posttranscriptional gene silencing process** that occurs in eukaryotic organisms from yeast to humans—and represents another adaptive immune system for eukaryotes, somewhat akin to CRISPR/Cas in prokaryotes in its use of **derived RNA sequences** to recognize "foreign" RNA. While acclaim and Nobel Prizes followed the description of the system in invertebrates, the basic principles of the system of silencing had in fact been worked out in plants using RNA plant viruses years earlier. These principles included the **sequence specificity** of the process, and the involvement of **dsRNA**, and amplification mechanisms for magnifying the signal. It is now well accepted that gene silencing by RNAi is a major defense system against **viruses** and **mobile genetic elements** (MGEs) in many eukaryotes without an adaptive immune system as mammals have, including fungi, plants, and invertebrates such as *Drosophila* sp. and the **nematode** *Caenorhabditis elegans*. The situation with mammals is less clear, however, as will be discussed below.

The pathways that generate and use RNAi all have the following attributes:

- Generation of **20–30 nt ssRNAs** from various forms of **dsRNA.**
- **Sequence-specific** complementary binding between these and "**target**" RNA sequences.
- Use of "**Argonaute family**" (**Ago**) proteins.

The sources of dsRNA can be widely different—foreign (=**viral**) **dsRNA genomic** segments or **replicative forms**, "**hairpin**" folds in host or viral ssRNA, partially base-paired **sense- and antisense transcripts** from one area of dsDNA. The **outcome** of siRNA generation can also vary, from **cleavage** and degradation of **target RNA**, to **modification** of gene expression, to long-term **expression changes** through **epigenetic** modification. Given that the components of RNAi pathways are distributed **very widely** across **eukaryotes**, it is possible that the **LECA** (last eukaryote common ancestor, see Chapter 3) **had a functional RNAi pathway**, some billion or so years ago.

RNAi as an adaptive antiviral immune system in plants and insects

It is obvious that in these eukaryotes, RNAi is used as a virus and a MGE defense. In these organisms, **dsRNA** is specifically recognized by a RNase-III family **ribonuclease ("Dicer")** and cut or diced into "**short interfering RNAs**" (**siRNA**; **viRNA** if derived from viral dsRNA) 21–24 nt in length. These are **dsRNAs** with 2-nt 3′ overhangs. These are then loaded into an **Ago-containing** complex also known as **RISC**, or RNA-induced silencing complex, and **one strand is degraded**. The RISC then **anneals to mRNA** or other ssRNA that has the complementary sequence to the viRNA, and "**slices**" it within the bound sequence. This is the **core siRNA pathway**: however, many organisms are also able to **amplify si/viRNAs** using a **host RdRp**, which in **plants** is **recruited by the RISC complex** to synthesize long dsRNAs which are then processed into cognate **si/viRNAs** by a **Dicer**. In **nematodes**, the RdRp binds the **si/viRNA-target duplex** and **synthesizes secondary siRNAs** without added components being necessary.

RNAi made in one cell of both plants and nematodes **can move to adjacent cells** and also be **trafficked more widely**. In plants, this would be via the **plasmodesmata**, and also the **phloem-associated** transport network (see earlier): in *Arabidopsis* 21-nt siRNAs can move cell-to-cell for up to **15 cells** away from the generating cell; it appears as though **dsRNA** may be loaded into **phloem cells** for longer-range transport. Nematodes are thought to have a **passive channel** between cells for RNA movement. This kind of transport means that **RNA silencing** can occur in cells **far removed** from the ones originally infected—which accounts for how plants may generate **new tissue** adjacent to virus infected areas, **without any detectable virus** in it. This is known as a "**recovery phenotype.**"

While RNAi was shown to have an antiviral function in plants by the late 1990s, its relevance in **insects** was limited to cultured cells until 2004, when it was shown to be part of a natural defense system in *Anopheles gambiae* **mosquitoes** against the **ssRNA+ togavirus** O'Nyong-nyong virus. By 2008 RNAi had also been shown to be an antiviral defense in **fungi**.

Given that a virus-host "arms race" has been going on in land plants and terrestrial insects and other arthropods and fungi for at least 450 million years (see Fig. 3.11), it is natural to wonder whether or not viruses have evolved **antisilencing defenses**. It is increasingly obvious that they have, and that these counter-measures are highly varied. **Viral suppressors of RNAi (VSRs)** are almost universal among RNA and indeed **ssDNA** viruses of plants—with some viruses having **multiple VSRs** (e.g., **closteroviruses**). The first **animal virus** VSR discovered was the **B2 protein** of the ssRNA+ **nodavirus** Flock house virus—which was initially shown to suppress RNAi in plants, which FHV can infect, before it was shown to work in the beetle host too. While many VSRs are **dsRNA-binding proteins**—presumably to sequester dsRNA away from **Dicer** complexes—others target Dicer (**potyvirus** HC-Pro, **FHV** B2), or **Ago proteins** (**cucumovirus** 2b; **polerovirus** P0), and some others may even be RNA molecules that interfere in viRNA binding in RISC complexes. Other mechanisms include blocking long-distance transport of siRNAs in plants (potexvirus P25 protein), or encoding a **transcriptional regulator** that may **modify expression of host genes** responsible for RNAi activity (ssDNA **begomovirus**, AC2 ORF protein).

Given such evidence of the varied involvement of RNAi in both normal eukaryote cell function as well as in defense against viruses, researchers were quick to apply **recombinant RNAi technology** for use in studying **gene function and regulation** in many organisms—generally by transient expression of **synthetic siRNAs** targeting **specific gene sequences**—as well as employing it to **engineer resistance to viruses**. In one of the simplest manifestations of this, it has been shown that **synthetic dsRNA** derived from **insect** gene sequences can be applied **topically to plant leaves** ("Spray Induced Gene Silencing"), and can either be taken up directly by insects, or **get into plant cells** and then be taken up via piercing mouthparts of **feeding insects**. The dsRNAs are processed in the insect into **siRNAs**, and then target **mRNAs of essential genes** such as acetylcholinesterase or actin (Cagliari et al., 2019) (Fig. 6.13).

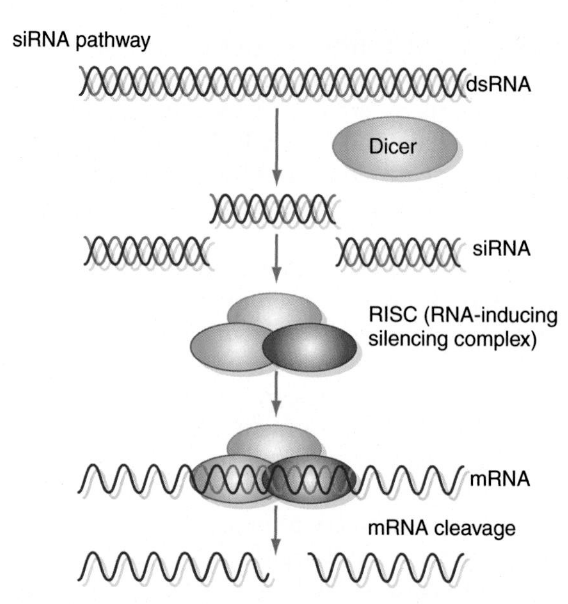

FIG. 6.13 Mechanism of RNAi action. siRNAs have base complementarity to their target RNA molecules. The resulting double-stranded RNAs are processed by various enzymes, notably Dicer, to produce a complex (RISC) which carries out cleavage of the target mRNAs.

RNAi in mammals

In mammals, small RNAs include **small interfering RNAs (siRNAs)** and **miRNAs**. siRNAs, with perfect base complementarity to their targets, activate **RNAi-mediated cleavage** of the target **mRNAs**, while miRNAs generally induce **RNA decay and/or translation inhibition** of target genes.

While there has been some skepticism about the existence of viRNAs in **mammals**, as they did not seem to be made anywhere except in **stem cells** or in **mouse embryos**, evidence is accumulating that RNAi does play some role in antiviral defenses in mammals. While this is mainly limited to **cultured cells** from humans, mice and hamsters, **viRNAs** have also been found in **suckling mice**. Moreover, **VSRs** have been found in many **mammalian viruses**: these include **influenza A** (NS1) and **Ebolavirus** (VP35) proteins that are **dsRNA-binding**, and suppress both **IFN** activation as well as the **RNAi** pathway. **SARS-CoV N protein** binds dsRNA, and is also an IFN antagonist; **vaccinia virus** VP55 **polyadenylates miRNAs** (see below), causing their degradation. It is speculated that VSRs may in fact **mask antiviral RNAi** in mammals, for which there is evidence from experiments with VSR-deficient viruses (**human enterovirus 71** with mutated **3A protein**): in this case, the mutated virus had severe replication defects, which were **partially rescued** in cells that **did not express Dicer**. It is speculated that RNAi may be the antiviral defense system that is preferred over the IFN response in **stem cells**, which is why siRNAs are more readily found in them. This could be because IFN may interfere with maintenance of the **pluripotent state** in stem cells, while the RNAi machinery may be important for maintenance of stem cell properties—especially as **MGE activity** is high in rapidly-dividing undifferentiated cells, and **RNAi is involved in MGE suppression** as well (Schuster et al., 2019).

While viRNAs may not be regarded as being important in developed animals, **miRNAs** most certainly are. Mammals, including humans, encode **hundreds or thousands of miRNAs**. Some **viruses** with eukaryotic hosts also encode miRNAs. **Herpesviruses** in particular encode multiple miRNAs; most other nuclear DNA viruses encode one or two miRNAs. RNA viruses and cytoplasmic DNA viruses appear to lack any miRNAs. Virus miRNAs may serve two major functions. Several have been shown to **inhibit the expression of cellular factors** that play a role in cellular **innate** or **adaptive** antiviral immune responses, so reducing the effectiveness of the immune response. Alternatively, virus miRNAs may downregulate the expression of virus proteins, including key immediate-early or early regulatory proteins. In **HSV**, miRNAs are expressed at high levels during **latency**, but not during productive replication, so their action is thought to **stabilize latency**.

In mammals miRNA is a powerful **regulator of gene expression**, including virus genes. Many viruses use miRNA to control their own gene expression and that of their host cells. On infection of a host cell, viruses encounter a range of miRNA species, many of which have been shown to **restrict virus gene expression**. Thus they have had to evolve a range of mechanisms to evade miRNA restriction is the same way that they have evolved other mechanisms to mitigate the impact of innate immunity. These include:

- **Blocking** miRNA function
- **Avoiding** 3′UTR targets complementary to cellular miRNAs
- Evolving **very short** 3′UTRs
- Evolving **structured** 3′UTRs

RNAi expression can be induced by dsRNA, and this approach has been used to investigate gene function in a variety of organisms including plants and insects. However, this method cannot be applied to mammalian cells as **dsRNAs longer than 30 nt** induce the **IFN response** (see earlier), which results in the degradation of mRNAs and causes a **global inhibition of translation**. To circumvent this problem, chemically synthesized siRNAs or plasmid vectors manipulated to produce **short hairpin RNA** molecules can be used to investigate gene function in mammals. In the future it may be feasible to treat virus diseases by shutting off gene expression by **directing the degradation of specific mRNAs**, and many clinical trials are currently underway. Although RNA interference has been used widely in cultured cells to inhibit virus replication and to probe biological pathways, considerable problems must be overcome before it becomes a useful **therapy**, including the development of suitable delivery and targeting systems and solving the issue of stability in vivo.

Virus-host interactions

Viruses do not set out to kill their hosts. Virus pathogenesis is an abnormal situation of no value to the virus—**the vast majority of virus infections are asymptomatic**. However, for pathogenic viruses, a number of critical stages in replication determine the nature of the disease they produce. For all viruses, pathogenic or nonpathogenic, the first factor that influences the course of infection is the **mechanism and site of entry into the body** (Fig. 6.14):

- **The skin:** Mammalian skin is a highly effective barrier against viruses. The outer layer (epidermis) consists of dead cells and therefore does not support virus replication. Very

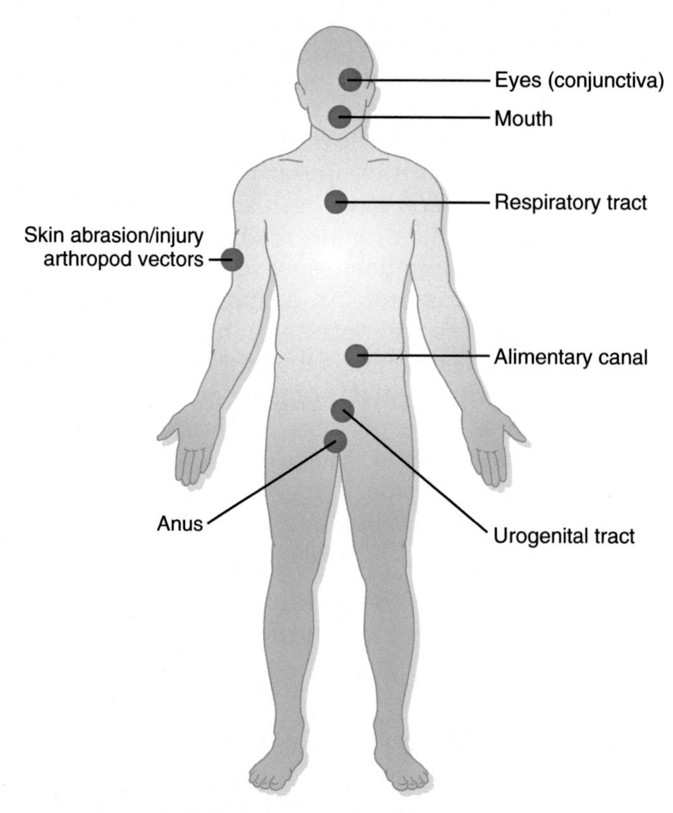

FIG. 6.14 Sites of virus entry into the human body. The course a virus infection follows depends on the biology of the virus and the response to infection by the host, but is also influenced by the site at which the virus enters the body. The diagram depicts the most common sites of access to the human body by viruses.

few viruses infect directly by this route unless there is **prior injury** such as minor trauma or **puncture of the barrier**, such as insect or animal **bites** or subcutaneous **injections.** Some viruses that do use this route include **HSV** and **papillomaviruses**, although these viruses probably still require some form of disruption of the skin such as small abrasions or eczema.

- **Mucosal membranes:** The "external" mucosal membranes of the **eye** and **genitourinary** (GU) tract are much more favorable routes of access for viruses to the tissues of the body. This is reflected by the number of viruses that can be sexually transmitted (e.g., **HIV, HPVs, human herpesvirus 2, HBV, HCV**); virus infections of the **eye** are also quite common (e.g., **HHV-1**; Table 6.3).
- **Alimentary canal:** Viruses may infect the alimentary canal via the **mouth, oropharynx, gut,** or **rectum**, although viruses that **infect the gut via the oral route** must survive passage through the **stomach**—an extremely hostile environment with a **low pH (pH 1.5–3.5)**—and then the **small intestine**, with high concentrations of **digestive enzymes**. Nevertheless, the **gut** is a highly valued prize for viruses—the **intestinal epithelium** is constantly replicating and a good deal of **lymphoid tissue** is associated with the gut

TABLE 6.3 Viruses and their routes of infection.

Virus	Site of infection
Mucosal surfaces	
Adenoviruses	Conjunctiva
Picornaviruses—enterovirus 70	Conjunctiva
Papillomaviruses	Genitourinary tract
Herpesviruses	Genitourinary tract
Retroviruses—HIV, human T-cell leukemia virus (HTLV)	Genitourinary tract
Internal mucosal membranes	
Herpesviruses	Mouth and oropharynx
Adenoviruses	Intestinal tract
Caliciviruses	Intestinal tract
Coronaviruses	Intestinal tract
Picornaviruses—enteroviruses	Intestinal tract
Reoviruses	Intestinal tract
Respiratory tract	
Adenoviruses	Upper respiratory tract
Coronaviruses	Upper respiratory tract
Orthomyxoviruses	Upper respiratory tract
Picornaviruses—rhinoviruses	Upper respiratory tract
Paramyxoviruses—parainfluenza, respiratory syncytial virus	Lower respiratory tract

which provides many opportunities for virus replication. In fact, it has recently been realized that the **gut-associated lymphoid tissue (GALT)** is possibly the largest collection of such tissue in the mammalian body. Moreover, the constant intake of food and fluids provides ample opportunity for viruses to infect these tissues (Table 6.3). To counteract this problem, the gut has many **specific** (e.g., secretory antibodies) and **nonspecific** (e.g., stomach acids and bile salts) defense mechanisms.

- **Respiratory tract:** The respiratory tract is probably the **most frequent site of virus infection**. As with the gut, it is constantly in contact with external virions which are taken in during respiration. As a result, the respiratory tract also has defenses aimed at virus infection—**filtering** of particulate matter in the sinuses and the presence of **cells and antibodies** of the immune system in the lower regions. Viruses that infect the respiratory tract usually come directly from the respiratory tract of others, as **aerosol spread**—the main route of transmission of **SARS-CoV-2**, and also of **measles virus**—is very efficient: the old adage that "**coughs and sneezes spread diseases**" has never been reinforced as much recently as by the example of SARS-CoV-2 (Table 6.3).

The natural environment is a considerable barrier to virus infection. Most viruses are relatively sensitive to **heat, drying, UV light** (sunlight), and other environmental factors, although a few types are quite resistant to them. **TMV**, for example has a **$T_{1/2}$ of 30 min at 80 °C**; many plant viruses and some insect viruses are tolerant to their virions being **dried out**, and many (generally nonenveloped) virions can be **lyophilized** for storage. This is particularly important for viruses that are spread via **contaminated water** or **foodstuffs**—not only must they be able to survive in the environment until they are ingested by another host, but, as most are spread by the **fecal-oral route**, they must also be able to pass through the **stomach** to infect the **gut** before being shed in the **feces**.

One way of overcoming environmental stress in between hosts is to take advantage of a **secondary vector** for transmission between the **primary hosts** (Fig. 6.15). As with plant viruses, the virus **may or may not replicate** while in the vector. Viruses without a secondary vector must rely on **continued host-to-host transmission** and have evolved various strategies to do this (Table 6.4):

- **Horizontal transmission:** The direct **host-to-host** transmission of viruses. This strategy relies on a **high rate of infection** to maintain the virus population, as is the case with **measles virus, influenza** and **common cold viruses,** and **SARS-CoV-2**.
- **Vertical transmission:** The transmission of the virus from one generation of hosts to the next. This may occur by **infection of the fetus** before, during, or shortly after birth (e.g., during **breastfeeding**). More rarely, it may involve direct transfer of the virus **via the**

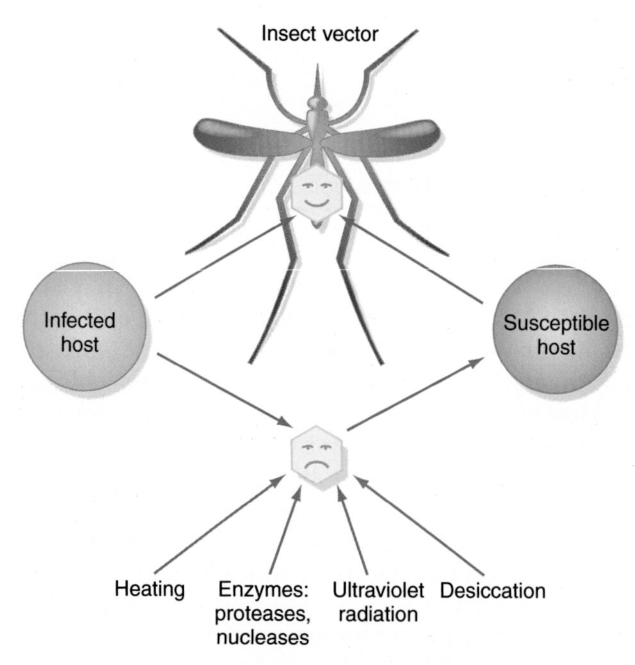

FIG. 6.15 Transmission of viruses through the environment. Some viruses have adopted the use of vectors such as insects or other arthropods to avoid environmental stresses when outside their host organism. This may or may not involve replication of the virus in the vector, depending on the virus and the vector.

TABLE 6.4 Virus transmission patterns.

Pattern	Example
Horizontal transmission	
Human-human (aerosol)	Influenza, SARS-CoV-2, measles
Human-human (fecal-oral)	Rotaviruses, polioviruses, hepatitis A
Animal-human (direct)	Rabies (bite), Nipah virus (contamination of palm sap by bats feeding)
Animal-human (arthropod vector)	Bunyaviruses (tick, mosquito); flaviviruses (mosquitoes)
Vertical transmission	
Placental-fetal	Rubella, CMV
Mother-child (birth)	HSV, HIV
Mother-child (breastfeeding)	HIV, HTLV
Germ line	In mice, retroviruses; in humans (?)

germ line itself (e.g., **retroviruses**). In contrast to horizontal transmission, this strategy relies on **long-term persistence** of the virus in the host rather than rapid propagation and dissemination of the virus.

Having gained entry to a potential host, the virus must initiate an infection by **entering** a susceptible cell (**primary replication**). This initial interaction frequently determines whether the infection will remain **localized** at the site of entry or spread to become a **systemic infection** (Table 6.5). In some cases, virus spread is controlled by infection of **polarized epithelial cells** and the preferential **release** of virus from either the **apical** (e.g., **influenza virus**—a localized infection in the upper respiratory tract) or **basolateral** (e.g., **rhabdoviruses**—a systemic infection) surface of the cells (Fig. 6.16).

Following primary replication at the site of infection, the next stage may be spread throughout the host. In addition to **direct cell-cell contact**, there are two main mechanisms for spread throughout the host. These are:

- **Via the bloodstream:** Virions may get into the bloodstream by **direct inoculation**—for example, by **arthropod vectors, blood transfusion**, or **intravenous drug use** (sharing of nonsterilized needles). The virus may travel **free in the plasma** (e.g., **togaviruses, enteroviruses**) or in association with **red cells (orbiviruses), platelets (HSV), lymphocytes (EBV, CMV)**, or **monocytes (lentiviruses)**. **Primary viremia**—the first appearance of virions in the blood—usually precedes and is necessary for the spread of virions to other parts of the body via the bloodstream, and is followed by a more generalized, **higher titer secondary viremia** as the virus reaches the other target tissues or replicates directly in **white blood cells** (mammals) or all blood cells (birds).
- **Via the nervous system:** As above, spread of virus to the nervous system is usually preceded by primary viremia. In some cases, spread occurs directly by contact with neurons at the primary site of infection; in other cases, it occurs via the bloodstream. Once in peripheral nerves, the virus can spread to the central nervous system (CNS) by

TABLE 6.5 Examples of localized and systemic virus infections.

Virus	Primary replication	Secondary replication
Localized infections		
Papillomaviruses	Dermis	–
Rhinoviruses	Upper respiratory tract	–
Rotaviruses	Intestinal epithelium	–
Systemic infections		
Enteroviruses	Intestinal epithelium	Lymphoid tissues, CNS
Herpesviruses	Oropharynx or GU tract	Lymphoid cells, CNS
Measles virus	Upper respiratory tract	Lymphoid tissue, skin

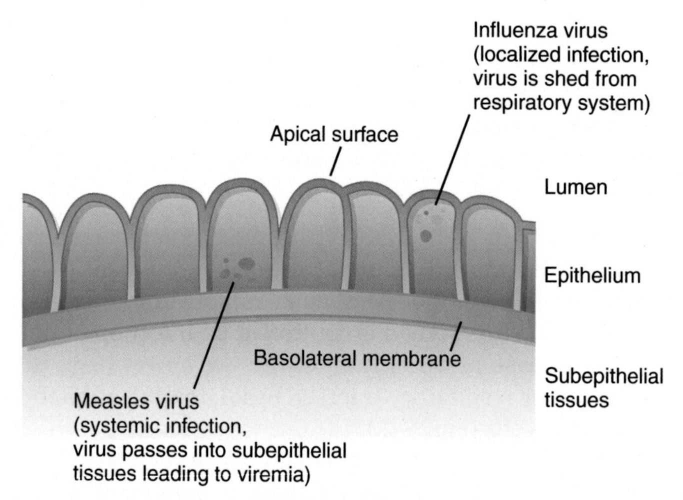

FIG. 6.16 Virus infection of polarized epithelial cells. Some viruses which infect epithelial cells are released from the apical surface (e.g., influenza virus) while others are released from the basolateral surface of the cells (e.g., rhabdoviruses). This affects the way in which the virus spreads through the body and the subsequent course of the infection: apical release puts virions back into the environment they came from (e.g., respiratory passages); basolateral release puts them into tissue and the circulation underlying the epithelium.

axonal transport along **neurons, or "nerve cells."** The classic example of this is **HSV** (see the section "Latent infection"). Viruses can cross **synaptic junctions** as these frequently contain virus **receptors,** allowing the virus to jump from one cell to another.

The spread of the virus to various parts of the body is controlled to a large extent by its favored cell type, or its **tissue tropism**. Tissue tropism is controlled partly by the route of infection, but largely by the interaction of a **virus-attachment protein** (VAP) with a specific receptor molecule on the surface of a cell (as discussed in Chapter 4) and has considerable effect on pathogenesis.

At this stage, following significant virus replication and the production of virus antigens, the **host immune response** comes into play: this obviously has a major impact on the outcome of an infection. To a large extent, the efficiency of the immune response determines the amount of **secondary replication** that occurs and, hence, the spread to other parts of the body. If a virus can be prevented from reaching tissues where secondary replication can occur, **generally no disease results**, although there are some exceptions to this. The immune response also plays a large part in determining the amount of **cell and tissue damage** that occurs as a result of virus replication. As described above, the production of IFNs is a major factor in preventing virus-induced tissue damage.

The immune system is not the only factor that controls cell death, the amount of which varies considerably for different viruses. Viruses may replicate widely throughout the body **without any disease symptoms** if they **do not cause significant cell damage or death**. **Retroviruses** do not generally cause cell death, for example, as they are released from the cell by **budding** rather than by **cell lysis**, and can cause **persistent infections**, even being passed vertically to the offspring if they infect the germ line. All vertebrate genomes, including humans, are littered with retrovirus **genomes** (ERVs) that have been with us for millions of years (Chapter 3). At present, these ancient ERVs are not firmly linked to any common disease in humans, although there are examples of tumors caused by them in rodents, and there is increasing evidence of their involvement in a variety of human diseases. Conversely, **picornaviruses** cause lysis and death of the cells in which they replicate, leading to **fever and increased mucus secretion**, in the case of **rhinoviruses**, and **paralysis or death** in the case of polioviruses. Death in polio cases is usually due to **respiratory failure** due to damage to the CNS, resulting in part from virus replication in these cells, as well as damage due to the immune response. This type of disease was responsible for the mass hospitalization of children in the mid-1900s in "**iron lungs,**" or tank ventilators, to enable them to breathe.

The eventual outcome of any virus infection depends on a balance between two processes. **Clearance** is mediated by the immune system (as discussed previously); however, the virus is a moving target that responds rapidly to pressure from the immune system by **altering its antigenic composition** (whenever possible) by **mutation**. The classic example of this phenomenon is **influenza virus**, which displays two genetic mechanisms that allow the virus to alter its antigenic constitution. These are:

- **Antigenic drift:** This involves the gradual accumulation of minor mutations (e.g., nucleotide substitutions) in the virus genome which result in subtly altered coding potential and therefore altered antigenicity, leading to decreased recognition by the immune system. This process occurs in all viruses all the time but at greatly different rates; for example, it is much more frequent in RNA viruses than in DNA viruses. In response, the immune system constantly adapts by recognition of and response to novel antigenic structures—but it is always one step behind. In most cases, however, the immune system is eventually able to overwhelm the virus, resulting in clearance.
- **Antigenic shift:** In this process, a sudden and dramatic change in the antigenicity of a virus occurs owing to reassortment of the segmented virus genome with another genome of a different antigenic type (see Chapter 3). This results initially in the failure of the immune system to recognize a new antigenic type, giving the virus the upper hand (Fig. 6.17).

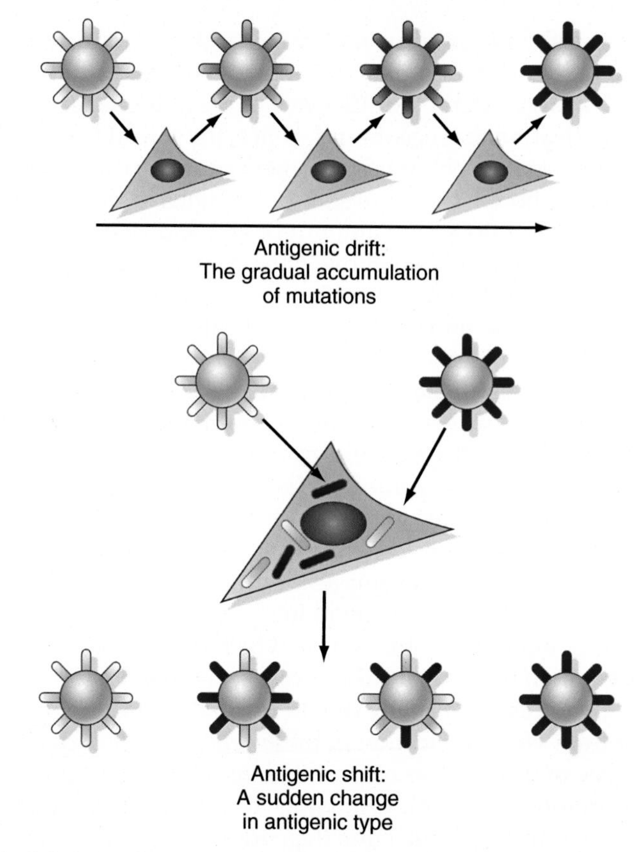

FIG. 6.17 Antigenic shift and drift in influenza viruses. Variation in the antigenicity of influenza viruses occurs through two mechanisms: these are gradual antigenic drift and sudden antigenic shifts. These are shown by gradual color shift in the HA/NA spikes with passage by the virions at the top of the figure, and by the acquisition of new HA/NA spikes by virions after coinfection with another strain of virus, at the bottom.

The occurrence of past **antigenic shifts** in influenza virus populations is recorded by analysis of **historical pandemics** (worldwide epidemics; Fig. 6.18). These events are marked by the sudden introduction of a **new antigenic type of HA and/or NA** into the circulating virus, or **complete replacement** of the circulating virus, as possibly happened in 1918 (see Chapter 1): this overcomes previous immunity in the human population. Previous HA/NA types become **resurgent** when a sufficiently high proportion of the people who have "immunological memory" of that type have died, or are outnumbered by new generations, thus overcoming the effect of "**herd immunity**." This is possibly what happened in 1977, when **influenza A H1N1** re-emerged into the human population, after being absent since its replacement by **H2N2** in 1957 (possibly with human help).

The other side of the relationship that determines the eventual outcome of a virus infection is the **ability of the virus to persist in the host**. Long-term persistence of viruses results from two main mechanisms. The first is the **regulation of lytic potential**. The strategy followed here is to achieve the continued survival of a **critical number** of virus-infected cells, so that

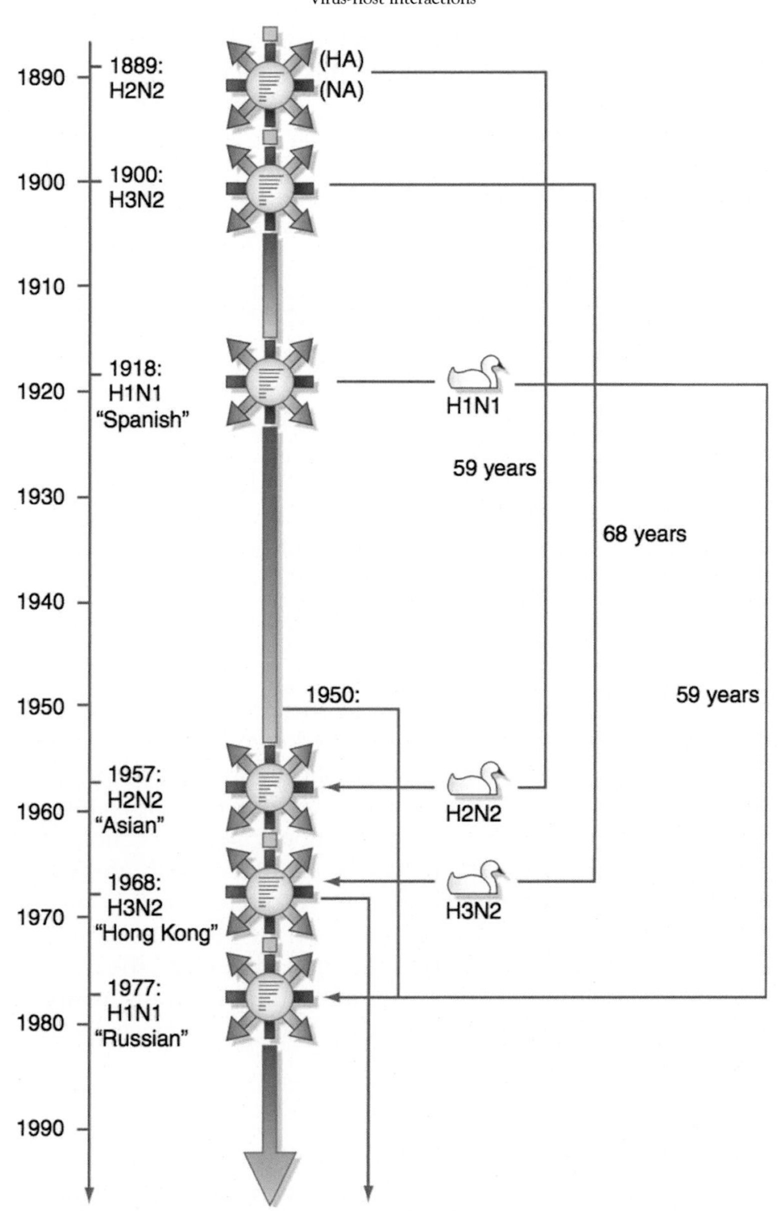

FIG. 6.18 Historical influenza pandemics. This chart shows the history of influenza pandemics throughout the 20th century, with subtype (e.g., H2N2, H3N2) shown where this is known. The eight genome segments are shown inside virions: several of these usually remain in the new virus after reassortment, although complete replacement cqan also occur (e.g., 1918 Flu, that probably emerged as is from birds). The first pandemic of the 21st century occurred in 2009 and was caused by an H1N1 type virus emerging out of pigs, although this was not as damaging as earlier pandemics. This was possibly due to a high proportion of people have partial immunity to it because of the continuous circulation of H1N1 since 1977.

this is **sufficient to continue the infection** without killing the host organism. For viruses that do not usually kill the cells in which they replicate, this is not usually a problem; hence, these viruses tend naturally to cause persistent infections (e.g., **retroviruses**). For viruses that undergo **lytic infection** (e.g., **herpesviruses**), it is necessary to develop mechanisms that **restrict virus gene expression** and, consequently, cell damage. The second aspect of persistence is the evasion of immune surveillance, discussed above.

The course of virus infections

Patterns of virus infection can be divided into a number of different types.

Abortive infection

Abortive infection occurs when a virus infects a cell (or host) but cannot complete the full replication cycle, so this is a **nonproductive infection**. The outcome of such infections is not necessarily insignificant: for example, **SV40** infection of nonpermissive **rodent cells** sometimes results in **transformation** of the cells (see Chapter 7).

Acute infection

This pattern is familiar for many common virus infections (e.g., "colds"). In these **relatively brief infections**, the virus is usually **eliminated completely** by the immune system. Typically, in acute infections, much of the virus replication occurs before the onset of any symptoms (e.g., fever), which are the result not only of virus replication but also of the **activation of the immune system**; therefore, acute infections present a serious problem for the epidemiologist and are the pattern most frequently associated with **epidemics** (e.g., influenza, measles). In this sort of disease outbreak, there is rapid spread to susceptible people, which can become **exponential** if there is no immunity in the population, until a great enough number have recovered that **herd immunity** comes into play. Here, the chances of infecting susceptibles goes down as more and more are immune, until the transmission rate or **reproduction number** (R_0) drops below sustainable levels: that is, one infected person on average infects fewer than one susceptible ($R_0 < 1$).

Chronic infection

These are the converse of acute infections (i.e., they are **prolonged** and **stubborn**). To cause this type of infection, the virus must **persist** in the host for a significant period. To the clinician, there is no clear distinction among **chronic**, **persistent**, and **latent** infections, and the terms are often used interchangeably. They are listed separately here because to virologists, there are significant differences in the events that occur during these infections. We use the term chronic to mean infections characterized by the continued presence of infection and of infectious virions long after **primary infection,** and also the possibility of **chronic** or **recurrent** disease. Hepatitis B virus (HBV) infection provides an interesting case study for a chronic infection. This blood-borne and sexually-transmitted virus has infected up to **one-third** of all people presently alive, with most (up to ~**150 million per annum**) having acute disease, with the liver as the site of secondary replication, which clears after several months. However, a proportion of infected people go on to develop chronic infections, which may last **lifelong**: in 2017, up to **390 million** people were estimated to be chronically infected, and carriers of the live virus. While many who are chronically infected are **asymptomatic, 10%** or so of the chronically-infected can go on to develop **hepatocellular carcinoma**, or liver cancer—and approximately **300,000** of these die every year. This is largely as a result of ongoing damage to the cells making up the liver by **CTLs**, which kill infected cells and stimulate a **chronic inflammatory state** that worsens the damage. The need for prevention of infection by this virus becomes starkly apparent when one considers that **90% of babies** who get infected at or near the time of birth develop chronic hepatitis B infections, while the figure for those infected after the age of five is **less than 10%.**

Other viruses that establish chronic infections in humans include hepatitis C (HCV; see the section "Chemotherapy of virus infections") and HIV-1 and -2. A recent investigation has shown that in these two cases, there are molecular signatures of immune dysfunction in chronically infected people that are similar to those seen in aging. One feature of chronic infections is a state of **chronic inflammation**, which is highly disruptive of the immune system: this is also seen in aged people, and plays an important role in diseases associated with aging. Features that are shared include T-cell "**memory inflation,**" upregulation of **intracellular signaling pathways** of inflammation, and **reduced sensitivity** to **cytokines** in lymphocytes and other cells of the immune system. While people infected with HIV-1 do not show recovery of normal immune responses even after a decade of virus suppression with antiretrovirals, if HCV is **cured** (with sofosbuvir; see below), there is **partial recovery of sensitivity** to interferon-A (see References).

Persistent infection

These infections result from a delicate balance between the virus and the host organism, in which ongoing virus replication occurs but the virus adjusts its replication and pathogenicity to avoid killing the host. In chronic infections, the virus is usually eventually cleared by the host (unless the infection proves fatal), but in persistent infections the virus may continue to be present and to replicate in the host for its entire lifetime.

The best-studied example of such a system is **lymphocytic choriomeningitis virus** (LCMV; an **arenavirus**) infection in mice (Fig. 6.19). Mice can be experimentally infected with this virus either at a peripheral site (e.g., a footpad or the tail) or by **direct inoculation into the brain (CNS inoculation)**. Adult mice infected in the latter way are killed by the infection, but among those infected by a peripheral route there are two possible outcomes to the infection: some mice die but others **survive**, having **cleared** the virus from the body completely. It is not clear what factors determine the survival or death of LCMV-infected mice, but the outcome is related to the immune response to the virus. In **immunosuppressed** adult mice infected via the CNS route, a **persistent infection** is established in which the virus is not cleared (due to the nonfunctional immune system), but, remarkably, these mice **are not killed** by the virus. If, however, **syngeneic** LCMV-specific T-lymphocytes (i.e., of the same MHC type) are injected into these persistently

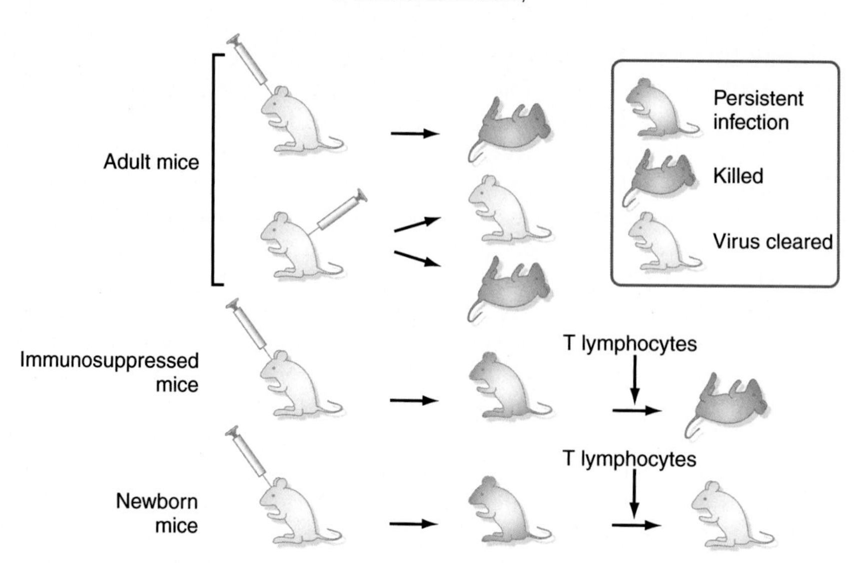

FIG. 6.19 Persistent infection of mice with LCMV. Lymphocytic choriomeningitis virus (LCMV) is an arenavirus. The course of infection in mice depends in part on the immune response of the host to the virus, and the route of infection.

infected mice, the animals develop the full pathogenic symptoms of LCMV infection and die. When newborn mice with immature immune systems are infected via the CNS route, they also develop a persistent infection. However, if they are subsequently injected with syngeneic LCMV-specific T-lymphocytes, they **clear the virus** and survive the infection. The mechanisms that control these events are not completely understood, but evidently there is a delicate balance between the virus and the host animal, and the immune response to the virus is partly responsible for the **pathology** of the disease and the death of the animals.

Not infrequently, persistent infections may result from the production of **defective-interfering (DI) particles** (see Chapter 3). Such particles contain a **partial deletion** of the virus genome and are **replication defective**, but they are maintained and may even tend to accumulate during infections because they can **replicate in the presence of replication-competent helper virus**. The production of DI particles is a common consequence of virus infection of animals, particularly by RNA viruses, but also occurs with DNA viruses and plant viruses and can be mimicked in vitro by continuous **high-titer passage** of virus. Although not able to replicate themselves independently, DI particles are not necessarily genetically inert, and may alter the course of an infection by **recombination** with the genome of a replication-competent virus. The presence of DI particles can profoundly influence the course and the outcome of a virus infection. In some cases, they appear to **moderate pathogenesis**—possibly by competing with viable genomes for replication machinery—whereas in others they **potentiate it**, making the symptoms of the disease much more severe. Moreover, as DI particles effectively cause restricted gene expression (because they are genetically deleted), they may also result in a persistent infection by a virus that normally causes an acute infection and is rapidly cleared from the body.

Latent infection

When it is in a **latent state**, the virus is able to **downregulate** its gene expression and enter an inactive state with strictly limited gene expression and **without ongoing virus replication**. Latent virus infections typically persist for the **entire life** of the host. Examples of such infection in humans are **HSV** and other **herpesviruses**. Infection of **sensory nerves** serving the mucosa during localized primary replication results in the virus travelling **via axon transport mechanisms** further into the nervous system. There, it "hides" or becomes latent in **dorsal root ganglia**, such as the **trigeminal ganglion** next to the human ear, establishing a truly latent infection. The nervous system is an **immunologically privileged site**: it is isolated away from and is not patrolled by mobile cells of the immune system in the same way as the rest of the body, so there is no ongoing virus clearance. The major factor in HSV latency, however, is the ability of the virus to restrict its gene expression. This eliminates the possibility of recognition of infected cells by the immune system via MHC-I presentation. Restricted gene expression is achieved by tight regulation of **α-gene expression** (see Chapter 5), which is an essential control point in herpesvirus replication. In the latent state, HSV makes an 8.3 kb RNA transcript called the **latent RNA** or **latency-associated transcript (LAT)**. The LAT is broken down into even smaller strands called **microRNAs (miRNAs)**, and these block the production of proteins which reactivate the virus. Drugs which block production of these miRNAs could in theory reactivate all the dormant virus genomes, making them vulnerable to the immune system and to antiviral therapy, which raises the eventual possibility of a cure for herpes infections. Expression of the LAT promotes **neuronal survival** after HSV infection by inhibiting **apoptosis**. This antiapoptosis function could promote reactivation by the following mechanisms:

- Providing more **latently infected neurons** for future reactivations.
- Protecting neurons in which **reactivation** occurs.
- Protecting **previously uninfected neurons** during a reactivation.

When reactivated by some provocative stimulus—generally thought to be some sorts of **stress stimuli**, including local **inflammation**—HSV travels down the sensory nerves to cause **peripheral lesions** due to local virus multiplication, such as **"cold sores" or genital ulcers**. It is not altogether clear what constitutes a provocative stimulus, but there are many possible alternatives, including psychological and physical factors. **Periodic** reactivation—such as of cold sores due to **HSV**, or **shingles** due to **EBV**—establishes the pattern of infection, with **sporadic**, sometimes very painful reappearance of disease symptoms for the rest of the host's life. Even worse than this, **immunosuppression** later in life can cause the latent infection to flare up: this indicates that the immune system normally has a role in helping to suppress these latent infections. This can result in a very severe, **systemic**, and sometimes life-threatening infection.

In a manner somewhat similar to herpesviruses, infection by **retroviruses** may result in a latent infection. Integration of the **provirus** into the host genome certainly results in the persistence of the virus for the lifetime of the host organism, and may lead to an **episodic pattern of disease**. In some ways, AIDS—which results from HIV infection—shows aspects of this pattern of infection. The pathogenesis of AIDS is discussed in detail in Chapter 7.

Prevention and therapy of virus infection

There are two aspects of the response to the threat of virus diseases: first, **prevention** of infection, and second, **treatment** of the disease. The former strategy relies on two approaches. The first is **public and personal hygiene**, which perhaps plays the major role in preventing virus infection. This includes provision of **clean drinking water** and safe treatment and disposal of **sewage**, and good medical practice such as the **sterilization of surgical instruments**. The second, and most important intervention in the last century, is **vaccination**, which makes use of the immune system to combat virus infections. Most of the damage to cells during virus infections occurs very early, often before the clinical symptoms of disease appear. This makes the treatment of virus infection very difficult; therefore, in addition to being less expensive, **prevention** of virus infection is undoubtedly better than cure.

Prophylactic viral vaccines

To design effective **vaccines**, it is important to understand both the **immune response** to virus infection and the **stages of virus replication** that are appropriate targets for immune intervention (see also Entrance, Entertainment, and Exit, Box 4.1). To be effective, vaccines must stimulate as many of the body's defense mechanisms as possible. In practice, this usually means trying to **mimic the disease** without causing **pathology**—for example, the use of **live attenuated viruses** as vaccines such as nasally administered **influenza** vaccines and orally administered **poliovirus** vaccines. To be effective, it is not necessary to get 100% uptake of vaccine. "**Herd immunity**" results from the break in transmission of a virus that occurs when a sufficiently high proportion of a population has been vaccinated. This strategy is most effective where there is **no alternative host for the virus** (e.g., **measles, poliovirus, smallpox**) and in practice is the situation that usually occurs, as it is impossible to achieve 100% coverage with any vaccine. However, this is a risky business; if protection of the population falls below a critical level, **epidemics** can easily occur.

The types of virus vaccines that are well established are listed below. These are:

- **Live attenuated vaccines**: these are viruses that have been genetically altered by **repeated passage** in cells or even in animals, to become nonvirulent. Examples include vaccinia virus (VV), the **smallpox** vaccine; the **yellow fever vaccine** (see Chapter 1); **measles** and **mumps** and the **Sabin polio** vaccines. These require only low doses, and no adjuvants.
- **Killed whole-virion vaccines**: these are purified or semipurified whole infectious virions that have been "killed" or inactivated, by means that include chemical (**formaldehyde, β-propiolactone**) and physical (UV light) treatments. The most successful of these has to be the inactivated **rinderpest** vaccine, made in 1960: this was quite cheap to manufacture, elicited lifelong immunity—and its use resulted in the **complete eradication of rinderpest by 2011**. Other examples include the original **Salk polio** vaccine, **hepatitis A** vaccine, some **rabies** vaccines and **tick-borne encephalitis flavivirus** vaccine, as well as some of the **SARS-CoV-2** candidate vaccines. These generally require quite high doses, possibly repeated, with coadministration of an adjuvant.
- **Subunit vaccines** consist of only some **purified components** of the virus, sufficient to induce a protective immune response. They are generally safe, except for very rare cases in which adverse immune reactions may occur. It is also necessary to administer

relatively large quantities of antigen to get a response compared to live or even whole-killed virion preparations, because of their relatively poor immunogenicity—a problem inherent in the use of soluble proteins as immunogens. **New delivery systems**, such as **synthetic nanoparticles** and adjuvants can help. The best viral example is **influenza virus vaccines** (see below).

The effectiveness of **attenuated vaccines** relies on the fact that a **complete spectrum** of virus proteins is expressed, including **nonstructural proteins**, and gives rise to potent **cell-mediated** immune responses. The vaccine strain may (rarely) be a **naturally occurring virus** (e.g., the use of cowpox virus by Edward Jenner to vaccinate against smallpox) or **artificially attenuated** in vitro (e.g., the **oral poliomyelitis** vaccines produced by Albert Sabin) or even in vivo (the original **rabies** vaccines). The advantage of attenuated vaccines is that they are **good immunogens** and induce long-lived, appropriate immunity. Set against this advantage, however, are their many disadvantages. They are often biochemically and genetically **unstable** and may either lose infectivity or **revert to virulence** unexpectedly. Despite intensive study, it is not possible to produce an attenuated vaccine to order, and there appears to be no general mechanism by which different viruses can be reliably and safely attenuated. Contamination of the vaccine stock with other, possibly pathogenic viruses is also possible—this was the way in which **SV40** was first discovered in **oral poliovirus vaccine** in 1960 (see Chapter 1). Inappropriate use of live virus vaccines, for example in immunocompromised hosts or during pregnancy, may lead to **vaccine-associated disease**, whereas the same vaccine given to a healthy individual may be perfectly safe.

Despite these difficulties, vaccination against virus infection has been one of the great triumphs of medicine during the 20th century. Most of the success stories result from the use of live attenuated vaccines—for example, the use of **VV against smallpox**. On May 8, 1980, the World Health Organization (WHO) officially declared **smallpox to be completely eradicated**, the first virus disease to be eliminated from the world. The WHO aims to eradicate a number of other virus diseases such as **poliomyelitis** and **measles**, but targets for completion of these programs have undergone much slippage due to the formidable difficulties involved in a worldwide undertaking of this nature.

The example of **egg-made influenza vaccines**, that essentially rely on technology from the 1930s to the 1940s (see Chapter 1), is worth discussing further. For routine manufacture of **seasonal** influenza vaccines, **reassortant** high-producer strains of virus that have the **HA and NA** proteins of interest are grown in chicken eggs—typically **1 egg for 1 dose**—and the virions semipurified, then inactivated by **chemical treatment** (formaldehyde or β-propiolactone), "**split**" into subunits by **detergent** treatment, and purified further by chromatography. The seasonal vaccines are made by mixing subunits from relevant isolates of **influenza A H1N1, H3N2, and one or two influenza B subtypes.** About **5 μg** of HA protein per type is used per dose.

Newer vaccine technologies, mostly without candidates yet on the market, include **synthetic peptide** vaccines, **recombinant protein** or **virus-like particle (VLP)** vaccines, **DNA** and **RNA** "genetic" vaccines, and **recombinant virus-vectored** vaccines. These are discussed individually below.

Synthetic vaccines in use presently are limited to "glycoconjugate" bacterial polysaccharides, whose structure is derived from **bacterial capsular** material. While **short, chemically**

synthesized peptides containing **B- and often T-cell epitopes** have been trialed for a variety of viruses over a couple of decades now, including for **HIV** for example, there are none that are licensed for general use. The major disadvantage with these molecules is that they are not usually very effective immunogens and are very costly to produce. However, because they can be made to order for any desired sequence, they have great theoretical potential, but this is as yet untapped.

Recombinant protein vaccines are produced by genetic engineering. A few such vaccines are already in general use; they are better than synthetic or purified subunit vaccines because production by **yeast fermentation** or **insect cell culture** is probably cheaper and more scalable than peptide synthesis and safer than **whole live virion** preparation. Moreover, the natural tendency of virion proteins to **self-assemble** to produce **virus-like particles** makes immunogens that elicit a more effective immune response. For example, vaccination against **hepatitis B virus (HBV)** used to rely on the use of the so-called "**Australia antigen"—22 nm lipoprotein particles** of the HBV **S** or **surface** protein **(HBsAg)**—that were obtained from the **serum of chronic HBV carriers**. This was a very risky practice indeed—not least because HBV carriers used as donors were often also infected with **HIV**. A completely safe HBV vaccine produced in **yeast** is now used: this relies on the natural assembly of recombinantly expressed **S protein** with plasma membrane-derived **lipids** to form 22 nm **nanoparticles** that contain no nucleic acids, and are also considerably more immunogenic than soluble versions of the protein alone. This type of vaccine has largely **halted the spread of HBV**, as it is included in most countries' extended programs of immunization (EPI) vaccine bundle that is given free to babies and children. In the long term, this will bring down the incidence of lifelong carriage of HBV, and the development of hepatocellular carcinoma due to HBV. Another major success in this regard are the **HPV vaccines**. The two different brands of vaccine are both based on **virus-like particles (VLPs)** made by recombinant expression of the **major capsid protein L1.** This is purified as **pleomorphic particles** from yeast or insect cell lysates, that are then chemically disassembled and allowed to **reassemble** in vitro: this allows more uniform particle formation (~50 nm isometric particles), and gets rid of adventitiously-encapsidated **nucleic acids** from the manufacturing cell systems. It is also possible to make HBV and HPV VLPs in plant expression systems, which may revolutionize applications of these vaccines.

While recombinant **VLP-based nanoparticle vaccines** seem an obvious choice for viruses with nonenveloped virions, there has also been some success with making **envelope protein-based VLPs** or nanoparticles for viruses with enveloped virions. This includes **HIV**, where particles resembling **immature virions** (see Fig. 4.21) containing **uncleaved Gag** and embedded **Env**, with no genomes, can be produced by **budding** in cell cultures. Moreover, recombinant **influenza virus HA-based vaccines** that form VLPs can be produced by budding in a variety of culture systems. Several different nanoparticle-based recombinant protein vaccines have also been trialed for **SARS-CoV-2** (see Chapter 8).

DNA vaccines are another type of vaccine that is long on promise, but short on delivery: there are presently no licensed DNA vaccines, despite more than 20 years of development. These generally consist of only a **plasmid-like DNA molecule** with an appropriate promoter sequence or sequences, encoding the antigen(s) of interest—and possibly also **costimulatory molecules** such as **cytokines**. The concept behind these vaccines is that the DNA component will be taken up into suitable cells after injection, typically into muscle tissue; be transported to the local cell nuclei, and **expressed** by transcription in vivo. This results in the production of small amounts of antigenic protein in a context that is very similar to what happens in natural infection, in that this is processed in cells for MHC-I and possibly MHC-II presentation. This serves to **"prime"** the immune response by **CD8+** and **CD4+** T-cell binding, and possibly also by **release** of whole protein or **presentation on cell surfaces** (e.g., of HIV Env) for **B-cell** activation, so that a protective response can be rapidly generated when the real antigen is encountered. In theory, these vaccines could be manufactured quickly and locally, and should elicit **cell-mediated** and possibly **humoral** immunity. In practice, however, many clinical trial studies have indicated that DNA vaccines are disappointingly weak immunogens that are best used in "heterologous prime-boost" immunization regimes. Thus, there is still some way to go until this experimental technology becomes a practical proposition.

A spin-off from the concept of "genetic" or DNA-based vaccines that has gained much attention in the last year or so is **RNA vaccines**. Most surprisingly, the **SARS-CoV-2** vaccines that were quickest out of the lab to testing and trial included two RNA-based candidates. These were **ssRNA+ molecules** synthesized in vitro using some nucleoside modifications to enhance mRNA stability, with an **efficient context for translation** of a single ORF encoding a **modified version of the full-length SARS-CoV-2 S protein**, delivered in **liposomes**, or a bubble of synthetic lipid-based envelope. As with several other SARS-CoV-2 vaccine candidates, the technology had been in development for several years for other vaccines, including for **Ebola** and **SARS-CoV** and **MERS-CoV**, and a quick pivot to including the SARS-CoV-2 S gene was quite a simple thing to do. The vaccines work by being taken up by cells, **uncoated** in the cytoplasm, and then immediately being used for translation of the message, just as a ssRNA+ Class IV virus genome does. The S protein is then processed through the **ER and Golgi**, and is displayed on the surface of cells containing the mRNA. These vaccines elicit **good cell-mediated responses** due to the mimicking of the natural virus protein production and MHC-I presentation, as well as the whole protein being localized on cell surfaces for antibody and B-cell recognition. The apparently better immunogenicity compared to DNA vaccines could be due to there being no need for the vaccine molecule to get into the nucleus for expression.

Virus-based vectors are **recombinant virus genomes** from "safe" nonpathogenic viruses, that are genetically manipulated to express **protective antigens** from other, pathogenic viruses. The idea here is to utilize the genome of a well-understood, **attenuated** virus to express and present antigens from other viruses to the immune system. Many different viruses offer possibilities for this type of approach. One of the most highly developed systems so far is based on the **vaccinia virus (VV)** genome. This virus was used to vaccinate hundreds of millions of people worldwide in the campaign to eradicate smallpox, and is generally considered a safe and effective vaccine. **VV-rabies** recombinants have been used in ongoing campaigns to eradicate rabies in **European fox** populations—scattering chicken head "**baits**" impregnated with the very stable vector. VV-based vaccines have advantages and disadvantages for use in humans—a high percentage of the human population has already been vaccinated during the smallpox eradication campaign, and this lifelong protection may result in poor response to recombinant vaccines. Although generally safe, **VV is dangerous in immunocompromised hosts**, thus it cannot be used in HIV-infected individuals, for example. A possible solution to these problems may be to use avipoxvirus vectors (e.g., fowlpox or canarypox) as "suicide vectors" that can only establish **abortive infections** of mammalian cells and offer the following advantages:

- Expression of **high levels** of foreign proteins.
- No danger of **pathogenesis** (abortive infection).
- No **natural immunity** in humans (avian virus).

Further development of VV, however—the production of **modified vaccinia Ankara (MVA)**, that does not replicate to completion in human cells—made VV an even more suitable vehicle for antigen delivery to humans in particular, as it has most of the advantages of the vectors listed above, with the added factor being that its production is routine and very well established, unlike some of the avian viruses. The virus has been used in numerous clinical trials, including of **HIV** vaccines in **mixed prime-boost regimens**.

Two other vaccine vectors that came to prominence recently, due to **emergency use** in the West Africa Ebola outbreak from 2014 to 2016 (see Chapter 8), were **vesicular stomatitis virus (VSV)** and two **adenoviruses**: these were a **chimpanzee** adenovirus (**cAd3**) and human adenovirus 26 (**Ad26**). All contained the **Ebola Zaire virus envelope glycoprotein, GP**. While the outbreak was mostly over before the vaccines were fully deployed, **emergency use authorizations (EUAs)** allowed the successful trial of VSV-vectored Ebola virus GP protein, as well as of cAd3, and a combination of MVA prime/Ad26 boost GP vaccines. All of these, together with **human adenovirus 5 (Ad5)** vector, have been used for **SARS-CoV-2** vaccine candidates. While the Oxford University chimpanzee AdV (**ChAdOx**) vectoring the S protein was one of the first to be licensed as a two-shot vaccine, an Ad26 single-shot version is also approved, and an **Ad5/Ad26 prime-boost** combination is also available. Other Ad5-based versions are still in trial.

Most existing virus vaccines are directed against viruses which are relatively **antigenically invariant**, for example, **measles, mumps, rubella** and **yellow fever** viruses, where there is in fact only one serotype of the virus. Viruses whose antigenicity alters continuously are a major problem in terms of vaccine production, and the classic example of this is **influenza** virus (see earlier). In response to this problem, new technologies such as **reverse genetics** could be used to improve and to shorten the lengthy process of preparing vaccines. RNA virus genomes can be easily manipulated as DNA clones to contain nucleotide sequences which match currently circulating strains of the virus. Infectious virions are rescued from the DNA clones by introducing these into cells. Seed viruses for distribution to vaccine manufacturers can be produced in as little as 1–2 weeks, a much shorter time than the months this process takes in conventional vaccine manufacture. Using the same technology, universal influenza vaccines containing crucial virus antigens expressed as fusion proteins with other antigenic molecules could feasibly be produced, making the requirement for constant production of new influenza vaccines obsolete. Although this has not yet been achieved, advances toward these goals are being made. The explosion of molecular techniques described in earlier chapters is now being used to inform vaccine design (as well as the design of antiviral drugs) rather than simply relying on trial-and-error approaches. However, developing safe and effective vaccines remains one of the greatest challenges facing virology.

Therapeutic vaccines

Although prevention of infection by prophylactic vaccination is much the preferred option, **postexposure therapeutic vaccines** can be of great value in modifying the course of some virus infections. Examples of this include **rabies** virus, where the course of infection may be very long and there is time for postexposure vaccination to generate an effective immune response and prevent the virus from carrying out the secondary replication in the CNS that is responsible for the pathogenesis of rabies. Other potential examples can be found in **virus-associated tumors**, such as **HPV**-induced cervical carcinoma, where therapeutic vaccines have been trialed quite intensively over the last 20 years.

Viruses as therapeutics

Phage therapy, the use of bacteriophages to treat or prevent disease, stretches back a century to the earliest days of the discovery of phages (see Chapter 1). Long before the discovery of antibiotics, the thought that viruses which lyse bacteria could be used to treat diseases was highly attractive—yet this idea has never become a widespread practical reality, despite its **routine medical application in Eastern Europe** up until now. Devotees of phage therapy have over the years defended their cherished belief with almost religious fervor, but there are serious obstacles to be overcome, such as the **narrow host range** of most phages (generally only a few strains of bacteria) and the speed at which **bacteria develop resistance** to infection. As the spectrum of clinically useful antibiotics dwindles, and as the ease of genetic modification increases with new technologies, phage therapy increases in attractiveness. However, it is unlikely ever to replace the antibiotic golden era of disease treatment we are now leaving behind, given the **very specific targeting of bacteria** that is inherent to phage biology.

Another aspect of "**virotherapy**" is the growing interest in **oncolytic viruses**—viruses used outside of their normal hosts, or which have been engineered, to **kill only cancer cells**. The usefulness of many different types of virus has been investigated, including **adenoviruses, herpesviruses, reoviruses,** and **poxviruses**. Although safety is a concern even in patients with terminal illnesses, this is one area of medical research where optimism is considerable. Many clinical trials are underway at it seems certain that this approach to cancer treatment will eventually become more common, possibly as an **adjunct** to other forms of therapy such as surgery, drugs, and radiotherapy.

Viruses have also developed as **gene delivery systems** for the **treatment of inherited and acquired diseases**. Virus-based gene therapy offers:

- Delivery of **large biomolecules** to cells.
- The possibility of **targeting delivery** to a specific cell type.
- **High potency** of action due to **replication** of the vector.
- Potential to treat certain diseases (such as **cervical** and **head and neck** cancers and **brain** tumors) that **respond poorly to other therapies** or may be inoperable.

The very first **retroviral** and **adenoviral** vectors were characterized in the early 1980s. The first human trial to treat children with immunodeficiency resulting from a **lack of the enzyme adenosine deaminase** began in 1990 and showed encouraging although not completely successful results. Like most of the initial attempts, this trial used **recombinant retrovirus genomes** as vectors. In 1995, the first successful gene therapy for **motor neurons** and **skin cells** was reported, while the first phase three (widespread) gene therapy trial was begun in 1997. In 1999, the first successful treatment of a patient with **severe combined immunodeficiency**

disease (SCID) was reported, but, sadly, the first death due to a virus vector also occurred, and in 2002 the occurrence of **leukemias** due to oncogenic insertion of a retroviral vector was seen in some SCID patients undergoing treatment. Several different viruses are being tested as potential vectors (Table 6.6). After initial optimism, gene therapy involving virus vectors has fallen from favor, and nonvirus methods of gene delivery including liposome/DNA complexes, peptide/DNA complexes, and direct injection of recombinant DNA are also under active investigation. It is important to note that such experiments are aimed at augmenting defective cellular genes in the somatic cells of patients to alleviate the symptoms of the disease and not at manipulating the human germ line, which is a different issue.

Chemotherapy of virus infections

The alternative to **vaccination** is to attempt to treat virus infections **using drugs that block virus replication** (see Chapter 4, Box 4.1, and Table 6.7). Historically, the discovery of **antiviral drugs** was largely down to luck. Spurred on by successes in the treatment of bacterial infections with antibiotics, drug companies launched huge blind-screening programs to identify

TABLE 6.6 Virus vectors in gene therapy.

Virus	Advantages	Possible disadvantages
Adenoviruses	Relatively easily manipulated in vitro (cf. retroviruses); genes coupled to the major late promoter are efficiently expressed in large amounts	Possible pathogenesis associated with partly attenuated vectors (especially in the lungs); immune response makes multiple doses ineffective if gene must be administered repeatedly (virus does not integrate)
Parvoviruses (AAV)	Integrate into cellular DNA at high frequency to establish a stable latent state; not associated with any known disease; vectors can be constructed that will not express any viral gene products	Only ~5 kb of DNA can be packaged into the parvovirus capsid, and some virus sequences must be retained for packaging; integration into host-cell DNA may potentially have damaging consequences
Herpesviruses	Relatively easy to manipulate in vitro; grows to high titers; long-term persistence in neuronal cells without integration	(Long-term) pathogenic consequences?
Retroviruses	Integrate into cell genome, giving long-lasting (lifelong?) expression of recombinant gene	Difficult to grow to high titer and purify for direct administration (patient cells must be cultured in vitro); cannot infect nondividing cells—most somatic cells (except lentiviruses?); insertional mutagenesis/activation of cellular oncogenes
Poxviruses	Can express high levels of foreign proteins. Avipoxvirus vectors (e.g., fowlpox or canarypox) are "suicide vectors" that undergo abortive replication in mammalian cells so there is no danger of pathogenesis and no natural immunity in humans	A high proportion of the human population has already been vaccinated—lifelong protection may result in poor response to recombinant vaccines (?). Dangerous in immunocompromised hosts

TABLE 6.7 Some antiviral drugs in current use.

Drug	Viruses	Chemical type	Target
Vidarabine	Herpesviruses	Nucleoside analog	Virus polymerase
Acyclovir	HSV	Nucleoside analog	Virus polymerase
Gancyclovir	CMV	Nucleoside analog	Virus polymerase (requires virus UL98 kinase for activation)
Nucleoside-analog reverse transcriptase inhibitors (NRTI)—zidovudine (AZT), didanosine (ddI), zalcitabine (ddC), stavudine (d4T), lamivudine (3TC)	Retroviruses (HIV)	Nucleoside analog	RT
Nonnucleoside reverse transcriptase inhibitors (NNRTI)—nevirapine, delavirdine	Retroviruses: HIV	Nucleoside analog	RT
Protease inhibitors—saquinavir, ritonavir, indinavir, nelfinavir	HIV	Peptide analog	HIV protease
Ribavirin	Broad-spectrum: HCV, HSV, measles, mumps, Lassa fever	Triazole carboxamide	RNA mutagen
Amantadine/rimantadine	Influenza A	Tricyclic amine	Matrix protein/ hemagglutinin
Neuraminidase inhibitors—oseltamivir, zanamivir	Influenza A and B	Ethyl ester prodrug requiring hydrolysis for conversion to the active carboxylate form	Neuraminidase

chemical compounds with antiviral activity, with relatively little success until recently. The key to the success of any antiviral drug lies in its **specificity**: almost any stage of virus replication can be a target for a drug, but the drug must be **more toxic to the virus than the host**. This is measured by the **chemotherapeutic index**, given by the formula:

$$\frac{\text{Dose of drug that inhibits virus replication}}{\text{Dose of drug that is toxic to host}}$$

The **smaller** the value of the chemotherapeutic index, the better. In practice, a difference of several orders of magnitude between the two toxicity values is usually required to produce a **safe and clinically useful** drug. Modern technology, including detailed knowledge of the molecular biology of virus replication and expression (Chapters 4 and 5), as well as increasingly sophisticated structural biology and computer-aided design of chemical compounds, allows the **deliberate design of drugs**. However, it is necessary to "know your enemy"—to understand the **key steps** in virus replication or expression that might be inhibited. Any of the stages of virus replication or expression can be a target for antiviral intervention. The only requirements are:

- The process targeted must be **essential** for either process.
- The drug is **active against the virus** but has "**acceptable toxicity**" (preferably very low) to the **host** organism.

What degree of toxicity is "acceptable" clearly varies considerably—for example, between a cure for the common cold, which might be sold over the counter and taken by millions of people, and a drug used to treat fatal acute virus infections such as **Ebola virus disease**, or potentially fatal chronic virus diseases such as **AIDS** or **HBV** (Box 6.2).

The development of antiviral chemotherapies has been one of the success stories of the last 30 or so years, with a huge expansion in the availability of effective drugs that can limit virus replication and even cure disease. If one understands in detail the various and often very different mechanisms viruses employ to enter our cells (Chapter 4), to express and replicate their genomes once within (Chapter 5), and to disperse themselves again, it is possible to develop drugs targeted at the various processes. Thus, one may interfere specifically with **attachment** of viruses to cells; one can attempt to prevent **entry and uncoating** of virus components inside the cell; one can target specific viral-encoded enzymes that are essential for **expression and processing** of viral proteins; one can use drugs that block up the **replication machinery** they use to make copies of their own genomes, and possibly the **maturation and release** of virions—all hopefully without adversely affecting the essential work of the cell (Table 6.7) (Box 6.3) (Fig. 6.20).

BOX 6.2

The drugs don't work

Pharmaceutical companies have a love/hate relationship with vaccines—and mostly hate. They are expensive and difficult to produce and save millions of lives, but if one child is harmed by an alleged bad reaction to a vaccination, the company suffers terrible publicity. Antiviral drugs however, now that's a different story: after suitable clinical trials antivirals are very safe, and they make money—lots of money. People like the idea of popping pills to cure diseases. Which is a shame, because the truth is in spite of all the effort put in, we have pitifully few effective antiviral drugs available, except for HIV. Got a cold? Hard luck. And as far as most developing countries are concerned, pricing puts most drugs out of reach of the people who need them. Antiretroviral therapy can keep AIDS patients alive for decades (if you or your country can afford it), but what about the millions who die each year from respiratory infections or diarrhea? What about Lassa fever in West Africa, that emerges every year—and can be treated with an old drug, but one that is not generally available there (see below)? These things must change!

BOX 6.3

Why can't we use antibiotics against viruses?

People often ask why **antibiotics** used for **bacterial or fungal** infections cannot be used against viruses too. The answer is that antibiotics are targeted at a number of **bacterial- or fungal-specific functions**, which occur **inside their cells**, removed from the essential functions of the host organisms or its cells by the bacterial or fungal **plasma membrane and cell wall**. Thus, one can specifically interfere in bacterial cell wall biosynthesis (e.g., **penicillin**) without affecting the host organism; similarly, one could use **ciprofloxacin** which specifically targets the **DNA gyrase** in **bacteria** to stop DNA replication, but does not affect the animal host.

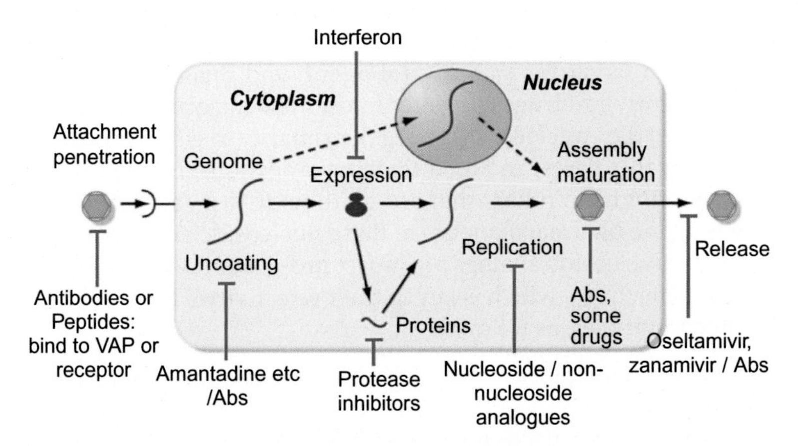

FIG. 6.20 Targets for therapeutic intervention in the virus replication cycle. Antibodies—naturally-derived or monoclonal—can interfere at a number of stages in the virus life cycle, from attachment, through uncoating, to assembly and release. Other important stages that can be blocked with existing or future therapies are shown by the *red bars* (Box 6.5).

Interfering with attachment

Other than by binding of **specific antibodies**, the **attachment phase** of replication can be inhibited in two ways: these are by agents that mimic the **virus attachment protein (VAP)** and bind to the **cellular receptor,** or by agents that **mimic the receptor** and **bind to the VAP**, so as to block its attachment. Custom-designed synthetic peptides are one of the most logical compounds to use for these purposes. However, while this is a very promising line of research, there are problems with the clinical use of these substances: these are primarily the high cost of synthetic peptides, and the poor pharmacokinetic properties of many of these synthetic molecules. The latter problem is potentially solvable by better and more sophisticated design and formulation, and the former to rapid advances in technology in this area of synthesis.

Preventing penetration and uncoating

The **penetration/uncoating** stages of virus replication may also be specifically targeted, although uncoating is largely mediated by the structure of specific virions and/or cellular enzymes, and is therefore a poor target for general intervention. **Amantadine** and **rimantadine** are two well-established drugs that are active against cell entrance by **influenza A virions**. The action of these closely related agents is to block **cellular membrane ion channels**: the target for both drugs is the **influenza virus M2** protein—a **viroporin**—but resistance to the drug may also map to the **HA** gene. This biphasic action results from the inability of drug-treated cells to **lower the pH** of the endosomal compartment—a function normally controlled by M2—which is essential to induce conformational changes in the HA protein to permit membrane fusion (see Chapter 4). Nonneutralizing "**binding**" **Abs** can also act at this stage, by preventing VAP-mediated fusion or other conformational change.

Blocking viral replication

Many viruses have evolved their own specific enzymes to replicate virus nucleic acids preferentially, at the expense of cellular molecules. There is often sufficient specificity in virus polymerases to provide a target for an **antiviral agent**, and this method has produced the majority of the specific antiviral drugs currently in use. The majority of these drugs function as **polymerase substrates** (i.e., **nucleoside** analogs), primarily to either **terminate chain elongation** or insert **noncopyable** bases, and their **toxicity** varies considerably, from some that are well tolerated (e.g., **acyclovir**) to others that are quite **toxic** (e.g., **azidothymidine** or **AZT**). There is a problem with the pharmacokinetics of these nucleoside analogs in that their typical serum half-life is 1–4 h. Nucleoside analogs are in fact **pro-drugs**, as they must be **phosphorylated** before becoming effective—which is key to their selectivity. There are a number of such drugs specific for **herpesviruses**, as follows:

- **Acyclovir** is phosphorylated by **HSV thymidine kinase** 200 times more efficiently than by cellular enzymes.
- **Ganciclovir** is 10 times more effective against **CMV** than acyclovir but must be phosphorylated by a **kinase encoded by CMV gene UL97** before it becomes pharmaceutically active.
- Other nucleoside analogs derived from these drugs and active against other **herpesviruses** have been developed (e.g., **valciclovir** and **famciclovir**). These compounds have improved pharmacokinetic properties, such as better oral bioavailability and longer half-lives.

The drug **cidofovir** is a nucleoside analog that is a potent agent for treatment of **poxvirus** infections, probably because the viruses **do not use cell nuclear machinery** for replication of their DNA.

Targeting the replication machinery of **RNA viruses** is an increasingly popular strategy, as this is a function which is **not present in host cells**. A nucleoside analog that was developed as a broad-spectrum antiviral drug, and was used with some success in West Africa in 2014–16 to treat Ebola virus infections—**remdesivir**—was recently repurposed to treat **COVID-19** patients. This was chiefly useful in serious cases, where it shortened the period of severe illness if treated early. **Ribavirin** is a **guanosine analog** with a very wide spectrum of activity against many different viruses, especially against **many (–)sense RNA viruses**. This drug, which is one of the oldest **antivirals**, acts as an **RNA mutagen**, causing a 10-fold increase in mutagenesis of RNA virus genomes and a **99% loss in virus infectivity** after a single round of virus infection in the presence of ribavirin. Ribavirin is thus quite unlike the other nucleoside analogs described above, and its use might become much more widespread in the future if it were not for the frequency of adverse effects associated with this drug. Paradoxically, while it has been known for decades that it is effective against **Lassa fever** that emerges annually in West Africa in rainy seasons (see Box 6.2), it is not often used there because of the cost of the drug.

There are a number of **nonnucleoside analogs** that **inhibit virus polymerases**: for example, **foscarnet** is an analog of **pyrophosphate** that interferes with the binding of incoming nucleotide triphosphates by **virus DNA polymerases** (Box 6.4).

BOX 6.4

The case of hepatitis C

A highly successful story of antiviral chemotherapy is that of **hepatitis C virus (HCV)**: while some infections with this **ssRNA+ flavivirus** are **acute** and often **asymptomatic**, and are resolved by the immune system, around 80% of cases go on to develop **chronic** infections. About **1.5%** of the global adult population **(62–89 million** people) have chronic disease. Between 5% and 20% of these people develop **cirrhosis** over a period of 20 years or more, and for these there is a 25% risk of progression to end-stage liver disease or **hepatocellular carcinoma**. There are seven or more genotypes of the virus, which can differ in genome sequence by up to 50%—and which have different susceptibility to therapeutics. The only accepted therapy for many years was treatment with **polyethylene glycol-conjugated interferon (PEG-IFN)** in combination with **ribavirin** for up to 48 weeks, which had only 40%–50% **sustained virological response (SVR)** rates, and many unpleasant or dangerous side effects.

This changed in 2011, with regulatory approval of two **direct-acting antivirals** (DAAs)—oral **boceprevir** and **telaprevir**—which could be used in combination with the older therapy, and increased SVR rates in patients to as much as **70%**, albeit still with some adverse side effects. In 2013 two new drugs—**simeprevir** and **sofosbuvir**—were released within weeks of each other, and the oral once-daily combination treatments were well tolerated and produced SVR rates **greater than 90%** with certain genotypes of HCV. Other genotypes still required codosing with PEG-IFN and RBV. Newer DAA drugs soon followed, and now it is possible to treat **all genotypes of HCV** without using the older drugs (Kish et al., 2017). The main targets for chemotherapy are the **viral NS3/4A protease** that is essential for viral polyprotein processing (drugs have the suffix "-previr"), the **NS5B polymerase** component of the RdRp (drugs with the suffix "-buvir"), and the **NS5A replicase component** (names end in "-asvir").

Thus, chronic infections of hepatitis C may now be effectively treated using a variety of drugs—in fact, it is now possible to **cure chronic hepatitis C infections**, something the Egyptian government embarked upon as a national strategy recently because of the very high number of people there who have been infected as a result of lax injection hygiene in clinics.

Blocking viral gene expression

Virus gene expression is less amenable to chemical intervention than **genome** replication, because viruses are much more dependent on the cellular machinery for transcription, mRNA splicing, cytoplasmic export than for replication—and are by their definition absolutely dependent on **cellular ribosomes** for **translation**. **Interferon** acts to stop **translation** (see above) but has its own side effects. To date, no clinically useful drugs that discriminate between virus and cellular gene expression have been developed.

Interfering with virion maturation and release

For the majority of viruses, the detailed processes of **assembly**, **maturation**, and **release** are poorly understood and therefore have not yet become specific targets for antiviral intervention, other than by use of monoclonal Abs. An exception here are the **antiinfluenza** drugs **oseltamivir** and **zanamivir**, which are inhibitors of influenza virus **neuraminidase** action. NA is involved in the **release** of virions budding from infected cells, and these drugs are believed to reduce the spread of virus to other cells (Box 6.5).

BOX 6.5

Blocking HIV replication and expression

The development of drugs to combat HIV infections started with the application of what few antiviral therapies there were at the time—drugs known to inhibit **reverse transcriptase**—and has mushroomed ever since. There are now, nearly 40 years since the pandemic was recognized, quite a wide variety of drugs that are given to people with **HIV** infections to keep their virus loads low, or even as **preexposure prophylactics**. These are collectively known as **antiretrovirals (ARVs)**, and their routine use has converted a diagnosis of HIV/AIDS from being a death sentence, to being almost just another chronic, manageable disease.

Drugs that block virus particles from **attaching** to cells are known as **attachment inhibitors**, and include very expensive monoclonal antibodies (**mAbs**) such as **ibalizumab** which blocks **CD4 receptors** on cells; compounds such as **fostemsavir** which bind to the **virion gp120** and block its binding to **cell-surface CD4** protein as the first stage of entry; the **CCR5** antagonist **maraviroc**, which binds to and blocks viral attachment to the **second** (CCR5) of two cell proteins vital for viral entry. **Fusion inhibitors** work at the next stage of infection, and block the virus from entering the cell by fusing membranes: the best-known is **enfuvirtide**, a **peptide** whose sequence is derived from the HIV-1 **gp41** protein.

Agents that specifically interfere with the **replication machinery** of the virus are **nucleoside** and **nonnucleoside reverse transcriptase inhibitors** (NRTIs and NNRTIs). These include well-known NRTIs such as **zidovudine** or **azidothymidine** (AZT) which was one the first agents licensed in 1987; **lamivudine** (3TC), **tenofovir** (TDF), and **emtricitabine** (FTC). NNTRIs include **nevirapine** (NVP), **efivarenz** (EFV), and **rilpirivine** (RPV). Agents that interfere with the **virus protease** that is vital for proper processing of virus proteins include **ritonavir** (RTV) and **saquinavir** (SQV). **Integrase** inhibitors such as **dolutegravir** (DTG) and **raltegravir** (RAL) block the action of the **viral integrase** enzyme, which is necessary for the HIV dsDNA form to integrate into host chromosomes (see Fig. 6.21). ARVs have since been recognized as being useful for the blocking of replication of HBV, and for combatting the activation of HERVs (see Chapter 3) where this leads to pathology.

Continued

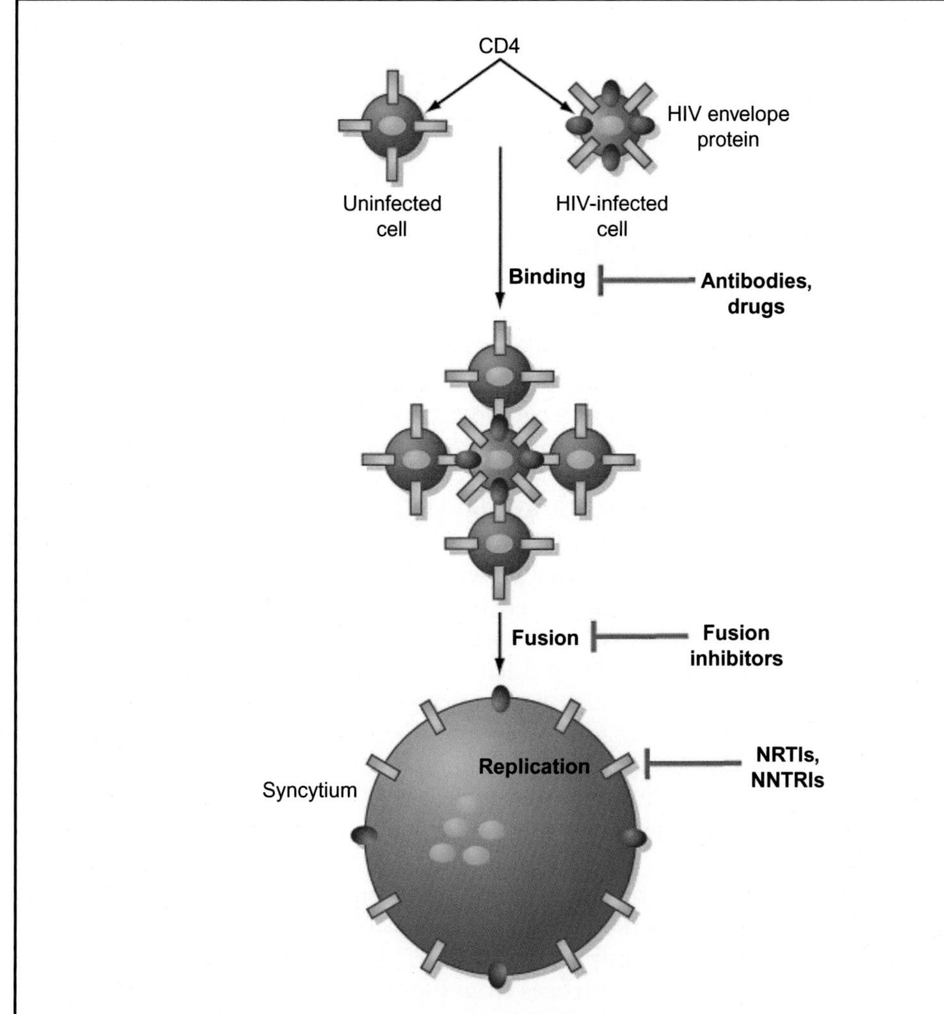

FIG. 6.21 Inhibition of HIV entry, fusion, and replication. The virus envelope glycoprotein (Env) plays a major role in receptor binding and subsequent membrane fusion, and is expressed on the surface of infected cells. Uninfected CD4+ cells coming into contact with these infected cells are fused together to form a multinucleate syncytium. The points at which the various classes of antiviral agents that inhibit HIV infections act are shown with *red bars*.

The most striking aspect of antiviral chemotherapy is how few clinically useful drugs are available for anything other than **herpes**- and **retroviruses**. As if this were not bad enough, there is also the problem of **drug resistance** to consider. In practice, the speed and frequency with which resistance arises when drugs are used to treat virus infections varies considerably and depends largely on the **biology of the virus** involved rather than on the chemistry of the compound. To illustrate this, two extreme cases are described here.

Acyclovir, the nucleoside analog used to treat **HSV infections**, is easily the most widely used antiviral drug. This is particularly true in the case of **genital herpes**, which causes painful recurrent ulcers on the genitals. It is estimated that 40–60 million people suffer from this condition in the United States alone. Fortunately, resistance to acyclovir arises infrequently. This is partly due to the **high fidelity** with which the dsDNA genome of HSV is copied (Chapters 3 and 4). Mechanisms that give rise to acyclovir resistance include:

- HSV ***pol*** **gene mutants** that **do not incorporate** acyclovir.
- HSV **thymidine kinase (TK) mutants** in which TK activity is absent (TK^-) or reduced, or shows altered substrate specificity.

Strangely, it is possible to find mutations that give rise to each of these phenotypes with a frequency of 1×10^{-3} to 1×10^{-4} in clinical HSV isolates. The discrepancy between this and the very low frequency with which resistance is recorded clinically is probably explained by the observation that most *pol*/TK mutants appear to be **attenuated** (e.g., TK^- mutants of HSV do not reactivate from the latent state).

Conversely, the use of the nucleoside analog **azidothymidine** (AZT) for treatment of **HIV** infection is much less effective. In untreated HIV-infected individuals, AZT produces a rise in the numbers of $CD4^+$ cells within **2–6 weeks**. However, this beneficial effect is **transient**; after **20 weeks**, $CD4^+$ T-cell counts generally revert to baseline. This is due partly to the development of **AZT resistance** in treated HIV populations, and also to the **toxicity** of AZT on **hematopoiesis**—the production of all of the cellular components of blood and plasma—as the chemotherapeutic index of AZT is much worse than that of acyclovir. AZT resistance is initiated by the acquisition of a mutation in the HIV reverse transcriptase (RT) gene at codon 215. In conjunction with two to three additional mutations in the RT gene, a fully **AZT-resistant phenotype** develops. After 20 weeks of treatment, **40%–50%** of AZT-treated patients develop at least one of these mutations. This high frequency is due to the error-prone nature of reverse transcription (Chapter 3).

Because of the large number of replicating HIV genomes in infected patients and the high mutation rate (Chapters 3 and 7), the mutations that confer resistance **already potentially exist in untreated virus populations.** Thus, treatment with AZT does not cause but merely **selects these resistant viruses** from the total pool. However, some combinations of resistant mutations may make it difficult for HIV to replicate, and **resistance to one RT inhibitor may counteract resistance** to another. The current strategy for therapy of HIV infection is known as **HAART** (for **highly active antiretroviral therapy**) and employs combinations of different drugs such as a protease inhibitor plus two nucleoside RT inhibitors.

South Africa presently has nearly **8 million** people living with HIV-1, and with over 4 million on treatment, has about **20% of the world's total** of people who are on HIV therapeutics. The initial or **first-line therapy regimens** for previously untreated patients in 2017 were chosen from among the following combinations:

- a mixture of the NRTIs TDF and FTC or 3TC, plus the NNTRI EFV
- NRTIs TDF + FTC/3TC plus integrase inhibitor DTG
- TDF + FTC/3TC plus NNTRI RPV (if viral load >100,000 copies/mL)

Second-line regimens—given when first-line treatment is not working—preferably include **two NRTIs and a RTV-boosted protease inhibitor**: this involves using low-dose RTV, which inhibits the breakdown of the added inhibitor. Molecular mechanisms of resistance and drug interactions are both important to consider when designing such combination regimes:

- Combinations such as AZT + ddI or AZT + 3TC have antagonistic patterns of resistance and are effective.
- Combinations such as ddC + 3TC that show cross-reactive resistance should be avoided.

Certain protease inhibitors affect **liver function** and can favorably affect the pharmacokinetics of RT inhibitors taken in combination. Other potential benefits of combination antiviral therapy include **lower toxicity profiles** and the use of drugs that may have **different tissue distributions** or cell **tropisms**. Combination therapy may also prevent or delay the development of drug resistance. Combinations of drugs that can be employed include not only small synthetic molecules, but also "**biological response modifiers**" such as **interleukins** and **IFNs** (Fig. 6.21).

Summary

Virus infection is a complex, multistage interaction between the virus and the host organism. The course and eventual outcome of any infection are the result of a balance between host and virus processes. Host factors involved include exposure to different routes of virus transmission, and the control of virus replication by the immune response. Virus processes include the initial infection of the host, spread throughout the host, and regulation of gene expression to evade the immune response. Medical intervention against virus infections includes the use of vaccines to stimulate the immune response and drugs to inhibit virus replication. Molecular biology is stimulating the production of a new generation of antiviral drugs and vaccines, and the process is snowballing.

References

Cagliari, D., Dias, N., Galdeano, D., et al., 2019. Management of pest insects and plant diseases by non-transformative RNAi. Front. Plant Sci. 10, 1319. https://doi.org/10.3389/fpls.2019.01319.

Dietzgen, R.G., Mann, K.S., Johnson, K.N., 2016. Plant virus-insect vector interactions: current and potential future research directions. Viruses 8 (11), 303. https://doi.org/10.3390/v8110303.

Ghequire, M.G.K., de Mot, R., 2015. The Tailocin tale: peeling off phage tails. Trends Microbiol. 23 (10), 587–590. https://doi.org/10.1016/j.tim.2015.07.011.

Kish, T., Aziz, A., Sorio, M., 2017. Hepatitis C in a new era: a review of current therapies. Pharm. Ther. 42 (5), 316–329. https://www.ncbi.nlm.nih.gov/pmc/articles/PMC5398625/.

Rath, D., Amlinger, L., Rath, A., Lundgren, M., 2015. The CRISPR-Cas immune system: biology, mechanisms and applications. Biochimie 117, 119–128. https://www.sciencedirect.com/science/article/pii/S0300908415001042.

Schuster, S., Miesen, P., van Rij, R., 2019. Antiviral RNAi in insects and mammals: parallels and differences. Viruses 11 (5), 448. https://doi.org/10.3390/v11050448.

Solovyev, A.G., Savenkov, E.I., 2014. Factors involved in the systemic transport of plant RNA viruses: the emerging role of the nucleus. J. Exp. Bot. 65 (7), 1689–1697. https://doi.org/10.1093/jxb/ert449.

Recommended reading

Aliyari, R., Ding, S.W., 2009. RNA-based viral immunity initiated by the Dicer family of host immune receptors. Immunol. Rev. 227 (1), 176–188.

Black, S., 2020. Phages Containing Huge Amounts of DNA Are Found Around the Globe. The Science Advisory Board. https://www.scienceboard.net/index.aspx?sec=ser&sub=def&pag=dis&ItemID=497.

Cullen, B.R., 2010. Five questions about viruses and microRNAs. PLoS Pathog. 6 (2), e1000787. https://doi.org/10.1371/journal.ppat.1000787.

Cullen, B.R., 2013. How do viruses avoid inhibition by endogenous cellular microRNAs? PLoS Pathog. 9 (11), e1003694. https://doi.org/10.1371/journal.ppat.1003694.

Grens, K., 2016. Giant Virus Has CRISPR-Like Immune Defense. https://www.the-scientist.com/the-nutshell/giant-virus-has-crispr-like-immune-defense-33938.

Loc-Carrillo, C., Abedon, S.T., 2011. Pros and cons of phage therapy. Bacteriophage 1 (2), 111–114. https://doi.org/10.4161/bact.1.2.14590.

Marraffini, L.A., Sontheimer, E.J., 2010. CRISPR interference: RNA-directed adaptive immunity in bacteria and archaea. Nat Rev Genet. 11 (3), 181–190. https://doi.org/10.1038/nrg2749.

Marshall, M., 2009. Timeline: The Evolution of Life. https://www.newscientist.com/article/dn17453-timeline-the-evolution-of-life/.

Roossinck, M.J., 2013. Plant virus ecology. PLoS Pathog. 9 (5), e1003304. https://doi.org/10.1371/journal.ppat.1003304.

CHAPTER

7

Viral pathogenesis

INTENDED LEARNING OUTCOMES

On completing this chapter, you should be able to:

- Discuss the link between virus infection and disease.
- Explain how virus infection may injure the body, including how HIV infection causes AIDS and how some viruses may cause cancer.

Pathogenicity, or the capacity of **one organism to cause disease in another**, is a complex and variable process. At the simplest level, there is the question of defining what disease is. An all-embracing definition would be that **disease is a departure from the normal physiological parameters of an organism**. This could range from a temporary and very minor condition, such as a slightly raised temperature or lack of energy, to **chronic pathologic conditions** that eventually result in death. Any of these conditions may result from a tremendous number of internal or external sources. There is rarely one single factor that "causes" a disease; most disease states are multifactorial at some level.

In considering virus diseases, two aspects are involved—the **direct effects** of virus replication and **the effects of the body's responses** to the infection. The course of any virus infection is determined by a delicate and dynamic balance between the host and the virus, as described in Chapter 6. The extent and severity of virus pathogenesis is determined similarly. In some virus infections, most of the pathologic symptoms observed are not directly caused by virus replication but are to the **side effects of the immune response**. Inflammation, fever, headaches, and skin rashes are not usually caused by viruses themselves but by the cells of the immune system due to the release of potent chemicals (generally ***cytokines***) such as **interferons** and **interleukins**. In the most extreme cases, it is possible that none of the pathologic effects of certain diseases is caused directly by the virus, except that its presence stimulates activation of the immune system.

In the past few decades, and especially in these days of "**multiomics**" research, molecular analysis has contributed enormously to our understanding of virus pathogenesis. Nucleotide sequencing and site-directed mutagenesis have been used to explore **molecular determinants of virulence** in many different viruses. Specific sequences and structures found only in disease-causing strains of virus and not in closely related **attenuated** or **avirulent** strains have been identified. Sequence analysis has also led to the identification of T-cell and

Cann's Principles of Molecular Virology
https://doi.org/10.1016/B978-0-12-822784-8.00007-6

B-cell epitopes on virus proteins responsible for their recognition by the immune system. Unfortunately, these advances do not automatically lead to an understanding of the mechanisms responsible for pathogenicity.

This chapter is largely about viruses that cause disease in animals, as viruses that cause disease in plants have already been considered in Chapter 6. Three major aspects of virus pathogenesis are considered: **direct cell damage** resulting from virus replication, **damage resulting from immune activation or suppression**, and **cell transformation** caused by viruses (Box 7.1.).

BOX 7.1

Don't blame the viruses!

Virus pathogenesis is an abnormal and fairly rare situation. The majority of virus infections are **silent** and do not result in any outward signs of disease. It is sometimes said that viruses would disappear if they killed their hosts. This is not necessarily true: it is possible to imagine viruses with a **hit-and-run strategy**, moving quickly from one dying host to the next and relying on continuing circulation for their survival. Nevertheless, there is a clear tendency for viruses not to injure their hosts if possible. A good example of this is the **rabies** virus. The symptoms of human rabies virus infections are truly dreadful, but thankfully rare. In its normal hosts (e.g., **foxes, bats**), rabies virus infection produces a much milder disease that does not usually kill the animal. Humans are an unnatural, **dead-end host** for this virus, and the severity of human rabies is as extreme as the condition is rare. Another example is **Marburg virus**: this generally causes very severe **hemorrhagic disease** in humans, but is a transient, cold-like infection in fruit bats in central Africa. Ideally, a virus would not even provoke an immune response from its host, or at least would be able to hide to avoid the effects. **Herpesviruses** and some **retroviruses** have evolved complex lifestyles that enable them to get close to this objective, remaining silent for much of the time.

Of course, fatal infections such as rabies and acquired immune deficiency syndrome (AIDS) always grab the headlines. Much less effort has been devoted to isolating and studying the many viruses that have not (yet) caused well-defined diseases in humans, domestic animals, or economically valuable crop plants, although metagenomic studies are rapidly filling this gap.

Mechanisms of cellular injury

Virus infection often results in a number of changes that are detectable by visual or biochemical examination of infected cells. These changes result from the production of virus proteins and nucleic acids, but also from alterations to the biosynthetic capabilities of infected cells. Virus replication sequesters cellular apparatus such as ribosomes and raw materials that would normally be devoted to synthesizing molecules required by the cell. Eukaryotic cells must carry out **constant macromolecular synthesis**, whether they are growing and dividing or in a state of quiescence, if they are to stay "alive." A growing cell needs to manufacture more proteins, more nucleic acids, and more of all of its components to increase its size before dividing. However, there is another reason for such continuous activity. The function of all cells is regulated by controlled expression of their genetic information and the subsequent degradation of the molecules produced. Such control relies on a delicate and dynamic balance between synthesis and decay which determines the intracellular levels of all the important molecules in the cell. This is particularly true of the **control of the cell cycle**, which determines the behavior of dividing cells (see the section "Cell transformation by DNA viruses"). In general terms, a number of common phenotypic changes can be recognized in virus-infected cells. These changes are often referred to as the **cytopathic effects** (**c.p.e.**) of a virus, and include:

- **Altered shape:** Adherent cells that are normally attached to other cells (in vivo) or an artificial substrate (in vitro) may assume a rounded shape different from their normal flattened appearance. The extended "processes" (extensions of the cell surface resembling tendrils) involved in attachment or mobility are withdrawn into the cell.
- **Detachment from the substrate:** For adherent cells, this is the stage of cell damage that follows that above. Both of these effects are caused by partial degradation or disruption of the cytoskeleton that is normally responsible for maintaining the shape of the cell.
- **Lysis:** This is the most extreme case, where the entire cell breaks down. Membrane integrity is lost, and the cell may swell due to the absorption of extracellular fluid and finally break open. This is an extreme case of cell damage, and it is important to realize that not all viruses induce this effect, although they may cause other cytopathic effects. Lysis is beneficial to a virus in that it provides an obvious method of releasing new virion s from an infected cell; however, there are alternative ways of achieving this, such as **release** by **budding** (Chapter 4).
- **Membrane fusion:** The membranes of adjacent cells fuse, resulting in a mass of cytoplasm containing more than one nucleus, known as a **syncytium**, or, depending on the number of cells that merge, a giant cell. Fused cells are short lived and subsequently lyse—apart from direct effects of the virus, they cannot tolerate more than one nonsynchronized nucleus per cell.
- **Membrane permeability:** A number of viruses cause an increase in membrane permeability, allowing an influx of extracellular ions such as sodium. Translation of some virus mRNAs is resistant to high concentrations of sodium ions, permitting the expression of virus genes at the expense of cellular messages.
- **Inclusion bodies:** These are areas of the cell where virus components have accumulated. They are frequent sites of virus assembly, and some cellular inclusions consist of crystalline arrays of virion s. It is not clear how these structures damage the cell, but they are frequently associated with viruses that cause cell lysis, such as herpesviruses and rabies virus.
- **Apoptosis:** Virus infection may trigger apoptosis ("programmed cell death"), a highly specific mechanism involved in the normal growth and development of organisms (see Chapter 6).

In some cases, a great deal of detail is known about the molecular mechanisms of cell injury. A number of **viruses that cause cell lysis** exhibit a phenomenon known as **shutoff** early in infection. Shutoff is the sudden and dramatic cessation of most host-cell macromolecular synthesis. In **poliovirus**-infected cells, shutoff is the result of production of the **virus**

2A protein. This molecule is a protease that cleaves the p220 component of **eIF-4F**, a complex of proteins required for cap-dependent translation of messenger RNAs by ribosomes. Because poliovirus RNA does not have a 5′ methylated cap but is modified by the addition of the **VPg protein**, virus RNA **continues to be translated**. In poliovirus-infected cells, the dissociation of mRNAs and polyribosomes from the cytoskeleton can be observed, and this is the reason for the inability of the cell to translate its own messages. A few hours after translation ceases, **lysis of the cell** occurs.

In other cases, cessation of cellular macromolecular synthesis results from a different molecular mechanism. For many viruses, the sequence of events that occurs is not known. In the case of adenoviruses, the **penton protein** (part of the virus capsid) has a **toxic effect on cells**. Although its precise action on cells is not known, addition of purified penton protein to cultured cells results in their rapid death. Toxin production by pathogenic bacteria is a common phenomenon, but this is the only well-established case of a virus-encoded molecule with a toxin-like action. However, some of the normal contents of cells released on lysis may have toxic effects on other cells, and antigens that are not recognized as "self" by the body (e.g., **nuclear proteins**) may result in **immune activation and inflammation**. The **adenovirus E3–11.6K** protein is synthesized in small amounts from the E3 promoter at early stages of infection and in large amounts from the major late promoter at late stages of infection (Chapter 5). It has recently been shown that E3–11.6K is required for the **lysis of adenovirus-infected cells** and the **release** of virion s from the nucleus.

Membrane fusion is the result of virus-encoded proteins required for infection of cells (see Chapter 4), typically, the **surface glycoproteins** of **enveloped viruses**. One of the best-known examples of such a protein comes from **Sendai virus** (a paramyxovirus), which has been used to induce cell fusion during the production of monoclonal antibodies (Chapter 1). At least 9 of the 11 known **herpes simplex virus (HSV/HHV-1)** glycoproteins have been characterized regarding their role in virus replication. Several of these proteins are involved in fusion of the virus envelope with the cell membrane and also in **cell penetration**. Production of **syncytia, or fused cells** (a.k.a. multinucleate giant cells) is a common feature of HSV infection.

Another virus that causes cell fusion is **human immunodeficiency virus (HIV)**. Infection of $CD4^+$ cells with some but not all isolates of HIV causes cell-cell fusion and the production of syncytia or giant cells (Fig. 6.21). The protein responsible for this is the **transmembrane envelope glycoprotein of the virus (gp41)**, and the domain near the amino-terminus responsible for this fusogenic activity has been identified by molecular genetic analysis. Because HIV infects $CD4^+$ cells and it is the reduction in the number of these crucial cells of the immune system that is the most obvious defect in AIDS, it was initially believed that direct killing of these cells by the virus was the basis for the pathogenesis of AIDS. Although direct cell killing by HIV undoubtedly occurs in vivo, it is now believed that the pathogenesis of AIDS is considerably more complex (see the section "Viruses and immunodeficiency"). Many animal retroviruses also cause cell killing and, in most cases, it appears that the envelope protein of the virus is required, although there may be more than one mechanism involved.

Viruses and immunodeficiency

At least two major groups of viruses—**herpesviruses** and **retroviruses**—directly infect **cells of the immune system**. This has important consequences for the outcome of the infection and for the immune system of the host. HSV establishes a **systemic infection**, spreading via the bloodstream in association with **platelets**, but it does not show particular **tropism** for cells of the immune system. However, **Herpesvirus saimirii**—which is **T-lymphotropic**—and **Marek's disease virus** are herpesviruses that cause **lymphoproliferative diseases** (but not clonal tumors) in monkeys and chickens, respectively. The most recently discovered **human** herpesviruses (HHVs)—HHV-6, HHV-7, and HHV-8—all infect lymphocytes (Chapter 8).

Epstein-Barr virus (EBV; HHV-4) infection of **B-cells** leads to their immortalization and proliferation, resulting in "**glandular fever**" or **mononucleosis**, a debilitating but benign condition. EBV was first identified in a lymphoblastoid cell line derived from **Burkitt's lymphoma:** in rare instances, and often in association with other infections such as malaria, EBV infection may lead to the formation of a **malignant tumor** (see the section "Cell transformation by DNA viruses"). While some herpesviruses such as HSV are highly **cytopathic**—cause damage to cells—most of the lymphotropic herpesviruses do not cause a significant degree of cellular injury. However, infection of the delicate cells of the immune system may perturb their normal function. Because the immune system is internally regulated by complex networks of interlinking signals, relatively small changes in cellular function can result in its collapse. Alteration of the normal pattern of production of **cytokines** could have profound effects on immune function. The ***trans*-regulatory proteins** involved in the control of herpesvirus gene expression may also affect the **transcription of cellular genes**; therefore, the effects of herpesviruses on immune cells are more complex than just cell killing. Incidentally, such proteins may cause activation of HIV-1 expression in CD4+ lymphocytes.

Retroviruses cause a variety of pathogenic conditions including **paralysis, arthritis, anemia**, and malignant cellular **transformation**. A significant number of retroviruses infect the **cells of the immune system**. Although these infections may lead to a diverse array of diseases and **hematopoietic abnormalities** such as anemia and **lymphoproliferation**, the most commonly recognized consequence of retrovirus infection is the formation of **lymphoid tumors**. However, some degree of **immunodeficiency**, ranging from very mild to quite severe, is a common consequence of the interference with the immune system resulting from the presence of a **lymphoid** or **myeloid tumor**.

Acquired immunodeficiency syndrome (AIDS)

The most prominent aspect of virus-induced immunodeficiency is acquired immunodeficiency syndrome (AIDS), a consequence in humans of **infection with HIV-1 or -2**, members of the genus *Lentivirus* of the family *Retroviridae*. A number of similar lentiviruses cause immunodeficiency diseases in animals: these include many **simian** as well as **bovine** and **feline** immunodeficiency viruses (SIVs, BIV and FIV). Unlike infection by other types of retrovirus, HIV infection does not directly result in the formation of tumors. Some tumors such as **B-cell lymphomas** are sometimes seen in AIDS patients, but these are a consequence of the **lack of immune surveillance** that is responsible for the destruction of tumors in healthy individuals.

The **clinical course of AIDS** is long and very variable. A great number of different abnormalities of the immune system are seen in AIDS. As a result of the biology of lentiviruses and their infections, the pathogenesis of AIDS is highly complex (Fig. 7.1). It is still not clear how much of the pathology of AIDS is caused directly by the virus and how much is caused by the

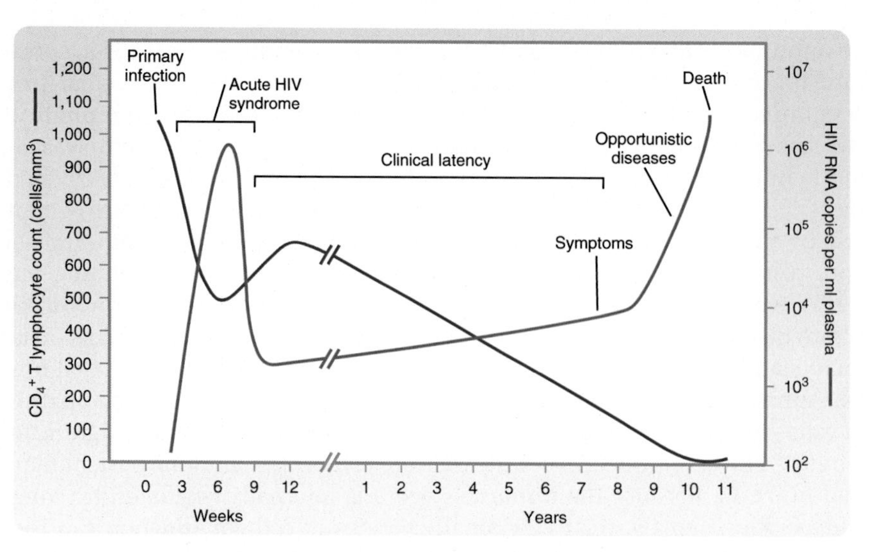

FIG. 7.1 Time course of HIV infection. This diagram shows a typical sequence of events in an HIV-infected person during the interval between infection with the virus and the development of AIDS.

immune system. Numerous models have been suggested to explain how HIV causes immunodeficiency. These mechanisms are not mutually exclusive, and indeed it is probable that the **underlying loss of CD4⁺ cells** (see Chapter 6) in AIDS is complex and multifactorial. **AIDS is defined as the presence of HIV infection**, plus **one or both of the following**:

- A **CD4⁺ T-cell count** of **less than 200 cells/ml** of blood (the normal count is 600–1000 per mL).
- Development of an **opportunistic infection** that occurs when the immune system is not working correctly, such as *Pneumocystis carinii* pneumonia (PCP), certain eye diseases, encephalitis, and some specific tumors such as Kaposi's sarcoma (KS).

It is facile to say that the best way to avoid AIDS is not to become infected with HIV, but that is not much help to the over **38 million people** worldwide who already are infected with the virus—or the **1.7 million** people who became **newly infected with HIV** in 2019, or the **690,000 people who died** from AIDS-related illnesses in 2019.

If we are to find a cure for AIDS, we need to understand the mechanisms by which the virus causes the disease. Although the basic biology of HIV is well understood (see References and Recommended Reading at the end of this chapter), scientists have never had a complete understanding of the processes by which **CD4⁺ T-helper cells are depleted** in HIV infection, and therefore have never been able to fully explain why HIV destroys the body's supply of these vital cells.

There have been many theories about how HIV infection results in AIDS. Soon after HIV-1 was discovered in the 1980s, it was shown that the virus could kill CD4⁺ cells in culture. Early experiments suggested there might not be enough virus present in AIDS patients to account for all the cell loss seen. Later, sensitive **PCR techniques** (Chapter 1) suggested that with the amount of virus present in infected individuals, the CD4⁺ cell count should in fact decline much faster and AIDS develop much earlier than it does after HIV infection. Researchers have used a "**tap and drain**" analogy to describe $CD4^+$ cell loss in HIV infection. In this description of the disease, $CD4^+$ cells (like water in a sink) are constantly being **eliminated** by HIV (**the drain**), while the body is constantly **replacing** them with new ones (**the tap**). Over time, the tap cannot keep up with the drain, and CD4 counts begin to drop, leaving the body susceptible to the infections that define AIDS. $CD4^+$ cells that are activated in response to invading microbes (including HIV itself) are highly susceptible to infection with the virus, and following infection these cells may produce many new copies of HIV before dying. One explanation for $CD4^+$ cell loss is the "**runaway**" hypothesis, in which $CD4^+$ cells infected by HIV produce more virion s, which activate more $CD4^+$ cells that in turn become infected, leading to a **positive feedback cycle** of **$CD4^+$ cell activation, infection, HIV production,** and **cell destruction.** Unfortunately, mathematical models consisting of a series of equations to describe the processes by which $CD4^+$ cells are produced and eliminated suggest that if the runaway hypothesis was correct, then $CD4^+$ cells in HIV-infected individuals would fall to low levels over a few months, not over several years as usually happens. This implies that the hypothesis cannot explain the **slow pace of $CD4^+$ cell depletion** in HIV infection. That leaves open the question of what exactly is going on between the time someone becomes infected with HIV and the time that they develop AIDS. While **rapid virus adaptation** and **antigenic variation** as a result of a high **mutation frequency** (see Chapter 3) is important in the biology of HIV, this alone cannot explain the whole story.

In general, HIV is regarded as an incurable infection, although in many cases doctors are able to stave off the onset of AIDS by giving patients sustained courses of **antiretroviral** drugs (see Box 6.6, Chapter 6). As retrovirus biology includes **integration of the virus genome** into host cell chromosomes, this constitutes a major obstacle to **eradicating the virus** from the body as happens in normal **acute** infections (see Chapters 3 and 6). In HIV-infected people receiving **antiviral therapy** there is a **reservoir** of **latently infected resting $CD4^+$ T cells**. Many HIV patients can manage their infection with cocktails of **antiretrovirals** (ARVs) which can reduce their "**viral load**"—the amount of virus circulating in the blood plasma—to undetectable levels, and incidentally, also render them **effectively noninfectious**. However, even in such people HIV is still lurking in **gut** and other **lymphoid** tissues, and still infecting other immune cells in the blood. Mathematical modeling and clinical observations suggest it might not ever be possible to completely eradicate the virus from the body with current therapies. The hope is that new approaches such as **RNAi** or even **CRISPR/Cas** (see Chapter 6) might 1 day be able to tackle this latent virus pool and completely eliminate the virus from the body, curing the infection.

Recent work has focused on the development of therapies to **disrupt virus latency**, and expose the virus to **replication-blocking drugs** and to the **immune response** which still exists in HIV-positive people. Even if this is possible, the cost of these advanced therapies would be beyond the reach of developing countries where the majority of HIV-infected people live. It may be that **adaptation** of HIV to **replication in humans** may eventually temper the pathogenicity of the virus by lowering its replication capacity, or by lessening the state of **chronic immune activation** that presently occurs with HIV. This is in fact what is supposed to have happened in **chimpanzees**, which were postulated to have originally been infected thousands of years ago with a **vervet** (="African green") **monkey SIV** to which they became adapted, over generations of genetic selection, so that **SIV-cpz** now no longer causes AIDS in chimps.

BOX 7.2

Stealthy does it

The more we study viruses, the more examples we find of viruses interacting with the immune system. Not interacting as in the "Argh! I'm dead" sense, but interacting as in a "I wonder what happens if I twist this knob?" sense. **Almost all viruses moderate the immune responses directed against them**. This makes sense—if they could not do this, while they might be able to replicate, they would not survive long enough to establish lasting infections. Some viruses are masters of the art, subtly tweaking and muting and subverting strands of the immune system to make life easier for them. **Herpesviruses** and **poxviruses** spring to mind: these viruses both encode numerous proteins and even nucleic acid sequences that are solely engaged in suppressing or subverting host immune responses. So too human papillomaviruses: metaviromic investigations of the healthy-looking skin of human adults or even on young babies turns up a multitude of HPVs—which do not appear to be doing any damage at all.

In comparison to them, viruses which appear go for an all-out assault on the body or on the immune system seem like amateurs—and may in fact be, in terms of infecting that class of host. Here, **Ebola virus** infections in humans spring to mind. The consequences on their hosts are devastating, which is bad for both the virus and the host. So let's hear it for the true masters of the craft of sneaking around, of getting on with things quietly!

However, this process will be speeded up by antiviral therapies in many different forms, including "**genetic therapies,**" **drugs**, and—eventually—**vaccines** (Box 7.2).

Virus-related diseases

Virus infections are believed to be a necessary prerequisite for a **number of human diseases which are not directly caused by the virus**. In some instances, the link between a particular virus and a pathological condition is well established, but it is clear that the pathogenesis of the disease is complex and also involves the immune system of the host. In other cases, the pathogenic involvement of a particular virus is less certain and, in a few instances, rather speculative.

Although the incidence of **measles virus** infection has been reduced sharply by **vaccination** (Chapter 6), measles still causes a few thousands of deaths worldwide each year—although, thanks to vaccination, this is down from over **800,000 deaths per annum** in the 1940s. The normal course of measles virus infection is an **acute febrile illness** during which the virus spreads throughout the body, infecting many tissues, with the most obvious manifestation being a widespread **skin rash**. The vast majority of people spontaneously recover from the disease without any lasting harm. In **rare cases (about 1 in 2000)**, measles may progress to a **severe encephalitis**. This is still an acute condition that either regresses or kills the patient within a few weeks; however, there is another, much rarer late consequence of measles virus infection that occurs **many months or years** after initial infection of the host. This is the condition known as **subacute sclerosing panencephalitis (SSPE)**. Evidence of **prior measles virus infection** (antibodies or direct detection of the virus) is found in **all patients with SSPE**, whether they can recall having a symptomatic case of measles or not. In about **1 in 300,000 cases** of measles, the virus is not cleared from the body by the immune system but establishes a **persistent infection in the CNS**. In this condition, virus replication continues at a low level, but **defects in the envelope protein genes** prevent the production of **extracellular infectious virions**. The lack of envelope protein production causes the **failure of the immune system to recognize and eliminate infected cells**; however, the virus is able to **spread directly from cell to cell**, bypassing the usual route of infection. It is not known to what extent damage to the cells of the brain is caused directly by virus replication or whether there is any contribution by the immune system to the pathogenesis of SSPE. Vaccination against measles virus and the prevention of primary infection should ultimately eliminate this condition forever.

Another well-established case where the immune system is implicated in pathogenesis concerns **dengue virus infections**. Dengue virus is a **flavivirus** that is transmitted from one human host to another via **mosquitoes**. The primary infection may be asymptomatic or may result in dengue fever. Dengue fever is normally a self-limited illness from which patients recover after 7–10 days without further complications. Following primary infections, patients carry antibodies to the virus. Unfortunately, there are **four serotypes of dengue virus** (DEN-1, 2, 3, and 4), and the presence of antibody directed against one type **does not give cross-protection** against the other three; worse still is the fact that antibodies can **enhance the infection** of peripheral blood mononuclear cells by **Fc-receptor-mediated uptake of antibody-coated dengue virions** (antibody-dependent enhancement or ADE; see Chapter 4). In a few cases, the consequences of dengue virus infection are much more severe than the usual fever, as the **dengue hemorrhagic fever** (DHF) that may result from such enhancement is a life-threatening disease. In the most extreme cases, so much **internal hemorrhaging** occurs that **hypovolemic shock** (dengue shock syndrome or DSS) occurs. DSS is frequently fatal. The cause of shock in dengue and other hemorrhagic fevers is partly due to the virus, but largely due to **immune-mediated damage of virus-infected cells** (Fig. 7.3). DHF and DSS following primary dengue virus infections occur in approximately **1 in 14,000** and **1 in 500 patients**, respectively; however, after **secondary dengue virus infections**, the incidence of DHF is **1 in 90** and DSS **1 in 50**, as cross-reactive but **nonneutralizing antibodies** to the virus are now present. These figures show the problems of cross-infection with different serotypes of dengue virus, and the difficulties that must be faced in developing a **safe vaccine** against the virus. Dengue virus is discussed further in Chapter 8 (Fig. 7.2).

Another instance where virus vaccines have resulted in increased pathology rather than the prevention of disease is the occurrence of **postvaccination Reye's syndrome**. Reye's syndrome is a **neurological condition** involving **acute cerebral edema** or "fluid on the brain," and occurs almost exclusively in children. It is well known as a rare postinfection complication of a number of different viruses, but most commonly **influenza virus** and **varicella-zoster virus** (human alphaherpesvirus 3, HHV-3), the virus causing **chicken pox**. Symptoms include **frequent vomiting, painful headaches, behavioral changes, extreme tiredness,** and

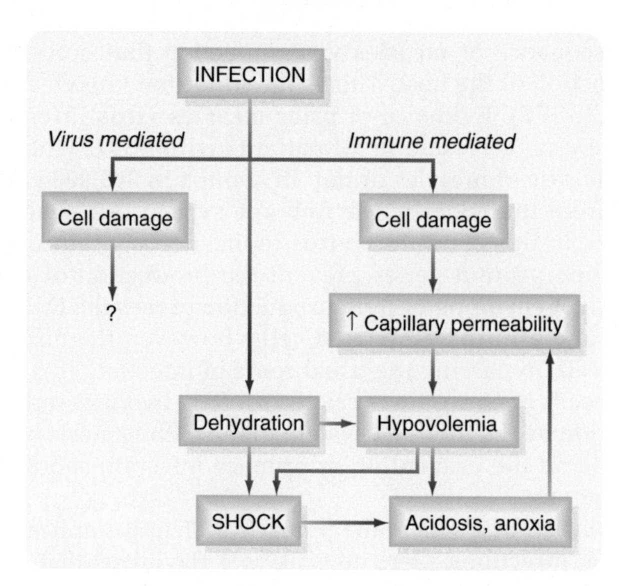

FIG. 7.2 Causes of shock in hemorrhagic fevers. The cause of hypovolemic shock in dengue and other hemorrhagic fevers is partly due to the virus, but largely due to immune- and inflammation-mediated damage of virus-infected cells.

disorientation. The chances of contracting Reye's syndrome are increased if **aspirin** is administered during the initial illness. The basis for the pathogenesis of this condition is completely unknown, but some of the most unfortunate cases have followed the administration of **experimental influenza virus vaccines**.

Guillain-Barré syndrome is another mysterious condition in which **demyelination of nerves** results in **partial paralysis** and muscle weakness. The onset of Guillain-Barré syndrome usually follows an **acute "virus-like" infection**—including **respiratory** or a **gastrointestinal** infections or more recently, **Zika virus**—but no single agent has ever been firmly associated with this condition. **Kawasaki disease (KD)** is similar to Reye's syndrome in that it occurs in children but is distinct in that it results in serious damage to the **heart**. Like Guillain-Barré syndrome, KD appears to follow acute infections—such as with **influenza virus A H1N1pdm** in 2009. The disease itself is not infectious but does appear to occur in **epidemics**, which suggests an infectious agent as the cause. A large number of bacterial and virus pathogens have been suggested to be associated with the induction of KD, but once again the underlying cause of the pathology is unknown. A novel **Kawasaki-like hyperinflammatory syndrome**, named **multisystem inflammatory syndrome in children (MIS-C)** by the WHO, was recently reported in children with high titers of IgG and IgM against SARS-CoV-2, but with no detectable virus. The syndrome shares a number of similarities with KD, toxic shock syndrome and macrophage activation syndrome, but differences to KD include higher age of incidence, gastrointestinal involvement, myocarditis and other cardiac involvement (see Gkoutzourelas et al., 2020). It would appear that **acute infection itself**—and possibly the **immune reaction** to it—may be responsible for the onset of these diseases, rather than a particular pathogen.

In recent years, there has been a search for an agent responsible for the disease called **chronic fatigue syndrome (CFS)** or **myalgic/myalgia encephalomyelitis (ME)**. Unlike the other conditions described above, CFS is a rather ill-defined disease and is not recognized by all physicians. Recent research has discounted the earlier idea that **EBV** might cause CFS, but a variety of other possible virus causes, including other **herpesviruses, enteroviruses,** and **retroviruses** have also been suggested. In October 2009 it was reported that a novel retrovirus—named **xenotropic murine leukemia virus-related virus (XMRV)**—might be a possible cause. However, subsequent research findings about XMRV proved to be contradictory and confusing, and there is **no strong evidence** for the role of XMRV in CFS, or in fact **that it exists at all in humans**: much of the evidence for it was **PCR-based sequence amplification**, and it has been found that XMRV-related sequences occur commonly in the human genome as repeated sequence elements—and not as **whole provirus sequences** (see Panelli et al., 2017).

Bacteriophages and human disease

Can **bacteriophages**, viruses that are only capable of infecting **prokaryotic** cells, play a role in human disease? Surprisingly, the answer is yes. **Shiga toxin (Stx)-producing *Escherichia coli* (STEC)** are able to cause intestinal foodborne diseases such as **diarrhea** and **hemorrhagic colitis**. STEC serotype O157:H7, the "**hamburger bug**," has received much attention in recent years. STEC infections can lead to fatal complications, such as **hemolytic-uremic syndrome**, as well as **neurological disorders**. The major virulence characteristics of these strains of bacteria are the ability to colonize the bowel (a natural trait of *E. coli*) and the **production of secreted "Shiga toxins,"** which can damage endothelial and tubular cells and may result in acute kidney failure. At least 100 different *E. coli* serotypes produce Stx toxins, and STEC bacteria occur frequently in the bowels of cattle and other domestic animals such as sheep, goats, pigs, and horses. Meat is infected by **fecal contamination**, usually at the time of slaughter. Ground meat such as hamburger is particularly dangerous as **surface bacterial contamination** may become buried deep within the meat where it may not be inactivated by cooking.

What has this got to do with bacteriophages? Various types of Stx are known, but they fall into two main types: **Stx1** and **Stx2**. The Stx1 and Stx2 toxin genes are **encoded in the genome of lysogenic lambda-like prophages** within the bacteria. Stimuli such as **UV light** or the antibiotic **mitomycin C** are known to **induce** these prophages to release a crop of virions which can infect and lysogenize other susceptible bacteria within the gut, accounting for the high prevalence of STEC bacteria (up to **50% of cattle** in some herds). Recent research has shown that the scandalous overuse of antibiotics as "growth promoters" in animal husbandry, and even antibiotic treatment of infected people, can stimulate the production of phage particles and contributes to the increased prevalence of STEC bacteria and growing human death toll. Other **bacterial virulence determinants** are also encoded by lysogenic phages: these include **diphtheria, cholera, botulism** and **Shiga toxins**, *Streptococcus* **erythrogenic** toxins and *Staphylococcus* **enterotoxins** (see ViralZone, n.d.), although the selective pressures that maintain these arrangements are not yet understood. Emerging bacterial genome sequence data strongly indicate that **phages have been responsible for spreading virulence determinants across a wide range of pathogens**.

The other area where bacteriophages may influence human illness is **phage therapy**—the use of **bacteriophages as antibiotics**. This is not a new idea, with initial experiments having been performed—albeit largely unsuccessfully—since shortly after the discovery of bacteriophages over 100 years ago (Chapter 1). However, with **increasing resistance of bacteria to antibiotics** and the emergence of "**superbugs**" immune to all effective treatments, this idea has experienced a resurgence of interest.

Cell transformation by viruses

Transformation is a change in the **morphological, biochemical,** or **growth parameters** of a cell. Transformation may or may not result in cells able to produce tumors in experimental animals, which is properly known as **neoplastic transformation**; therefore, transformed cells do not automatically result in the development of "**cancer**." **Carcinogenesis** (or more properly, **oncogenesis**) is a complex, multistep process in which cellular transformation may be only the first, although essential, step along the way. Transformed cells have an **altered phenotype**, which is displayed as one (or more) of the following characteristics:

- **Loss of anchorage dependence:** Normal (i.e., nontransformed) adherent cells such as **fibroblasts** or **epithelial** cells require a surface to which they can adhere. In the body, this requirement is supplied by adjacent cells or structures; in vitro, it is met by the glass or plastic vessels in which the cells are cultivated. Some transformed cells lose the ability to adhere to solid surfaces and float free (or in clumps) in the culture medium without loss of viability.
- **Loss of contact inhibition:** Normal adherent cells in culture divide and grow until they have coated all the available surface for attachment. At this point, when adjacent cells are touching each other, cell division stops—the cells do not continue to grow and pile up on top of one another. Many transformed cells have lost this characteristic. Single transformed cell in a culture dish becomes visible as **small thickened areas of growth** called "**transformed foci**"—clones of cells all derived from a single original cell.
- **Colony formation in semisolid media:** Most normal cells (both adherent and nonadherent cells such as lymphocytes) will not grow in media that are partially solid due to the addition of substances such as agarose or hydroxymethyl cellulose; however, many transformed cells will grow under these conditions, forming colonies since movement of the cells is restricted by the medium.
- **Decreased requirements for growth factors:** All cells require multiple factors for growth. In a broad sense, these include compounds such as ions, vitamins, and hormones that cannot be manufactured by the cell. More specifically, it includes regulatory peptides such as **epidermal growth factor** and **platelet-derived growth factor** that regulate the growth of cells. These are potent molecules that have powerful effects on cell growth. Some transformed cells may have decreased or may even have lost their requirement for particular factors. The production by a cell of a growth factor required for its own growth is known as **autocrine stimulation** and is one route by which cells may be transformed.

Cell transformation by viruses is a **single-hit process**; that is, a single virus transforms a single cell. This contrast with **oncogenesis**, which is the formation of tumors and is a **multistep process**. **All or part** of the virus genome **persists** in the transformed cell and is usually (but not always) **integrated into the host-cell chromatin**. Transformation is usually accompanied by continued expression of a limited repertoire of virus genes, or rarely by **productive infection**. Virus genomes found in transformed cells are frequently **replication defective** and contain substantial deletions. Examples here are **HBV genomes**, frequently found in transformed liver cells, and the **HPV genomes** found integrated in cervical and other **HPV-associated carcinomas**.

Transformation is mediated by proteins encoded by **oncogenes**. These **regulatory genes** can be grouped in several ways—for example, by their origins, biochemical function, or subcellular locations (Table 7.1). Cell-transforming viruses may have RNA or DNA genomes, but **all have at least a DNA stage in their replication cycle**; that is, the **only RNA viruses** directly capable of cell transformation are the **retroviruses** (Table 7.2). Certain retroviruses carry homologues of **c-*oncs*** derived originally from the cellular genes, and now known as **v-*oncs***. In contrast, the oncogenes of cell-transforming DNA viruses are **unique to the virus**

TABLE 7.1 Categories of oncogenes.

Type	Example
Extracellular growth factors (homologues of normal growth factors)	c-*sis*: Encodes the PDGF B chain (v-*sis* in simian sarcoma virus)
	int-2: Encodes a fibroblast growth factor (FGF)-related growth factor (common site of integration for mouse MMTV)
Receptor tyrosine kinases (associated with the inner surface of the cell membrane)	c-*fms*: Encodes the colony-stimulating factor 1 (CSF-1) receptor—first identified as a retrovirus oncogene
	c-*kit*: Encodes the mast cell growth factor receptor
Membrane-associated nonreceptor tyrosine kinases (signal transduction)	c-*src*: v-*src* was the first identified oncogene (Rous sarcoma virus)
	lck: Associated with the CD4 and CD8 antigens of T cells
G-protein-coupled receptors (signal transduction)	*mas*: Encodes the angiotensin receptor
Membrane-associated G-proteins (signal transduction)	c-*ras*: Three different homologues of c-*ras* gene, each identified in a different type of tumor and each transduced by a different retrovirus
Serine/threonine kinases (signal transduction)	c-*raf*: Involved in the signaling pathway; responsible for threonine phosphorylation of mitogen-activated protein (MAP) kinase following receptor activation
Nuclear DNA-binding/transcription factors	c-*myc* (v-*myc* in avian myelocytomatosis virus): Sarcomas caused by disruption of c-*myc* by retroviral integration or chromosomal rearrangements
	c-*fos* (v-*fos* in feline osteosarcoma virus): Interacts with a second proto-oncogene protein, Jun, to form a transcriptional regulatory complex

TABLE 7.2 Cell-transforming retroviruses.

Virus type	Time to tumor formation	Efficiency of tumor formation	Type of oncogene
Transducing (acutely transforming)	Short (e.g., weeks)	High (up to 100%)	c-*onc* transduced by virus (i.e., v-*onc* present in virus genome; usually replication defective)
cis-Activating (chronic transforming)	Intermediate (e.g., months)	Intermediate	c-*onc* in cell genome activated by provirus insertion—no oncogene present in virus genome (replication competent)
trans-Activating	Long (e.g., years)	Low (<1%)	Activation of cellular genes by *trans*-acting virus proteins (replication competent)

genome—there are **no homologous sequences present in normal cells**. Genes involved in the formation of tumors can be grouped by their biochemical functions:

- **Oncogenes and proto-oncogenes:** Oncogenes are mutated forms of proto-oncogenes, cellular genes whose normal function is to promote the normal growth and division of cells.
- **Tumor suppressor genes:** These genes normally function to inhibit the cell cycle and cell division: relevant examples are the **p53** and **retinoblastoma** (RB) genes in mammals, both involved in cell cycle regulation (see Fig. 7.6), and both targeted by viruses such as **HPV**.
- **DNA repair genes:** These genes ensure that each strand of genetic information is **accurately copied** during cell division of the cell cycle. Mutations in these genes lead to an increase in the frequency of other mutations (e.g., in conditions such as ataxia-telangiectasia and xeroderma pigmentosum).

The function of **oncogene products** depends on their cellular location (Fig. 7.3). Several classes of oncogenes are associated with the process of **signal transduction**—the transfer of information derived from the binding of **extracellular ligands to cellular receptors to the nucleus** (Fig. 7.4). Many of the kinases in these groups have a common type of structure with conserved functional domains representing the **hydrophobic transmembrane** and **hydrophilic intracellular kinase regions** (Fig. 7.5). These proteins are **associated with the cell membranes** or are present in the **cytoplasm**. Other classes of oncogenes located in the **nucleus** are normally involved with the control of the cell cycle (Fig. 7.6). The products of these genes overcome the restriction between the **G1 and S phases** of the cell cycle, which is the key control point in **preventing uncontrolled cell division**. Note, though, that **some viruses cause transition to S or pre-S phase** in order to allow cell synthesis of proteins such as DNA polymerases involved in their own replication.

Some virus oncogenes are not sufficient on their own to produce a fully transformed phenotype in cells; however, in some instances, they may **cooperate** with another **oncogene of complementary function** to produce a fully transformed phenotype—for example, the **adenovirus E1A gene** plus either the **E1B gene** or the **c-*ras* gene** transforms NIH3T3 cells (a mouse fibroblast cell line). This further underlines the fact that **oncogenesis is a complex, multistep process**.

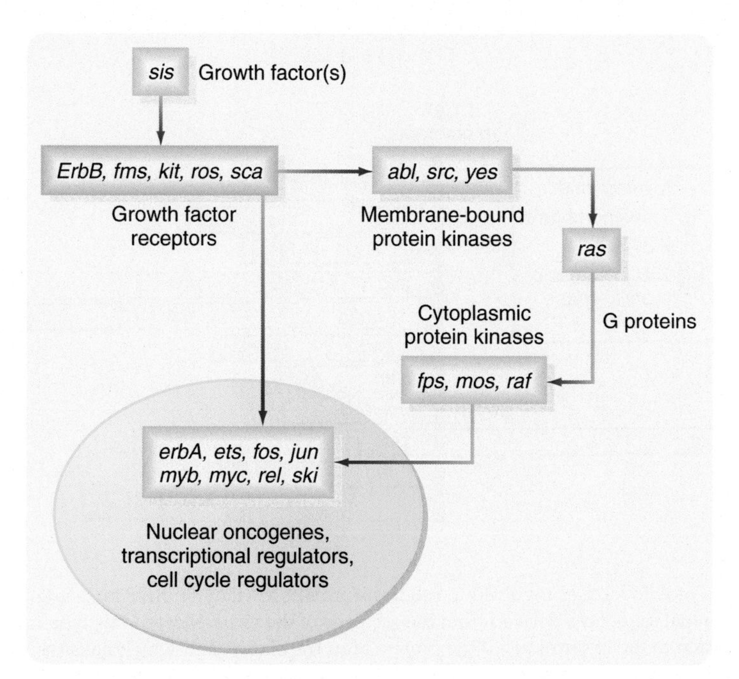

FIG. 7.3 Subcellular location of oncoproteins. The function of most oncogene products depends on their cellular location, for example, signal transduction proteins in the plasma membrane and cytoplasm, transcription factors in the nucleus, etc.

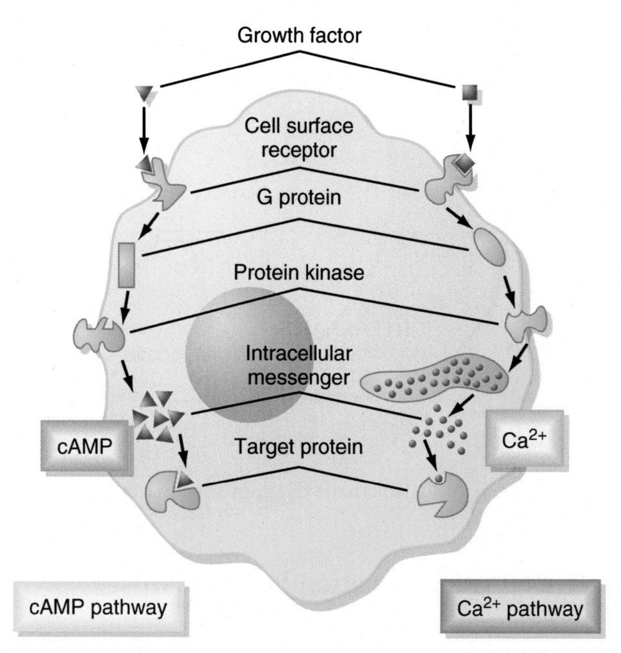

FIG. 7.4 Cellular mechanism of signal transduction. Several classes of oncogenes are associated with the process of signal transduction—the transfer of information derived from the binding of extracellular ligands to cellular receptors to the nucleus.

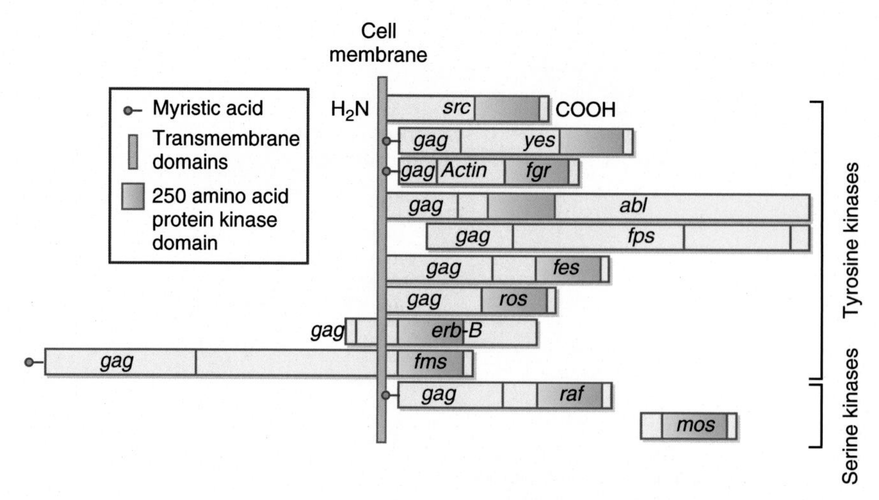

FIG. 7.5 Retrovirus protein kinases involved in cell transformation. Many of these molecules are fusion proteins containing amino-terminal sequences derived from the *gag* gene of the virus. Most of this type contain the fatty acid myristate which is added to the N-terminus of the protein after translation and which links the protein to the inner surface of the host-cell cytoplasmic membrane. In a number of cases, it has been shown that this posttranslational modification is essential to the transforming action of the protein.

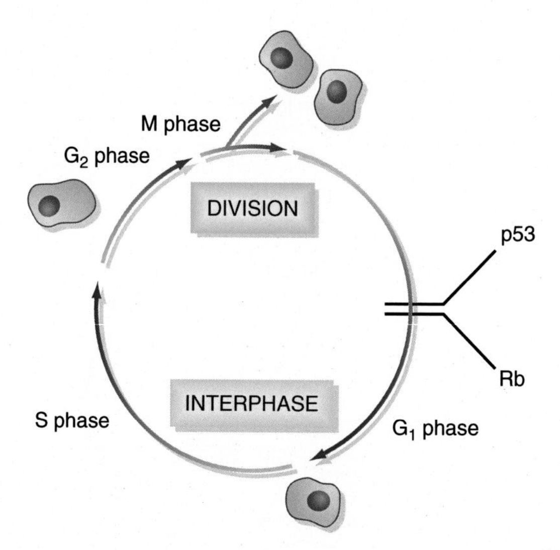

FIG. 7.6 Phases of the eukaryotic cell cycle. Schematic diagram showing the paths of the eukaryotic cell cycle discussed in the text.

Cell transformation by retroviruses

Not all retroviruses are capable of transforming cells—for example, **lentiviruses** such as HIV do not transform cells, although they are **cytopathic**. The retroviruses that can transform cells fall into three groups: these are **transducing**, ***cis*-activating**, and ***trans*-activating**. The characteristics of these groups are given in Table 7.2. If oncogenes are present in all cells, why does transformation occur as a result of virus infection? The reason is that **oncogenes may become activated in one of two ways**, either by **subtle changes to the normal structure of the gene** or by **interruption of the normal control of expression**. The transforming genes of the acutely transforming retroviruses (**v-*oncs***) are derived from and are highly homologous to **c-*oncs*** and are believed to have been **transduced** by viruses; however, most v-*oncs* possess slight alterations from their c-*onc* progenitors. Many contain **minor sequence alterations** that alter the **structure** and the **function** of the oncoprotein produced. Others contain **short deletions** of part of the gene. Most oncoproteins from **replication-defective, acutely transforming retroviruses** are **fusion proteins**, containing additional sequences derived from virus genes, most commonly **virus *gag* sequences at the amino-terminus of the protein**. These additional sequences may alter the function or the cellular localization of the protein, and these abnormal attributes result in transformation.

Alternatively, retroviruses may result in **abnormal expression** of an unaltered oncoprotein. This might be either the **overexpression** of an oncogene under the control of a **virus promoter** rather than its normal promoter in the cell, or it may be the **inappropriate temporal expression** of an oncoprotein that disrupts the cell cycle. **Chronic transforming retrovirus genomes** do not contain oncogenes. These viruses activate c-*oncs* by a mechanism known as **insertional activation**. A **provirus** that integrates into the host-cell genome close to a c-*onc* sequence may indirectly activate the expression of the gene in a way analogous to that in which v-*oncs* have been activated by transduction (Fig. 7.7). This can occur if the provirus is integrated upstream of the c-*onc* gene, which might be expressed via **a read-through transcript** of the virus genome plus downstream sequences; however, insertional activation can also occur when a provirus integrates **downstream** of a c-*onc* sequence or upstream but **in an inverted orientation**. In these cases, activation results from **enhancer elements** in the virus promoter (see Chapter 5). These can act even if the provirus integrates at a distance of several kilobases

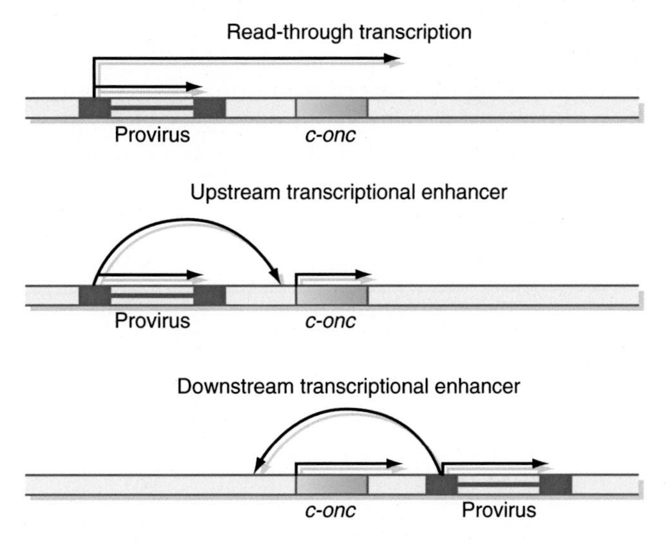

FIG. 7.7 Transcriptional activation of cellular oncogenes by insertional mutagenesis. Mechanisms by which cellular oncogenes can be transcriptionally activated by retrovirus insertional mutagenesis.

from the c-*onc* gene. The best-known examples of this phenomenon occur in **chickens**, where insertion of **avian leukosis virus** activates the ***myc*** gene, and in **mice**, where **mouse mammary tumor virus (MMTV)** insertion activates the ***int*** gene.

Transformation by the third class of retroviruses operates by quite a different mechanism. **Human T-cell leukemia viruses (HTLVs)** and related animal viruses (**bovine** leukemia, **simian** leukemia) encode a **transcriptional activator protein** in the virus ***tax*** gene. The Tax protein acts in *trans* to stimulate **transcription from the virus long terminal repeat**. It is believed that the protein also activates transcription of many cellular genes by interacting with cellular transcription factors (Chapter 5); however, **HTLV oncogenesis** (i.e., the formation of a leukemic tumor) has a **latent period of some 20–30 years**. Therefore, **cell transformation** (which can be mimicked in vitro) and **tumor formation** (which cannot) are not one and the same—additional events are required for the development of leukemia. It is thought that chromosomal abnormalities that may occur in the population of HTLV-transformed cells are also required to produce a malignant tumor, although because of the difficulties of studying this lengthy process, this is not completely understood.

Cell transformation by DNA viruses

In contrast to the **oncogenes** of retroviruses, the **transforming genes** of **DNA tumor viruses** (see Chapter 1) have no cellular counterparts. Several families of DNA viruses are capable of transforming cells (Table 7.3). In general terms, the functions of their oncoproteins are much less diverse than those encoded by retroviruses. They are **mostly nuclear proteins involved in the control of DNA replication** which directly affect the **cell cycle**. They achieve their effects by interacting with cellular proteins which normally appear to have a **negative regulatory role** in cell proliferation. Two of the most important cellular proteins involved are known as **p53** and **Rb** (for retinoblastoma).

p53 was originally discovered by virtue of the fact that it **forms complexes with SV40 T-antigen**. It is now known that it also interacts with other DNA virus oncoproteins, including those of **adenoviruses** and **papillomaviruses**. The gene encoding p53 is mutated or altered in the majority of tumors, implying that loss of the normal gene product is associated with the emergence of malignantly transformed cells. Tumor cells, when injected with the native protein in vitro, show a **decreased rate of cell division** and **decreased tumorigenicity** in vivo. **Transgenic** or **knockout mice** that do not possess an intact p53 gene are developmentally

TABLE 7.3 Transforming proteins of DNA tumor viruses.

Virus	Transforming protein(s)	Cellular target
Adenoviruses	E1A+E1B	Rb, p53
Polyomaviruses (SV40)	T-antigen	p53, Rb
Papillomaviruses:	E5	PDGF receptor
BPV-1	E6	p53
HPV-16, 18	E7	Rb

normal but are susceptible to the formation of **spontaneous tumors**; therefore, it is clear that p53 plays a central role in controlling the cell cycle. It is believed to be a tumor suppressor or "**antioncogene**" and has been called the "guardian of the genome." p53 is a transcription factor that activates the expression of certain cellular genes, notably **WAF1**, which encodes a protein that is an **inhibitor of G1 cyclin-dependent kinases**, causing the cell cycle to **arrest at the G1 phase** (Fig. 7.6). Because these viruses **require ongoing cellular DNA replication for their own propagation**, this explains why their transforming proteins target p53.

Rb was discovered when it was noticed that the gene that encodes this protein is always damaged or deleted in a **tumor of the optic nerve** known as **retinoblastoma**; therefore, the normal function of this gene is also thought to be that of a tumor suppressor. The Rb protein forms complexes with a **transcription factor called E2F**. This factor is required for the transcription of **adenovirus** genes, but E2F is also involved in the transcription of **cellular genes which drive quiescent cells into S phase**. The formation of inactive E2F-Rb complexes thus has the same overall effect as the action of p53—**arrest of the cell cycle at G1**. Release of E2F by replacement of E2F-Rb complexes with E1A-Rb, T-antigen-RB, or E7-RB complexes therefore **stimulates cellular and virus DNA replication**. Incidentally, **plant geminiviruses** also target the **RB homologue** in their hosts, in order to induce DNA synthesis so that their genomes can replicate.

The **SV40 T-antigen** is one of the known virus proteins that binds p53. Chapter 5 describes the role of large T-antigen in the regulation of SV40 transcription. Infection of cells by SV40 or other polyomaviruses can result in two possible outcomes. These are:

- Productive (**lytic**) infection
- Nonproductive (**abortive**) infection

The outcome of infection appears to be determined primarily by the cell type infected; for example, **mouse polyomavirus** establishes a **lytic** infection of mouse cells but an **abortive infection** of rat or hamster cells, while **SV40** shows **lytic** infection of monkey cells but **abortive infection** of mouse cells. However, in addition to transcription, T-antigen is also involved in genome replication. SV40 DNA replication is initiated by binding of large T-antigen to the origin region of the genome (see Fig. 5.27). The function of T-antigen is controlled by **phosphorylation** by a **kinase**, which decreases the ability of the protein to bind to the SV40 origin.

The SV40 genome is very small and does not encode all the information necessary for DNA replication; therefore, it is essential for the host cell to **enter S phase**, when cell DNA and the virus genome are replicated together. **Protein-protein interactions** between T-antigen and **DNA polymerase** α directly stimulate replication of the virus genome. The precise regions of the T-antigen involved in binding to **DNA, DNA polymerase** α**, p53,** and **Rb** are all known (Fig. 7.8). Inactivation of tumor suppressor proteins bound to T-antigen **causes G1-arrested cells to enter S phase and divide**, and this is the mechanism that results in transformation; however, the **frequency** with which abortively infected cells are transformed is low (**about** $\mathbf{1 \times 10^{-5}}$). Therefore, the function of T-antigen is to alter the cellular environment to permit virus DNA replication. Transformation is a **rare and accidental consequence** of the sequestration of tumor suppressor proteins.

The immediate-early proteins of **adenoviruses** are analogous in many ways to SV40 T-antigen. **E1A** is a **trans-acting** transcriptional regulator of the adenovirus early genes (see

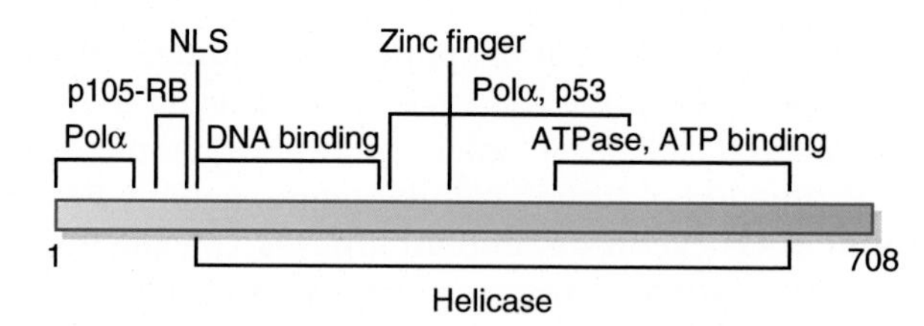

FIG. 7.8 Regions of SV40 T-antigen involved in protein-protein interactions. Other functional domains of the protein involved in virus DNA replication are also shown, including the helicase, ATPase, and nuclear location signal (NLS) domains.

Chapter 5). Like T-antigen, the **E1A protein binds to Rb**, inactivating the regulatory effect of this protein, permitting virus DNA replication, and **accidentally stimulating cellular DNA replication**. **E1B binds p53** and reinforces the effects of E1A. The combined effect of the two proteins can be seen in the phenotype of cells transfected with DNA containing these genes (Table 7.4). However, the interaction of these transforming proteins with the cell is more complex than simple induction of DNA synthesis. Expression of **E1A alone** causes cells to undergo **apoptosis**. Expression of **E1A and E1B together** overcomes this response and permits transformed cells to survive and grow.

Human papillomavirus (HPV) genital infections are very common, occurring in **more than 50%** of young, sexually active adults, and are usually **asymptomatic**. This is in fact the case for most of the **cutaneous HPVs**, which are only known about because of their detection via PCR amplification of viral DNA from **swabs of healthy skin**—including from newborn babies. However, certain HPV **types** (older taxonomic term based on genome homology) are **high risk**, and are associated with a risk of subsequent development of anogenital cancers such as **cervical, penile** and **anal carcinoma**, after an incubation period of several decades.

HPV is the primary cause of cervical cancer: **99.7%** of all **cervical squamous cell cancer** cases worldwide test positive for one or more high-risk type of HPV. Of the **200**-odd HPV types currently recognized, only **14** are associated with a high risk of tumor formation (**HPVs 16, 18, 31, 33, 35, 39, 45, 51, 52, 56, 58, 59, 66, and 68**). HPVs 16 and 18 together cause around 70% of cervical cancer cases, hence their inclusion in the first HPV vaccines. However, the new Gardasil-9 vaccine also includes types **31, 33, 45, 52, and 58**, which could prevent another 20% of cases. Nearly **600,000 new cases** of cervical cancer were diagnosed in 2018, with over **300,000 deaths**, more than **90%** of which were in low and middle-income countries. This makes cervical cancer the **fourth most common cause of cancer death in women** globally. However, high-risk HPVs are also strongly implicated in a variety of other human cancers:

TABLE 7.4 Role of the adenovirus E1A and E1B proteins in cell transformation.

Protein	Cell phenotype
E1A	Immortalized but morphologically unaltered; not tumorigenic in animals
E1B	Not transformed
E1A+E1B	Immortalized and morphologically altered; tumorigenic in animals

these include **vulvar, vaginal, anal, oropharyngeal,** and other **head and neck cancers**, with the last three types also occurring in men along with **penile** cancers.

Once again, transformation is mediated by the **early gene products of the virus**. However, the transforming proteins appear to vary from one type of papillomavirus to another, as shown in Table 7.3. In general terms, it appears that **two or more early proteins often cooperate** to give a transformed phenotype. Although some papillomaviruses can transform cells on their own (e.g., bovine papillomavirus 1, BPV-1), others appear to require the cooperation of an **activated cellular oncogene** (e.g., HPV-16/*ras*). In **bovine papillomavirus**, it is the **E5 protein** that is responsible for transformation. In HPV-16 and HPV-18, the **E6 and E7 proteins**—now well-characterized **oncogenes**—are involved (Fig. 7.9).

While the circular dsDNA HPV genome is maintained **episomally** in infected cells and in cells at an early stage of transformation, in **most tumors all or part of the papillomavirus genome**, including the putative transforming genes, is maintained as an **integrated linear form** in the tumor cells. This integration **disrupts the E2 gene**, leading to a **loss of repression** of the expression of the **E6 and E7 oncogenes**, the products of which interfere with the function of **p53** and **Rb**, respectively. However, in some cases (e.g., **BPV-4**) the virus DNA may be lost after transformation, which may indicate a possible **hit-and-run mechanism** of transformation. Different papillomaviruses appear to use slightly different mechanisms to achieve genome replication, so cell transformation may proceed via a slightly different route depending on the virus. It is imperative that a better understanding of these processes is obtained.

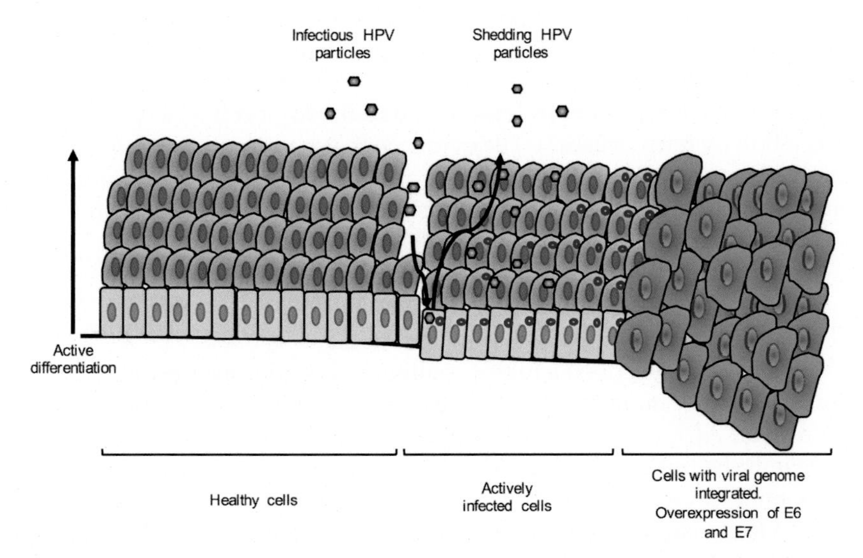

FIG. 7.9 The life cycle of a typical high-risk HPV. Infection occurs at basal epithelial cells through anatomically accessible points such as microlesions. The genomes of HPVs stay as episomes in the host cells nuclei. Cells proliferate and differentiate. The expression of structural proteins, L1 and L2, viral assembly and release only occur at late stages of the cell life cycle. Integration of the viral genome into the host's genome leads to overexpression of E6 and E7 which disrupt the cell life cycle regulation which promotes prolonged cell life leading to genomic instability and cancer. At this stage no viral structural proteins are expressed. *From Chabeda, A., Yanez, R.J.R., Lamprecht, R., Meyers, A.E., Rybicki, E.P., Hitzeroth, I.I., 2018. Therapeutic vaccines for high-risk HPV-associated diseases. Papillomavirus Res. 5, 46–58. https://doi.org/10.1016/j.pvr.2017.12.006.*

In recent years, evidence has emerged that p53 and Rb are major cellular sensors for **apoptosis**. Loss of these protein functions triggers apoptosis—**the major anticancer mechanism in cells**; thus, viruses that interfere with these proteins must have evolved mechanisms to counteract this effect (see discussion in Chapter 6).

Viruses and cancer

There are numerous examples of **viruses that cause tumors in experimental animals**—for example, **SV40** and **mouse mammary tumor virus** (see Chapter 1)—which stimulated a long search for viruses that might be the cause of cancer in humans. For many years, this search was unsuccessful; so much so that a few scientists categorically stated that viruses did not cause human tumors. Like most rash statements, this was wrong: **up to 20% of all human cancers worldwide may be caused by viruses**.

Although it is convenient to consider human tumor viruses as a discrete group of viruses, the **seven viruses** so far known to cause human cancers have very different genomes, replication cycles, and come from **six different virus families**. The viruses are **EBV/HHV-4** and **Kaposi's sarcoma associated (HHV-8)** herpesviruses, **HBV** (hepadnavirus), **HCV** (hepacivirus), **HPVs**, **HTLVs** (retroviruses), and **Merkel cell polyomavirus**. The path from virus infection to tumor formation is slow and inefficient. Only a minority of infected individuals progress to cancer, usually years or even decades after primary infection. Virus infection alone is generally not sufficient for cancer, and additional events and host factors, such as **immunosuppression, somatic mutations, genetic predisposition,** and **exposure to carcinogens**, must also play a role.

The role of the **HTLV tax protein in leukemia** has already been described (see the section "Cell transformation by retroviruses"). The evidence that **papillomaviruses** may be involved in human tumors is now well established. There are almost certainly many more viruses that cause human tumors, but the remainder of this chapter describes two examples that have been intensively studied: these are **EBV** or **HHV-4**, and **HBV**.

Epstein-Barr virus

In 1962, Dennis Burkitt described a highly malignant **lymphoma**, the distribution of which in Africa paralleled that of **malaria**. EBV was first identified in 1964 in a **lymphoblastoid cell line** derived from an African patient with Burkitt's lymphoma. Burkitt recognized that this tumor was rare in India but occurred in Indian children living in Africa and therefore looked for an environmental cause. Initially, he incorrectly thought that the tumor might be caused by a **virus spread by mosquitoes**. To be fair, the association between EBV and Burkitt's lymphoma was not entirely clear cut:

- **EBV is widely distributed** worldwide but Burkitt's lymphoma is **rare**.
- EBV is found in **many cell types** in Burkitt's lymphoma patients, **not just in the tumor cells**.
- Rare cases of **EBV-negative Burkitt's lymphoma** are sometimes seen in countries where malaria is not present, suggesting there may be **more than one route** to this tumor.

EBV has a **dual cell tropism** for **human B-lymphocytes** (generally a **nonproductive** infection) and **epithelial cells**, in which a **productive infection** occurs. The usual outcome of EBV infection is **polyclonal B-cell activation** and a benign **proliferation** of these cells which is frequently asymptomatic but sometimes produces a relatively mild disease known as **infectious mononucleosis** or **glandular fever**. This was also once known as the "kissing disease," because of its high incidence in people attending colleges and universities for the first time in the US prior to the 1960s. In 1968, it was shown that EBV could efficiently **transform** (i.e., immortalize) **human B-lymphocytes** in vitro. This observation clearly strengthens the case that EBV is involved in the formation of tumors. It is also a very useful feature of the virus that allows its now-routine use for this purpose in the **production of human mAbs** (see Chapter XX). There is now epidemiological and/or molecular evidence that EBV infection is associated with **at least five human tumors**. These are:

- Burkitt's lymphoma.
- **Nasopharyngeal carcinoma** (NPC), a highly malignant tumor seen most frequently in China. There is a strong association between EBV and NPC. Unlike Burkitt's lymphoma, **the virus has been found in all the tumors that have been studied**. Environmental factors, such as the consumption of nitrosamines in salted fish, are also believed to be involved in the formation of NPC (cf. the role of malaria in the formation of Burkitt's lymphoma).
- **B-cell lymphomas** in immunosuppressed individuals (e.g., AIDS patients).
- Some clonal forms of **Hodgkin's lymphoma**.
- **X-linked lymphoproliferative syndrome (XLP)**, a rare condition usually seen in males where infection with EBV results in a **hyperimmune response**, sometimes causing a fatal form of glandular fever and sometimes cancer of the lymph nodes. XLP is an inherited defect due to a faulty gene on the X chromosome.

Cellular transformation by EBV is a complex process involving the cooperative interactions between several viral proteins. Three possible explanations for the link between EBV and Burkitt's lymphoma are:

1. EBV **immortalizes a large pool of B-lymphocytes**; concurrently, **malaria** causes **T-cell immunosuppression**. There is thus a large pool of target cells in which a **third event** (e.g., a **chromosomal translocation**) results in the formation of a malignantly transformed cell. Most Burkitt's lymphoma tumors contain translocations involving **chromosome 8**, resulting in **activation of the c-*myc* gene**, which supports this hypothesis.
2. **Malaria** results in **polyclonal B-cell activation**. EBV subsequently **immortalizes a cell** containing a **preexisting c-*myc* translocation**. This mechanism would be largely indistinguishable from the above.
3. EBV is just a **passenger virus**! Burkitt's lymphoma also occurs in Europe and North America although it is very rare in these regions; however, **85% of these patients are not infected with EBV**, which implies that there are other causes for Burkitt's lymphoma.

Although it has not been formally proved, it seems likely that either (1) and/or (2) is the true explanation for the origin of Burkitt's lymphoma.

EBV/HHV-4 may also be associated with the **causation of multiple sclerosis (MS)**: a major recent study conducted between 1993 and 2013 on more than 10 million US military personnel found a much higher incidence of prior EBV infection in people who developed MS than in controls, and no association with any other human virus (see Doctrow, 2022). This indicates that even this recalcitrant disease could perhaps be prevented by an EBV vaccine.

Hepatitis B virus

Another case where a virus appears to be associated with the formation of a human tumor is that of **HBV and hepatocellular carcinoma (HCC)**. The term "hepatitis" simply means an **inflammation of the liver**, and as such is not a single disease. Because of the central role of the liver in metabolism, many virus infections may involve the liver; however, **at least seven viruses** seem specifically to **infect and damage hepatocytes**, the most common cell type in the liver. No two of these belong to the same family, although all except HBV are **ssRNA+** viruses: **hepatitis A virus** (HAV) is a picornavirus; **HCV** is a hepacivirus (family *Flaviviridae*); **HDV**—or hepatitis delta—is a **satellite virus** whose genome resembles a viroid; **HEV** is a calicivirus; **HGV** is a flavivirus.

HBV is the prototype member of the family *Hepadnaviridae* and causes the disease formerly known as "**serum hepatitis**." This disease was distinguished clinically from "**infectious hepatitis**" (caused by other hepatitis viruses) in the 1930s. HBV infection a few decades ago was often the result of inoculation with **human serum** (e.g., via **inadvertent needle contamination** in clinics; via **blood transfusions**), or obtained via **organ transplants**, and was common among **intravenous drug abusers**. The virus is also commonly transmitted **sexually**, and by **oral ingestion** in children and from **mother to child**, which accounts for **familial clusters** of HBV infection. However, **all blood, organ, and tissue donations** in many countries are now tested for HBV, and risk of transmission by these means is extremely low.

The virus **does not replicate in tissue culture**, and the only animal model was the **chimpanzee**, which has seriously hindered investigations into its pathogenesis. HBV infection has three possible clinical outcomes:

1. An **acute infection** followed by complete recovery and **immunity from reinfection** (**>90%** of cases).
2. **Fulminant hepatitis**, developing quickly and lasting a short time, causing liver failure and a **mortality rate of approximately 90% (<1%** of cases).
3. **Chronic infection**, leading to the establishment of a **carrier state** with virus persistence (about **10%** of cases).

There are approximately **350 million chronic HBV carriers** worldwide. The total population of the world is approximately **7 billion**; therefore, about **5% of the world population** is persistently infected with HBV. All of these chronic carriers of the virus are at **100–200 times** the risk of noncarriers of **developing HCC**. This is a rare tumor in the Western hemisphere, where it represents **<2% of fatal cancers**. Most cases that do occur in the West are **alcohol related**, and this is an important clue to the pathogenesis of the tumor. In **Southeast Asia and in China, HCC is the most common fatal cancer**, resulting in about **500,000 deaths** every year.

The virus could cause the formation of the tumor by three different pathways: these are **direct activation** of a cellular oncogene(s), ***trans*-activation** of a cellular oncogene(s), or **indirectly** via **persistent tissue regeneration** (Fig. 7.10). As with EBV and Burkitt's lymphoma, the relationship between HBV and HCC is not clear cut:

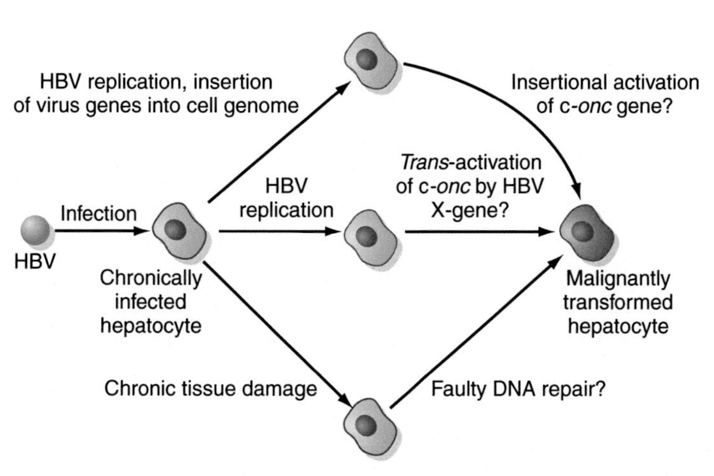

FIG. 7.10 Possible mechanisms of HCC formation due to HBV infection. The complex relationship between HBV infection and HCC means that it is not certain whether any or all of these possible mechanisms are involved.

- **Cirrhosis** is a hardening and loss of organization of the liver tissue, which is the result of infections or various toxins, such as alcohol, and appears to be a **prerequisite for the development of HCC**. Chronic liver damage induces **tissue regeneration**, and if this happens often enough, **faulty DNA repair mechanisms** result eventually in **malignant cell transformation**. Unrelated viruses that cause **chronic active hepatitis**, such as **HCV**, are also associated with HCC after a long latent period.
- A number of **cofactors**, such as **aflatoxins** and **nitrosamines** (both potent carcinogens), can induce HCC-like tumors in experimental animals without virus infection; therefore, such substances may also be involved in human HCC (cf. nitrosamines and NPC, above).

For many years, it was thought that **HBV genome integration events** were random with regard to their sites within the human genome, but when the relationship between *"**fragile sites**"* in the host genome and virus integration events are compared, HBV DNA is found to integrate **within or near** many of these fragile regions. In most cases, integration at a particular site has only been reported for a single or small number of tumors. However, a closer look shows that individual integration sites alter the expression of different components **in the same or redundant biochemical or signaling pathways** that support the **hepatocellular growth and survival** that is important for tumor development. Most (but not all) HBV integration events retain the open reading frame encoding the **HBx antigen (HBxAg;** see Fig. 5.24), which suggests that this protein contributes to HCC, **possibly via miRNAs**. It is in fact possible that all the mechanisms shown in Fig. 7.10 might operate in vivo. The key risk factor, however, is the **development of a chronic as opposed to an acute HBV infection**. This in itself is determined by a number of other factors. These are:

- **Age**—the frequency of chronic infections **declines with increasing age** at the time of infection.
- **Sex**—for chronic infection, the **male:female ratio is 1.5:1**; for cirrhosis, the **male:female ratio is 3:1**.

- **HCC**—The male:female ratio is **6:1**.
- **Route (and possibly dose) of infection**—Oral or sexual infections give rise to fewer cases of chronic infection than infection via **serum**.

Until there is a much better understanding of the pathogenesis and normal course of HBV infection, it is unlikely that the reasons for these differences will be understood. There may be a happy ending to this story, however. A safe and effective **recombinant subunit vaccine** that prevents HBV infection has been available since 1986, and has been widely used in the areas of the world where HBV infection is **endemic** as part of the WHO Expanded Programme on Immunization. This means that it has been given free to most of the children in developing countries for years now: this will prevent **a million deaths annually** from HCC and HBV disease in the future.

Other viruses in human cancers

As mentioned earlier, there are seven viruses involved in the causation of human cancers. We have covered EBV and HBV; here we will summarize what is known about the other five.

Kaposi sarcoma-associated herpesvirus, or HHV-8

Kaposi's sarcoma (KS) is caused by the eponymous virus (abbreviated **KSHV**), and is an especially common cancer in **HIV-infected individuals** who are not on ARVs, and are **immunosuppressed**. The virus infects endothelial cells, and interferes with pathways controlling cell proliferation, and gene expression and metabolism. The tumors that form appear dark because of the mass of tiny blood vessels they contain, and can appear on the skin, on mucosal surfaces, or in **the lungs, liver, stomach, intestines,** and **lymph nodes. ARV treatment** in HIV+ people can in fact help the lesions regress.

Hepatitis C virus

HCV is in fact **the most common cause of HCC**, in which it has a multifactorial role. **Chronic infections** can activate pathways that lead to **liver fibrosis**, that interfere with **induction of apoptosis** and **promote cellular survival**. Progression of disease through to HCC can take decades: it involves **progressive fibrosis** of the liver by **damage and regeneration**, as well as **irreversible genetic and epigenetic alterations** at the cellular level. Malignant transformation of hepatocytes involves several pathways, including mutations of **p53, PIKCA,** and **beta-catenin genes**. HCC risk is partly dependent on **viral genotype**, with HCV genotypes **3, 25, 26,** and **27** being more carcinogenic than others.

Human papillomaviruses

HPVs and causation of cancers have been dealt with to some extent above, in the context of both **interfering with p53 and Rb function**, and **dysregulation of the viral genome** by integration. More importantly, however is the fact that there has been a steep decline (~70% from 1955 to 1992) in cervical cancer mortality in developed countries, due largely to the widespread application of cervical cancer screening in the form of **Papanicolaou smears** (Pap tests), which use **cytology** to diagnose the presence of abnormal cells. Further, the HPV vaccination campaigns that began in the **early 2000s** have also caused a **marked**

decrease in the incidence of cervical intraepithelial neoplasias, the cervical lesions that are the precursors of cervical cancers. It is in fact predicted that high HPV vaccination coverage in low and middle income countries, even of girls only, **could lead to cervical cancer elimination by the end of the 21st century** (see Brisson et al., 2020). It is expected that other HPV-related cancers will also decline drastically—and especially if boys are included in vaccination drives.

Human T-cell leukemia

While HTLV infections (**HTLVs I, II, and IV**) rarely cause serious disease, **HTLV-I** infection (**~10 million** worldwide infections) may lead to **adult T-cell leukemia (ATL) in between 2% and 5% of infected people**. Transmission occurs from mother-to-child via **breastfeeding**, by **sexual interaction**, and via **blood transfusion** or potentially, contaminated needles. It is difficult to work with HTLVs in cell culture; however, **infected T cells** undergo permanent **genetic** and ***epigenetic*** changes. The disease is classed as a highly aggressive mature T-cell neoplasm, and **acute and lymphoma** subtypes of ATL have a poor prognosis.

Merkel cell polyomavirus

This virus (MCV) has recently been **firmly linked to the rare skin cancer Merkel cell carcinoma**—which is the first and **only cancer** so far linked to a **human polyomavirus** (or any polyomavirus—see Chapter 1). The cancer is **neuroendocrine** in origin, and occurs **in immune-suppressed** or partially suppressed **sun-exposed people**, and usually only the elderly. MCVs can be found associated with the **skin microbiome** in the absence of any disease, as is also true for many HPVs.

Adeno-associated virus

A recent report in Nature Genetics (Nault et al., 2015) implicates **adeno-associated virus type 2 (AAV2)** in causation of **human hepatocellular carcinoma (HCC)**—specifically by means of insertional mutagenesis in "cancer driver" genes. Clonal integrations—with the same genome at the same site(s)—were found in 11 of 193 HCCs sampled, and the authors noted multiple insertions in some tumors, with significant effects on gene expression. It is especially interesting that tumors with integrated viral genomes were found in non-cirrhotic liver (9 of 11 cases) and tumors in patients without known risk factors (6 of 11 cases). The authors suggest a clear pathogenic role for AAV2 in these cases, and conclude that *"AAV2 is a DNA virus associated with oncogenic insertional mutagenesis in human HCC."*

Summary

Virus pathogenesis is a complex, variable, and relatively rare state. Like the course of a virus infection, **pathogenesis is determined by the balance between host and virus factors**. Not all of the pathogenic symptoms seen in virus infections are caused directly by the virus—the **immune system** also plays a significant role in causing cell and tissue damage. Viruses can **transform** cells in the course of their normal infections, so that they continue to grow indefinitely. In some but not all cases, this can lead to the formation of **tumors**. There are some well-established cases where certain viruses provoke human tumors, and possibly many others that we do not yet understand. The relationship between the virus and the formation of the tumor is not a simple one, but **prevention of infection** undoubtedly significantly reduces the risk of tumor formation.

References

Brisson, M., Kim, J.J., Canfell, K., Drolet, M., Gingras, G., Burger, E.A., Martin, D., Simms, K.T., Bénard, É., Boily, M.C., Sy, S., Regan, C., Keane, A., Caruana, M., Nguyen, D.T.N., Smith, M.A., Laprise, J.F., Jit, M., Alary, M., Bray, F., Fidarova, E., Elsheikh, F., Bloem, P.J.N., Broutet, N., Hutubessy, R., 2020. Impact of HPV vaccination and cervical screening on cervical cancer elimination: a comparative modelling analysis in 78 low-income and lower-middle-income countries. Lancet 395 (10224), 575–590. https://doi.org/10.1016/S0140-6736(20)30068-4 (Epub 2020 Jan 30).

Doctrow, B., 2022. Study Suggests Epstein-Barr Virus May Cause Multiple Sclerosis. NIH Research Matters. February 1 2022 https://www.nih.gov/news-events/nih-research-matters/study-suggests-epstein-barr-virus-may-cause-multiple-sclerosis.

Gkoutzourelas, A., Bogdanos, D.P., Sakkas, L.I., 2020. Kawasaki disease and COVID-19. Mediterr. J. Rheumatol. 31 (Suppl 2), 268–274. https://doi.org/10.31138/mjr.31.3.268.

Nault, J.C., Datta, S., Imbeaud, S., Franconi, A., Mallet, M., Couchy, G., Letouzé, E., Pilati, C., Verret, B., Blanc, J.F., Balabaud, C., Calderaro, J., Laurent, A., Letexier, M., Bioulac-Sage, P., Calvo, F., Zucman-Rossi, J., 2015. Recurrent AAV2-related insertional mutagenesis in human hepatocellular carcinomas. Nat. Genet. 47 (10), 1187–1193. https://doi.org/10.1038/ng.3389.

Panelli, S., Lorusso, L., Balestrieri, A., Lupo, G., Capelli, E., 2017. XMRV and public health: the retroviral genome is not a suitable template for diagnostic PCR, and its association with Myalgic encephalomyelitis/chronic fatigue syndrome appears unreliable. Front. Public Health 5, 108. https://doi.org/10.3389/fpubh.2017.00108.

ViralZone, Bacteriophage-Encoded Exotoxin. https://viralzone.expasy.org/3967#:~:text=Bacteriophage%2Dencoded%20toxins%20(e.g.%20botulism,symptoms%20associated%20with%20human%20diseases.

Recommended reading

Angel, R.M., Valle, J.R., 2013. Dengue vaccines: strongly sought but not a reality just yet. PLoS Pathog. 9 (10), e1003551.

Brubaker, J., 2019. The 7 Viruses That Cause Human Cancers. American Society for Microbiology. https://asm.org/Articles/2019/January/The-Seven-Viruses-that-Cause-Human-Cancers.

Durer, C., Babiker, H.M., 2021. Adult T-cell leukemia. In: StatPearls. StatPearls publishing, Treasure Island, FL. [updated 2021 Jun 30]. Available from: https://www.ncbi.nlm.nih.gov/books/NBK558968/.

Kaplan, J.E., 2020. Kaposi's Sarcoma (KS). WebMD. https://www.webmd.com/hiv-aids/guide/aids-hiv-opportunistic-infections-kaposis-sarcoma.

Thorley-Lawson, D.A., Allday, M.J., 2008. The curious case of the tumour virus: 50 years of Burkitt's lymphoma. Nat. Rev. Microbiol. 6 (12), 913–924. https://doi.org/10.1038/nrmicro2015.

CHAPTER

8

Panics and pandemics

INTENDED LEARNING OUTCOMES

On completing this chapter, you should be able to:

- Define and differentiate between emerging and reemerging viruses.
- Describe how a virus could emerge to cause disease in humans.
- Understand the factors that are important in recognizing and dealing with a potentially pandemic virus.

New, emerging, and re-emerging viruses

What constitutes a "new" infectious agent? Are these just viruses that were never seen previously, or are they previously known viruses that have reappeared or changed their behavior? This chapter will describe and attempt to explain current understanding of a number of agents that meet the above criteria.

In the last century, massive and unexpected ***epidemics***—a disease outbreak affecting a community, population or limited geographical region—and ***pandemics***—epidemics that spread over multiple regions—have been caused by certain viruses. For the most part, these epidemics have not been caused by completely new (i.e., previously unknown) viruses but by viruses that were well known in certain geographical areas—in other words, were ***endemic*** in certain limited areas—in which they may cause **recurrent epidemics or outbreaks** of disease. Such viruses are known as **emerging or re-emerging viruses** (Table 8.1), depending on whether or not they have previously broken out of their geographical and/or climatic restraints. Such emergences may be climate-related, as in heavy rainy seasons providing pools of water for mosquitoes to lay eggs in, or related to movement of vectors or hosts into new areas due to loss of habitat, or the result of intrusion of humans into areas that were previously uninhabited or uncultivated. There are also numerous examples of viruses that appear to have **altered their behavior with time**, with significant effects on their pathogenesis, and possibly on their ability to spread. The World Health Organization (WHO) published an updated list of **priority diseases** in 2021—all caused by viruses—that needed urgent research

Cann's Principles of Molecular Virology
https://doi.org/10.1016/B978-0-12-822784-8.00008-8

TABLE 8.1 Some examples of emerging viruses.

Virus	Type	Comments
Cacao swollen shoot	Badnavirus	Emerged in 1936 and is now the main disease of cacao in Africa, seriously affecting cocoa and chocolate production. Deforestation increases population of mealy bug vectors and disease transmission
Hendra virus	Paramyxovirus	Emerged in Brisbane, Australia, September 1994. Causes acute respiratory disease in horses with high mortality and a fatal encephalitis in humans—with several deaths so far. The disease, normally carried by fruit bats (with no pathogenesis), has reemerged in humans in Queensland several times since 1994
Nipah virus	Paramyxovirus	Emerged in Malaysia in 1998. Closely related to Hendra virus; a zoonotic virus transmitted from animals (pigs, bats) to humans. Mortality rate in outbreaks of up to 70%
Phocine distemper	Paramyxovirus	Emerged in 1987 and caused high mortalities in seals in the Baltic and North Seas. Similar viruses subsequently recognized as responsible for cetacean (porpoise and dolphin) deaths in Irish Sea and Mediterranean. The virus was believed to have been introduced into immunologically naive seal populations by a massive migration of harp seals from the Barents Sea to northern European coasts
Rabbit hemorrhagic disease (RHD), also known as rabbit calicivirus disease (RCD) or viral hemorrhagic disease (VHD)	Calicivirus	Emerged in farmed rabbits in China in 1984, spread through the United Kingdom, Europe, and Mexico. Introduced to Wardang Island off the coast of South Australia to test potential for rabbit population control, the disease accidentally spread to Australian mainland, causing huge kill in rabbit populations. A vaccine is available to protect domestic and farmed rabbits. In August 1997, RHD was illegally introduced into the South Island of New Zealand and escaped into the United States in April 2000

and development attention, both for understanding their potential to cause major outbreaks, and for development of vaccines. These are the following:

- COVID-19
- Crimean-Congo hemorrhagic fever
- Ebola and Marburg virus diseases
- Lassa fever
- Middle East respiratory syndrome (MERS-CoV) and severe acute respiratory syndrome (SARS) coronaviruses
- Nipah and other henipaviral diseases
- Rift Valley fever
- Zika
- "Disease X: *a pathogen currently unknown to cause human disease*"

https://www.who.int/activities/prioritizing-diseases-for-research-and-development-in-emergency-contexts

"Pathogen X" for the period 2019–22 seems to be **SARS-CoV-2,** which has recently usurped pole position in terms of priority for vaccine and other research.

Newly **emerging viruses** are generally ***zoonotic*** in origin: that is, they have emerged out of wildlife either directly, or via domestic animals. The increasing incidence of such diseases—viral as well as bacterial, protozoan and other parasites—has sparked the growth of the **One Health initiative**, which "*...recognises that the health of people is closely connected to the health of animals and our shared environment,*" and aims to promote collaboration across the human and animal and plant health and environmental research sectors, to achieve the best health outcomes for all (https://www.onehealthcommission.org/).

The importance of the approach can be seen when one realizes that **6 out of every 10** infectious diseases in humans are **zoonotic**, and **7 out of 10** of **emerging or re-emerging infections** are **vector-borne or zoonotic.** The kinds of emerging zoonotic disease agent that are of concern to One Health include:

- those that are transmitted directly from wild animals to humans (e.g., **hantaviruses;** rodent borne **arenaviruses** such as **Lujo** and **Lassa** viruses**; rabies** virus**; SARS** and **SARS2** coronaviruses);
- agents that **originate in wild animals and then spread human to human** (e.g., **HIV-1** and -2; **Ebola** and **Marburg** viruses; **SARS** and **MERS** and **SARS2**);
- agents that are **transmitted from wild to domestic animals to humans** (e.g., **Nipah** and **Hendra** viruses; **MERS** and potentially **SARS2**); and
- those that **move from wild to domestic animals** then go on to be **transmitted long-term between humans** (e.g., **pandemic influenza viruses, SARS-CoV-2**).
- **Vector-borne viruses** that use an **animal amplification host** and are transmitted by **arthropod vectors** such as mosquitoes and ticks, *Culicoides* midges and sandflies to sensitive animals and humans, where they can cause severe disease. This includes **West Nile virus, Rift Valley fever virus (RVFV),** and **Crimean Congo hemorrhagic fever virus (CCHFV),** emerging arboviruses such as **Zika virus,** and **endemic mosquito transmitted viruses.**

Re-emerging viruses

Polioviruses

One of the better-known examples of this phenomenon is **poliovirus**. It is known that poliovirus and poliomyelitis have existed in human populations for at least 4000 years. For most of this time, the pattern of disease was **endemic** rather than **epidemic**: that is, there was a low, continuous level of infection in particular geographical areas. During the first half of the 20th century, the pattern of occurrence of poliomyelitis in Europe, North America, and Australia changed to an epidemic one, with vast annual outbreaks of infantile paralysis. Although we do not have samples of polioviruses from earlier centuries, the clinical symptoms of the disease give no reason to believe that the virus changed substantially. Why, then, did the pattern of disease change so dramatically? It is believed that the reason is as follows. In rural communities with primitive sanitation facilities, poliovirus circulated freely. Serological surveys in similar contemporary situations reveal that **more than 90% of children of 3 years of age** have antibodies to at least one of the three serotypes of poliovirus. We note that even the **most virulent strains of poliovirus** cause **100–200 subclinical infections** for each case of **paralytic poliomyelitis** seen. In such communities, infants experience subclinical immunizing infections while still protected by maternal antibodies—a form of natural **vaccination**. The relatively few cases of paralysis and death that do occur are likely to be overlooked, especially in view of high infant mortality rates.

During the 19th century, industrialization and urbanization changed the pattern of poliovirus transmission. Dense urban populations and increased traveling afforded opportunities for rapid transmission of the virus. In addition, **improved sanitation** broke the natural pattern of virus transmission. Children were likely to encounter the virus for the first time **at a later age** and without the protection of **maternal antibodies**. These children were at far greater risk when they did eventually become infected, and it is believed that these social changes account for the altered pattern of disease. Much the same is true for another picornavirus disease, **hepatitis A**: people in developing countries are far more likely to have had hepatitis A as a small child—when it is almost always a **mild disease**—than people in developed economies. The result of this is that adults in the latter are more likely have severe disease, which can kill **2% of infected adults**.

Fortunately, the widespread use of poliovirus vaccines has since brought the situation under control in industrialized countries (Chapter 6). In 1988, the World Health Organization committed itself to wiping out polio completely ("**eradication**") by the year 2000. However, the disease has proved to be troublingly resilient in a few of the poorest and most dangerous countries, and is still hanging on. This is despite the fact that **wild-type polioviruses types 1 and 3 have been eradicated**: all that is left are a few pockets of wild-type poliovirus type 2 in Pakistan and Afghanistan, and—worryingly—a wider distribution of **vaccine-derived poliovirus type 2** that circulates uncontrolled, and can cause **poliomyelitis**. Polio eradication is no longer a technical challenge; rather, it is a political and economic one. The same is true for hepatitis A: **effective killed vaccines** are available for the one serotype of the virus that is found worldwide; however, there have recently been outbreaks in major US cities due to fecal exposure among homeless people—because of local politics affecting sanitation, and the lack of universal vaccination.

Measles and smallpox

There are many examples of the **epidemic** spread of viruses caused by movement of human populations. For example, measles and smallpox were not known to the ancient Greeks. Both of these viruses are maintained by direct person-to-person transmission and have no known alternative hosts; therefore, it has been suggested that it was not until human populations in **China** and the **Roman Empire** reached a critical density that these viruses were able to propagate in an epidemic pattern and cause recognizable outbreaks of disease. Before this time, the few cases that did occur could easily have been overlooked.

It is suggested that the ssRNA- **measles virus** (*Paramyxoviridae*) diverged from the closely related cattle pathogen **rinderpest virus** sometime around **600 BCE**: this resulted from the increased animal husbandry that accompanied the rise of larger settlements and cities leading to the "**spillover**" into humans (see Düx et al., 2020).

Smallpox reached Europe from the Far East in **710 CE**, and in the 18th century it achieved "plague" proportions—five reigning European monarchs died from smallpox. However, the worst effects occurred when these viruses were transmitted to the New World. Smallpox was inadvertently transferred to the Americas by Hernando Cortés in 1520, by transport of infected soldiers. In the next 2 years, **3.5 million Aztecs died** from the disease—thus, the Aztec empire was decimated by disease rather than conquest. Although not as highly pathogenic as smallpox, epidemics of measles subsequently finished off the Aztec and Inca civilizations. More recently, the first contacts with isolated groups of Inuit and tribes in New Guinea and South America have had similarly devastating results, although on a smaller scale. These historical incidents illustrate the way in which a known virus can suddenly cause illness and death on a catastrophic scale following a **change in human behavior**—and especially migration.

Smallpox was of course declared to have been **eradicated** by the World Health Assembly in 1980, following a concerted worldwide campaign of education and vaccination. A similar campaign has been attempted for measles, given that there is no other host for the virus and there has been a very effective vaccine since 1963: the WHO estimates that from 2000 to 2018, **vaccination prevented over 23 million deaths**, and that measles mortality **had decreased by 73%.** However, vaccine resistance and occasional nonavailability means that there are repeated re-emergences of the virus, with **140,000 people**—mostly children under 5—having died in 2018 alone.

Influenza viruses

Influenza viruses, and especially the pandemics they have caused, are excellent teaching aids for understanding how serious human disease can result from **repeated reintroductions into humans** of viruses, or even only reassorted components of them, derived from wild animals. While they were not recognized as such at the time, major or **pandemic outbreaks** of influenza disease—presumably caused by **influenza type A viruses**—have occurred throughout recorded history. Medical historians have used contemporary reports to identify probable influenza epidemics and pandemics from as early as **412 BCE**—and the term "**influenza**" was first used in 1357 CE in Italy, describing the supposed "influence" of the stars on the disease. The first convincing report of an epidemic of the disease was from 1694, and reports of epidemics and pandemics in the 18th century increased in quality and quantity (Fig. 8.1).

The first pandemic that historians agree on was in **1580**: this started in Asia, and spread to Africa, took in the whole of Europe in 6 months, and even got to the Americas. Subsequent pandemics with significant death rates occurred in **1729** and **1781–2**; there was a major pandemic in **1880–1883** that attacked up to **25% of affected populations**, and another in **1898–1900 which was probably H2N2**. It is speculated that the so-called "**Russian Flu**" of 1889–1894 that preceded the H2N2 outbreak may in fact have been a **bovine coronavirus**—similar to the emergence of SARS-CoV-2—given the high attack rate, and tendency to relapse (see Berche, 2022).

The "Spanish flu" pandemic 1918–20

While the first reports of this pandemic were from Spain, this was largely because theirs was possibly the only uncensored press in Europe at the time because of the 1914–18 World War. It is possible that the virus was **brought to the US from China**, given contemporary accounts of disease in that region; however, a recent report indicates that a "**influenza pneumococcal purulent bronchitis**" reported in 1916–17 in military camps in Europe may have been a precursor to the 1918 outbreak (see Oxford and Gill, 2018). In any case, it is generally accepted that the disease **was in France** by **April 1918**, and subsequently spread quickly across

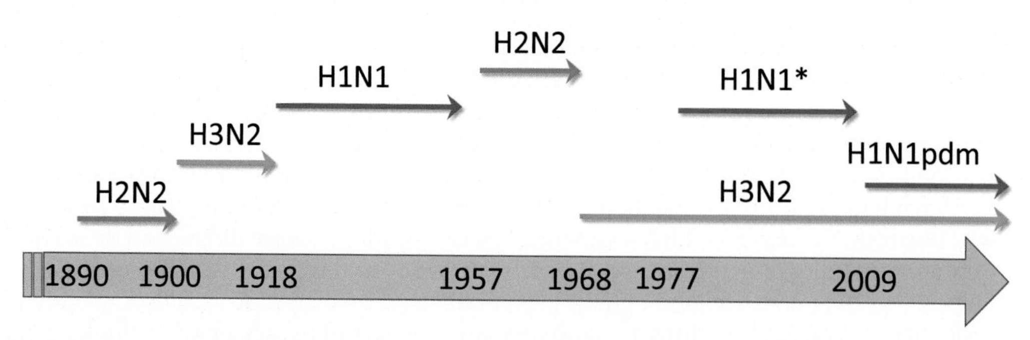

FIG. 8.1 Influenza A pandemics. Known influenza A pandemics in modern times. *Probably reintroduced in 1977 from a laboratory from the H1N1 circulating from 1918 until 1957.

Europe, and via troop transports again to **northern Russia, north Africa,** and **India**. Further spread then occurred, to **China, New Zealand**, and **The Philippines**, all by **June 1918**.

Initially, there was nothing unusual: infections spread quickly for a while and then declined, and **death rates were not higher** than in previous pandemics. However, from **August 1918**—marked by a ship-borne outbreak in **Sierra Leone** in West Africa—the virus seemed to have become **markedly more virulent**, and the death rate is supposed to have increased **10-fold**. The virus quickly spread through Europe, to the United States, to India by **October 1918**, and to Australia by **January 1919,** all the while spreading through and around Africa. Introduction of the virus to South Africa occurred in two waves, in September 1918 via Durban in the west, and in October 1918 via Cape Town in the south. The later introduction was of the most severe variant: over the next several months, up to 300,000 people died—making South Africa the fifth worst affected country in the world in terms of **case fatality rate** (Fig. 8.2).

Some countries had **second and even third waves** of infection, in **1918–19** and **1919–20**. The pandemic was initially calculated as having killed some **20 million people**: however, later estimates which took into account in particular the African, Indian and Chinese death tolls have increased the death toll to **at least 50 million**, and possibly up to 100 million. The virus probably infected **over one-third of the humans alive at the time (~1.5 billion)**, with a **case fatality rate of up to 15%**. Some regions, like Alaska and parts of Oceania, had death rates of **up to 25%** of the total population. By contrast, the normal mortality rate for seasonal flu is **0.05%–0.15%** of those infected.

The pandemic was unusual in that it seemed to affect **mainly young adults:** The graph **in** Fig. 8.3 shows case fatality rates in percent for pneumonia and influenza combined for **1918–19**, and for seasonal influenza for **1928–29**, for different age groups. **The "W" shape** for the 1918–19 figures is most unusual; the later seasonal data show a **far more usual "U" curve.** The green line shows what could have happened if—as is suspected—people over 40 had not had some immunity to the virus, **due to prior exposure** to the virus or components of it—possibly during **the 1880 pandemic.**

Although **secondary bacterial infections** of the lungs were common in fatal cases in 1918, and contributed significantly to mortality, there were also many cases of rapid death where bacterial infection could not be demonstrated—so these were due to a so-called "**abacterial pneumonia.**" Incidentally, the archiving of pathology specimens from especially military cases in the United States proved invaluable in "viral archeology" studies as late as 1997, which resulted in sequencing and reconstruction of the 1918 virus.

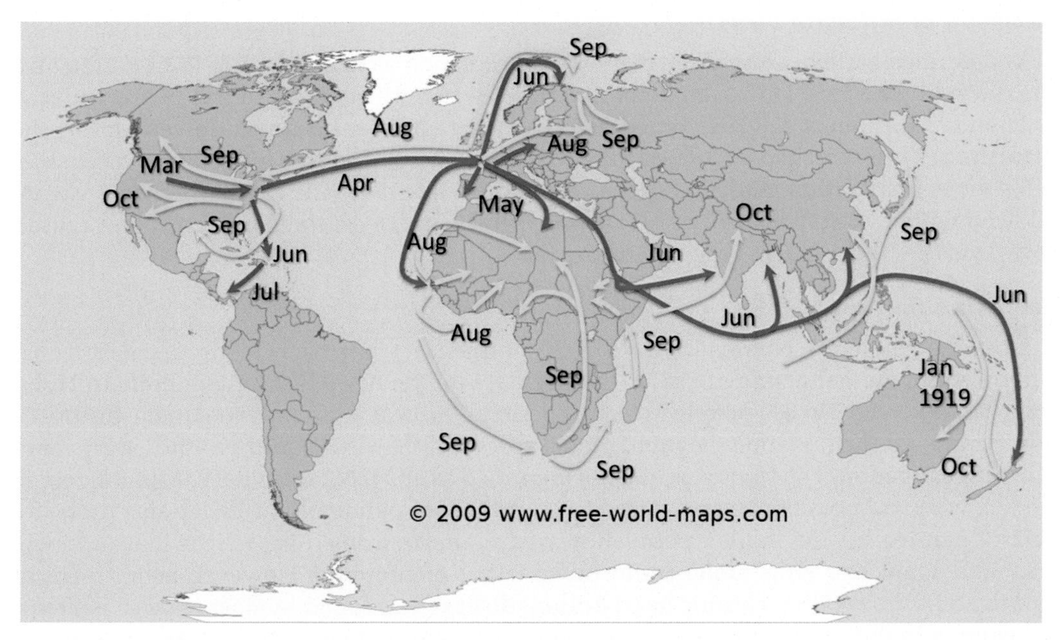

FIG. 8.2 Spread of 1918 flu around the world. The virus had essentially reached all parts of the planet by early 1919—and it had traveled first by ship, and then by train or other land transport. First wave is shown in *red*, and second in *yellow*.

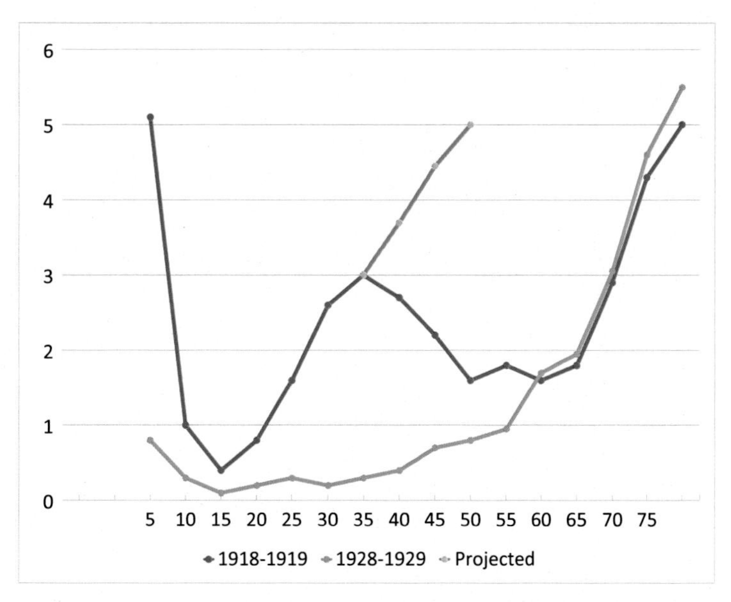

FIG. 8.3 Impact of 1918 flu. Combined age-specific case-fatality rates (CFRs) for influenza and pneumonia in the United States for the periods 1918–19 and 1928–29. *Graph redrawn from data from Taubenberger, J.K., Morens, D.M., 2006. 1918 Influenza: the mother of all pandemics. Emerg. Infect. Dis. 2006 12 (1), 15–22. https://doi.org/10.3201/eid1201.050979.*

Discovery of influenza virus

As described earlier (Chapter 1), there was speculation as early as 1918–19 that influenza was caused by a virus. However, definitive proof was only obtained in the 1930s, with the isolation of "**influenza A,**" later **typed as H1N1**: this virus was a **direct descendant of the Spanish flu virus,** and had circulated in humans since **1918**. The first influenza A vaccine—**a killed virus preparation made in eggs**—was made in late 1943. The first **influenza B isolate** followed with a vaccine made by 1945. It was then clear that seasonal influenza was caused by two viruses: the **A H1N1 type** and **influenza B.**

The "Asian flu" of 1957–58

After the influenza pandemic of 1918–20, influenza went back to its usual seasonal pattern—until the **pandemic of 1957**. This started with the news that an **epidemic in Hong Kong** had involved **250,000 people** in a short period. This was a unique event in the history of influenza, as for the first time the rapid global spread of the virus **could be studied by laboratory investigation.** The virus was quickly identified as an **H2N2 subtype**. Except for people over 70, who had possibly been exposed to an **influenza pandemic in 1898**—also probably a **H2N2 pandemic**—the human population was again confronted by a virus that was new to it—and again, **the virus alone could cause lethal pneumonia**. However, better medical investigation showed that **chronic heart or lung disease** was found in most of these patients, and **women in the third trimester of pregnancy** were also vulnerable (Fig. 8.4).

The 1957 pandemic was the first opportunity for medical people to observe the vaccination response in the many people who had not previously been exposed to the novel virus. This was very different to the 1918 virus that had been circulating ever since, meaning that **most**

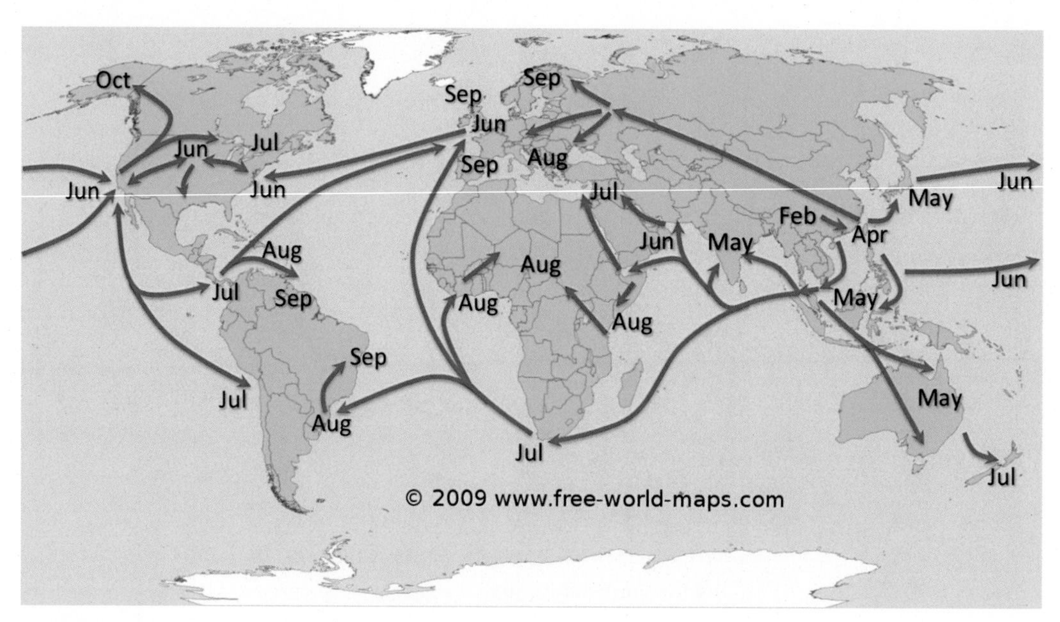

FIG. 8.4 Spread of the 1957 flu. Improved transport meant the virus traveled very swiftly—although again mainly by ship, followed by land transport—around the world.

people had no immunity to it at all. More vaccine was initially needed to give protective immunity than with the earlier type A vaccines. However, by 1960 as the virus recurred as a **seasonal infection**, immunity levels in the general population increased and vaccine responses were better, due to "**priming**" of the response by natural infection or first immunization. The death toll for this pandemic was around 2 million people—even though a vaccine was available by late 1957. Infections were most common among **school children, young adults, and pregnant women** in the early pandemic. **Elderly people** had the highest death rates, even though this was the **only group that had any prior immunity,** and there was a second wave in this group in 1958. The new H2N2 virus completely replaced the previous H1N1 type, and became the **new seasonal influenza type**.

The "Hong Kong Flu" of 1968–69

This pandemic started in mid-1968 in **Hong Kong**, and rapidly spread in a few months to **India, the Philippines, Australia, Europe,** and the **United States**. By 1969, it had reached **Japan, Africa,** and **South America**. Worldwide, the **death toll peaked in December-January**. However, although around 1 million people died, the death rate was lower than in 1957–58 for a number of reasons, including the following:

1. The virus was similar in some respects to the Asian Flu variant—it was an H3N2 isolate, similar to the pre-1918 seasonal type, sharing N2—meaning people infected then had partial immunity.
2. The better availability of antibiotics meant secondary bacterial infections were less of a problem (Fig. 8.5).

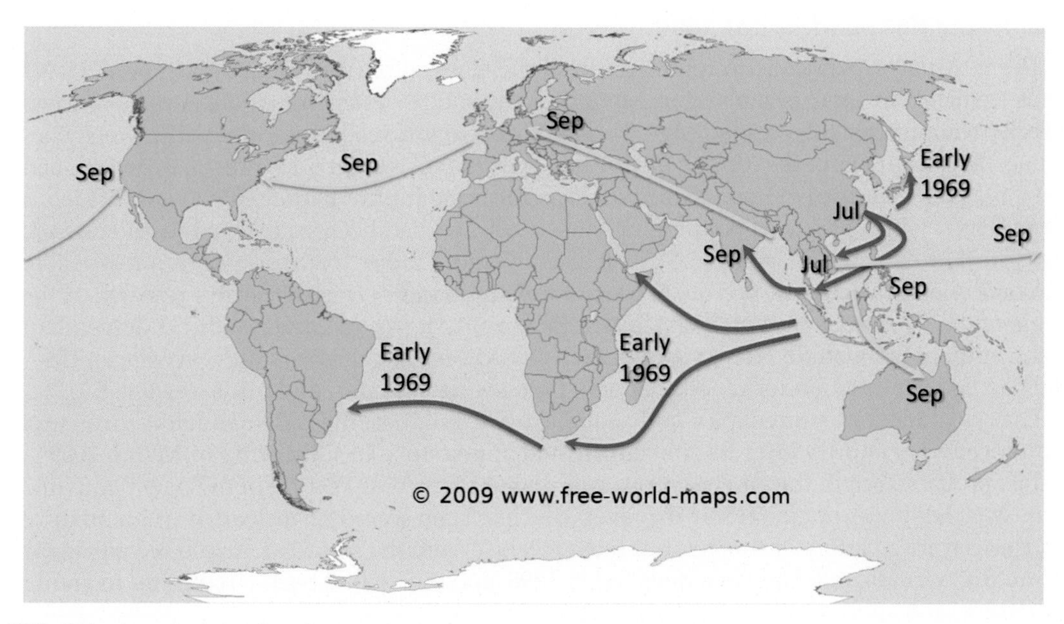

FIG. 8.5 Spread of 1968 flu. This was the first pandemic where air travel was a major factor, and spread was initially influenced by US military traffic to and from Southeast Asia. Subsequent traffic was mainly by ships and road transport, as before. Spread in 1968 is shown in *yellow*, and spread in 1969 in *red*.

A vaccine to the new virus became available a month after the epidemic peaked in the United States—following a trend which had started with the 1958 pandemic, of **vaccines becoming available only after the peak of the pandemic had passed**.

An interesting development soon after this was the finding that **waterfowl are the natural hosts of all influenza A viruses**—and that there was a greater diversity of viruses in birds than in humans. Thus, all influenza A outbreaks are at their root **zoonotic**, or derived from animals.

The "Russian flu" of 1977

Between May and November of 1977, an epidemic of influenza spread **out of north-eastern China and the former Soviet Union**—hence the name "Russian Flu." The disease was, however, **limited to people under the age of 25**—and was generally mild. It was soon found that virus responsible was effectively identical to the **H1N1 that had circulated from 1918 through to 1958**, and which had been replaced by the Asian flu (H2N2), which was in turn supplanted by the Hong Kong flu (H3N2). **This was a most unlikely scenario**, given that it was already known that influenza A viruses mutated rapidly as they multiplied—and it had been 20 years since the Spanish or H1N1 flu had been seen in humans. It also explained why infections were limited to young people: anyone who had caught the **seasonal flu prior to 1958** was protected.

There has been speculation that the pandemic was due to an **inadequately-inactivated or attenuated vaccine** released in a trial; there has even been mention of **escape from a freezer** in a biological warfare lab. There is no firm evidence for either possibility; however, the result is that the virus that had reappeared **then cocirculated with the H3N2 as a seasonal virus,** continuously until the next pandemic. This was unusual, as a pandemic virus usually becomes the next seasonal strain.

The "Swine flu" pandemic of 2009

The next major pandemic to follow on from the 1968 outbreak was **again a type A H1N1 virus**—which this time, originated in Mexico or the south-western USA, and probably **came directly from intensively-farmed pigs**. This had been an unusually long interval between pandemics, and warnings of the coming plague had been issued regularly for years: however, it had been expected that the next pandemic would involve the **highly pathogenic avian influenza virus H5N1**, which had been popping up since 1997, and had been established as an **endemic virus in farmed chickens since 2004**. This was therefore rather a surprise—but a reasonably welcome one, as the virus turned out to be **relatively mild in its effects**. Intensive research on the origin of the virus threw up some very interesting results: it was effectively a **direct descendant of the original Spanish flu H1N1 virus**, but which had been circulating in pigs ever since 1918—and had had contributions of genetic material from **swine, humans, and birds** (see Fig. 8.2).

Thanks to modern surveillance and sequencing techniques, the 2009 pandemic virus **was sequenced very shortly after its first confirmed appearance in California in March 2009**—and its probable origin traced very soon after that. A disturbing feature of the new pandemic virus was that it was not the **H5N1** that everyone had been scared of: instead, it was a mixture of viruses from a variety of sources, but most recently from the so-called "**triple recombinant swine flu**" viruses, that had been detected in 1998, and which had been **circulating in swine ever since**, and sporadically infecting humans (Fig. 8.6).

Interestingly, the new virus incorporated **two structural components** (M1 and NA proteins) from an "avian-like" Eurasian swine flu; **two replication components** (PB2 and PA)

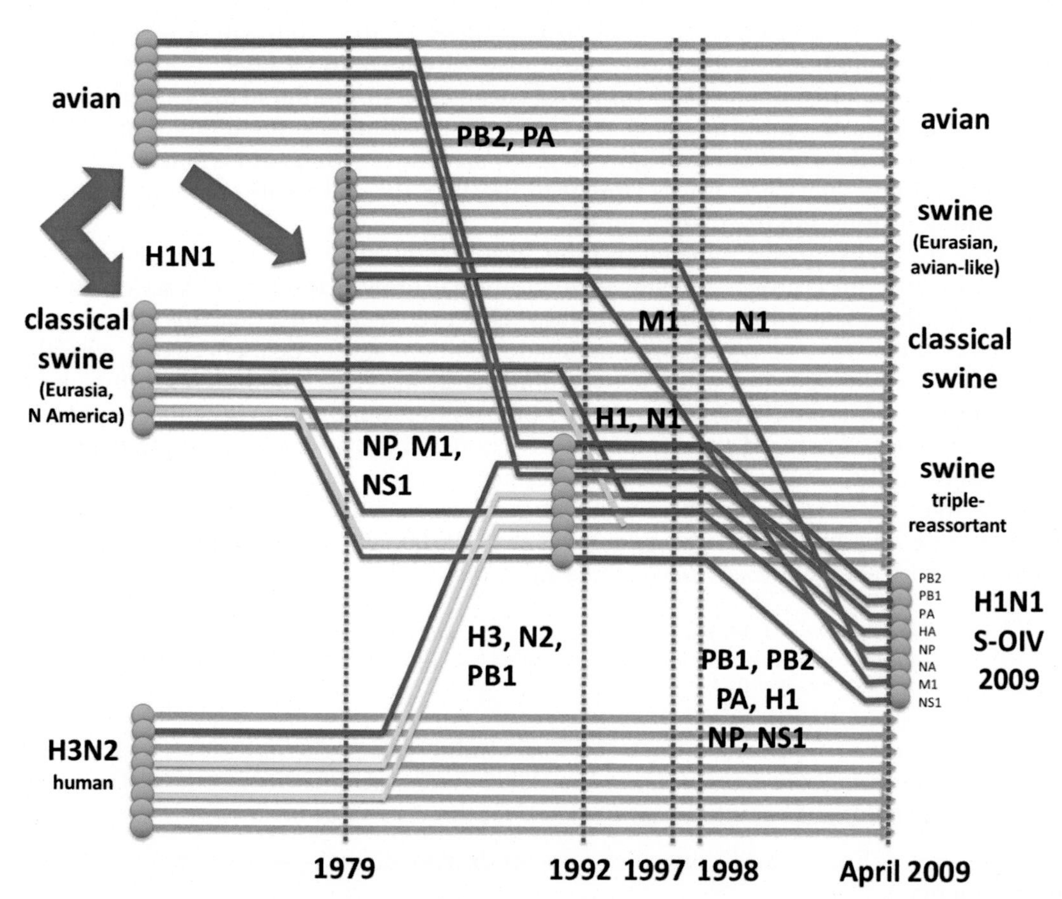

FIG. 8.6 "Swine flu" pandemic virus. Origin of Swine Origin Influenza (SOIV) H1N1pdm 2009. The graphic shows the **evolution by reassortment** of the pandemic virus: *blue circles* grouped together represent complete influenza A virus genomes; *gray lines* are genes remaining with the parent virus; *yellow* are those that exchanged but did not get into the H1N1 A2009; *red lines* are the genes that ended up in the new virus. *Image redrawn from Neumann, G., Noda, T., Kawaoka, Y., 2009. Emergence and pandemic potential of swine-origin H1N1 influenza virus. Nature 459, 931–939. http://www.nature.com/nature/journal/v459/n7249/full/nature08157.html.*

from an avian H1N1 virus, and **NP and NS1** from classical H1N1 swine flu via the **triple-recombinant swine lineage**; **HA** directly from classical swine H1N1, and **PB1** from human H3N2 via the triple reassortant complex. What is important to notice here is the **central role of domestic pigs**: **all of the ancestor viruses were in swine**, even though some components came recently from birds and one from humans, and some had been circulating continuously in swine **since 1918**.

By June 2009 the World Health Organization (WHO) had raised **the pandemic alert level to Phase 6**—the highest level, indicating that the **virus had spread worldwide and that there were infected people in most countries**. The "swine flu" pandemic was not as serious as had been feared, however: symptoms of infection were similar to seasonal influenza, albeit with a greater incidence of diarrhea and vomiting. The virus was also found to preferentially bind to cells deeper in the lungs than seasonal viruses: this explained both why it was generally mild—it did not often get that far down—**but also why it could be fatal**, as it could cause severe and sudden pneumonia if it did penetrate deep enough, **similar to the 1918 influenza. Binding to cells in the intestines** also explained the unusual nausea and vomiting. It was also found that there were **distinct high-risk groups, including pregnant women and obese individuals**. In these respects it was **similar to the 1918 flu**, as this also predominantly affected **young people, and pregnant mothers**.

Vaccine manufacture was initiated in **June 2009** by the WHO and manufacturers: while there was some concern over the **slower-than-normal growth rate** of the vaccine strains of the virus, this was rectified in a few months. However, as also happened with the other pandemics, there was not enough vaccine made soon enough to deal effectively with the pandemic—even though similarities between the pandemic virus and the 1977 outbreak virus meant that **most middle-aged people had preexisting immunity to it,** which either prevented infection, or reduced the severity of infections. This also meant **a single dose was sufficient in adults**, similar to the seasonal vaccine.

While the disease may have been mild in most cases, and initially the death toll was thought to be low, **by 2012 it was calculated that** 300,000 or more people probably died, mainly in Africa and Southeast Asia. A sobering quote:

> "*... since the people who died were much younger than is normally the case from influenza, in terms of years of life lost the H1N1 pandemic was significantly more lethal than the raw numbers suggest.*" https://www.nybooks.com/articles/2011/06/23/who-died-flu/

The virus has now become a normal **seasonal strain**, replacing the previously-circulating **H1N1,** but interestingly, has **not replaced the H3N2** that has circulated since 1968. The Mayo Clinic lists the following complications from H1N1pdm 2009 virus infection:

- Worsening of **chronic conditions**, such as heart disease, diabetes, and asthma
- Pneumonia
- **Neurological signs** and symptoms, ranging from confusion to seizures
- Respiratory failure

A recent study also showed that in experimentally infected animals, the **level of replication in the lungs was higher** than for the previous seasonal viruses, and **autopsies showed that the 2009 virus mainly targeted the lower respiratory tract**, which resulted in "severe acute respiratory distress syndrome."

A matter for some concern following the pandemic was that a vaccine used in Europe—a monovalent inactivated H1N1pdm 2009 vaccine, with an oil-in-water adjuvant—was associated with an increase in the incidence of narcolepsy, which is a "... chronic neurological disorder caused by the brain's inability to regulate sleep-wake cycles normally." While this was still a **rare complication**, the CDC stated that this did not occur in the United States, where no **adjuvants** were used in pandemic or seasonal vaccines. This appeared to implicate the adjuvant; however, subsequent analyses showed that a very similar vaccine made in Canada did not cause narcolepsy, and that the **specific viral antigen** used in the 37 million doses distributed in Europe may have been the cause. This was supported by reports from China that indicated that natural infection with the H1N1pdm 2009 virus was associated with a **3–4-fold increase in narcolepsy diagnoses**. A 2013 report confirmed that **narcolepsy was in fact an**

autoimmune disease: sufferers have a special group of CD4+ T cells that target the **hormone hypocretin**, which are only found in people with narcolepsy—and these same cells are stimulated by a component of the 2009 H1N1 virus.

As described earlier (Chapter 6, Fig. 6.19), "**antigenic drift**" in influenza A and B viruses is caused by gradual accumulation of mutations in HA and NA surface proteins, and can result in partial escape of circulating viruses from protective immunity. While influenza B viruses are apparently less susceptible to such drifting, the **B types** seen globally have in fact diverged in sequence to the point that distinct "**B Victoria**" and "**B Yamagata**" lineages (named for locale of first isolation) now cocirculate, which do not cross-protect very well. This necessitates regular **updates of vaccines** to cope with the most recent versions of the viruses: thus, the US Food & Drug Administration's recommendations for the US 2020–21 influenza season are, for quadrivalent egg-made split vaccines:

- an A/Guangdong-Maonan/SWL1536/2019 **(H1N1) pdm09-like** virus;
- an A/HongKong/2671/2019 (**H3N2**)-like virus;
- a B/Washington/02/2019- like virus (**B/Victoria lineage**); and
- a B/Phuket/3073/2013-like virus (**B/Yamagata lineage**).

Antigenic shift, resulting from acquisition of new HA and usually NA components as well, is completely unpredictable, both in timing and in nature—and the resultant **emerging virus** is almost impossible to protect against with existing vaccines, hence the urgency in researching possible **universal flu vaccines**.

Influenza A viruses with pandemic potential

The WHO maintains a constantly-updated list of viruses it considers potential threats, as well as vaccine status for these (https://www.who.int/news-room/fact-sheets/detail/influenza-(avian-and-other-zoonotic)). While there are **hundreds of strains of influenza A viruses, only four**—that is, **H5N1, H9N2, H7N3,** and **H7N7**—are known to have repeatedly caused human infections, with another four—**H7N9, H10N8, H6N1, and H5N6**—occurring less often. In addition, the following A subtypes have **sporadically** infected humans, generally **directly from animals**, and may have pandemic potential:

1. H5N1: 1997 and 2003 emergences in **poultry, occasionally** infecting humans; last deaths reported in Egypt and Indonesia in 2017.
2. H7N2: two human cases associated with **outbreak in cats** in the United States from 2016.
3. H9N2: from 1998 in humans exposed to poultry, mostly in Hong Kong and nearby mainland **China**.
4. **H7N7**: from 1996, in laboratory and farm workers in United Kingdom and Netherlands, from birds.
5. H3N2v: from 2012, farm workers and fair visitors in the United States, from pigs.
6. H7N9: 2012–13, people exposed to chickens in China; continues to be detected in birds in China.
7. H10N8 and H6N1 viruses: March 2015, caused sporadic human infections on Chinese mainland and Taiwan.
8. H5N6: April-May 2014, China; a case of death by severe pneumonia. In Viet Nam, **H5N6** is **epizootic, or widely circulating**.

Arboviruses

Measles and smallpox viruses are transmitted **exclusively from one human host to another**. For viruses with more complex cycles of transmission (e.g., those with secondary hosts and insect vectors), control of infection becomes much more difficult (Fig. 8.7). This is particularly true of the families of viruses known collectively as "**arboviruses**": these are **arthropod-borne viruses**, and include **bunyaviruses** (mosquitoes, ticks), **flaviviruses** (mosquitoes, ticks), **orbiviruses** (ticks, mosquitoes, gnats, and midges), and **togaviruses** (mosquitoes). As human territory has expanded, this has increasingly brought people into contact with the type of environment where these viruses are found—**warm, humid, vegetated areas** where **insect vectors occur in high densities**, such as swamps and jungles. For example, unusually heavy rains in East Africa can lead to outbreaks of **Rift Valley fever** (bunyavirus) due to explosions in the mosquito vector populations.

Of more than 500 arboviruses known, at least 100 are pathogenic for humans and at least 20 would meet the criteria for **emerging viruses**. Attempts to control these diseases rely on twin approaches: these are the **control of insect vectors** responsible for transmission of the virus to humans, and the development of **vaccines** to protect human populations. However, both of these approaches present considerable difficulties, the former in terms of avoiding environmental damage and the latter in terms of understanding virus pathogenesis and developing appropriate vaccines (see discussion of dengue virus pathogenesis in Chapter 7).

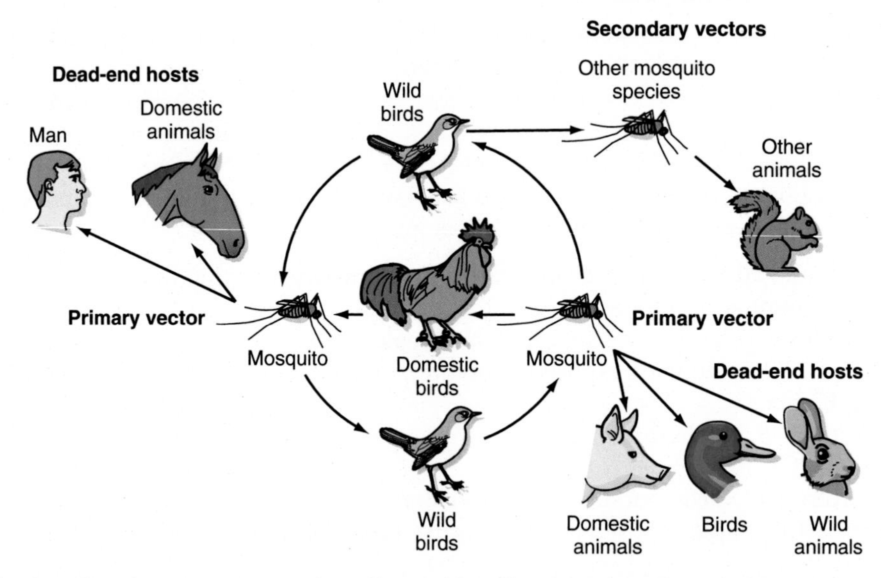

FIG. 8.7 Complex transmission pattern of an arbovirus like yellow fever virus. Because of their complex transmission patterns involving multiple host species, arthropod-borne viruses are difficult to control, let alone to eradicate. They generally have a sylvan cycle—between insects and wild animals—which results in human or domestic animal outbreaks when these come into contact with the arthropod vectors as a result of climate-induced vector movement, or invasion of pristine wilderness areas.

Flaviviruses

A classic example of the effects of a re-emerging virus is the major disease mortality caused by ssRNA+ **yellow fever virus** (YFV; type member of genus *Flavivirus*) during the building of the Panama Canal at the end of the 19th century. The virus had previously reached this region due to transport of people and mosquitoes from Africa during the slave trade (see Chapter 1), and had played an important role in the success of the Haiti rebellion against their French colonial occupiers. More recently, the increasing pace of ecological alteration in tropical areas has resulted in the resurgence of yellow fever in Central America, particularly an urban form of the disease transmitted **directly from one human to another by mosquitoes**.

Dengue fever, caused by **four serotypes of dengue virus** (genus *Flavivirus*) is also primarily an **urban disease of the tropics**, transmitted by *Aedes aegypti*, a domestic, day-biting mosquito that prefers to feed on humans (see Chapter 7). Some outbreaks of dengue fever have involved more than a million cases, with **attack rates of up to 90% of the population**. There are believed to be over 40 million cases of dengue virus infection worldwide each year. This disease was first described in 1780. By 1906, it was known that the virus was transmitted by mosquitoes, and the **virus** was isolated in 1944; therefore, this is not a new virus, but the frequency of dengue virus infection has increased dramatically in the last 30 years due to changes in human activity, and **inadvertent carriage** of mosquitoes around the world via cargo ships.

Another re-emerging flavivirus that caused a major impact recently was **Zika virus (ZIKV)** (**Fig. 8.8**). This was originally isolated in 1947 in the **Yellow Fever Research Institute in Uganda** from a rhesus macaque that had been caged in the Zika Forest near Lake Victoria, as a "sentinel" for sampling primate-infecting arboviruses. Isolation from an *Aedes africanus* mosquito from the same site followed in 1948, and a serological survey in 1952 showed around **6%** of those sampled had antibodies to the virus. It is believed that the virus was maintained in this area by a "sylvatic cycle" in forest-dwelling nonhuman primates and mosquitoes (Fig. 8.7).

FIG. 8.8 Zika virus. Semitranslucent view showing core and envelope proteins. The envelope (*E*) proteins are obtained by budding, but associate so tightly as dimers in a $T=3$-like structure that virions appear isometric. *Courtesy of Russell Kightley Media.*

Subsequent serosurveys from the mid-1950s showed that antibodies to the virus were distributed widely in people in Africa, as well as in India, Pakistan, Malaysia, Thailand and north Vietnam, Philippines, and Indonesia. However, it was only in 1966 that the virus was isolated from *Aedes aegyptii* mosquitoes in Malaysia (Wikan and Smith, 2016). Given that the earliest reported distribution of the virus and antibodies to it closely matches where British Imperial forces derived from Africa (the "King's African Rifles") were stationed during and after World War II, it is tempting to speculate that the virus was transported to south-east Asia by Africa-to-Asia military traffic—however, there is no evidence of this.

While there was only evidence of 13 naturally-infected cases of Zika virus in humans up to the 2000s, in 2007, 49 confirmed and 59 probable cases of ZIKV infection were recorded in **Yap State,** 1 of the 4 states of the Federated States of Micronesia in the western Pacific, north of the equator. By this time PCR testing was common, and sequences of isolates showed 90% homology with the original ZIKV—and demonstrated that there were **two African subclades** as well as an "**Asian lineage,**" with the Yap State virus being a relative that had spread from SE Asia, in the Asian lineage. In 2008 two scientists working in Senegal returned to the United States and developed a febrile illness—and the wife of one also contracted ZIKV, possibly by **sexual transmission**: this was noticed in later outbreaks too. From 2010 to 2012, two children with fevers—one in Cambodia, and one in Philippines—were also shown to be infected with Asian lineage ZIKV. Subsequently, there were **sporadic** identifications of ZIKV in travelers to and from SE Asia—until 2013, when a major outbreak in several islands of French Polynesia Tahiti, Moorea, Bora Bora, and others) in the south-central Pacific Ocean. Within a year of the first confirmed cases, it is estimated that there had been **19,000 cases** of disease, with the virus apparently being derived directly from the **Asian lineage** rather than from Yap State, and only around the time of detection of the first case. Further cases were confirmed in travelers returning from visiting the area, but these were again sporadic. In 2013 the Pacific Island of New Caledonia had an outbreak—1385 laboratory-confirmed cases—with importation being traced to French Polynesia and other Pacific islands. In nearly all the cases reported up until this date, there was only **low viremia** and it was difficult to isolate ZIKV from patients.

In 2015 the French Polynesian lineage had reached **Brazil**, with very similar viruses being isolated in Camaçari in Bahia State and in Natal in Rio Grande do Norte, over 1000 km apart. It has been speculated that the virus could have entered Brazil from a Pacific territory with soccer World Cup spectators in 2014—however, it is not known for certain. Since then, Brazil has had between **440,000 and 1,300,000 cases**, and nearby ZIKV was reported from South American countries of Colombia, Paraguay, Venezuela, Suriname, French Guiana, Ecuador, Guyana, and Bolivia; then Mexico and much of Central America, and Caribbean territories such as Martinique, Puerto Rico, and Barbados.

The virus evidently changed in its travels, as pre-Polynesian outbreaks were of a **mild, short duration febrile illness with skin rash**, and no hospitalization. By the time of the French Polynesia outbreak, **cases were more severe**—including cases of **Guillain-Barré syndrome**. In Brazil, there was a dramatic increase in cases of **microcephaly** in newborns, or lack of proper brain development, leading to a small and misshapen head and cognitive deficits in infants. There was a **20-fold increase** in incidence of the syndrome: this led to general warnings to women in the affected areas to try to avoid getting pregnant, as infection in the first trimester of pregnancy was significantly associated with the syndrome: the same was found in retrospective studies in Polynesia. ZIKV was isolated in **amniotic fluid** of pregnant women with microcephalic fetuses, and in the brains of fetuses aborted because of microcephaly, and ZIKV was seen to have a strong tissue tropism for **neural progenitor cells**. Subsequent deep sequencing studies revealed that there were certain **mutations** associated with microcephaly cases, in the **nonstructural protein NS3** that interfaces with NS5 in replication machinery—possibly making replication more efficient. Other mutations in **the virion surface glycoproteins** could be involved in changes in **tropism** (see Borucki et al., 2019).

The difference between Asian and African lineages of ZIKV, and between patterns of disease emergence in Africa and elsewhere, are puzzling: however, recent work has shown that low-passage strains representing **recent African ZIKV diversity**—still separated by a deep phylogenetic divide from Asian and New World isolates—are **more transmissible by mosquitoes** than Asian lineage viruses in direct comparisons, and exhibit **higher lethality** in both adult and fetal mice than Asian isolates. However, African populations of *A. aegyptii* mosquitoes are significantly **less susceptible to ZIKV infection by any variant**, and are **not as efficient vectors** as populations elsewhere—presumably because of longer exposure to the virus in nature. Thus, Zika outbreaks in Africa may not be noticed as much as elsewhere, partly because there **is less transmission** (and so smaller outbreaks), and because the African viruses may be causing **fetal death** rather than neural damage—limiting the occurrence of microcephaly. It is speculated that African lineages of the virus could have significant epidemic potential if vectored by more susceptible mosquitoes—which are out-competed in Africa (Aubry et al., 2021).

West Nile virus (WNV) is a member of the Japanese encephalitis antigenic complex of the **ssRNA+ family *Flaviviridae***. All known members of this complex are transmissible by **mosquitoes**, and many of them can cause febrile, sometimes fatal, illnesses in humans. WNV was first isolated in the West Nile district of Uganda in 1937 but is in fact the most widespread of the flaviviruses, with geographic distribution including **Africa and Eurasia**. It is also possible that the virus spread out of Africa as yellow fever is supposed to have, by movement of infected people and possibly mosquitoes. Unexpectedly, an outbreak of **human encephalitis** caused by WNV occurred in the **United States** in New York and surrounding states in 1999. In this case, the virus appears to have been transmitted from wild, domestic, and exotic birds by *Culex* mosquitoes (an urban mosquito that flourishes under dry conditions)—a classic pattern of arbovirus transmission. WNV RNA has been detected in overwintering mosquitoes and in birds, and the disease is now **endemic** across the United States, causing outbreaks each summer, and is considered a serious problem. This rapid spread into a new territory shows that spread did not rely on environmental factors such as climate change—the North American environment was already suitable for the virus once it had been introduced, probably via air travel from the Middle East. There have also been sporadic outbreaks of WNV in southern Europe recently, probably due to the **increasing geographical range** of vector mosquitoes due to climate change. There are no accepted human vaccines for WNV, although vaccines for use especially in horses are available.

Bunya- and alphaviruses

The ssRNA− **Rift Valley fever virus (RVFV)** (genus *Phlebovirus*, family *Phenuiviridae*, order *Bunyavirales*) was first isolated from sheep in 1930, but has caused repeated epidemics in sub-Saharan Africa during the last few decades, with human infection rates in epidemic areas

as high as 35%. This is an **epizootic** disease, transmitted from sheep and other small animals to humans by a number of different mosquitoes. The construction of dams which increase mosquito populations, increasing numbers of sheep and goats, and the movement of these and human populations are believed to be responsible for the upsurge in this disease. RVFV **continues to extend its range in Africa and the Middle East**, and is a significant health and economic burden in many areas of Africa, remaining a serious threat to other parts of the world. While there are vaccines for livestock, there is no generally licensed human vaccine for RVFV.

A **tick-transmitted** virus—**Crimean-Congo hemorrhagic fever virus, CCHFV** (genus *Orthonairovirus*, family *Nairoviridae*, order *Bunyavirales*)—is also listed as an agent of concern by the WHO and CDC, as it is expanding its geographical range with global warming. It is presently endemic in Africa, the Balkan region, the Middle East and south Asia in regions south of 50° North latitude, which delimits the geographical range of its principal vector—hard-bodied ticks in the genus *Hyalomma*, which has long, distinctly red and white (or yellow) striped legs. It is transmitted to people via ticks that are found in the wild or on domestic animals, or by contact with animal flesh in abattoirs. It causes **severe hemorrhagic fever**, with a case fatality rate of **up to 40%**—and like Ebola or Marburg infections, can be spread between people by close contact with blood or secretions from infected people. There are regular outbreaks—generally of only a few cases—in Pakistan, sporadic cases all over the endemic regions—and a 2020 outbreak in Turkey, with 480 human cases and 15 deaths (Outbreak News Today, 2020). There are generally available vaccines for CCHFV, although the Bulgarian armed forces use a **locally-developed killed vaccine**.

Its potential for spread into Western Europe was shown by a fatal case in 2016 in Spain caused by an **African lineage of CCHFV** after a tick bite, possibly as a result of airborne carriage of the tick by a migrating bird. A survey of ticks in Spain had found evidence of CCHFV infection in ticks from red deer 5 years previously, over 300 km away (Negredo et al., 2017). Ticks collected from migrating birds in Greece in 2009 had previously been found to be positive for CCHFV, and in 2019 a *H. rufipes* tick nymph—mainly found in **sub-Saharan Africa**, and infected with an **African genotype III virus**—was found on a *trans*-Saharan migrating whinchat (Mancuso et al., 2019).

Chikungunya virus (CHIKV) is transmitted by *Aedes* mosquitoes and was first isolated in 1953 in what is now Tanzania. CHIKV is a member of the ssRNA+ genus *Alphavirus* in the family *Togaviridae*. The disease caused by this virus typically consists of an **acute illness** characterized by fever, rash, and incapacitating joint pain. The word chikungunya means "**that which bends up**" in the Makonde language from the south of East Africa, and refers to the effect of the **joint pains** that characterize this dengue-like infection. Chikungunya is a specifically **tropical disease**, but was previously geographically restricted and outbreaks were relatively uncommon. The virus remained largely unknown until a major outbreak in 2005 and 2006 on islands across the Indian Ocean. Plausible explanations for this outbreak—and subsequent spread, which has continued into the **Caribbean region** and neighboring countries—include increased tourism, CHIKV introduction into a **naive population**, and **virus mutation**. It is the last of those three factors which seems to be most significant in this case, with the outbreak strain showing a **single amino acid change** in the **envelope glycoprotein** which allows more effective transmission due to more efficient crossing of the **mosquito gut membrane barrier**. There is every possibility that CHIKV will continue to extend its territory. There is presently no vaccine.

Mayaro virus is another emerging member of the *Togaviridae,* first isolated in 1954 in Trinidad and Tobago. It has three genotypes, and is endemic in South and Central America and Mexico, and sporadically emerges in Haiti and other Caribbean states. It can cause high fevers and intense joint pain in acute episodes of 3–7 days, and **sporadic emergence** is linked to exposure to **rural** or **semirural forested areas**. The virus is maintained in a **sylvan cycle**—as is true for chikungunya and Zika—involving wild vertebrates and birds, and generally being vectored by *Haemogogus* spp. mosquitoes. However, populations of the nonnative invading *A. aegyptii* can also be infected and transmit the viruses—which could lead to "breakout" into the areas preferred by these mosquitoes, such as more urban settings. There are no vaccines for the viruses, so the only level of control that could be used is eradication or control of the vector mosquitoes.

Plant viruses

Plant viruses can also be responsible for emerging diseases. The ssDNA **geminiviruses** in the genus *Begomovirus* are transmitted by sap-sucking insect vectors called **whiteflies** (mainly *Bemisia* spp., Chapter 6). These viruses cause a great deal of crop damage in plants such as tomatoes, beans, squash, cassava, and cotton, and their spread may be directly linked to the inadvertent worldwide dissemination of a particular **biotype** of the whitefly *Bemisia tabaci*. This vector is an indiscriminate feeder, encouraging the rapid and efficient spread of viruses from indigenous plant species to neighboring crops. The viruses are said to have a **persistent nonpropagative** mode of carriage: that is, host proteins and factors such as heat shock proteins and chaperonins encoded by bacterial endosymbionts assist in transporting the viruses along the food canal to the esophagus and midgut, across the filter chamber and the midgut into the hemolymph, and translocation into the primary salivary glands from where they are injected into plants. However, tomato yellow leaf curl virus (TYLCV-Is) is now known—after many years of speculation—to be transcribed and to replicate in the insects' midgut and filter chamber. This triggers the insect autophagic response, which inhibits replication and virion destruction. Ovaries and fat cells may also support replication, but with the same result (Czosnek et al., 2017). Thus, while geminiviruses may technically replicate in their insect vector(s), it is not as important a factor in their transmission as it is for animal-infecting arboviruses.

The emergence of geminiviruses infecting **maize** in Africa—mostly **maize streak virus** (MSV), genus *Mastrevirus*—occurred after the introduction of maize to the continent, probably by the Portuguese in the 1500s (see Chapter 1). The virus is transmitted in the same manner as begomoviruses by **leafhoppers** in the family Cicadellidae. Severe "**streak**" disease in maize was first noticed and investigated in the late 1800s-early 1900s in Kenya and South Africa (see Shepherd et al., 2010), and was subsequently proposed to be due to the emergence of a new mastrevirus that was a **recombinant** between two or more distinct mastreviruses that **infected grasses**. This recombination resulted in a virus that both **replicates and spreads** better in maize plants than grass-infecting viruses, and moreover causes **striking symptoms** that attract the leafhopper vectors. The genetic range of viruses causing severe symptoms in maize is very narrow, which indicates a recent origin, unlike the plethora of widely different mastreviruses infecting native grasses in Africa. MSV and

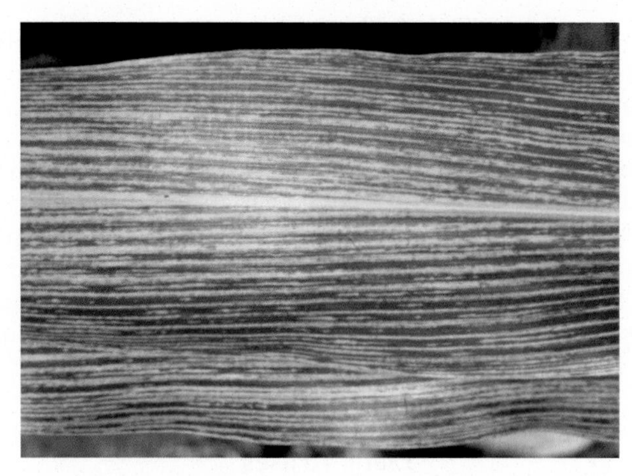

FIG. 8.9 Symptoms of severe MSV infection on maize. Photograph taken by EP Rybicki in Githungiri District, central Kenya, in 1997. Note the very distinct striated yellow streaks in the leaf.

its close relatives occur **only in Africa** and neighboring **Indian Ocean territories** such as Mauritius and La Réunion, possibly as a result of vectors being **blown in from mainland Africa** by prevailing winds (Fig. 8.9).

Occasionally, there appears an example of an emerging virus that has **acquired extra genes** and as a result of this new genetic capacity has become capable of infecting a **completely new range of hosts**. A possible example of this phenomenon is seen in **tomato spotted wilt virus (TSWV)**. TSWV is a ssRNA− bunyavirus with a very wide plant host range, infecting over 600 different species from 70 families. In recent decades, this virus has been a major agricultural pest in Asia, the Americas, Europe, and Africa. Its rapid spread has been the result of dissemination of its insect vector (the **thrip** *Frankinellia occidentalis*) and diseased plant material. TSWV is the type species of the *Tospovirus* genus and has a morphology and genomic organization similar to the other bunyaviruses (Chapter 3). However, TSWV undergoes **propagative transmission**, and it has been suggested that it may have acquired an extra gene in the M segment via **recombination**, either from a plant or from another plant virus. This new gene encodes a **movement protein** (Chapter 6), conferring the capacity to **move in plants**, and cause extensive **systemic** damage.

Given the close association between insects and plants on land from over 400 million years ago (see Fig. 3.11), it is not surprising that viruses associated with insects can replicate in plants. One example of virus that is possibly transitioning from being an insect virus to also being one of plants, like TSWV, is **flock house virus (FHV),** first isolated in New Zealand from "grass grubs" (*Costelytra zealandica*). This is a ssRNA+ **alphanodavirus** (family *Nodaviridae*) with a two-component genome that replicates in a number of insect species, including fruit flies (*Drosophila* spp.), mosquitoes (*Anopheles gambiae*), the tsetse fly, and the triatomine bugs that vector Chagas disease. The virus can also **replicate in plant tissues** into which it is introduced by its insect hosts: however, it does not have a movement protein, and so **remains within the inoculated cells**, from where it can be taken up again even weeks later. Thus, plants can act as a "**propagative noncirculative vector**" for infecting insects, in a turnaround from the normal situation. It is also possible, incidentally, for plants to act as nonpropagative vectors: the aphid-infecting ***Rhopalosiphum padi* virus** (a dicistrovirus, similar to picornaviruses) can be injected into, and remain stable in plant tissues for long enough to be picked up 3 weeks after injection, **without replicating**.

Novel human viruses

In addition to viruses whose ability to infect their host species appears to have changed, new human viruses are being discovered continually.

Human herpesviruses

After many years of study, three new **human herpesviruses (HHVs)** have been discovered comparatively recently:

- **HHV-6:** First isolated in 1986 in lymphocytes of patients with lymphoreticular disorders; **tropism** for $CD4^+$ lymphocytes. HHV-6 is now recognized as being an **almost universal human infection**. Discovery of the virus solved a longstanding mystery: the primary infection in childhood causes "roseola infantum" or "fourth disease," a common childhood rash of previously unknown cause. **Antibody titers** are highest in children and decline with age. The consequences of childhood infection appear to be mild. Primary infections of adults are rare but have more severe consequences—mononucleosis or hepatitis—and infections may be a severe problem in transplant patients.
- **HHV-7:** First isolated from human $CD4^+$ cells in 1990. Its **genome organization** is similar to but distinct from that of **HHV-6**, and there is limited antigenic cross-reactivity between the two viruses. Currently, there is no clear evidence for the direct involvement of HHV-7 in any human disease, but might it be a cofactor in HHV-6-related syndromes.
- **HHV-8 (Kaposi's sarcoma herpesvirus):** In 1995, sequences of a unique herpesvirus were identified in DNA samples from AIDS patients with KS and in some non-KS tissue samples from AIDS patients. There is a strong correlation (>95%) with KS in both HIV^+ and HIV-2-infected patients. HHV-8 can be isolated from lymphocytes and from tumor tissue and appears to have a less ubiquitous world distribution than other HHVs; that is, it may only be associated with a specific disease state (cf. HSV, EBV). However, the virus is not present in KS-derived cell lines, suggesting that **autocrine** or paracrine factors may be involved in the formation of KS. There is some evidence that HHV-8 may also cause other tumors such as B-cell lymphomas (± EBV as a "helper"), and it appears to be involved in multiple sclerosis as well (see Chapter 7, Epstein-Barr Virus).

Human hepatitis viruses

Although many different virus infections may involve the **liver**, at least six viruses seem specifically to infect and damage hepatocytes (see Chapter 7). No two of these belong to the same family! The identification of these viruses has been a long story:

- Hepatitis B virus (HBV) (hepadnavirus; ds circular DNA): 1963
- Hepatitis A virus (HAV) (picornavirus; ssRNA+): 1973

- Hepatitis delta virus (HDV) (deltavirus; see Chapter 9—circular ssRNA): 1977
- Hepatitis C virus (HCV) (flavivirus; ssRNA+): 1989
- Hepatitis E virus (HEV) (*Orthohepevirus, Hepeviridae*—ssRNA+): 1990
- GBV-C/HGV (flavivirus): 1995

Reports continue to circulate about the existence of other hepatitis viruses. Some of the agents are reported to be sensitive to chloroform (i.e., **enveloped**) while others are not. This may suggest the existence of **multiple viruses, as yet undescribed**, although this is still uncertain. Although the viral causes of the majority of cases of infectious human hepatitis have now been identified, it is possible that further hepatitis viruses will be described in the future.

An interesting finding in Japan in 1997 was of a small nonenveloped virus with a circular ssDNA genome, unrelated to **circoviruses**, that was implicated in **transfusion-related hepatitis**. However, it and related sequences were subsequently found in up to 90% of human sera in a number of investigations, without being firmly associated with any disease. Viruses have naked $T=1$ isometric virions 30–32nm in diameter, a monopartite genome of 3–4kb in length, up to 4 overlapping or partially overlapping ORFs encoded in only one strand, and a noncoding region with one to two 80–110nt sequences with high GC content. The original isolate is now named "**torque teno virus**" (Latin for "thin necklace"; TTV) (genus *Alphatorquevirus*, family *Anelloviridae*). TTV and related viruses are found worldwide, and prevalence can reach 100% in some populations; the virus can be transmitted by transfusion, respiratory droplets and fecal-oral route. Incidence in children increases with age. While not associated with disease, **viral load** can be used a **marker for lack of regular immune function**.

Human retroviruses

The very early discovery of **avian viruses associated with cancer**, and the subsequent failure for many years to isolate similar viruses from mammals, gave some researchers the idea that possibly birds were unique in this regard. However, "**RNA tumor viruses**" or **oncornaviruses**, as they were known for a time (see Chapter 1), were first demonstrated to affect mammals when **mouse mammary tumors** were shown to be due to a virus in 1936, **by transmission in milk** and **vertically**, from mother to pups. By 1951 it was known that **leukemia could be passaged in mice** using cell-free extracts, and in the 1960s a **mouse sarcoma virus** and a **feline leukemia virus** had been isolated, and **bovine leukemia** was shown to be a viral disease. In 1971 the first **primate leukemia virus**—from **gibbons**—was described, and **the first retrovirus (foamy virus)** described from **humans**. **Bovine leukemia virus (BLV)** was characterized as a retrovirus in 1976.

It is not surprising, therefore, that many labs tried to find cancer-causing disease agents in humans. However, such effort had been put into finding oncornaviruses associated with human tumors, with such lack of success, that it led to people talking of "human rumor viruses"—a useful list of which can be seen in Voisset et al. (2008). Nevertheless, by 1980 Robert Gallo's group had succeeded in finding characteristic "type C retrovirus particles" from lymphocytes of a patient with cutaneous T-cell lymphoma, which they called **human T-cell leukemia virus (HTLV)**. The breakthrough was made possible by their prior discovery of "**T cell growth factor,**" now called **interleukin 2 (IL-2)**, which meant human T cells could be successfully cultured for the first time. A group of Japanese researchers described an "**adult T cell leukemia virus**" (ATLV) in 1982: this proved to be the same as what became **HTLV-1**, given the description also in 1982 by Gallo's group of another retrovirus associated with a T-cell variant of hairy cell leukemia, which they dubbed **HTLV-2.**

HTLV-1 is associated with the rare and genetically-linked adult T-cell leukemia, found mainly in southern Japan, as well as with a demyelinating disease called "HTLV-I associated myelopathy/tropical spastic paraparesis (HAM/TSP)" and HTLV-associated **uveitis** and infective dermatitis. The areas of highest prevalence are Japan, Africa, the Caribbean islands and South America. **HTLV-2** had a mainly Amerindian and African pygmy distribution, although it is now found worldwide, and causes a milder form of HAM/TSP, as well as arthritis, bronchitis, and pneumonia. It is also frequent among injecting drug users. However, except for rare incidences of **cutaneous lymphoma** in people coinfected with **HIV**, and the fact of its origin in a hairy cell leukemia, **there is no good evidence that HTLV-2 causes lymphoproliferative disease**. The two viruses infect between 15 and 20 million people worldwide. HTLV-1 infections can lead to an often rapidly fatal leukemia.

By 2005 another two viruses had joined the family: **HTLV-3 and HTLV-4** were described from samples from Cameroon that were presumably **zoonoses**—being associated with bushmeat hunters—and which are not associated with disease. Interestingly, **all the HTLVs have simian counterparts**—indicating species cross-over at some point in their evolution. Collectively they are known as the **primate T-lymphotropic viruses (PTLVs)** as they constitute an evolutionarily related group. Another relative is BLV. The **HTLV-1/STLV-1** and **HTLV-2/STLV-2** relationships are relatively ancient, at more than **20,000 years since divergence**. However, their evolution differs markedly in that STLV-I occurs in Africa and Asia among at least 19 species of **Old World primates**, while STLV-2 has only been found in **bonobos**, or *Pan paniscus* dwarf chimpanzees from DR Congo. It is therefore quite possible that there are other HTLVs undiscovered in primates in Africa and elsewhere, that may yet emerge into the human population (Fig. 8.10).

Human immunodeficiency virus type 1 (HIV-1) was for a time after its discovery in 1983 called **HTLV-III** by the Gallo group and **lymphadenopathy virus (LAV)** by the **Montagnier** group in France; however, evidence later obtained from sequencing and genome organization showed by 1986 that it was in fact a **lentivirus**, related to viruses such as **feline immunodeficiency virus (FIV)** and the **equine infectious anemia virus** (EIAV) discovered in 1904, and it was renamed.

HIV is indirectly implicated in cancer because it creates an environment through immunosuppression that allows the development of **opportunistic tumors** that would normally be controlled by the immune system: these include **HPV-related cervical cancer**, and **Kaposi's sarcoma** caused by HHV-8 (see above). It is also possible that HIV may directly cause lymphoma development in AIDS patients by **insertional activation of cellular oncogenes**, although this appears to be rare.

Zoonoses

Many **emerging virus diseases** are **zoonoses** (i.e., transmitted from animals to humans): a large percentage of all newly identified infectious diseases as well as many existing ones are zoonoses. This emphasizes the importance of the "species barrier" in preventing transmission of infectious

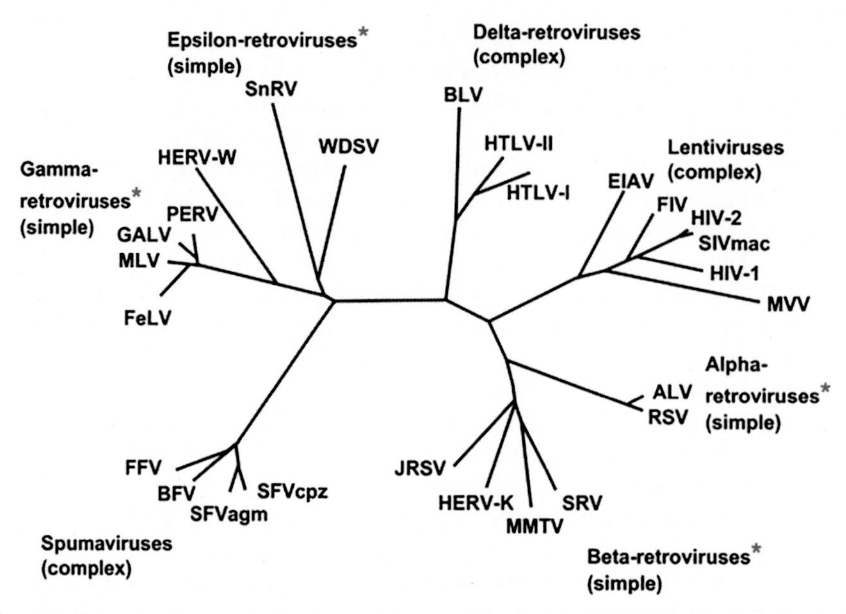

FIG. 8.10 Phylogeny of retroviruses. *From* "**Phylogeny of Retroviruses**" *by Weiss, R.A., 2006. The discovery of endogenous retroviruses. Retrovirology 3, 67. Review. PMID 17018135. https://doi.org/10.1186/1742-4690-3-67. Licensed under CC BY 3.0 via Wikimedia Commons—http://commons.wikimedia.org/wiki/File:Phylogeny_of_Retroviruses.jpg#/media/File:Phylogeny_of_Retroviruses.jpg.*

diseases, and several recent examples (e.g., HIV, SARS) illustrate the potentially disastrous consequences that can occur when this is breached. Strictly speaking, many of the "arboviruses" discussed earlier are zoonotic in humans, but their transmission involves an **insect vector**. On occasions, viruses such as **HIV** or **influenza A viruses** can spread from **animals into the human population** and then be transmitted from **one person to another** without the involvement of a vector or the original animal host. Important examples of emerging zoonoses are given below.

Hantaviruses

Viruses classified in the *Hantavirus* genus of family *Bunyaviridae* are a particular cause for concern. Hantaviruses cause **millions of cases of hemorrhagic fever** each year in many parts of the world. Unlike arboviruses, hantaviruses are transmitted directly **from rodent hosts** to humans—via **fecal and urine** contamination of food, clothing and bedding, or even as dried matter in dust or **aerosols**—rather than by an invertebrate host. Hantaviruses cause two acute diseases: **hemorrhagic fever with renal syndrome (HFRS)** and **hantavirus pulmonary syndrome (HPS)**. HFRS was first recognized in 1951 after an outbreak among US troops stationed in Korea. In 1993, HPS was first recognized in the Four Corners region of the US Southwest, and a new virus, **Sin Nombre** (meaning "without a name," chosen because of political pressure), was identified as the cause. This was found to spread by the **deer mouse**, and research found evidence of its presence across the United States in mouse populations. It is now known that at least three different hantaviruses cause HFRS and four different viruses cause HPS. By 1995, HPS had been recognized in 102 patients in 21 states of the United

States, in seven patients in Canada, and in three in Brazil, with an **overall mortality rate** of approximately **40%**. These statistics illustrate the disease-causing potential of emerging viruses. There are also no vaccines for any of these viruses.

Filoviruses: Marburg and Ebola

These viruses are classic "newly emerging" viruses in their appearance and impact upon humans, and all known natural outbreaks appear linked to either impingement of humans upon as yet **unknown vectors** in forest settings (Ebola viruses), or inadvertent exposure to an animal species that hosts the virus without overt disease (Marburg virus). The taxonomy of the viruses was dealt with in Chapter 2: suffice it to say here that there is one **Marburg virus** (genus *Marburgvirus*), and five **Ebola viruses—Ebola (EBOV), Sudan (SUDV), Reston (RESTV), Taï Forest (TAFV),** and **Bundibugyo (BDBV) viruses** (genus *Ebolavirus*) in family *Filoviridae*, of enveloped ssRNA− viruses (see Fig. 2.14).

Marburg virus

In 1967, the world was rudely introduced to a new virus: 31 people in Marburg and Frankfurt in Germany, and Belgrade in the then Yugoslavia, became infected in a linked outbreak with a **novel hemorrhagic fever agent**. Twenty-five of them were **laboratory workers** associated with research centers, and were directly infected via contact with infected **vervet monkeys** (*Circopithecus aethiops*) imported to all three centers from Uganda. Seven people died. In what was a remarkably short period of time for that era—given that this was **presequencing and cloning of nucleic acids**, let alone viruses, so that classical virological techniques had to be used—**it took less than 3 months** for scientists from Marburg and Hamburg to isolate and characterize what was being called "**green monkey virus.**" The new agent was named **Marburg virus (MARV)**, after the city with the greatest number of cases.

The first electron micrograph of the virus clearly exhibits the filamentous nature of the particles, complete with the **now-famous "shepherd's crook"** (see later). The virions were typically 790nm long and about 80nm wide.

The virus disappeared until 1975, when an Australian hitchhiker who had traveled through what is now **Zimbabwe** was hospitalized in Johannesburg, South Africa, with symptoms reminiscent of Marburg disease. He died, and his female companion and then a nurse also became infected with what was suspected to be either of **yellow fever** or **Lassa viruses**. In an example for later outbreaks, this led to rapid implementation of **strict barrier nursing** and **isolation** of the patients and their contacts, which resulted in quick containment of the outbreak—with recovery of the two secondary cases. MARV was later identified in all three patients. It is noteworthy that Uganda—the place of origin of the first known outbreak—and Zimbabwe are separated by **over 3000km**, meaning there was no link between the two events.

Subsequent natural outbreaks of what is now known as "**Marburg virus disease**" (MVD)—there were also several laboratory infections—occurred as follows:

- 1980, Nairobi (Kenya): 2 cases, 1 fatal, associated with Kitum Cave in Mt Elgon National Park
- 1987, Nairobi: 1 fatal case, associated with visit to same cave

- 1998–2000, Durba and Watsa areas in DRC: 128 fatalities among 154 cases in several overlapping outbreaks associated with gold mining in caves
- 2004–2005, Uíge, Angola: 252 cases, 227 fatalities; no known associations
- 2007, Kamwenge, Uganda: 1 fatality among 4 cases, associated with gold mining in Kitaka Cave
- 2008, tourists from USA and Netherlands: 2 cases, 1 fatality; linked to visit to Python Cave, Maramagambo Forest, Uganda
- 2012, Kabale, Uganda: 4 fatalities among 15 infected; no obvious associations
- 2014, Kampala, Uganda: single fatal case; no known links

This gives a total of 373 fatalities out of 468 diagnosed cases, or a CFR of 80%—although later outbreaks had improved survival rates to around 50% due to improved barrier nursing and palliative care. The disease is now thought to originate in **exposure to *Rousettus aegyptiacus* fruit bats** in caves or other roosts—presumably via faces and/or urine droplets—and then spread in humans via **direct contact** with blood or secretions or bodily fluids, although sexual transmission also occurs (WHO, 2021). The bat has a wide distribution across Africa, so further unlinked outbreaks are possible. Exposure via organs of infected animals is also possible, as in the original outbreak. The fruit bats that are infected with Marburg virus **show no obvious signs of illness**: in fact, it appears as though the virus causes only **transient common cold-like symptoms** in young bats, after which they are immune. Thus, human exposures may be **seasonal**, linked to the bat breeding cycles. Updated Marburg disease outbreaks are listed by the USA CDC, here: https://www.cdc.gov/vhf/marburg/outbreaks/chronology.html.

While recovery is mostly associated with complete clearance of the virus, it may **persist** in people who have recovered from disease, in "**immune privileged**" sites such as inside the **eye**, in **testicles**, or in the **placenta and fetus** in infected pregnant women. This persistence may result in systemic reinfection after several months, although this is rare. There are presently no available vaccines to MARV, although experimental candidates are being developed alongside Ebola virus vaccines (see below).

Ebola viruses

Ebola viruses burst into public notice in 1976, with two spectacular **outbreaks of severe hemorrhagic fever** in people—both in Africa. In the better-known outbreak for which the viruses were later named, **Ebola virus (EBOV)** was first associated with an outbreak that eventually totaled **318 cases**, starting in September 1976. This was in the Bumba Zone of the Equateur Region in the north of what was then Zaire, and is now the Democratic Republic of the Congo (DRC). The index case in the outbreak, as well as many of those subsequently infected, was treated in the **Yambuku Mission Hospital**. He was injected with chloroquine to treat his presumptive malaria: within a few days fever symptoms developed again; within a week, several others who had received injections around the same time also developed fevers which in several cases had hemorrhagic complications (Fig. 8.11).

Interestingly, **women 15–29 years of age** were most affected by the disease: this was strongly correlated with their attending **antenatal clinics** at the hospital, where they regularly received injections. Apparently the hospital had only five old-style syringes and needles, and these were reused without proper sterilization. Nearly all cases in this outbreak either received injections at the hospital, or had close contact with those who had. Most people were infected within the first 4 weeks of the outbreak, after which the hospital was closed because

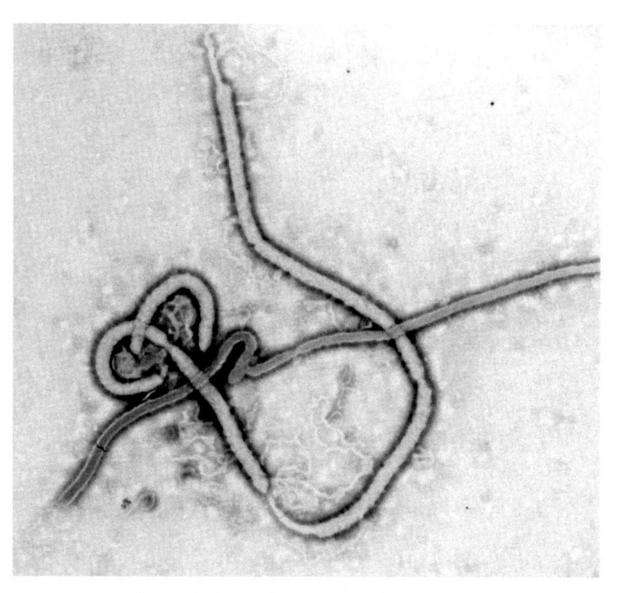

FIG. 8.11 Ebola virus. This electron micrograph was taken in 1976 by Frederick Murphy, a virologist then working at the Centers for Disease Control and Prevention (CDC) in Atlanta, Georgia, USA. This image is in the public domain.

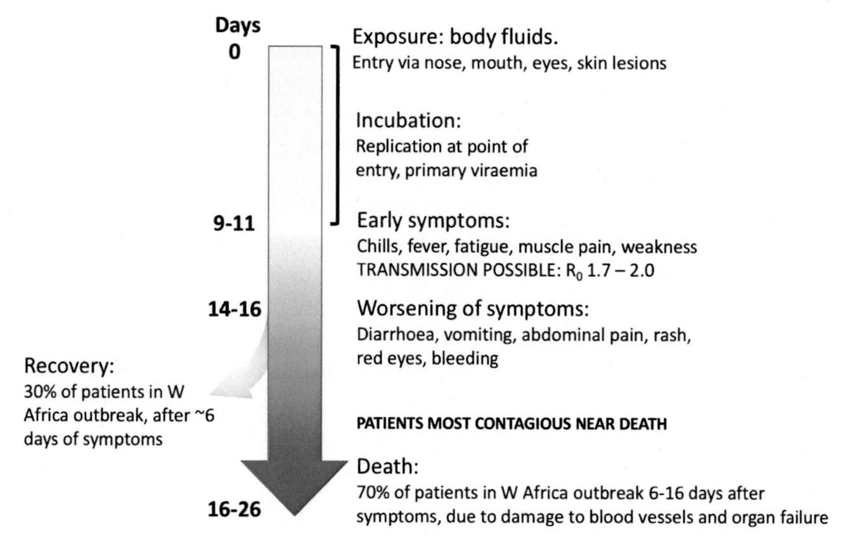

FIG. 8.12 The time course of Ebola virus disease. *Data from CDC Ebola virus disease advice for clinicians (https://www.cdc.gov/vhf/ebola/clinicians/evd/clinicians.html).*

11 of 17 staff had died. Another 269 people died, for a **total estimated case-fatality rate of 88%.** The **incubation period** for needle-transmitted Ebola virus was 5–7 days and that for person to person transmitted disease was 6–12 days. Interestingly, in postepidemic serosurveys in DRC, **antibody prevalence** to the "Zaire Ebola virus" has been **3%–7%:** this indicates that **subclinical infections** with the disease agent may well be reasonably common (Fig. 8.12).

The team that discovered the virus at the Antwerp Institute of Tropical Medicine in Belgium, did so after receiving **blood samples** in September 1976 from a sick Belgian nun with hemorrhagic symptoms who had been evacuated from Yambuku to Kinshasa in the DRC, for them to investigate a possible diagnosis of yellow fever. Following her death, **liver biopsy samples** were also shipped to Antwerp—where the team had already ruled out yellow fever and Lassa fever. Because of the severe nature of the disease, and its apparently novel agent, the World Health Organization (WHO) arranged that samples be sent to other reference centers for hemorrhagic viruses, including the Centers for Disease Control (CDC) in Atlanta, USA.

The Belgian team were the first to image the virus derived from cell cultures on an electron microscope—when it was obvious that the only thing it resembled was Marburg virus.

The CDC quickly confirmed that it was **Marburg-like**, with possibly **the most famous virus image** in the world (see Fig. 8.11). Particles were of very varied length, but had a constant diameter of 96–98nm—and it was a **distinct and new virus**. This meant it needed a name—and it was given one derived from the Ebola River near the town of Yambuku. Another, minor outbreak of the virus occurred in June 1979 in Tandala in north-western DRC: one young child died, and virus was recovered from her—and subsequent investigations showed that two clinical infections with Ebola virus had occurred in 1972 and that about 7% of the residents had antibodies to the virus. This further reinforced the idea that **subclinical infections** were possible.

Sudan virus

In June 1976—before the Yambuku epidemic in DRC—an outbreak of a hemorrhagic fever began in the town of **Nzara,** in what is now **South Sudan**. The presumed index case was a storekeeper in a cotton factory, who was hospitalized on June 30th, and died within a week. There were a total of **284 cases** in this outbreak: there were **67** in Nzara, where it is presumed to have originated, and where infection spread from factory workers to their families. There were also **213** in Maridi, a few hours drive away—where, as in Yambuku, the outbreak was amplified by "nosocomial" or hospital-acquired transmission in a large hospital. In this case, transmission seems to have been associated with nursing of patients. The incubation period in this outbreak was **7–14days**, with a case mortality rate of **53%.**

Two viral isolates were made from sera from Maridi hospital patients in November 1976. Antibodies to the now-identified "Ebola virus" from DRC were detected in **42 of 48** patients clinically-diagnosed patients from Maridi—but in only **6 of 31** patients from Nzara. However, it was subsequently shown that the Sudan and DRC Ebola viruses were different enough from one another to be **separate viral species** (see later), which undoubtedly affected the results.

Interestingly, **19%** of the Maridi case contacts had **antibodies** to the virus—with very few of them with any history of illness. This strongly indicates that the Sudan virus can cause **mild** or even **subclinical infections**.

An indication of the possible origin of the epidemic is the fact that **37% of the workers** in the Nzara cotton factory appeared to have been infected, with six independently-acquired infections—and that this was concentrated in the cloth room, where there were numerous rats as well as thousands of **insectivorous bats** in the roof. However, subsequent study of antibodies in the bats failed to detect evidence of infection, and no virus was isolated from bat tissue. There was another outbreak of the same type of Ebola hemorrhagic fever in the area of Nzara in July-October 1979: this resulted in **34 cases, 22 of them fatal**, with the index patient working at the cotton factory and all others being infected via the hospital he was admitted to. It is interesting that **antibodies** to the Sudan virus were detected in **18% of adults** not associated with the outbreak, leading the report's authors to speculate that the **virus was endemic in this region**.

It was thought that the Sudan and DRC outbreaks were linked: the original WHO Bulletin report on the Sudan outbreak even speculates that extensive truck-borne commercial goods traffic between Bumba in DRC and Nzara in what is now South Sudan could have caused the DRC outbreak. However, comparisons between the viruses isolated from the two epidemics later showed that they were distinct, both in terms of **virulence**, and antigenicity—meaning the **Sudan virus** (SUDV) got its own name.

It gives a sense of scale of the separation, given that many people think that the Sudan and Congo filovirus outbreaks in 1976 occurred near to each other, to note that Yambuku in the DRC and Nzara in what is now South Sudan, are 1100 km apart: this is as far away from each other as **Belgium and Poland**, or **Kansas and Kentucky**. They were thus entirely separated in terms of geography and movement of people, and had **completely separate origins**. A recent Sudan virus outbreak in Uganda was still ongoing at time of writing, and was the 5th worst Ebola outbreak at the time (https://www.who.int/emergencies/disease-outbreak-news/item/2022-DON421)

Later Ebola virus disease outbreaks

After a relatively long gap after 1979, there was a major scare in 1989 in **Reston, Virginia** (USA), when **crab-eating macaques** in an animal facility were diagnosed with **simian hemorrhagic fever**—which was very quickly assumed to be caused by EBOV. This proved incorrect, as the virus was antigenically different, and caused **no disease in exposed facility workers**, although some did develop **antibodies** to it. This was fortunate, as the new virus apparently spread quite efficiently via the air conditioning system. The incident sparked a great deal of media attention at the time, and a highly popular TV series was made recently detailing a slightly fictionalized account of the outbreak (The Hot Zone). The virus was subsequently found in two more primate research facilities (in Pennsylvania, and in Siena, Italy), and was traced back to a facility in the Philippines that **raised macaques for laboratory use**. It has been given its own name (RESTV).

"Ebola virus disease" (**EVD**) outbreaks subsequently occurred in several regions of Africa, starting in 1994. It became apparent that the first of these was also a new virus (**Taï Forest ebolavirus**, TAFV), while EBOV also broke out in a number of places subsequently. Table 8.2 recounts salient features of these outbreaks.

A major Ebola outbreak occurred in the town of **Kikwit** in the DRC in 1995, starting with nosocomial transmission in the Kikwit General Hospital from a patient who had had recent exposure to fresh and smoked "bushmeat" from antelope and monkeys. This went on to infect **317** people, and to kill **245** of them for a **CFR of near 80%.** The virus was effectively identical to the one that first emerged in 1976, prompting intense speculation as to where it had been hiding—and **in which host**, given the apparently very low degree of **sequence drift** that had occurred. A step-by-step account of the outbreak was chronicled in one of the very first set of web pages to be dedicated to the virus (see Rybicki, 2015). News of the outbreak was one of the first things to be featured on the invaluable **ProMED-mail** service, which has gone on to cover every relevant human and veterinary disease outbreak worldwide ever since. A link to a highly useful history of Ebola virus disease is given below (see Muyembe-Tamfum et al., 2012).

TABLE 8.2 Ebola virus disease outbreaks.

Date(s)	Virus	Location	Human cases/ Deaths
1994	TAFV	Taï Forest, Côte d'Ivoire. Many dead chimpanzees found	1/0
1994	EBOV	Minkebé Forest, Gabon. Linked to killing of sick gorilla	52/31
1995	EBOV	Kikwit, Democratic Republic of Congo. Index case a farmer with contact with a forest area	315/250
Jan 1996; Jul-Mar 1996–97; 2001–02	EBOV	Gabon. Linked to nonhuman primate "bushmeat"	31/21; 60/45; 65/53
2000–01	SUDV	Gulu, Mbarara, Masindi in Uganda. This is the largest SUDV outbreak to date	425/224
2001-Jul 2002; Dec 2002-Apr 2003; Dec 2003; 2005	EBOV	Mbomo, Kellé in Republic of Congo (ROC). Linked to gorilla hunting initially	59/44; 143/128; 35/29; 12/10
2004	SUDV	Yambio, South Sudan (near Nzara)	17/7
2007	EBOV	Kampungu, Kaluamba; West Kasai Province, DRC. Index case a hunter, possibly associated with a fruit bat migration	264/187
2007–08	BDBV	Bundibugyo and Kikyo, Uganda. Associated with bushmeat	149/37
2008–09	EBOV	Kaluamba, DRC	32/14
2012	SUDV	Uganda	24/17
2012	BDBV	Province Orientale, DRC. Linked to bushmeat	77/36
Dec 2013-Jan 2016	EBOV	Liberia, Guinea, Sierra Leone; other cases in travelers in Nigeria, Mail, the United States, Senegal, Spain, United Kingdom, Italy (see below)	28,616/11,310
2014	EBOV	Boende, Équateur Province, north-west DRC. Associated with wife of a bushmeat hunter. Virus most similar to Kikwit isolate	66/49
May-Jul 2018	EBOV	Bikoro, Iboko, Wangata in Equateur Province, north-east DRC	54/33
Aug 2018-Jun 2020	EBOV	North Kivu and Ituri Provinces, north-east DRC, and one case imported to adjacent area in Uganda	DRC 3470/2280 Uganda 4/0
May-Nov 2020	EBOV	Mbandaka, Équateur Province, north-west DRC. Virus not derived from Kivu or previous Équateur outbreaks	130/55
2021	EBOV	Butembo, North Kivu Province, NE DRC	12/6
2021	EBOV	Guinea. Reactivation of latent virus infection in a survivor of the 2013–16 epidemic	23/12
Oct 2021	EBOV	Beni Health Zone, North Kivu Province, DRC. Linked by sequence to the 2018–20 outbreak in the same region, possibly due to persistent infection in an EVD survivor leading to transmission	No data

Derived from Muyembe-Tamfum, J.J., Mulangu, S., Masumu, J., Kayembe, J.M., Kemp, A., Paweska, J.T., 2012. Ebola virus outbreaks in Africa: past and present. Onderstepoort J. Vet. Res. 79 (2), 451. https://doi.org/10.4102/ojvr.v79i2.451. http://www.scielo.org.za/scielo.php?script=sci_arttext&pid=S0030-24652012000200003; and https://en.wikipedia.org/wiki/List_of_Ebola_outbreaks.

The use of **immune serum** as a **therapy** was pioneered quite early during the outbreak: after early experience with the disease in 1976, the Congolese physician **Dr Jean-Jacques Muyembe-Tamfum** led a study in 1995 on the use of convalescent antibodies. Seven of eight patients recovered, prompting further work that led eventually to **the isolation of mAbs** from B-cells of a disease survivor in 2006. These proved to be **protective** in monkeys, leading to the development of candidate therapies for use in infected patients (see below).

The recent Ebola virus disease outbreak in West Africa which began in 2013 and continued until 2016 is by far the largest to date, having killed **over 11,000 people** of the nearly **29,000** known to have been infected. It is supposed to have started with a single zoonotic transmission event to a child in Meliandou in Guinea, that may be linked to exposure to the roost of an **insectivorous bat colony** (*Mops condylurus*) in a hollow tree (Marí Saéz et al., 2015). The epidemic eventually involved the neighboring West African nations of Guinea, Liberia, and Sierra Leone, none of which had previously reported Ebola cases. While it was caused by the **same Ebola virus** (EBOV) that emerged in 1976 and has re-emerged several times since, this was fortunately not the most virulent virus outbreak as the death rate was around **40%** of those known to have been infected, compared with over **90%** in some smaller previous outbreaks. However, the lack of experience of local health workers with the disease, coupled with lack of healthcare infrastructure in all three countries and widespread mistrust of the authorities, led to rapid transmission with people escaping centers of disease outbreaks, and attending funerals without any distancing or hygienic practices being employed. As in earlier outbreaks, inexperienced medical staff were among the earliest casualties. A major international relief effort and training of medical staff ensued, and the outbreak was eventually brought under control by **barrier nursing, rigorous contact tracing**, and official interventions in **funerary practices**, such as preventing access to bodies of the deceased.

This epidemic also marked the first use (albeit late in the outbreak) of **monoclonal antibodies** (mAbs), such as the three-Ab cocktail called **ZMapp**, as well as of the experimental **RdRp antagonist** antiviral remdesivir (see Chapter 6), and of emergency-use vaccines based on **adeno-**, **rhabdo-**, and **poxvirus** vectors (see Chapter 9, Virus biotechnology). The only vaccine that was tested sufficiently—before case loads dropped too low for a trial—was the **recombinant rhabdovirus VSV, rVSV-ZEBOV,** expressing Ebola virus major surface glycoprotein. This was used in "**ring vaccination**" trials, where all suspected and proven contacts of infected people were vaccinated in a given area around the contact. This strategy had previously been employed successfully in the eradication of **smallpox**.

The world's second-most serious outbreak—in a conflict zone in **North Kivu** and **Ituri** provinces in the **north-east of the DRC**—was declared in August 2018, and officially only declared over in June 2020. EBOV infected at least 3470 people, and killed over 2000 of them, for a CFR of 66%. While the DRC had successfully controlled all outbreaks since Kikwit in 1995, this region was effectively ungoverned and plagued by various rival militias, resulting in a breakdown of healthcare services and the targeting of both EVD patients and healthcare workers by armed groups. However, this epidemic was also the first time that both **vaccines** and **therapeutics** were widely deployed—and with some significant successes. While both remdesivir and ZMapp proved ineffective as therapeutics, two **mAb-based therapies—mAb114 and REGN-EB3**—caused **dramatic reductions in mortality** among people hospitalized soon after infection, and subsequently became "standard of care" for consenting patients in treatment centers. A vaccine based on **rRSV-ZEBOV** that was first tested in West Africa, was deployed to more than **300,000 people** who had been in contact with Ebola patients or their contacts, as determined by rigorous contact tracing. More than **80% of vaccinees did not develop disease**, and those who did had only **mild cases**, indicating a **resounding success**. In fact, claims that **EVD was now curable** were made following the success of the mAb therapies: in infected people who arrived at treatment centers soon after becoming ill, just 6% and 11% of those treated with REGN-EB3 or mAb114 died, compared to 24% of those treated with ZMapp, and 33% with remdesivir (Boseley, 2019). The most recent EVD outbreaks—also caused by EBOV—have occurred in the DRC, at effectively opposite ends of the country, with viruses that were not linked epidemiologically to the previous outbreaks in both areas (see Table 8.2). The new therapies were quickly employed, along with rRSV-ZEBOV ring vaccination.

Transmission of Ebola viruses

Epidemics and outbreaks have resulted from person to person transmission, nosocomial or in-hospital spread, or laboratory infections. The mode of primary infection and the natural ecology of these viruses are unknown. **Association with bats** has been implicated directly in at least 2 episodes when individuals entered the same bat-filled cave in Eastern Kenya. Ebola infections in Sudan in 1976 and 1979 occurred in workers of a cotton factory containing thousands of bats in the roof. However, unlike the case with Marburg virus mentioned earlier, **no evidence of natural infection** of bats with Ebola virus was found.

Most Ebola virus outbreaks appear to be associated with contact with **infected primates**; however, extensive ecological surveys in Central Africa have failed to show any evidence that primates (or any of the thousands of animals, plants, and invertebrate species examined) are the natural reservoir for infection. No animal reservoir for the virus has been positively identified, but **fruit and insectivorous bats** tested in captivity support replication and can have high titers of Ebola virus without necessarily becoming ill: this includes the *M. condylurus* bats implicated in the West African epidemic. As with SARS, though, consumption of exotic wild meats (called "**bushmeat**"), particularly primates, may be the major risk factor.

Given that **over 17,000** infected people had survived infection and disease in the West African outbreak, a considerable number have subsequently reported continuing health problems, over several years. These included severe eye problems, or **uveitis**; headaches and joint pains. One survivor went on to have a flare-up of disease in the form of severe meningitis, 9 months after being cleared of infection—presumably because of **persistent, low-level infection** in the brain as an immune privileged site, as had also been seen with Marburg. Another male survivor was linked to a new cluster of infections nearly **500 days** after discharge from care, due to sexual transmission of EBOV: virus was subsequently found in his seminal fluid, 531 days following his recovery.

More alarmingly, an outbreak declared in January 2021 in Guinea appears to have been linked to "**resurgence**" from a survivor of the West African outbreak, **5 years later**: this spread to several health care workers and later to relatives and friends during her funeral, resulting in 23 cases and 12 deaths. It is speculated that several of the known outbreaks of EVD may have resulted from just such a phenomenon—and that Ebola virus may in fact be **endemic** in certain areas without causing overt disease. A discovery that gives credence to this, and reinforces the notion that **many Ebola virus infections may by asymptomatic** and therefore go undetected, was the news that **11% of patients** whose blood was sampled for other reasons prior to the EVD outbreak in North Kivu in 2018, were **seropositive** for Ebola viruses.

Twenty-nine had antibodies to EBOV, and one to **Bombali ebolavirus** (BOMV), previously found only in **insectivorous** free-tailed bats (*Chaerephon pumilus*) roosting in people's houses, that were collected in 2016 in Sierra Leone. These results not only reinforce the possibility that Ebola viruses may lurk undetected in areas where they have been seen before, but there are yet more filoviruses, **possibly with outbreak potential**, still out there in the African bush.

Hendra and Nipah viruses

These viruses are monopartite ssRNA- viruses in genus *Henipavirus*, family *Paramyxoviridae*; both characterized by unusually long genomes for paramyxoviruses, and both are listed as "**priority diseases**" by the WHO because of the risks for their emergence.

Hendra virus

In September 1994, 13 of 19 horses and their trainer in Hendra, a suburb of Brisbane in Australia, all died of what was at first called **equine morbillivirus** for the fact that the causative virus appeared to be typical of viruses in genus *Morbillivirus*, family *Paramyxoviridae*. There was another outbreak in Mackay, **1000 km north** of Brisbane, retrospectively diagnosed as occurring in August 1994. This resulted in the deaths of two horses and their owner: he was diagnosed 3 weeks later with **meningitis**, recovered, but developed neurological symptoms 14 months later, and died. Hendra virus was found in his **brain** after autopsy.

Large endemic fruit bats in the genus *Pteropus*—also known as "flying foxes"—were found to be the source of infection, which was assumed to result from **fecal or other contamination** of grass or fruit under bat roosts in trees in the horse pasture. The virus was found to have an unusually long genomic ssRNA– of 18.2 kb, and was reclassified as **Hendra virus** (HeV) in 1998. The bat hosts develop viremia after experimental infection, and shed the virus in **feces**, **urine** and **saliva** for around a week, with **no signs of illness**. Up to **47% of bats** in certain roosts were found to be seropositive for HeV. Symptoms of virus infection in **humans—seven cases** to date, four fatal; **CFR of 60%**—include respiratory illness, including hemorrhage and edema in the lungs, and occasionally meningitis. Infections in horses usually result in pulmonary edema—evidenced by horses blowing out frothy secretions—and neurological symptoms. One case of a **dog with anti-HeV antibodies** was recorded in 2011, prompting fears that the virus could spill over into domestic animals other than horses.

As of 2014 there had been a total of **50 outbreaks** of the virus, along or near the **Australian east coast** from north of Sydney in New South Wales to Cairns in north Queensland. All involved horses, with 87 dying or being euthanized (**CFR 75%**) with proven or suspected infection. Four of these outbreaks involved transmission to humans. Outbreaks seem to occur in the cooler months—**May to October**, when large numbers of bats congregate in SE Queensland to forage for fruit—and all human cases seem to result from exposure to horses, and not to bats. A vaccine was approved in 2012, then licensed in 2015, for horses only. This is a subunit vaccine consisting of a **soluble version of the virion surface G glycoprotein** (HeV-sG), and several hundred thousand doses have been given to more than 100,000 horses.

Nipah virus

In 1998 a research team at the University of Malaya identified a new disease in people from the village of Sungai Nipah, and followed up by isolating and identifying the virus in 1999.

This was identified as a **novel paramyxovirus**, named **Nipah virus (NiV)**, and was similar to HeV in having a longer than normal genome. The two viruses were found to **cross-react in serological assays**, and were subsequently grouped into a new genus—*Henipavirus*—in family *Paramyxoviridae*. In humans by mid-1999 there had been 265 cases of **encephalitis** and 105 deaths reported from Malaysia, and 11 cases and one death from Singapore. The outbreak was linked to pigs and pig farmers, and millions of pigs were culled in 1999 to limit spread of the disease—which measure appeared to work in that instance. Evidence of infection was found in pigs, but also in other domestic animals such as **horses, goats, sheep, cats,** and **dogs**. Subsequently, outbreaks have been identified **almost annually**, from Malaysia, Singapore, Bangladesh, and India. Regular outbreaks in Bangladesh have been traced to **contamination** of palm wine collection vessels in trees with **saliva of bats** coming at night to drink from them—and then contact of vessels directly with humans.

Initial symptoms in humans include **fever, headaches, myalgia, vomiting,** and **sore throat**. This is sometimes followed by dizziness, drowsiness, and other **neurological symptoms** indicating **acute encephalitis**. Some cases exhibit atypical pneumonia and **acute respiratory distress**. Encephalitis and **seizures** occurred in severe cases, **progressing to coma** within 24–48 h.

The incubation period is from 4 to 14 days, but could be as long as 45 days. The **CFR is 40%–75%** depending on the outbreak: a 2018 outbreak in Kerala in India **killed 17 of the 19** infected, for a **CFR of 88%.** Most survivors recover fully, although some who had acute encephalitis have **residual neurological complications**. Some **relapses** have been reported. The virus is unlike HeV in that it can **readily be transmitted from person to person** in hospital settings, probably via respiratory secretions and saliva and droplets expelled during coughing, from infected patients who usually had breathing problems. There is presently no vaccine available, but a candidate **subunit vaccine**—HeV-sG, which it is hoped will cross-protect against NiV—is presently in early-stage trials.

The animal reservoirs associated with NiV outbreaks are **fruit bats** in the family *Pteropodidae*, and particularly in genus *Pteropus*. The bats show **no apparent signs of infection**, but shed live virus similarly to bats with HeV. It is rather concerning that **evidence of past infection** with viruses that cross-react serologically has been found in *Pteropus* bats from northern Australia, Bangladesh, Cambodia, southern China, India, Indonesia, Madagascar, Malaysia, Papua New Guinea, Thailand and Timor-Leste. Fruit bats of the genus *Eidolon* in family *Pteropodidae* found in **Ghana** were **seropositive for henipaviruses**, indicating that these viruses might be far more widely spread than was suspected. The Global Alliance for Vaccines and Immunization (**GAVI**) started in March 2021 that Nipah virus could well be **the agent that causes the next pandemic** (GAVI, 2021): it is **endemic** in a wide geographical area where there is **increasing contact with bats**; it is **already transmissible person-to-person**—and any increase in transmissibility could result in just the kind of deadly pandemic as was very convincingly portrayed in the **2011 film Contagion**, where the cause was a **recombinant Nipah virus** resulting from infection of a **pig by a bat** displaced from a bulldozed forest.

Bats and paramyxoviruses

A 2011 article reporting a major international survey for paramyxoviruses (PVs) in rodents and bats (see Drexler et al., 2012) identified 66 putative new species viruses—more than **the 36 already characterized** by the ICTV—from worldwide sampling by RT-PCR of 86 bat and 33 rodent species (4954 and 4324 individual samples, respectively). Detection rates of PVs were near 3% for both types of animals, with bat viruses being considerably more genetically diverse. In addition to other **rubulaviruses**, a virus that is conspecific with human **mumps virus** was found in an African bat—and a serological survey of 52 fruit bats and 78 insect-eating bats showed that 42% had antibodies that cross-reacted with mumps virus proteins. **Morbilliviruses** distantly related to human **measles** and animal **rinderpest** viruses were found in bats, and morbillivirus-like viruses in both bats and rodents. **Metapneumoviruses** in a sister clade to **human and bovine respiratory syncytial viruses** (RSVs) in subfamily *Pneumovirinae* were also found in bats, although these were not as closely related as viruses in **birds** that are postulated to be ancestors of the mammalian viruses.

Of more relevance to concern over NiV as a possible pandemic virus was the finding of sequences from **26 distinct viral clades** related to **henipaviruses** in 6 bat species from **5 African countries.** These included **19 novel henipaviruses**. It is speculated that a lineage of viruses from Africa was ancestral to both HeV and NiV given the greater diversity of these viruses in Africa—meaning the potential for "spillover" of **potentially lethal** henipaviruses from bats to humans may be high across Africa. Interestingly, two widely distributed **New World bat species** also harbored henipa-like viruses.

Severe acute respiratory syndrome coronaviruses

Severe acute respiratory syndrome (SARS) is a type of viral pneumonia, with symptoms including fever, a dry cough, shortness of breath, and headaches. Death may result from progressive respiratory failure due to lung damage. The syndrome may be caused by a number of agents, including **influenza viruses**—and occasionally by the human "**common cold**" viruses from family *Coronaviridae*; namely, hCoVs 229E, NL63, OC43, and HKU1. All are enveloped ssRNA+ viruses with **genomes of ~30 kb**, and 120 nm diameter particles with prominent **spikes** of trimeric S protein—hence "corona," or **crown** (see Fig. 1.13). The first two are **alphacoronaviruses**, and the latter **betacoronaviruses**, and have been infecting humans at least since their discovery in the 1960s for 229E and OC43, 2004 for NL63, and 2005 for HKU1. The viruses appear to be ubiquitous, all typically infect the upper respiratory tract, usually cause symptoms such as sore throat, cough and stuffy nose, and cause a significant number—10%–30%—of common colds. Infections with HCoV-NL63and HCoV-HKU1 appear to cause more severe respiratory disease in young children, adults with comorbidities, and the elderly.

Interestingly, **hCoV-OC43** is postulated to have emerged into humans **from domestic cattle** in the 1890s—and the epidemic that had been named "the Russian flu" in 1889–90, and which killed **over 1 million people**, may well have actually been caused by this virus—which has now, as influenza viruses have done previously, become endemic in humans as a mild seasonal disease agent (Fielding, 2020).

However, the most prominent recent emerging coronaviruses are **three betacoronaviruses** (genus *Betacoronavirus*, family *Coronaviridae*); namely, **SARS-CoV**, **Middle East respiratory syndrome coronavirus (MERS-CoV)**, and **SARS-CoV-2**.

SARS-CoV

The original SARS outbreak originated in the south-eastern **Guangdong Province** of China in the beginning of November 2002, and was reported in February 2003. At least **300 people** became ill and at least **five** died in the region, sparking an intense research effort to determine the cause. This was found to be a **novel coronavirus**, soon named **SARS-CoV**. Symptoms were initially **flu-like**, and could include fever, muscle pain, lethargy, **diarrhea**, a dry cough, and sore throat. Fever with **temperatures above 38C** was the most common indicator of infection. **Severe disease** symptoms were shortness of breath, and **viral** or secondary **bacterial** pneumonia. The virus is believed to be spread by droplets produced by coughing and sneezing, but other routes of infection may also have been involved, such as **fecal droplet or aerosol contamination of circulated air** in a hotel, and **sexual** transmission. The virus spread out of Guangdong (5327 cases/349 deaths in China) with **business air travelers**, via Hong Kong in February 2003—where hotel transmission amplified the outbreak (1755 cases/299 deaths)—and thence in order of severity to Taiwan (346/ 37), Canada (251/43), Singapore (238/33), and Vietnam (43/5). In total around the world, **8422 people** were confirmed to be infected, and **916 died**, for a **CFR of 11%.** The mean incubation period was **6.4 days**, with a range of 2–10 before symptoms. It is thought that the relatively long incubation period for a respiratory virus allowed **asymptomatic air travelers** to spread the virus quickly and widely (Chan-Yeung and Xu, 2003). Most clinical cases were in healthy adults aged 25–70 years, with few cases reported in children under 15, with milder disease. Risk factors The WHO states that the CFR among people with symptoms meeting the WHO case definition for **probable and suspected cases** of SARS was **around 3%.** The epidemic was declared "contained" by the WHO in July 2003, and only another four cases were detected in China up to January 2004. **Three laboratory accidents** subsequently led to infections: one in Singapore in August 2003, and two in China, resulting in a cluster of six cases related to two researchers in Beijing who picked up the virus **independently** of one another.

The rapidity of international spread and the high mortality rate caused international panic, with announcement of a **global alert** by the WHO in March 2003 being followed by strict isolation of cases and their contacts, cancellation of flights to and from affected cities, and temperature screening at airports. The **economic impacts** of the epidemic were relatively severe in China and Hong Kong, and less so for Canada and Singapore, with the main sectors affected being investment, air travel, tourism, and hospitality. The outbreak provided a salutary teaching moment for everyone waiting for "**The Big One**"—which was always supposed to be an **influenza A virus** pandemic, which did in fact come around in 2009 (see above), but was nothing like as severe as expected.

One postoutbreak study indicated that survivors had **high antibody titers** against SARS-CoV S protein for **2 years** after recovery, but by **3 years** this had dwindled to **55%**, and to **zero by 6 years** (Sariol and Perlman, 2020). This is similar to what happens with the **common cold hCoVs**—which can infect people again after a few years. However, another study found **low levels of serum IgG persisting 13 years** after infection. By contrast, levels of **memory T-cells** were generally maintained, in **70%–100% of patients up to 6 years** after recovery, even in people with **no B-cell responses**. While lack of serum IgG could mean people could get reinfected at some time after recovery, the maintenance of T-cell responses probably means that they would not become ill (Fig. 8.13).

Where did the SARS virus come from? Coronaviruses with 99% sequence similarity to the surface spike protein of human SARS isolates were isolated in May 2003 from animals in the local **live food market** in Guangdong, China, from apparently healthy **masked palm civets**, a cat-like mammal closely related to the **mongoose**. The unlucky palm civet

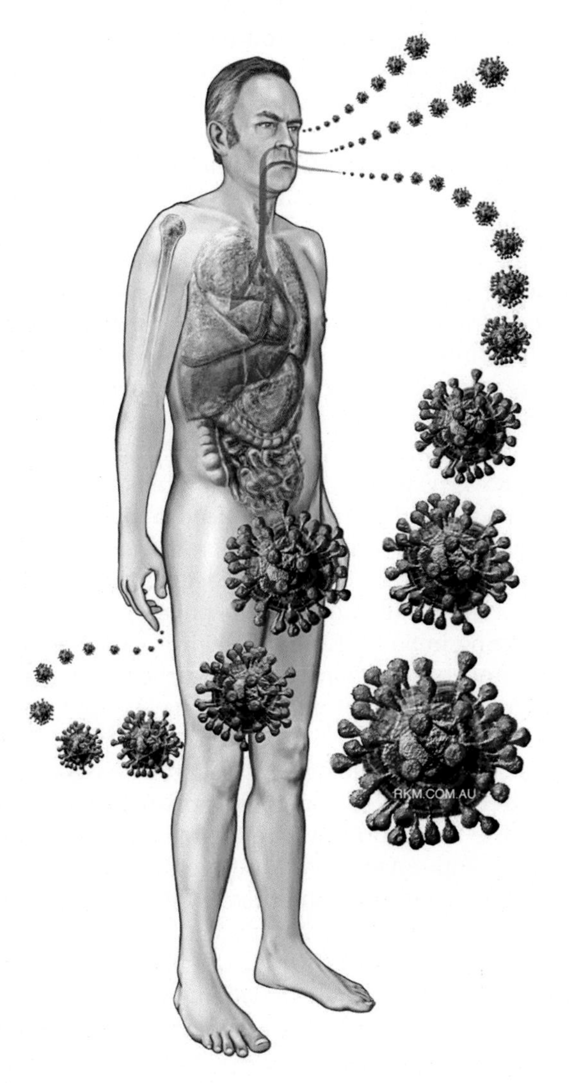

FIG. 8.13 Anatomical sites of infection and transmission for SARS-CoV. Virions may infect people via respiratory droplets or aerosols getting into the upper respiratory tract, or contacting the surface of the eyes, or being picked up on surfaces and conveyed to the mouth. While SARS-CoV is regarded as a respiratory virus, it can actually cause systemic infections. Virus may be transmitted via respiratory secretions, either directly, or indirectly via droplet or aerosol, or via fecal contamination or aerosolization. *Courtesy of Russell Kightley Media.*

is regarded as a delicacy in Guangdong and it is believed that humans became infected as they raised and slaughtered the animals rather than by consumption of infected meat. Later surveys detected virus in "raccoon dogs," "ferret badgers," and even domestic cats—and then in 2005, a number of SARS-CoV-like viruses were identified in **Chinese horseshoe bats** (*Rhinolophus* spp.). In 2017, "SARS-hunting" researchers found a population of horseshoe bats in a cave in the south-western Yunnan Province that harbored a mixture of SARS-CoV-like viruses. **Metaviromic sequencing** proved that of the **15 viruses** isolated from the bats, **no one virus** was identical to SARS-CoV—but the **sequence of the human virus** could be pieced together by **recombination** between members of the strain mixture. This is regarded as strong evidence for the **emergence of SARS-CoV from bats** via transmission to, and possible recombination in, other animals in the food market in Guangdong.

MERS-CoV

After 2003 SARS apparently went away, but in 2012 a related betacoronavirus—**Middle East respiratory syndrome coronavirus (MERS-CoV)**—popped up in Jeddah in Saudi Arabia in the Middle East, and has since been reported from the United Arab Emirates and South Korea. Like SARS, MERS is a **zoonotic** infection, probably originating in bats, but now found and transmitted among **dromedary camels**. MERS-CoV has a genome size of 30,119 bases, determined by next-generation sequencing, which has only 46% nucleotide sequence identity with SARS-CoV. Unlike SARS-CoV, which uses the **angiotensin converting enzyme 2 (ACE2)** as a receptor, MERS-CoV was found to use **dipeptidyl peptidase 4** (**DPP4**; also known as CD26). Camels have a transient common cold-like disease, and most human index cases are camel owners or people who work with camels. Subsequent **person-to-person transmission requires close contact**—and most outbreaks have been hospital-associated, with the largest such outbreak to date occurring in South Korea. The virus causes fever, coughing and shortness of breath, with pneumonia being common and diarrhea more rare. The average incubation period is **5.5 days** (range 2–15). Asymptomatic cases have been recorded during aggressive contact tracing. The disease is considerably more severe than that caused by SARS-CoV, as up to **35% of confirmed cases die**. Those who develop severe disease are generally older people, and people with **immune deficiencies** or **comorbidities** such as renal and chronic lung disease, cancer and diabetes. The true mortality rate is probably lower, as asymptomatic infections are more common than with SARS-CoV. Multiple therapies have been tried, including use of interferon, chloroquine, remdesivir and others, with no obvious efficacy. There is no licensed vaccine for MERS-CoV, although this may change in the near future with the very rapid redevelopment of SARS-CoV and MERS-CoV vaccine candidates into accepted SARS-CoV-2 vaccines (see below). Oxygenation of severely ill patients improved outcomes.

Since its identification in 2012, MERS-CoV cases have been reported in **28 countries**, including the United States, United Kingdom, and several European and Indian Ocean countries—mainly in travelers or cases linked to travelers from the Middle East. Over **2500 cases** worldwide had been reported as of January 2021, with **45 in 2020**. The largest outbreaks have been in Saudi Arabia in 2018 (147 cases) and South Korea in 2015: the index case here was a person who had visited Saudi Arabia. The outbreak spread to multiple hospitals with patients transferring themselves around, and 184 confirmed cases resulted, with 19 deaths. One positive note is that **antibody responses** could still be detected in **100% of MERS survivors** who had experienced **severe or moderate disease**, and in in **50%** who had had **mild disease**. This indicates that persistent immunity may be possible.

It is not clear whether the emergence of MERS is the result of a **single zoonotic event** with subsequent **human-to-human transmission**, or if the multiple geographic sites of infection across the Middle East and Africa represent **multiple zoonotic events** from a common source. Serosurveys of camels in the Middle East, Africa and South Asia have identified past

MERS-CoV infections—in **90% of evaluated animals** in one 2013 study in Saudi Arabia—and viruses that are identical to those isolated from people have been found in camels in Egypt, Oman, Qatar, and Saudi Arabia. Experimental infection of Jamaican fruit bats (*Artibeus jamaicensis*) showed that they were **readily infectible**, showed no signs of clinical disease, and **shed virus from respiratory and intestinal tracts** for up to 9 days. MERS-CoV has been isolated from a Saudi bat, and is closely related to *Tylonycteris* bat coronavirus HKU4 and *Pipistrellus* bat coronavirus HKU5. The supposition, then, is that the progenitor of MERS-CoV **entered camels in the mid-1990s or earlier**, and has circulated in them ever since.

SARS-CoV-2

The latest viral pandemic to afflict humankind was partially foreseen, in that researchers had been predicting for some years that an agent similar to that causing SARS could arise out of live meat markets in China—and indeed, it was at first thought that SARS-CoV had re-emerged, when a cluster of cases of **severe pneumonia**—three of five of whom were associated with the "**Huanan Seafood Market,**" which was also a live animal meat market—were identified in the city of **Wuhan in Hubei Province** in central China in December 2019. Bronchoalveolar lavage specimens were collected by local researchers from five patients with acute respiratory distress, and subjected to **next-generation sequencing** of total DNA, and of cDNA generated from RNA extracts (see Ren et al., 2020). The most prominent "microbial species" in three patients—representing 40%, 73%, and 80% of all sequence reads—and a more minor constituent in the other two (1.6% and 14%) was the same **betacoronavirus-related sequence**. Assembly of a consensus sequence from all reads gave a complete genome of 29,870 bases, with **79% identity with the type SARS-CoV genome** and **52% with MERS-CoV**. Moreover, the virus could be **cultured in Vero cells**, and its spike protein was detected by **antibodies in convalescent sera** from clinically defined cases. Accordingly, as the degree of identity fell below the species threshold of **90% amino acid homology** in the **conserved RdRp domain**, the virus deserved its own name. It was at first referred to as "**nCoV-19**" by the WHO for "new coronavirus 2019"; however, the **ICTV** soon named it **SARS-CoV-2** (see Chapter 2), for the virus that causes **COVID-19**, or "**coronavirus disease 2019.**" The rapid discovery and characterization of this virus was a triumph of modern techniques: **shotgun DNA and cDNA sequencing** from symptomatic patients; the use of **bioinformatics** to isolate and then identify common virus sequences in samples; then the use of more **conventional virus culture** and **immunofluorescence microscopy** to confirm the identification (Fig. 8.14).

The new virus soon proved to be **highly transmissible**, and to be spreading quickly outside of the area in which it was first described. While there was strong pressure to declare earlier, the WHO declared the outbreak a "**Public Health Emergency of International Concern**" (PHEIC) only on 30 January 2020, and then declared it to be a **pandemic** on **11 March 2020** as it spread more widely internationally. Since then, the world has seen inexorable spread—and **four distinct waves of disease**, depending on geographical locality and timing, despite worldwide measures aimed at restricting **travel** and **unnecessary social contact**, and **increasing hygienic measures** such as hand-washing and masking. By around mid-January 2021, the pandemic had surpassed the "Asian flu pandemic" of 1957 as the **second-worst known respiratory virus pandemic**, with **over 2 million deaths** attributed to COVID-19. The figure at time of writing was **over 6 million deaths**, and counting.

FIG. 8.14 Depiction of SARS-CoV-2 virion. The virion has characteristic faceting and a **corona** of spikes. The spikes—shown in reddish-orange—are trimers (three identical proteins bound noncovalently together) that allow the virion to attach to a cell surface. *Courtesy of Russell Kightley Media.*

While the SARS-CoV-2 case fatality rate (**CFR)** was initially estimated to be **above 3%,** it was repeatedly determined in independent investigations that there was a significant undercount of infections due to **asymptomatic infections**—which could account for as many as **a third of all infections**, especially among younger people. SARS-CoV-2 causes less severe disease and a greater proportion of asymptomatic infections than either SARS- or MERS-CoV: the estimated **infection fatality rate (IFR)** for SARS-CoV-2 is **0.5%–1%** (WHO, 2020), compared to CFRs of **11%** and nearly **35%**, respectively, for diseases with a lower proportion of asymptomatic cases. SARS-CoV-2 was obviously **more transmissible** than the previous emerging coronaviruses, as evidenced by the **speed** of worldwide spread: while it was thought initially that transmission was largely due to **droplets and surface contamination**, later determinations were that spread is primarily **via aerosols**: this means that one infected person can spread virus via **tiny droplets** that persist for much longer in the air than larger droplets, and **travel further**. An under-appreciated mode of transmission is also via the **mucous membranes of the eyes**: people wearing **glasses** or **eye protection** in hospitals were 30% less likely to become infected than those without, for the same use of other **personal protective equipment (PPE)**.

Children are significantly less likely to become ill than older people after infection, with risk of severe disease and death **increasing exponentially with age** after 20 years. Fig. 8.8 illustrates the difference between seasonal influenza and COVID-19 IFRs—COVID-19 is 5× up to 25× more severe—and that COVID-19 disproportionately affects older people (IFR of 4 at age 80 vs 0.02 at age 30) (Fig. 8.15).

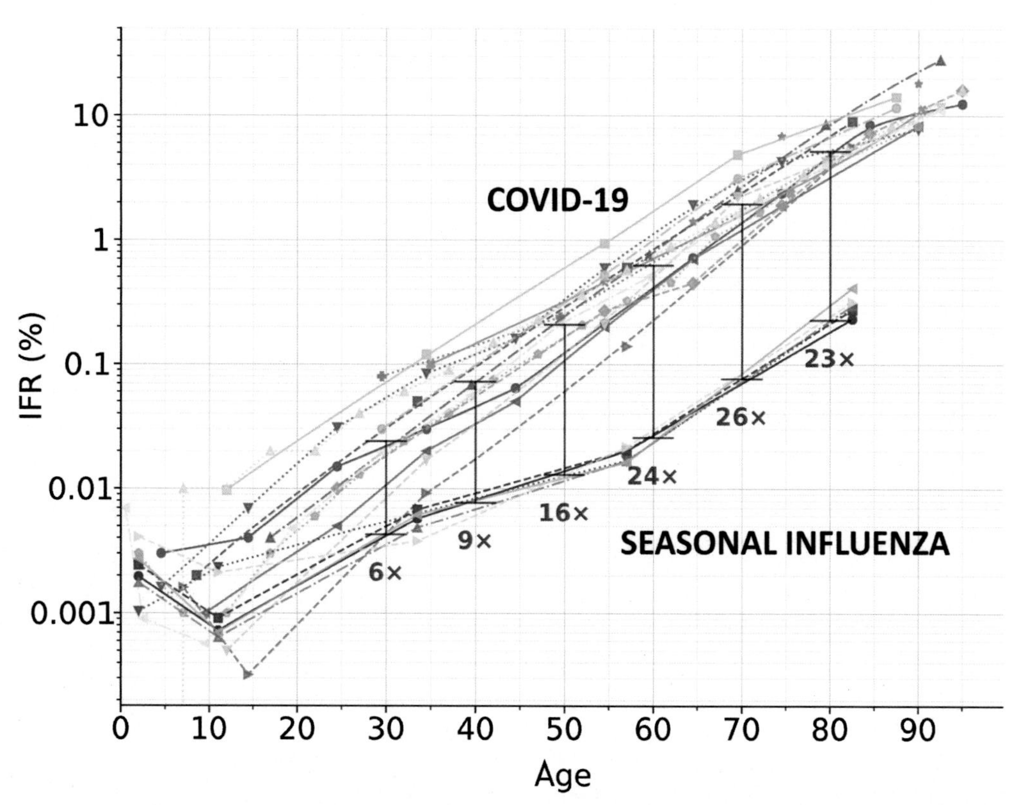

FIG. 8.15 Infection fatality rate of COVID-19 compared to seasonal influenza for different age groups. *From Bevand, M., 2021. The chart compares the IFR of COVID-19 to the IFR of seasonal influenza. COVID-19 is definitely significantly more fatal than influenza at all ages above 30 years. https://github.com/mbevand/covid19-age-stratified-ifr#comparing-covid-19-to-seasonal-influenza.*

Therapy for those infected was intensively researched, with early claims for the effectiveness of **hydroxychloroquine, remdesivir,** the antiparasite medicine **ivermectin** and various other compounds, as well as **convalescent serum** and (later) **mAbs**. The consensus is that **none of these treatments is effective**, and that the only therapeutic that had a definite, measurable outcome in treating severe disease was **dexamethasone**. This is an **antiinflammatory corticosteroid**, and in a large clinical trial in the United Kingdom reduced mortality of patients on respirators **by one-third** and among those on oxygen only, **by one-fifth**.

An issue that complicated the pandemic, and may have led to increased transmission frequency and possibly even disease severity in certain regions, was the appearance in late 2020 and early 2021 of several "**variants of concern**": these are viruses with collections of **mutations** (see Chapter 3) that affect the **virus phenotype**. The SARS-CoV-2 variants found in the United Kingdom, South Africa, and Brazil in late 2020—more properly referred to as the **B.1.1.7, B.1.135,** and **P.1** variants, but now known as **Alpha, Beta, and Gamma variants**—are examples, along with the newer **BA.1 variant** (branching into BA1.1 and BA1.2) now known as **Omicron** (CDC, 2022). All are reported to have significantly higher transmission rates than earlier lineages, and **Alpha** was reported to be more pathogenic as well, while Omicron appears less pathogenic. These variants very largely became the predominant viruses in their countries of supposed origin, and have spread to other countries. The **B.1.617 or Delta variant** was first described in March 2021, and was driving a surge in cases worldwide in April 2021 but has now been usurped worldwide by Omicron—with the BA1.2 subvariant of Omicron rapidly taking over owing to their successively higher transmission rates.

Another area of concern is the fact that human lineages can infect **domestic and farmed animals**: while reports that cats and dogs could be infected **were interesting**, it was far more serious that **mink and ferrets** farmed for their fur could be infected—and could then **generate their own unique variants**, and **give them back to human handlers**. Shortly before the time of writing, many million farmed mink in the Netherlands and in Denmark had been culled because of infection, at a time when experimental vaccines were available, and could possibly have been used.

The origins of the pandemic are certainly zoonotic, and the virus almost certainly originated in ***Rhinolophus*** **bats**—although an **intermediate host** may have been involved, as with SARS-CoV, possibly in live meat markets. Fig. 8.9 shows the relationships among betacoronaviruses found in bats and other animals in China and neighboring countries, derived from a WHO-China joint report published in February 2021 (see WHO, 2021b) (Fig. 8.16).

It is very clear that SARS-CoV-2 clusters closely with **several bat-derived CoVs**, in a monophyletic group distinct from pangolin-isolated viruses, "SARS-related" CoVs and SARS-CoV. Thus, the origin of SARS-CoV-2 was **almost certainly in bats**—however, phylogenetic analyses showed that the probable evolutionary distance between the closest-related bat viruses and SARS-CoV-2 **was several decades**, indicating that an **intermediate host** may have been involved and possibly one found in the farms that supply partially domesticated live animals to the live meat markets.

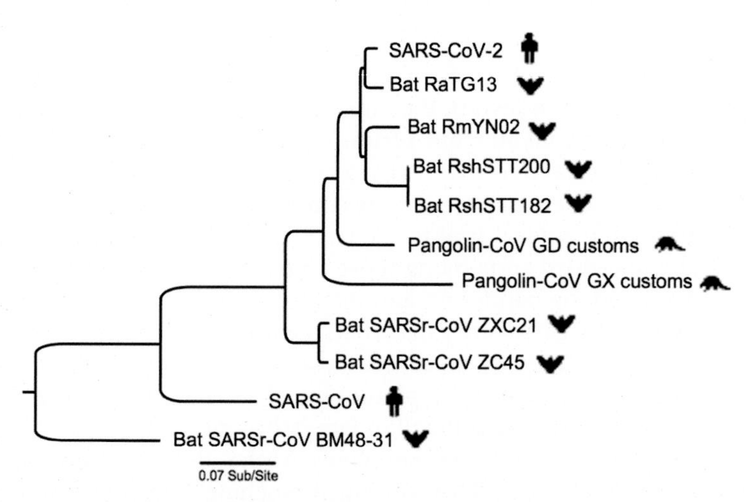

FIG. 8.16 Relationship of bat, human, and pangolin viruses. Phylogenetic tree of SARS-CoV-2 and other coronaviruses in bats and pangolins based on the concatenated protein sequences from all the genomes (WHO, 2021b).

Long COVID

Given the huge numbers of people infected during the pandemic, it was inevitable that rare and serious long-term "**sequelae**"—conditions that are the consequence of a previous disease—of SARS-CoV-2 infection would emerge, just as they had for the 1918 and other influenza epidemics and pandemics, and EVD outbreaks. Indeed, some survivors of the "Russian flu" epidemic of the 1890s that may have actually marked the emergence of a hCoV (see earlier), as well as of the 1918 Flu, were left with "nervous disorders" resulting from central nervous system involvement, possibly due to immune-mediated damage. These included severe headaches and extreme lethargy, which was worse in the 1890s outbreak. The link of this outbreak to the emergence of hCoV-OC43 which appears closely related to bovine coronaviruses (BCoVs) is interesting, as **both BCoV and the hCoV are known to be able to invade and damage the CNS in respective hosts**.

A phenomenon known as "Long COVID" has emerged in countries with the largest number of COVID-19 cases: this is also more formally known as "**post COVID syndrome (PCS),**" and is associated with several symptoms, the most prominent of which are **severe breathlessness, lasting fatigue**, and **cognitive effects**, also known as "brain fog." Other less frequently seen manifestations include the sudden development of **severe migraines** in people previously unaffected by these. Worryingly, long COVID could occur in people who had only mild symptoms of COVID, or who were asymptomatic and only diagnosed subsequently by antibody testing. Many of these symptoms are common to what is often termed "yuppie flu," or **chronic fatigue syndrome** (CFS), which could strengthen links between these syndromes and prior infection with a variety of viral or bacterial infections that now include SARS-CoV-2. Causes of some of the symptoms in this specific case could be **damage to the small blood vessels** around the lungs as well as to **lung tissue directly**, which could cause breathlessness; damage due to inflammation as a result of cytokine storms in severe cases; and autoimmune reactions affecting nerves and the CNS as a result of autoantibody induction during infection. One such manifestation of neurologic involvement is **anosmia**, or loss of sense of smell: this is quite a common symptom of early COVID-19, is more common with **mild disease**, and people generally recover this fully. This is linked to SARS-CoV-2 infection of the olfactory neuroepithelium, and accompanying **inflammation** (de Melo et al., 2021). However, some people have **long-term persistence of anosmia**—and have evidence of infected cells and viral transcripts in the same tissues, as well as continuing inflammation, months after "recovery." This could mean that viral persistence is more common than was realized, and could account for some of the symptoms of Long COVID as well as **relapsing symptoms** of COVID-19.

What is more concerning, and is something that can be linked directly to COVID-19, is the **significantly increased incidence of strokes** among people recovered from even mild COVID-19 infection, particularly **young people**. A major study published recently that involved 136 academic institutions in 32 countries around the world, found that of **380 patients** who had had strokes as well as COVID-19, **38% had no virus symptoms** and were only diagnosed after the stroke. Moreover, 75% of stroke patients had **acute ischemic strokes**, caused by blockage in an artery supplying blood to the brain, in the **absence of any of the classical risk factors** such as high blood pressure, high cholesterol levels, and clotting disorders. This worrying trend is almost certainly the result of blood clots caused by infection and damage of vascular endothelia by SARS-CoV-2, and points up the important fact that COVID-19 is a **systemic** and not just a respiratory disease—and that even **asymptomatic or mild disease** could have serious long-term consequences.

COVID-19 vaccines

The advent of COVID-19 as the much-feared and anticipated "Disease X" was fortuitously accompanied by very rapid pivoting of vaccine development research from experimental **SARS-** and **MERS-CoV** and from **Ebola virus disease vaccines**, to target **SARS-CoV-2**. Processes that can take many years—early laboratory development, preclinical animal trials, manufacture, and clinical trials from Phase 1 through to 3—were **compressed into less than a year**. The result was the very rapid authorization of **mRNA-** and **adenovirus-based vectors** for the expression of **SARS-CoV-2 S protein** as well as of **classical "killed" whole virion vaccines**, and their deployment around the world. Moreover, none of the frontrunner vaccine candidates had in fact had a predecessor licensed for routine use for another disease: the **mRNA-based S protein** vaccines had only ever been trialed in animals for other diseases; the **adenovirus-based S protein** expressing vaccines too had only had one prior human trial, as an "emergency-use-only" **adenovirus 26 expressing Ebola virus GP protein** during Ebola outbreaks in the late 2000s. Interestingly, the **rRSV** vector used so effectively to curb Ebola outbreaks was pulled after being trialed in 2020 for SARS-CoV-2, due to lack of effectiveness compared to the frontrunners. While protein-only vaccines were latecomers in the race, both an **adjuvanted nanoparticle-based spike trimer vaccine** and a **virus-like particle displaying S trimers made in plants** have been through Phase 3 clinical trial by early 2022, and are already being deployed.

A major concern for many people was the apparent ability of several of the variants of concern—**Alpha, Beta, Delta, and now Omicron**—to **avoid neutralization** to some extent by convalescent sera or mAbs directed against S protein, due to mutations in antibody-binding epitopes. While this could be a factor in people becoming **more susceptible to reinfection with SARS-CoV-2**, especially with one of the variants, it is reassuring to note that all of the nucleic acid-based vaccines—**mRNA and adenoviruses**—appear to elicit good **cellular immune responses** that are hardly affected by the mutations. This means that while reinfection could occur—which has now been demonstrated quite often—virus-infected cells would **quickly be cleared** by reactivation of memory T-cells, and only **very mild disease** might ensue. Moreover, vaccination following natural infection has been shown to elicit very strong B- and T-cell responses, **better than natural immunity or immunity elicited by double vaccination**. It does mean, however, that people who have recovered from COVID-19 and those who have been vaccinated should still observe infection hygiene so that they do **not inadvertently transmit virus** to unprotected people.

While there is concern that SARS-CoV-2 immunity **may wane rapidly**—as is the case with the **HCoVs causing common colds**, which means you can catch them again after a couple of years (Fig. 8.10)—it does appear as though **cellular immunity** to both SARS-CoV and SARS-CoV-2 is of longer duration (see above), and **antibody titers**—especially after severe disease and presumably vaccination—**are maintained as well**. Thus, it is less likely than it is with hCoVs that immunity will wane to the point of allowing reinfection, though use of SARS-CoV-2 vaccines **seasonally** may still happen (Fig. 8.17).

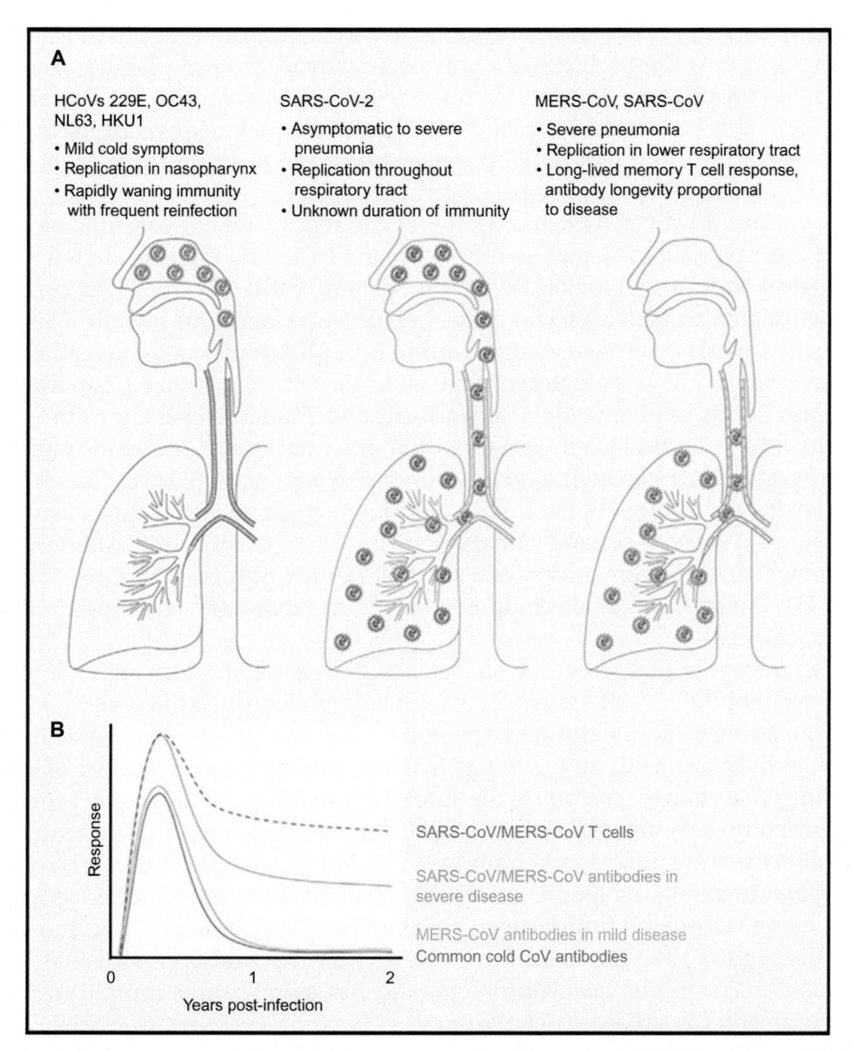

FIG. 8.17 Immunity to human coronaviruses. Schematic depicting sites of replication of human coronaviruses (A) and schematic of longevity of immune responses to common cold coronaviruses, SARS-CoV, and MERS-CoV (B) (not drawn to scale). Data not available for antibody longevity in patients following mild disease caused by SARS-CoV. *From Sariol, A., Perlman, S., 2020. Lessons for COVID-19 immunity from other coronavirus infections. Immunity 53 (2), 248–263. https://doi.org/10.1016/j.immuni.2020.07.005. Epub 2020 Jul 14.*

Biowarfare and bioterrorism

There have been anecdotal and apocryphal stories of the use of bioweapons generally, and virus diseases in particular, for many, many years. Some of these have effectively passed into the realm of myth—for example, firing saliva from **rabid dogs** from cannons in the 1600s, and distribution of **smallpox**-contaminated blankets to native Americans in 1763—but there is certainly an element of truth to some of them, such as a story about Confederates selling clothing used by **smallpox** and **yellow fever** victims to the Union troops in the United States in 1863. There was also the problem of inadvertent spread of viral plagues, that probably killed many more people. For example, the Spanish occupation of much of Central and South America was probably considerably eased by the fact that a **plague of smallpox** and possibly other viruses spread out through naïve populations in the lands in front of their soldiers, eventually killing **up to 90%** of native peoples there.

Closer to our time, however, there has been a lot about government-executed virus weapon development, starting with the Japanese military in the 1930s and then the United States in response. Biological weapons were not used in the Second World War; however, certainly their use was contemplated and research was done. In later years, accounts include claims that the United States had worked on **Venezuelan equine encephalitis virus** as a weapon pre-1969, and that former Soviet Union had stored large quantities of "weaponized" **smallpox** as well as experimenting with weaponization of **Marburg and Ebola viruses** up to the late 1980s. Reputedly, the Soviet Union also investigated the potential use of **foot-and-mouth disease virus, rinderpest,** and **African swine fever viruses** as weapons against cattle and other farm animals. There are also rumors of the Iraqi government working with viruses as weapons in the 1990s, including **rotaviruses** and **camelpox virus**. There is very little extant evidence for any of this; however, although at least one small **smallpox outbreak** and possibly the 1977 **influenza A H1N1 virus** pandemic could be blamed on laboratory "escapes" from possible bioweapons facilities.

It is highly relevant in this respect to sound a caution about a recent spate of popular articles on the desirability of "self-spreading vaccines." While these at first sight seem to be a good idea—spreading a vaccine through a population of animals or even people by inoculating just a few individuals, with huge savings in costs and logistics—a number of real-life examples should give us pause. The first is the spread of the **poxvirus causing myxomatosis**—a virulent rabbit pathogen—throughout Australia: this was intended as a **biocontrol agent**; however, rabbits there are now largely immune to it, and its uncontrolled spread could cause havoc with rabbit breeding in other countries. The second is the uncontrolled spread among humans of the **live attenuated poliovirus type 2 vaccine**, that has led to "**vaccine-associated paralytic poliomyelitis**" in a number of countries where the wild-type virus has been eliminated. This is a major public health problem, and has significantly complicated efforts to eradicate poliomyelitis as a disease of children.

The potential of **plant viruses as weapons** is also possibly underappreciated: the very high **yield** of certain viruses, their **lethality** in certain crop plants, and their **ease of transmission** by crop sprayer, for example, would make use of something like TMV both cheap and easy. However, their effects would not be quickly felt, so they are not much discussed as overt weapons.

Along with the threats from emerging viruses and virus weapons made by governments in wartime, the world currently faces the potential use of viruses **as terrorist weapons**. Although this issue has received much media attention, the reality is that the deliberate releases of such pathogens may have less medical impact than is generally appreciated. Many governments, for example, devoted considerable resources to the development of viruses as weapons of war as well as countermeasures to them, before deciding that their military usefulness was very limited. The **US Centers for Disease Control and Prevention (CDC)** only recognizes two types of virus as potentially dangerous terrorist weapons: these are **smallpox**, and agents causing **hemorrhagic fevers** such as **filoviruses** and **arenaviruses**. Emerging viruses such as **Nipah virus** and **hantaviruses** are also recognized as possible future threats.

While it is possible to isolate natural viruses for this purpose, it is also increasingly easy to synthesize whole virus genomes. While the first early efforts in this regard were of very small viruses from the 1970s on (see Chapter 1), a group of researchers in 2018 reported the **synthesis of an infectious horsepox virus genome, of 212kb in length** (Noyce et al., 2018). While they did this because the virus was probably extinct, and they wished as an expert lab to investigate its relationship to vaccinia virus and its potential as a vaccine, it is now feasible to do similar work on **smallpox**, for example. Indeed, there have been considerable international efforts for some years now to make sure that "custom gene" synthesis firms are much more circumspect in exactly which genes may be shipped to laboratories in which countries. It is also difficult to conceive of how a terrorist group could get all of the sophisticated equipment, chemicals and reagents necessary to do the job, let alone to also have all of the molecular biological and other expertise necessary to convert the sequence into anything dangerous—and then to manufacture it in quantities sufficient to be useful.

Moreover, the small number of viruses that could feasibly be considered as bioweapons is in contrast to a much larger number of **bacterial species and toxins**. The reason for this is that **bacterial pathogens** would be much easier for terrorist groups to prepare and disseminate than viruses. However, the potential threat from bioterrorism is in reality insignificant in relation to the **actual number of deaths caused by natural infections** worldwide each year. Nevertheless, this is an issue which governments are sensibly treating with great seriousness—and possibly why various militaries are so keen on developing vaccines for certain viruses.

Summary

New pathogenic viruses are being discovered all the time, and changes in human activities result in the re-emergence of known viruses, or the emergence of new or previously unrecognized diseases. Most of the viruses of concern are either arboviruses—transmitted by arthropods, in which they also multiply—or are derived from zoonotic infections, entering the human population from direct or indirect contact with wild animals. The potential of certain groups of viruses such as arbo-, hanta-, influenza A, filo-, paramyxo-, and coronaviruses to cause serious and unexpected outbreaks of disease is explored, together with the potential of viruses to be used as bioweapons.

References

Aubry, F., Jacobs, S., Darmuzey, M., Lequime, S., Delang, L., Fontaine, A., Jupatanakul, N., et al., 2021. Recent African strains of Zika virus display higher transmissibility and fetal pathogenicity than Asian strains. Nat. Commun. 12 (1), 916. https://doi.org/10.1038/s41467-021-21199-z.

Berche, P., 2022. The enigma of the 1889 Russian flu pandemic: a coronavirus? Presse Med. 51 (3), 104111. https://doi.org/10.1016/j.lpm.2022.104111.

Borucki, M.K., Collette, N.M., Coffey, L.L., Van Rompay, K.K.A., Hwang, M.H., Thissen, J.B., Allen, J.E., Zemla, A.T., 2019. Multiscale analysis for patterns of Zika virus genotype emergence, spread, and consequence. PLoS One 14 (12), e0225699. https://doi.org/10.1371/journal.pone.0225699.

Boseley, S., 2019. Ebola now Curable After Trials of Drugs in DRC, Say Scientists. The Guardian. 12 August 2019 https://www.theguardian.com/world/2019/aug/12/ebola-now-curable-after-trials-of-drugs-in-drc-say-scientists.

CDC, 2022. Omicron Variant: What You Need to Know. Centers for Disease Control and Prevention COVID-19. 29 March 2022 https://www.cdc.gov/coronavirus/2019-ncov/variants/omicron-variant.html.

Chan-Yeung, M., Xu, R.H., 2003. SARS: epidemiology. Respirology 8 (Suppl 1), S9–S14. https://doi.org/10.1046/j.1440-1843.2003.00518.x.

Czosnek, H., Hariton-Shalev, A., Sobol, I., Gorovits, R., Ghanim, M., 2017. The incredible journey of Begomoviruses in their whitefly vector. Viruses 9 (10), 273. https://doi.org/10.3390/v9100273.

de Melo, G.D., Lazarini, F., Levallois, S., Hautefort, C., Michel, V., Larrous, F., Verillaud, B., et al., 2021. COVID-19-related anosmia is associated with viral persistence and inflammation in human olfactory epithelium and brain infection in hamsters. Sci. Transl. Med. 13 (596), eabf8396. https://doi.org/10.1126/scitranslmed.abf8396 (Epub 2021 May 3).

Drexler, J.F., Corman, V.M., Müller, M.A., Maganga, G.D., Vallo, P., Binger, T., Gloza-Rausch, F., et al., 2012. Bats host major mammalian paramyxoviruses. Nat. Commun. 3, 796. https://doi.org/10.1038/ncomms1796.

Düx, A., Lequime, S., Patrono, L.V., Vrancken, B., Boral, S., Gogarten, J.F., Hilbig, A., Horst, D., Merkel, K., et al., 2020. Measles virus and rinderpest virus divergence dated to the sixth century BCE. Science 368 (6497), 1367–1370. https://doi.org/10.1126/science.aba9411.

Fielding, B.C., 2020. A Brief History of the Coronavirus Family – Including One Pandemic We Might Have Missed. The Conversation. 24 March 2020 https://theconversation.com/a-brief-history-of-the-coronavirus-family-including-one-pandemic-we-might-have-missed-134556.

GAVI, 2021. The Next Pandemic: Nipah Virus? GAVI Vaccines Work Pages. 15 March 2021 https://www.gavi.org/vaccineswork/next-pandemic/nipah-virus.

Mancuso, E., Toma, L., Polci, A., d'Alessio, S.G., Di Luca, M., Orsini, M., et al., 2019. Crimean-Congo hemorrhagic fever virus genome in tick from migratory bird, Italy. Emerg. Infect. Dis. 25 (7), 1418–1420. https://doi.org/10.3201/eid2507.181345.

Marí Saéz, A., Weiss, S., Nowak, K., Lapeyre, V., Zimmermann, F., Düx, A., Kühl, H.S., Kaba, M., et al., 2015. Investigating the zoonotic origin of the west African Ebola epidemic. EMBO Mol. Med. 7 (1), 17–23. https://doi.org/10.15252/emmm.201404792.

Muyembe-Tamfum, J.J., Mulangu, S., Masumu, J., Kayembe, J.M., Kemp, A., Paweska, J.T., 2012. Ebola virus outbreaks in Africa: past and present. Onderstepoort J. Vet. Res. 79 (2), 451. https://doi.org/10.4102/ojvr.v79i2.451.

Negredo, A., de la Calle-Prieto, F., Palencia-Herrejón, E., Mora-Rillo, M., Astray-Mochales, J., Sánchez-Seco, M.P., et al., 2017. Autochthonous Crimean-Congo hemorrhagic fever in Spain. N. Engl. J. Med. 377 (2), 154–161. https://doi.org/10.1056/NEJMoa1615162.

Noyce, R.S., Lederman, S., Evans, D.H., 2018. Construction of an infectious horsepox virus vaccine from chemically synthesized DNA fragments. PLoS One 13 (1), e0188453. https://doi.org/10.1371/journal.pone.0188453.

Outbreak News Today, 2020. Crimean-Congo Hemorrhagic Fever: Turkey Reports Hundreds of Cases. http://outbreaknewstoday.com/crimean-congo-hemorrhagic-fever-turkey-reports-hundreds-of-cases-31599/.

Oxford, J.S., Gill, D., 2018. Unanswered questions about the 1918 influenza pandemic: origin, pathology, and the virus itself. Lancet Infect. Dis. 18 (11), e348–e354. https://doi.org/10.1016/S1473-3099(18)30359-1 (Epub 2018 Jun 20).

Ren, L.L., Wang, Y.M., Wu, Z.Q., Xiang, Z.C., Guo, L., Xu, T., Jiang, Y.Z., et al., 2020. Identification of a novel coronavirus causing severe pneumonia in human: a descriptive study. Chin. Med. J. 133 (9), 1015–1024. https://doi.org/10.1097/CM9.0000000000000722.

Rybicki, E., 2015. Ebola Web Page. https://rybicki.blog/2015/07/21/ebola-on-the-web-20-years-on/.

Sariol, A., Perlman, S., 2020. Lessons for COVID-19 immunity from other coronavirus infections. Immunity 53 (2), 248–263. https://doi.org/10.1016/j.immuni.2020.07.005 (Epub 2020 Jul 14).

Shepherd, D.N., Martin, D.P., Van Der Walt, E., Dent, K., Varsani, A., Rybicki, E.P., 2010. Maize streak virus: an old and complex 'emerging' pathogen. Mol. Plant Pathol. 11 (1), 1–12. https://doi.org/10.1111/j.1364-3703.2009.00568.x.

Voisset, C., Weiss, R.A., Griffiths, D.J., 2008. Human RNA "rumor" viruses: the search for novel human retroviruses in chronic disease. Microbiol. Mol. Biol. Rev. 72 (1), 157–196. https://doi.org/10.1128/MMBR.00033-07.

WHO, 2020. Estimating Mortality From COVID-19—Scientific Brief. 4 August 2020 https://www.who.int/news-room/commentaries/detail/estimating-mortality-from-covid-19.

WHO, 2021a. Marbug Virus Disease. https://www.who.int/news-room/fact-sheets/detail/marburg-virus-disease.

WHO, 2021b. WHO-Convened Global Study of Origins of SARS-CoV-2: China Part. Joint WHO-China study: 14 January–10 February 2021. 30 March 2021 https://www.who.int/publications/i/item/who-convened-global-study-of-origins-of-sars-cov-2-china-part.

Wikan, N., Smith, D., 2016. Zika virus: history of a newly emerging arbovirus. Lancet Infect. Dis. 16 (7), e119–e126. https://doi.org/10.1016/S1473-3099(16)30010-X.

Recommended reading

Goldstein, T., 2020. Silent Spillover: Ebolavirus Antibodies Detected in People in DR Congo Before the 2018 Outbreak. BMC Blog Network. December 2020 https://blogs.biomedcentral.com/on-health/2020/12/01/silent-spillover-ebolavirus-antibodies-detected-in-people-in-dr-congo-before-the-2018-outbreak/.

Maxmen, A., 2020. World's Second-Deadliest Ebola Outbreak Ends in Democratic Republic of the Congo. Nature News, https://doi.org/10.1038/d41586-020-01950-0. 26 June 2020.

Recommended reading

CHAPTER

9

Subviral agents: Deltaviruses and prions

INTENDED LEARNING OUTCOMES

On completing this chapter, you should be able to:

- Discuss the minimal genome needed by a living entity.
- Explain what satellites and viroids are and how they differ from viruses.
- Describe how prions, infectious protein molecules seemingly with no genome at all, can cause disease.

What is the minimum **genome** size necessary to sustain an infectious agent? Could an entity with a genome of 1700 bases survive? Or a genome of 240 b? Could an infectious agent without any genome at all exist? We have already introduced the first two alternatives—viruses with really small genomes, viruses that appear to have been coopted by their hosts (polydnaviruses) as well as **viroids** and **satellite DNAs and RNAs**—in Chapter 1, but the idea of an infectious agent without a genome seems bizarre and ridiculous. Strange as it may seem, such agents as these do exist, are called **prions**, and cause disease in animals (including humans).

This Chapter will cover the case of **hepatitis delta virus (HDV)**, a highly unusual animal pathogen that has a circular ssRNA genome, as well as aspects of prion molecular biology and pathology. The characteristics of satellites and viroids are described in Table 9.1, for sake of comparison with HDV and its relatives.

Hepatitis delta virus

Hepatitis delta virus (HDV) was briefly introduced in Chapters 1 and 2, but some more detail on it would be useful to help in understanding the unique characteristics of this class of subviral agents. The HDV genome is a unique chimeric molecule—a circular ssRNA(−) molecule—with some of the properties of a **satellite** virus and some of a **viroid** (Table 9.2),

https://doi.org/10.1016/B978-0-12-822784-8.00009-X

TABLE 9.1 Satellites and viroids.

Characteristic	Satellites	Viroids
Helper virus required for replication	**Yes**	**No**
Protein(s) encoded	Yes	No
Genome replicated by	Helper virus enzymes	Host-cell RNA polymerase II
Site of replication	Same as helper virus (nucleus or cytoplasm)	Nucleus

TABLE 9.2 Properties of HDV.

Satellite-like properties	Viroid-like properties
Size and composition of genome—1640 nt (about four times the size of plant viroids)	Circular ssRNA genome with extensive secondary structure similar to viroids
Single-stranded circular RNA molecule, similar to some plant virus at RNAs	Sequence homology to the conserved central region involved in viroid replication
Dependent on HBV for replication—HDV RNA is packaged into coats consisting of lipids plus HBV encoded proteins	Replication by rolling circle mechanism mediated by host DNA-dependent RNA pol II
Encodes a single polypeptide, the δ antigen	

which causes disease in humans. HDV requires hepatitis B virus (HBV) as a helper virus for replication and is transmitted by the same means as HBV, benefiting from the presence of a protective coat composed of lipid plus HBV proteins. Virus preparations from HBV/HDV-infected animals contain heterologous particles distinct from those of HBV but with an irregular, ill-defined structure. These particles are composed of **HBV envelope glycoproteins M, L and S,** and contain the covalently closed circular 1.6 kb HDV RNA molecule in a **branched or rod-like configuration** similar to that of other viroids (Fig. 9.1). Unlike all viroids, **HDV encodes a protein**, the δ (D) antigen.

The virions infect cells by attaching to heparan sulfate on their surfaces via specific attachment of the M (major) HBSAg, followed by fusion of the virion membrane with the endocytotic vesicle membrane, and release of the HDAg-RNA nucleoprotein in the cytoplasm. This is then transported to the nucleus due to localization signals in the HDAg. In the nucleus the δ gene is transcribed so as to give mRNA encoding S-HDAg, the "early" form of the protein: this promotes further transcription and replication. The HDV genome is replicated by **host-cell RNA polymerase II** using a "**rolling circle**" mechanism, as are viroids, that produces **linear concatemers** that must be cleaved for infectivity. The cleavage is carried out by a **ribozyme domain present in the HDV RNA**, the only known example of a ribozyme in an animal virus genome. There is then **postreplicational genome editing** by the host enzyme complex ADAR1 (**a**denosine **d**eaminase **a**cting on **R**NA), which results in the production of L-HDAg (214 amino acids, extension by 19AA of L-HDAg), which is necessary for the assembly and release of HDV-containing particles. ADAR1 is incidentally involved in dampening interferon-triggered cytoplasmic innate immune responses to dsRNA by editing dsRNA structures (see Lamers et al., 2019, below). It is thought that edited genomic RNA is no longer infectious as it can no longer produce S-HDAg.

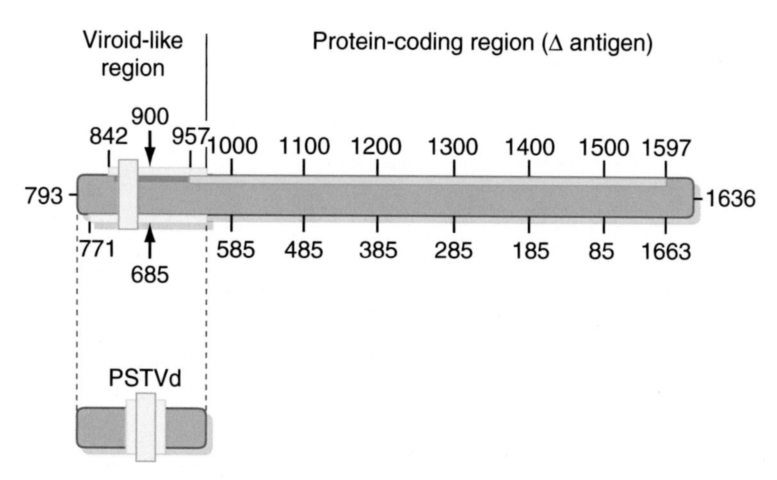

FIG. 9.1 Structure of hepatitis D virus RNA. A region at the left end of the genome strongly resembles the RNA of plant viroids such as PSTVd, which is shown for comparison.

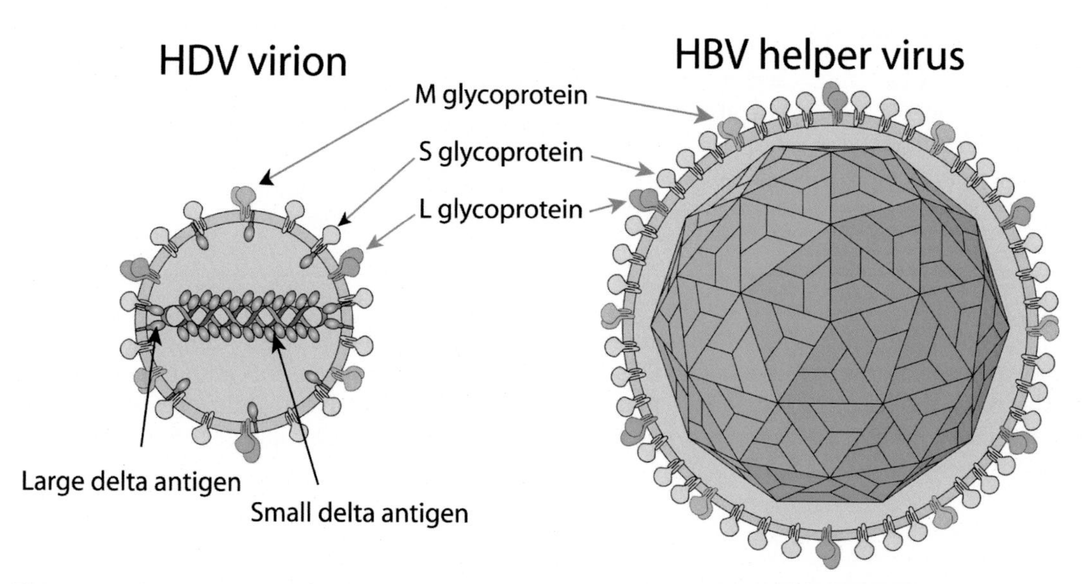

FIG. 9.2 The HDV virion. This is enveloped and spherical, and about 22 nm in diameter. The virion is covered by helper HBV membrane proteins. *Source: Philippe Le Mercier, ViralZone, Swiss Institute of Bioinformatics.*

HDV is found worldwide wherever HBV infection occurs. The interactions between HBV and HDV are difficult to study, but HDV seems to potentiate the pathogenic effects of HBV infection. **Fulminant hepatitis** (with a mortality rate of about 80%) is 10 times more common in coinfections than with HBV infection alone. Because HDV requires HBV for replication and because virions are covered in HBV proteins, it can be controlled by **HBV vaccination** (Chapter 6) (Fig. 9.2).

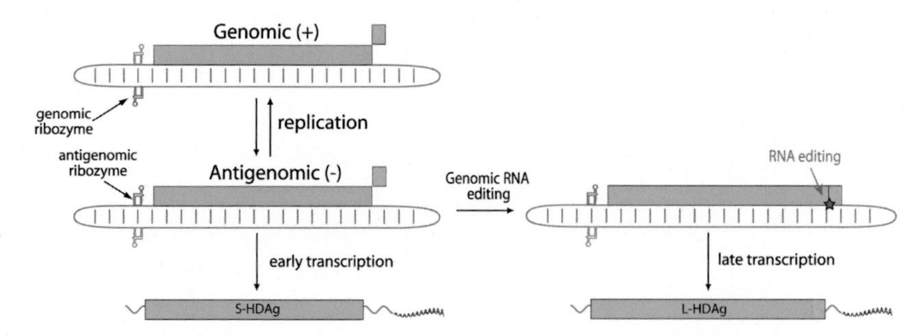

FIG. 9.3 HDV genome expression. ADAR1 editing of genomic RNA results in a frameshift that extends the S-HDAG ORF by 19 amino acids, to produce L-HDAg. The extension allows it to bind both the genomic RNA, and also into the site of budding on the inside of the cell membrane. *Source: Philippe Le Mercier, ViralZone, Swiss Institute of Bioinformatics.*

BOX 9.1

Is it a bird? Is it a plane? No, it's a—What exactly?

Hepatitis delta "virus" and its relatives are unique—there's nothing else like them. For a start, they're not viruses (capable of independent replication), even though we call them that. They look very, very like a viroid—except that they all encode a protein required for replication and nucleoprotein and virion formation—and viroids don't do that. So what are deltaviruses, exactly? HDV is one of the test cases that make virologists glad to be alive. Falling through the cracks between boring bacteria and monotonous mammals, HDV makes us think, and ask big questions about "life."

As mentioned in Chapter 2, it has recently been determined that HDV—long regarded as being a lone example of this type of virus—is in fact part of a **group of infectious agents now gathered in the Realm *Ribozyviria*,** named for the **ribozyme activity** that processes concatenated linear rolling circle replication products into individual, circular genomes (Box 9.1) (Fig. 9.3).

Prions

A particular group of **transmissible, chronic, progressive infections of the nervous system** show common pathological effects and are invariably fatal. Their pathology is similar to that of **amyloid diseases such as Alzheimer's syndrome,** and to distinguish them from such noninfectious (endogenous) conditions they are known as **transmissible spongiform encephalopathies (TSEs)**. The earliest record of any TSE dates from several centuries ago, when a disease called **scrapie** was first observed in sheep (see "TSE in animals," below). Originally thought to be caused by viruses, the first doubts about the nature of the infectious agent involved in TSEs arose in the 1960s. In 1967, Tikvah Alper was the first to suggest that the agent of scrapie might replicate without nucleic acid, and in 1982 Stanley Prusiner coined the term **prion** (proteinaceous infectious particle)—which, according to Prusiner, is pronounced **"pree-on."** The molecular nature of prions has not been unequivocally proved (see "Molecular biology of prions," below), but the evidence that they represent a new phenomenon outside the framework of previous scientific understanding is now generally accepted.

Pathology of prion diseases

All prion diseases share a similar underlying pathology, although there are significant differences between various conditions. A number of diseases are characterized by the deposition of **abnormal protein deposits** in various organs (e.g., kidney, spleen, liver, or brain). These "amyloid" deposits consist of accumulations of various proteins in the form of **plaques or fibrils** depending on their origin—for example, Alzheimer's disease is characterized by the deposition of plaques and "tangles" composed of **β-amyloid protein**. None of the "conventional" amyloidoses is an infectious disease, and extensive research has shown that they cannot be transmitted to experimental animals. These diseases result from endogenous errors in metabolism caused by a variety of largely unknown factors. Amyloid deposits appear to be inherently cytotoxic: although the molecular mechanisms involved in cell death are unclear, it is this effect that gives the "**spongiform encephalopathies**" their name owing to the characteristic holes in thin sections of affected brain tissue viewed under the microscope; these holes are caused by neuronal loss and gliosis, or **loss of glial cells**. Deposition of amyloid is the end stage of disease, linking conventional amyloidoses and TSEs and explaining the tissue damage seen in both types of disease. However, it does not reveal anything about their underlying causes. Definitive diagnosis of TSE cannot be made on clinical grounds alone and requires demonstration of prion protein (PrP) deposition by immunohistochemical staining of postmortem brain tissue, molecular genetic studies, or experimental transmission to animals, as discussed in the following sections.

TSE in animals

A number of TSEs have been observed and intensively investigated in animals. In particular, the **sheep disease scrapie** is the model for our understanding of human TSEs. Some of these diseases are naturally occurring and have been known about for centuries, whereas others have only been observed more recently and are almost certainly causally related to one another.

Scrapie

First described more than 200 years ago, scrapie is a naturally occurring disease of sheep found in many parts of the world, although it is not universally distributed. Scrapie appears to have originated in Spain and subsequently spread throughout Western Europe. **The export of sheep from Britain in the nineteenth century is thought to have helped scrapie spread around the world.** Scrapie is primarily a disease of sheep although it can also affect goats.

The scrapie agent has been intensively studied and has been experimentally transmitted to laboratory animals many times (see "Molecular biology of prions," below). Infected sheep show **severe and progressive neurological symptoms** such as abnormal gait; they often repeatedly scrape against fences or posts, a behavior from which the disease takes its name. The incidence of the disease increases with the age of the animals. Some countries, such as Australia and New Zealand, have eliminated scrapie by slaughtering infected sheep and by the imposition of rigorous import controls. Work has shown that **the land on which infected sheep graze may retain the condition and infect sheep up to 3 years later**. Thus, it was obvious that the causative agent must be highly stable.

The incidence of scrapie in a flock is related to the breed of sheep. Some breeds are relatively resistant to the disease while others are prone to it, indicating genetic control of susceptibility. In recent years **the occurrence of TSE in sheep in the United Kingdom closely parallels the incidence of bovine spongiform encephalopathy (BSE) in cattle** (Fig. 9.4). This is probably due to infection with the BSE agent (to which sheep are known to be sensitive) via infected feed. The natural mode of transmission between sheep is unclear. Lambs of scrapie-infected sheep are more likely to develop the disease, but the reason for

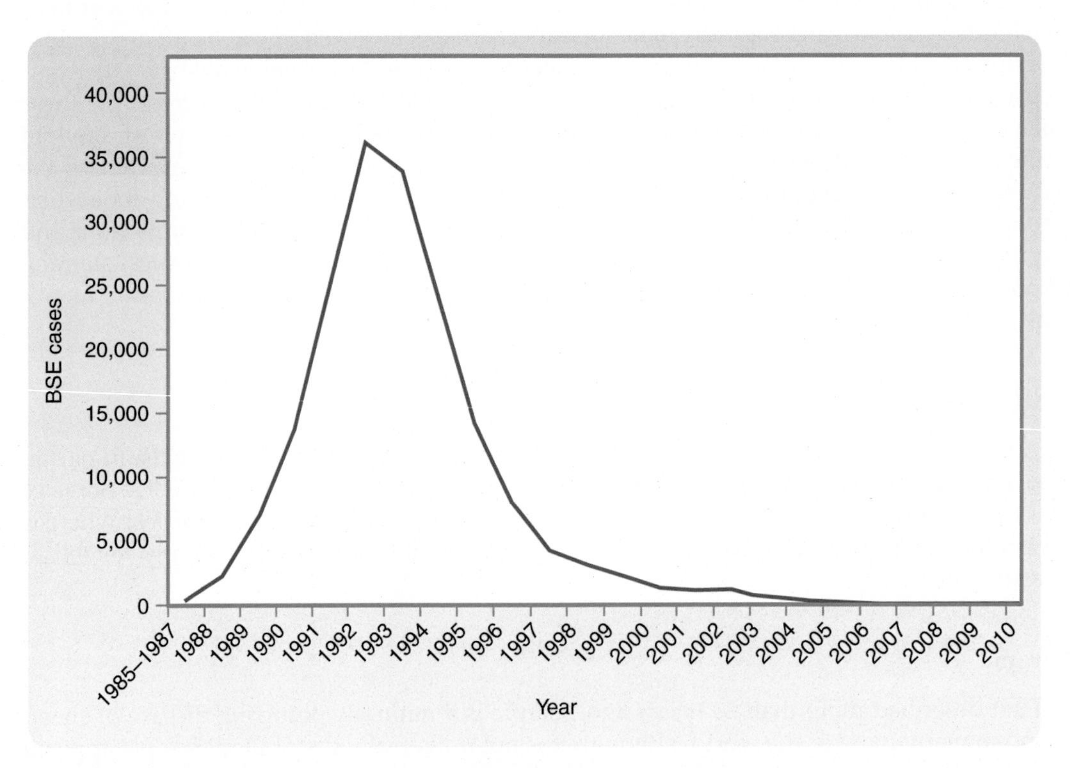

FIG. 9.4 Reported incidence of BSE in the United Kingdom. From the earliest recorded incidence in 1985, reported cases of BSE in the United Kingdom have now fallen back to single figures per year. BSE cases continue to be reported from many countries around the world.

this is unclear. Symptoms of scrapie are not seen in sheep less than one and a half years old which indicates that the incubation period of scrapie is at least this long. The first traces of infectivity can be detected in the tonsils, mesenteric lymph nodes, and intestines of sheep 10 to 14 months old which suggests an oral route of infection. The infective agent is present in the membranes of the embryo but it has not been demonstrated in colostrum or milk or in tissues of the newborn lambs.

Transmissible mink encephalopathy (TME)

TME is a rare disease of farmed mink caused by exposure to a scrapie-like agent in feed. The disease was first identified in Wisconsin in 1947 and has also been recorded in Canada, Finland, Germany, and Russia. Like other TSEs, TME is a slow progressive neurological disease. Early symptoms include changes in habits and cleanliness as well as difficulty in eating or swallowing. TME-infected mink become **hyperexcitable** and begin arching their tails over their backs, ultimately losing locomotor coordination. Natural TME has a minimum incubation period of 7 to 12 months, and, although exposure is generally through oral routes, **horizontal mink-to-mink transmission cannot be ruled out**. The origin of the transmissible agent in TME appears to be contaminated foodstuffs, but this is discussed further below (see "Bovine Spongiform Encephalopathy").

Feline Spongiform Encephalopathy (FSE)

FSE was recognized in the United Kingdom in May 1990 as a scrapie-like syndrome in domestic cats resulting in **ataxia** (irregular and jerky movements) and other symptoms typical of spongiform encephalopathies. By December 1997, a total of 81 cases had been reported in the United Kingdom. In addition, FSE has been recorded in a domestic cat in Norway and in three species of captive wild cats (cheetah, puma, and ocelot). Inclusion of cattle offal in commercial pet foods was banned in the United Kingdom in 1990, so the incidence of this disease is expected to decline rapidly (see "Bovine spongiform encephalopathy").

Chronic wasting disease (CWD)

CWD is a disease similar to scrapie which affects deer and captive exotic ungulates (e.g., nyala, oryx, kudu). CWD was first recognized in captive deer and elk in the western United States in 1967, and appears to be **endemic** in origin. Since its appearance in Colorado, the disease has spread to other states and has also been reported in Canada and South Korea. This disease seems to be more efficiently transmitted from one animal to another than other TSEs, so it seems unlikely that it will ever be eradicated from the regions in which it occurs. CWD prions taken from the brains of infected deer and elk are able to convert normal human prion to a protease-resistant form, a well-studied test for the ability to cause human disease, but the overall risk to human health from this disease remains unclear and **there is no evidence that this disease has ever been transmitted to humans**.

Bovine spongiform encephalopathy

BSE was first recognized in dairy cattle of the United Kingdom in 1986 as a typical spongiform encephalopathy. Affected cattle showed altered behavior and a staggering gait, giving the disease its name in the press of "**mad cow disease**." On microscopic examination, the brains of affected cattle showed **extensive spongiform degeneration**. It was concluded that BSE resulted from the use of contaminated foodstuffs. To obtain higher milk yields and growth rates, the nutritional value of feed for farmed animals was routinely boosted by the addition of **protein derived from waste meat products and bonemeal** (MBM) prepared from animal carcasses, including sheep and cows. This practice was not unique to the United Kingdom but was widely followed in most developed countries. By the end of 2010, a total of 184,607 cases of BSE had been reported in the United Kingdom, and thousands more cases in other countries (Fig. 9.4).

The initial explanation for the emergence of BSE in the United Kingdom was as follows. Because scrapie is **endemic** in Britain, it was assumed that this was the source of the infectious agent in the feed. Traditionally, MBM was prepared by a rendering process involving steam treatment and hydrocarbon extraction, resulting in two products: a **protein-rich fraction called "greaves"** containing about 1% fat from which MBM was produced and a **fat-rich fraction called "tallow"** which was put to a variety of industrial uses. In the late 1970s, the price of tallow fell and the use of expensive hydrocarbons in the rendering process was discontinued, producing an MBM product containing about 14% fat **in which the infectious material may not have been inactivated**. As a result, a ban on the use of ruminant protein in cattle feed was introduced in July 1988 (Fig. 9.4). In November 1989, human consumption of specified bovine offal products thought most likely to transmit the infection (**brain, spleen, thymus, tonsil, and gut**) was prohibited. A similar ban on consumption of offal from sheep, goats, and deer was finally announced in July 1996 to counter concerns about transmission of BSE to sheep. The available evidence suggests that **milk and dairy products do not contain detectable amounts of the infectious agent**. The total number of BSE cases continued to rise, as would be expected from the long incubation period of the disease, and the peak incidence was reached in the last quarter of 1992. Since then the number of new cases has started to fall; however, a number of false assumptions can be identified in the above reasoning.

It is now known that **none of the rendering processes used before or after the 1980s completely inactivates the infectivity of prions**—therefore cattle would have been exposed to scrapie prions in all countries worldwide where scrapie was present and MBM was used, not just in the United Kingdom in the 1980s. For example, the incidence of scrapie in the United States is difficult to determine, but in the 8 years after the level of compensation for slaughter of infected sheep was raised to $300 in 1977, the reported number of cases went up tenfold to a peak of about 50 affected flocks a year.

BSE is not scrapie. The biological properties of the scrapie and BSE agents are distinct—for example, transmissibility to different animal species and pattern of lesions produced in infected animals (see "Molecular biology of prions," below). There is no evidence to support the assumption that BSE is scrapie in cows. The only feasible interpretation based on present knowledge is that **BSE originated as an endogenous bovine (cow) prion** which was **amplified by the feeding of cattle-derived protein in MBM back to cows**. Thus, the emergence of BSE in the United Kingdom appears to have been due to a chance event compounded by poor husbandry practices (i.e., use of MBM in ruminant feed).

Important unanswered questions remain concerning BSE. Many of these are raised by the large number of infected cattle born after the 1988 feed ban. It is now generally acknowledged that the feed ban was initially improperly enforced and, moreover, only applied to cattle feed. The same mills that were producing cattle feed were also producing sheep, pig, and poultry food containing MBM, allowing many opportunities for contamination. As a result, in March 1996 the use of all mammalian MBM in animal feed was prohibited in the United Kingdom. It is now known that vertical transmission of BSE in herds can occur at a frequency of 1%–10%. Similarly, there is a possibility of environmental transmission similar to that known to occur with scrapie. Apart from the economic damage done by BSE, the main concern remains the possible risk to human health (discussed below).

Human TSEs

There are **four recognized human TSEs** (summarized in Table 9.3). Our understanding of human TSEs is derived largely from studies of the animal TSEs already described. Human TSEs are believed to originate from three sources:

- **Sporadic:** Creutzfeldt–Jakob disease (CJD) arises spontaneously at a frequency of about one in a million people per year with little variation worldwide. The average age at onset of disease is about 65, and the average duration of illness is about 3 months. This category accounts for 90% of all human TSE, but only about 1% of sporadic CJD cases are transmissible to mice.
- **Iatrogenic (acquired) TSE:** This occurs due to recognized risks (e.g., neurosurgery, transplantation). About 50 cases of TSE were caused in young people who received injections of human growth hormone or gonadotropin derived from pooled cadaver pituitary gland extracts, a practice that has now been discontinued in favor of recombinant DNA-derived hormone.
- **Familial:** Approximately 10% of human TSEs are familial (i.e., inherited). A number of mutations in the human PrP gene are known to give rise to TSE as an autosomal dominant trait acquired by hereditary Mendelian transmission (Fig. 9.5).

TABLE 9.3 TSEs in human

Disease	Description	Comments
CJD	Spongiform encephalopathy in cerebral and/or cerebellar cortex and/or subcortical gray matter, or encephalopathy with prion protein (PrP) immunoreactivity (plaque and/or diffuse synaptic and/or patchy/perivacuolar types)	Three forms: sporadic, iatrogenic (recognized risk, e.g., neurosurgery), familial (same disease in first degree relative)
Familial fatal insomnia (FFI)	Thalamic degeneration, variable spongiform change in cerebrum	Occurs in families with PrP_{178} asp-asn mutation
Gerstmann–Straussler–Scheinker disease (GSS)	Encephalo(myelo)pathy with multicentric PrP plaques	Occurs in families with dominantly inherited progressive ataxia and/or dementia
Kuru	Characterized by large amyloid plaques	Occurs in the Fore population of New Guinea due to ritual cannibalism, now eliminated

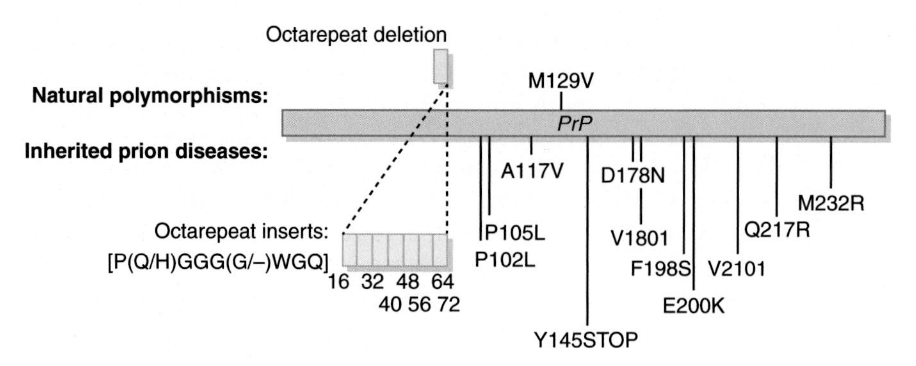

FIG. 9.5 Mutations in the human PrP gene. Approximately 10% of human TSEs are inherited. A number of mutations in the human PrP gene are known to give rise to TSEs.

Kuru was the first human spongiform encephalopathy to be investigated in detail and is possibly one of the most fascinating stories to have emerged from any epidemiological investigation. The disease occurred primarily in villages occupied by the Fore tribes in the highlands of New Guinea. The first cases were recorded in the 1950s and involved **progressive loss of voluntary neuronal control**, followed by death less than 1 year after the onset of symptoms. The key to the origin of the disease was provided by the profile of its victims—it was never seen in young children, rarely in adult men, and was **most common in both male and female adolescents and in adult women**. The Fore people practiced ritual cannibalism as a rite of mourning for their dead. Women and children participated in these ceremonies but adult men did not take part, explaining the age/sex distribution of the disease. **The incubation period for kuru can be in excess of 30 years** but in most cases is somewhat shorter. The practice of ritual cannibalism was discouraged in the late 1950s and the incidence of kuru declined dramatically. Kuru has now disappeared.

The above description covers the known picture of human TSEs which has been painstakingly built up over several decades. **There is no evidence that any human TSE is traditionally acquired by an oral route (e.g., eating scrapie-infected sheep).** There are good reasons why this should be (see "Molecular biology of prions," below). However, in April 1996 a paper was published that described a **new variant of CJD (vCJD)** in the United Kingdom. Although relatively few in number, these cases shared unusual features that distinguish them from other forms of CJD. The official U.K. Spongiform Encephalopathy Advisory Committee concluded that vCJD is "a previously unrecognized and consistent disease pattern" and that "although there is no direct evidence of a link, on current data and in the absence of any credible alternative the most likely explanation at present is that these cases are linked to exposure to BSE." **By 2010, 170 people had died of vCJD in the United Kingdom** and several more in other countries. Although three new deaths due to vCJD were recorded in the United Kingdom in 2010, the overall picture is that the vCJD outbreak in the United Kingdom is in decline, albeit with a pronounced tail. **One in 2000 people in the United Kingdom—over 25,000 people—is a carrier of vCJD, for which there is currently still no effective cure.** Apart from whether these people will ultimately become ill, another worrying aspect of this statistic is whether these silent carriers can pass the infection on to others through blood transfusions. There is no reliable test to screen for vCJD carriers. Many facts about human prion disease

BOX 9.2

How lucky did we get?

It seems wrong to say that we got "lucky" with vCJD when 177 people have died in the United Kingdom and nearly 300 have died worldwide. But some of the projections were much worse. At one stage, it was suggested that **14,000 people might die in the United Kingdom alone**. As time goes on, the chance of that happening becomes less (but there will still be more deaths for years to come). So is the relatively small number of deaths which actually happened just down to luck? No. It's down to the **species barrier** which protects each species from the prions of another. Millions of people were exposed to BSE in the United Kingdom alone, but fortunately, **humans are quite resistant to cattle prions**—more resistant than mice are to hamster prions for example. So it wasn't just luck. But it certainly wasn't down to good planning or judgment.

remain unknown, but what is clear is that decades after it started, mad cow disease has not gone away—the effects of the outbreak will rumble on for decades to come (Box 9.2).

Molecular biology of prions

The evidence that prions are not conventional viruses is based on the fact that nucleic acid is not necessary for infectivity, as they show:

- **Resistance to heat inactivation**: Infectivity is reduced but not eliminated by **high-temperature autoclaving** (135°C for 18 min). Some infectious activity is even retained after treatment at 600°C, suggesting that an inorganic molecular template is capable of nucleating the biological replication of the agent.
- **Resistance to radiation damage**: Infectivity was found to be resistant to shortwave ultraviolet radiation and to ionizing radiation. These treatments inactivate infectious organisms by causing damage to the genome. **There is an inverse relationship between the size of target nucleic acid molecule and the dose of radioactivity or ultraviolet light needed to inactivate them**; that is, large molecules are sensitive to much lower doses than are smaller molecules (Fig. 9.6). The scrapie agent was found to be highly resistant to both ultraviolet light and ionizing radiation, indicating that **if any nucleic acid were present, it must be less than 80 nt long**.
- **Resistance to DNAse and RNAse treatment**, to psoralens, and to Zn^{2+} catalyzed hydrolysis, all of which **inactivate nucleic acids**.
- **Sensitivity** to urea, detergents, phenol, and other **protein-denaturing chemicals**.

All of the above **indicate an agent with the properties of a protein rather than a virus**. A protein of 254 amino acids (PrP^{Sc}) is associated with scrapie infectivity. Biochemical purification of scrapie infectivity results in preparations highly enriched in PrP^{Sc}, and **purification of PrP^{Sc} results in enrichment of scrapie activity.** In 1984, Prusiner determined the sequence of 15 amino acids at the end of purified PrP^{Sc}. This led to the discovery that all mammalian

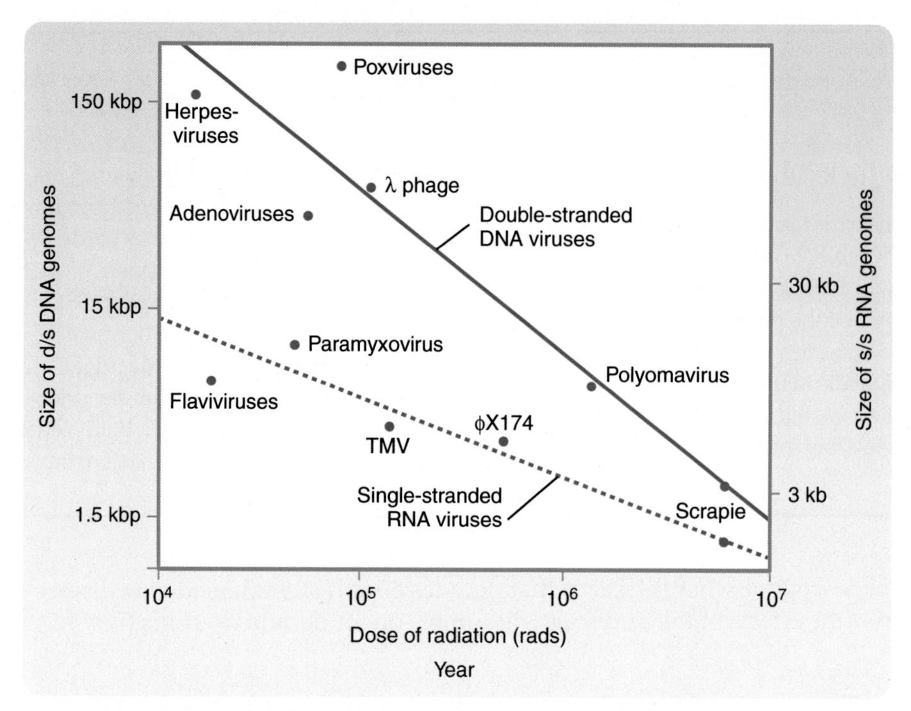

FIG. 9.6 Radiation sensitivity of infectious agents. The dose of ionizing radiation required to destroy the infectivity of an infectious agent is dependent on the size of its genome. Larger genomes (e.g., double-stranded DNA viruses, *upper line*) present a larger target and therefore are more sensitive than smaller genomes (e.g., single-stranded RNA viruses, *lower line*; N.B. ϕX174 has a single-stranded DNA genome). The scrapie agent is considerably more resistant to radiation than any known virus (note log scale on vertical axis).

cells contain a gene (*Prnp*) that encodes a protein identical to PrP^{Sc}, termed PrP^{C}. No biochemical differences between PrP^{C} and PrP^{Sc} have been determined, although, unlike PrP^{C}, PrP^{Sc} is partly resistant to protease digestion, resulting in the formation of a 141-amino-acid, protease-resistant fragment **that accumulates as fibrils in infected cells** (Fig. 9.7). Only a proportion of the total PrP in diseased tissue is present as PrP^{Sc}, but this has been shown to be the infectious form of the PrP protein, as highly purified PrP^{Sc} is infectious when used to inoculate experimental animals. Like other infectious agents, there is a dosage effect that gives a correlation between the amount of PrP^{Sc} in an inoculum and the incubation time until the development of disease.

Thus, TSEs, which behave like infectious agents, **appear to be caused by an endogenous gene/protein** (Fig. 9.8). Susceptibility of a host species to prion infection is codetermined by the prion inoculum and the *Prnp* gene. Disease incubation times for individual prion isolates vary in different strains of inbred mice, but for a given isolate in a particular strain they are remarkably consistent. These observations have resulted in two important concepts:

1. **Prion strain variation:** At least 15 different strains of PrP^{Sc} have been recognized. These can be determined from each other by the incubation time to the onset of disease and the type and distribution of lesions within the central nervous system (CNS) in inbred strains of mice. Thus, prions can be "fingerprinted," and BSE can be distinguished from scrapie or CJD.

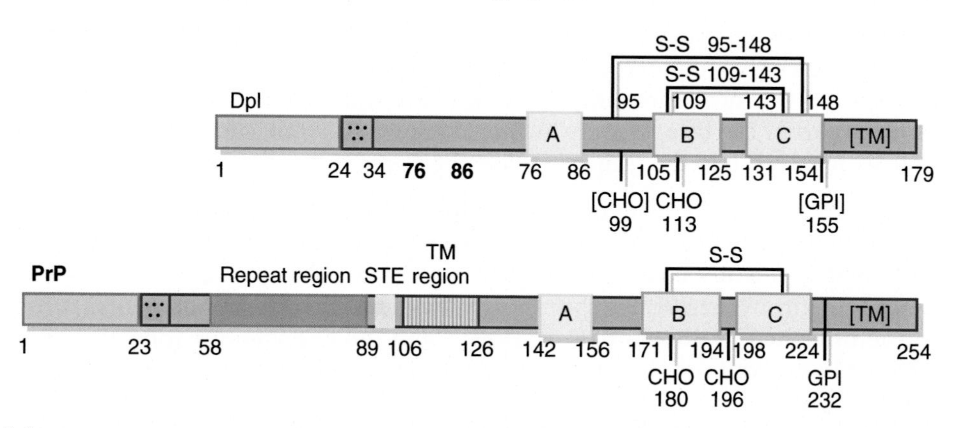

FIG. 9.7 Structure of the prion (PrP) and Doppel (Dpl) proteins. The PrP protein and the closely related Doppel (Dpl) protein, overexpression of which also causes neurodegeneration, are part of a family of genes have arisen by gene duplication.

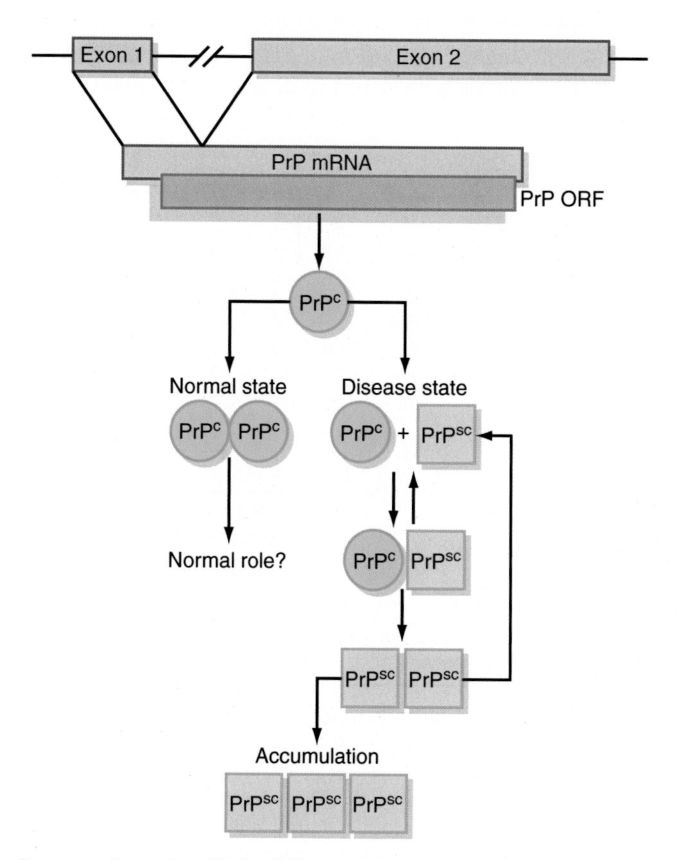

FIG. 9.8 Schematic diagram of the role of PrP in TSEs. TSEs behave like infectious agents but appear to be caused by a cellular gene/protein.

2. **Species barrier:** When prions are initially transmitted from one species to another, disease develops only after a very long incubation period, if at all. On serial passage in the new species, the incubation time often decreases dramatically and then stabilizes. This species barrier can be overcome by introducing a PrP transgene from the prion donor (e.g., hamster) into the recipient mice (Fig. 9.9).

PrP^C and PrP^{Sc}, however, **are not posttranslationally modified**, and the **genes that encode them are not mutated**, which is distinct from **Mendelian inheritance of familial forms of CJD**. How such apparently complex behavior can be "encoded" by a 254-amino-acid protein has not been firmly established, but there is evidence that **the fundamental difference between the infectious, pathogenic form (PrP^{Sc}) and the endogenous form (PrP^C) results from a change in the conformation of the folded protein**, which adopts a conformation rich in β-sheet (Fig. 9.10).

Transgenic "knockout" *Prnp*$^{0/0}$ mice that do not possess an endogenous prion gene are completely immune to the effects of PrP^{Sc} and do not propagate infectivity to normal mice, indicating that **production of endogenous PrP^C is an essential part of the disease process in TSEs** and that the infectious inoculum of PrP^{Sc} does not replicate itself. Unfortunately, these experiments have given few clues to the normal role of PrP^C. Most *Prnp*$^{0/0}$ mice are developmentally normal and do not have CNS abnormalities, suggesting that loss of normal PrP^C function is not the cause of TSE and that the accumulation of PrP^{Sc} is responsible for disease symptoms. However, one strain of *Prnp*$^{0/0}$ mouse was found to develop late-onset ataxia and neurological

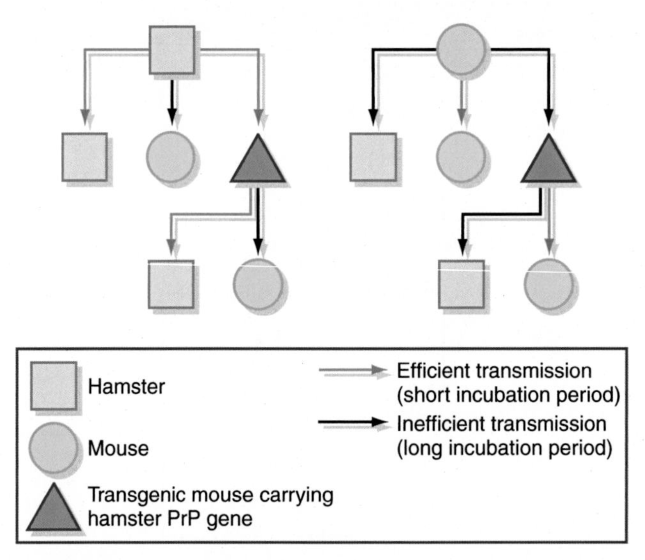

FIG. 9.9 Experimental transmission of scrapie to animals demonstrates a species barrier. Hamster-to-hamster and mouse-to-mouse transmission results in the onset of disease after a relatively short incubation period (75 days and 175 days, respectively). Transmission from one host species to another is much less efficient, and disease occurs only after a much longer incubation period. Transmission of hamster-derived PrP to transgenic mice carrying several copies of the hamster PrP gene (the *darker mice* in this figure) is much more efficient, whereas transmission of mouse-derived PrP to the transgenic mice is less efficient. Subsequent transmission from the transgenic mice implies that some modification of the properties of the agent seems to have occurred.

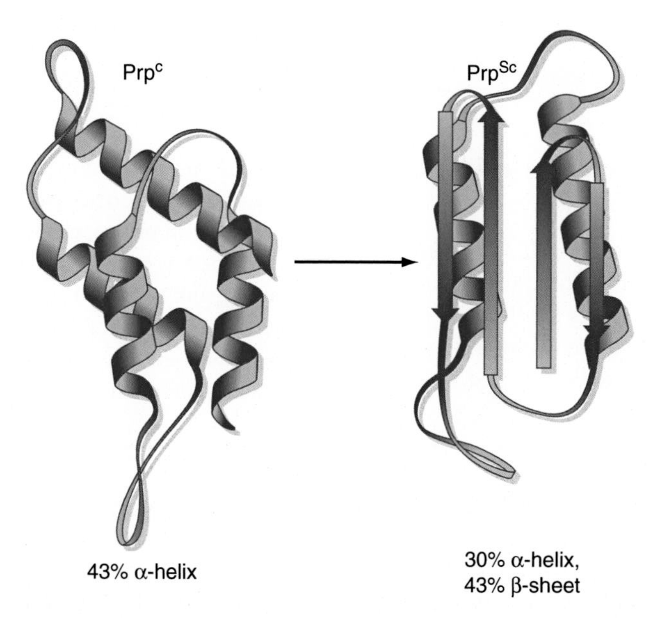

FIG. 9.10 Conformational changes in PrP. The fundamental difference between the infectious, pathogenic form (PrP^{Sc}) and the endogenous form (PrP^{C}) results from a change in the conformation of the folded protein, which adopts a conformation rich in β-sheet.

degeneration. This observation led to the discovery of another gene, called *Prnd*, that is close to the *Prnp* gene and encodes a 179-amino-acid, PrP-like protein designated Doppel (Dpl), overexpression of which appears to cause neurodegeneration (Fig. 9.7). Like *Prnp*, **this gene is conserved in vertebrates including humans and may have arisen from *Prnp* by gene duplication**. It is suspected that there may be other members of the *Prn* gene family.

The URE3 protein of the yeast *Saccharomyces cerevisiae* **has properties very reminiscent of PrP.** Other PrP-like proteins are also known (e.g., PSI in yeast and Het-s* in the fungus *Podospora*). URE3 modifies the cellular protein Ure2p, causing altered nitrogen metabolism; similarly, the PSI phenotype involves a self-propagating aggregation of Ure2p and the cellular protein Sup35p. **Cells "infected" with URE3 can be "cured" by treatment with protein-denaturing agents such as guanidium**, which is believed to cause refolding of URE3 to the Ure2p conformation. The explanation for the inherited familial forms of prion disease is therefore presumably that inherited mutations enhance the rate of spontaneous conversion of PrP^{C} into PrP^{Sc}, permitting disease manifestation within the lifetime of an affected individual. This concept also suggests that the sporadic incidence of CJD can be accounted for by somatic mutation of the PrP gene and offers a possible explanation for the emergence of BSE—**spontaneous mutation of a bovine PrP gene resulted in infectious prions which were then amplified through the food chain.** In mammalian cells, the evidence points to prion proteins playing a role in neurotransmission at synaptic junctions. This function of prions as part of the normal epigenetic inheritance of cells indicates their true reason for existence.

"Just as nucleic acids can carry out enzymatic reactions, proteins can be genes."
Reed Wickner, Recommended Reading, below.

BOX 9.3

Absence of evidence or evidence of absence?

Is there really no nucleic acid associated with prions? And if there isn't, how do you prove that? It took many years for the protein-only prion hypothesis to be generally accepted and for Stanley Prusiner to be awarded the Nobel Prize in 1997. But claims that there may be some sort of nucleic acid or conventional virus associated with prions, even if not a complete genome, just won't go away. How do you prove a negative hypothesis? In the end, scientists have to weigh the balance of evidence and opt for the most likely explanation. And always keep an open mind, just in case.

The prion hypothesis is revolutionary and justifiably met with a somewhat skeptical reception. In recent years, the construction of transgenic animals has cast further light on the above ideas. PrP is a very difficult protein to work with as it aggregates strongly and is heterogeneous in size, and even the best preparations require 1×10^5 PrP molecules to infect one mouse. This raises the question of whether or not some sort of unidentified infectious agent is "hiding" in the protein aggregates—although evidence that only the protein is required seems very strong. It is still possible, however, to construct numerous alternative theories of varying degrees of complexity (and plausibility) to fit the experimental data. Science progresses by the construction of experimentally verifiable hypotheses. For many years, research into spongiform encephalopathies has been agonizingly slow because each individual experiment has taken at least one and in some cases many years to complete. With the advent of molecular biology, this has now become a fast-moving and dynamic field. The next few years will undoubtedly continue to reveal more information about the cause of these diseases and will probably provide much food for thought about the interaction between infectious agents and the host in the pathogenesis of infectious diseases (Box 9.3).

References

Lamers, M.M., van den Hoogen, B.G., Haagmans, B.L., 2019. ADAR1: "editor-in-chief" of cytoplasmic innate immunity. Front. Immunol. 10, 1763. https://doi.org/10.3389/fimmu.2019.01763.

Recommended reading

Acevedo-Morantes, C.Y., Wille, H., 2014. The structure of human prions: from biology to structural models-considerations and pitfalls. Viruses 6 (10), 3875–3892. https://doi.org/10.3390/v6103875.

Aguzzi, A., Zhu, C., 2012. Five questions on prion diseases. PLoS Pathog. 8 (5), e1002651. https://doi.org/10.1371/journal.ppat.1002651 (Epub 2012).

Netter, H.J., Barrios, M.H., Littlejohn, M., Yuen, L.K.W., 2021. Hepatitis Delta virus (HDV) and Delta-like agents: insights into their origin. Front. Microbiol. 12, 652962. https://doi.org/10.3389/fmicb.2021.652962.

Pérez-Vargas, J., Pereira de Oliveira, R., Jacquet, S., Pontier, D., Cosset, F.L., Freitas, N., 2021. HDV-like viruses. Viruses 13 (7), 1207. https://doi.org/10.3390/v13071207.

Verma, A., 2016. Prions, prion-like prionoids, and neurodegenerative disorders. Ann. Indian Acad. Neurol. 19 (2), 169–174. https://doi.org/10.4103/0972-2327.179979.

Wickner, R.B., Shewmaker, F.P., Bateman, D.A., Edskes, H.K., Gorkovskiy, A., Dayani, Y., Bezsonov, E.E., 2015. Yeast prions: structure, biology, and prion-handling systems. Microbiol. Mol. Biol. Rev. 79 (1), 1–17. https://doi.org/10.1128/MMBR.00041-14.

Appendix 1

Glossary

Aminoacylatable	Referring to nucleic acids and in particular tRNAs or tRNA-like structures: the property of being able to be recognized by an aminoacyl-tRNA synthetase to covalently bind an amino acid to the RNA
Amoebae	Single-celled free-living eukaryotes
Anastomosis	A connection that arises between two tubular structures
Assay, assayed	An analytical procedure that measures the absolute or relative quantity of a given molecule
Attenuate, attenuated	A pathogenic agent that has been genetically altered and displays decreased virulence; attenuated viruses are the basis of live virus vaccines (see Chapter 6).
Autonomous	Independent of others: autonomously-replicating means replicating under its own set of instructions
Autoradiography, autoradiographic	A technique that uses X-ray or other film to record radioactive emissions—e.g., from nucleic acids labeled with ^{32}P in a Southern or northern blot
Biome	An area classified according to the species that live in that location, OR a description of a unique population of organisms in a particular environment (e.g., gut microbiome)
Centrifugation	A mechanical process that uses rotors spun at high speed to separate particles of different sizes and/or densities by centrifugal force
Chromatin	The ordered complex of DNA plus proteins (histones and nonhistone chromosomal proteins) found in the nucleus of eukaryotic cells.
Cis-, trans-acting	Cis: A genetic element that affects the activity of contiguous (i.e., on the same nucleic acid molecule) genetic regions Trans: A genetic element encoding a diffusible product which acts on regulatory sites whether or not these are contiguous with the site from which they are produced
Commensal	An organism that benefits by close association with another without adversely affecting or benefitting the other
Complement cascade	An enzyme-triggered cascade of molecular changes in the blood of mammals that results in a series of inflammatory events
Conjugation	The process of transfer of genetic information between bacteria in close contact
Culture	The process of multiplying eukaryotic or prokaryotic cells—or of viruses in such cells—under controlled, generally sterile conditions

Continued

Cytokine	A group of generally small, soluble proteins in mammals that are involved in cell signaling
Cytoskeletal motor protein	One of the three families of eukaryotic molecular motors that interact with the cell cytoskeleton fibrils (actin, myosin) to move cargoes around inside the cell
Encapsidate	The act of enclosing a nucleic acid or nucleoprotein inside a protein shell
Endemic	Viruses that regularly cause disease in certain restricted geographical areas
Epidemic, pandemic	Epidemic: a disease outbreak affecting a community, population or limited geographical region Pandemic–epidemic that spread over multiple regions of the world
Epigenetic	An effect that stably alters gene activity without changing the genome: this could be via DNA methylation, modification of histones in the chromatin, or even RNA silencing
Episome	An extrachromosomal entity—plasmid, virus genome—that is associated with host cell DNA but is not integrated, and can replicate and be expressed independently of it
Eukaryote	An organism whose genetic material is separated from the cytoplasm by a nuclear membrane and divided into discrete chromosomes
Exons, introns, RNA splicing	Exon: A region of a gene expressed as protein after the removal of introns by posttranscriptional splicing Intron: A region of a gene removed after transcription by splicing and consequently not expressed as protein Splicing: Posttranscriptional modification of primary RNA transcripts that occurs in the nucleus of eukaryotic cells during which introns are removed and exons are joined together to produce cytoplasmic mRNAs
Fragile sites	Chromosomal sites that are susceptible to breakage, and therefore to spontaneous recombination events
Immortalized	A cell capable of indefinite growth (i.e., number of cell divisions) in culture. On rare occasions, immortalized cells arise spontaneously but are more commonly caused by mutagenesis as a result of virus transformation
Inoculate	To introduce an infectious or noninfectious agent into another; to add virions or bacteria into suitable cultures or culture media
Integrin	Membrane proteins in animal cells that bind the outside of the cells to the extracellular matrix—and can link this to actin-based cytoskeleton inside the cells
Interferons	A group of cell signaling proteins that may be cytokines (e.g., gamma-Ifn) or which are secreted from virus-infected cells in order to induce an antiviral state in neighboring cells (alpha- and beta-Ifns)
Isometric	A particle displaying cubic symmetry, of which the icosahedron is the predominant form for viruses
Major histocompatibility complex	Class I and II MHCs play a major role in the adaptive immune system in vertebrates by displaying peptides derived from intracellular breakdown of either homologous or heterologous proteins for activation of cytotoxic T cells (Class I) or helper T cells (Class II)

Continued

Monocistronic; polycistronic	Monocistronic: A messenger RNA that consists of the transcript of a single gene and which therefore encodes a single polypeptide; a virus genome such as picornaviruses that acts as such an mRNA Polycistronic: A messenger RNA that encodes more than one polypeptide: these can all be expressed in prokaryotes, but generally only the 5′-proximal one is expressed in eukaryotes
Monophyletic; polyphyletic	Monophyletic: a class of organisms that derive from a common ancestor (e.g., picornaviruses) Polyphyletic: a group of organisms that derive from different evolutionary pathways (e.g., dsRNA viruses)
N-acetylneuraminic acid	Also known as Neu5Ac or NANA; the predominant sialic acid in most mammals: it is an important part of the glycans displayed on the outside of many cell types, and a receptor for influenza viruses
Neurotropic	Tending to preferentially infect cells of the nervous system in animals
Neurovirulent	The capacity of a virus to cause disease of the nervous system in animals
Nuclear localization signals	Short peptide sequences that are recognized by transport mechanisms that shuttle proteins or virions into the eukaryote cell nucleus
Nuclear matrix	A network of fibers that acts as the "skeleton" of the eukaryotic cell nucleus, and to which many proteins important for regulation of gene activity and functional organization of nuclear DNA attach
Oncogene	A gene that encodes a protein capable of inducing cellular transformation: these may be cellular or viral genes
Operator sequences	A genetic sequence (mainly in prokaryotes) that allows proteins involved in transcription to attach to cell and viral DNA
Open reading frame (ORF)	A region of a gene or mRNA that potentially encodes a polypeptide, bounded by an AUG translation start codon at the 5′ end and a termination codon at the 3′ end
Operon	A cluster of prokaryotic genes that are transcribed into one mRNA for coexpression
Organelle	Eukaryotic cellular substructures—e.g., mitochondria, peroxisomes, chloroplasts—that have membranes separating them from the cell cytoplasm and which have specific functions
Passage, passaging	Passaging or—more correctly—serially passaging viruses in culture or even in animals is the process of repeatedly serially transferring infectious virus from cultures or animals to new hosts. This is done to adapt viruses to new hosts or to attenuate their virulence
Plasmid	An extrachromosomal genetic element capable of autonomous replication—generally prokaryotic
Primary cell	A cultured cell explanted from an organism that is capable of only a limited number of divisions
Prion	A proteinaceous infectious particle, believed to be responsible for transmissible spongiform encephalopathies such as Creutzfeldt–Jakob disease (CJD) or bovine spongiform encephalopathy (BSE) (Chapter 9)

Continued

Prokaryote	An organism whose genetic material is not separated from the cytoplasm of the cell by a nuclear membrane (Archaea, Bacteria)
Prophage	The lysogenic form of a temperate bacteriophage genome integrated into the genome of the host bacterium
Protozoan	Single-celled and generally free-living eukaryotes, like amoebae or various ciliates
Provirus	The double-stranded DNA form of a retrovirus genome integrated into the chromatin of the host cell
Pseudoknots	An RNA secondary structure that causes "frame-shifting" during translation, producing a hybrid peptide containing information from an alternative reading frame
Quasispecies	A complex mixture of rapidly evolving and competing molecular variants of RNA virus genomes that occur in most populations of RNA viruses.
Receptor-mediated endocytosis	The process by which specific binding of particular proteins (including virions) to eukaryotic cell-surface receptors results in invagination of the plasma membrane, formation of clathrin-coated pits, and internalization of the protein/virion in a membrane-bound vesicle
Recombination: homologous, illegitimate	Homologous: interaction of similar DNA or RNA sequences that results in swapping or exchange or insertion of one sequence with another. Happens frequently with coronaviruses and geminiviruses Illegitimate: random recombination of one sequence with another (rare)
Replicon	A nucleic acid molecule containing the information necessary for its own replication; includes both viral genomes and other entities such as plasmids and satellites
Reverse transcriptase	A complex of enzymes encoded by the *pol* of retroviruses, retrotransposons and pararetroviruses, that has RNA-dependent DNA polymerase and DNA-dependent-DNA polymerase activities, as well as RNase H (DNA-dependent RNase). Retroviruses additionally incorporate integrase and site-specific endonuclease activities
Ribocell; virocell	Ribocell: organisms that encode the full suite of protein-synthesizing machinery and make their own ribosomes; cellular organisms Virocell: acellular organisms that do not encode ribosomes
Ribosome	A complex nucleoprotein "nanomachine" found in all cells that translates mRNAs into polypeptides
Serology, serological	Serology: the study of serum and other body fluids in animals Serological: the use of components of serum (typically antibodies) in assays or tests
Signal transduction	The translation of the binding of an extracellular messenger in eukaryote cells (cytokine, hormone) to a cell surface receptor into intracellular processes, including gene transcription
siRNA	Small interfering RNA or silencing RNA: a class of short dsRNAs (20–24 nt) involved in gene regulation, including silencing of particular mRNAs
Subclinical	An infection that is not detectable, or that does not present symptoms

Continued

Super virus hallmark genes	A set of genes that can be used for deep classification of viruses: these include RdRps, reverse transcriptases, superfamily 3 helicases, single- and double-jelly-roll capsid proteins, and rolling circle replication initiator proteins (Chapter 3)
Synchronous infection	Effectively simultaneous infection of all cells in a culture
TATA box	Also Goldberg–Hogness (Archaea, eukaryotes) or Pribnow (bacteria) box: a TA-rich DNA sequence that is part of the core promoter region for gene expression
Telomere	A feature of the two ends of eukaryotic linear chromosomes: these consist of ~300 bp (yeast)—many kbp (humans) of arrays of G-rich 6–8 bp repeat sequences, with complex end structures bound to telomerase enzymes and short RNA templates
Tissues	Four basic types of connected cell masses in animals: these are connective, muscle, nervous, and epithelial tissues
Transfection	Infection of cells mediated by the introduction of nucleic acid rather than by virus particles
Transform, transformation	Any change in the morphological, biochemical, or growth parameters of a cell; often accompanies genetic changes leading to immortalization of mammalian cells Also refers to introduction of recombinant plasmids into bacteria
Transforming retroviruses	These are retroviruses that can cause transformation of infected cells, although not necessarily: acutely transforming retroviruses are generally defective, and induce much more frequent transformation because they encode cell-derived oncogenes
Transgenesis, transgenic	Genetically manipulating eukaryotes to contain additional genetic information from another species. The additional genes may be carried and/or expressed only in the somatic cells of the transgenic organism or in the cells of the germ line, in which case they may be inheritable by any offspring
Vaccination	The administration of a vaccine
Virulent	Extremely poisonous or injurious—in the context of virus infections it means "causes severe and damaging infections"
Zoonosis, zoonotic	Zoonosis: disease of humans that originates in animals. Zoonotic: description of a disease originating in animals

Index

Note: Page numbers followed by "*f*" indicate figures, "*t*" indicate tables, and "*b*" indicate boxes.

I

K

L

M

N

W

X

Y

Z